Gui-Qiang G. Chen • Helge Holden
Kenneth H. Karlsen
Editors

Hyperbolic Conservation Laws and Related Analysis with Applications

Edinburgh, September 2011

Springer

Editors
Gui-Qiang G. Chen
Mathematical Institute
University of Oxford
Oxford, United Kingdom

Helge Holden
Department of Mathematical Sciences
Norwegian University of Science
and Technology
Trondheim, Norway

Kenneth H. Karlsen
Centre of Mathematics for Applications
University of Oslo
Oslo, Norway

ISSN 2194-1009 ISSN 2194-1017 (electronic)
ISBN 978-3-642-39006-7 ISBN 978-3-642-39007-4 (eBook)
DOI 10.1007/978-3-642-39007-4
Springer Heidelberg New York Dordrecht London

Library of Congress Control Number: 2013949498

Mathematics Subject Classification (2010): 35-06, 35L65, 35L40, 35Q35, 76J20, 35L67, 35R35, 76N10

Printed on acid-free paper

Springer is part of Springer Science+Business Media (www.springer.com)

Preface

This book collects together a selection of refereed expository articles and original research papers which stem from the Workshop on Hyperbolic Conservation Laws and Related Analysis with Applications hosted at the International Centre for Mathematical Sciences (ICMS), Edinburgh, UK, on September 19–23, 2011. There were 28 plenary lectures given by 24 distinguished mathematicians, including three introductory crash courses which surveyed and summarized the state of the art in the area of hyperbolic conservation laws and its important applications/connections to other areas, 18 research lectures which presented some recent important results, and 1 public lecture entitled *What Can Mathematics Say about Liquid Crystals?* delivered by Prof. Sir John Ball (Oxford). Most of the papers appearing in this book are authored by the leading mathematicians who spoke at the workshop.

Thirteen papers comprise this work, representing a cross section of the most significant recent advances and current directions in nonlinear hyperbolic conservation laws and related analysis with applications. The general theory of hyperbolic conservation laws emerged just 50 years ago, even though the seeds of the field were originally planted in the eighteenth and nineteenth centuries, especially by the leading scientists of the times: Euler, Cauchy, Poisson, Challis, Stokes, Rayleigh, Kelvin, Riemann, Clausius, Rankine, Hugoniot, and Mach, among many others. In recent years, the field has experienced a vigorous growth, and the research is marching on at a brisk pace.

This book presents a survey of recent analytical and numerical advances, as well as phenomena and theories likely to be important for future developments in the field, through discussing fundamental mathematical problems in nonlinear hyperbolic conservation laws arising in fluid mechanics, elasticity, solid mechanics, and differential geometry, including questions of existence, uniqueness, regularity, formation of singularities, and asymptotic behaviour of solutions. It contains two introductory papers:

- *Multi-dimensional Systems of Conservation Laws: an Introductory Lecture*
  by Denis Serre

- *The Nash-Moser Iteration Technique with Application to Characteristic Free-Boundary Problems*
  by Ben Stevens

In addition, the book's contributions offer two other perspectives:

Papers on the general analytical treatment of the theory and related analysis. These include:

- *The Semigroup Approach to Conservation Laws with Discontinuous Flux*
  by Boris Andreianov
- *SBV Regularity Results for Solutions to 1D Conservation Laws*
  by Laura Caravenna
- *Existence and Stability of Global Solutions of Shock Diffraction by Wedges for Potential Flow*
  by Gui-Qiang G. Chen & Wei Xiang
- *Some Well-Posedness Results for the Ostrovsky-Hunter Equation*
  by Giuseppe Maria Coclite, Lorenzo di Ruvo & Kenneth Karlsen
- *Divergence-Measure Fields on Domains with Lipschitz Boundary*
  by Hermano Frid

Papers on applications originating from significant realistic mathematical models of natural phenomena. These include:

- *On Numerical Methods for Hyperbolic Conservation Laws and Related Equations Modelling Sedimentation of Solid-Liquid Suspensions*
  by Fernando Betancourt, Raimund Büger, Ricardo Ruiz-Baier, Héctor Torres & Carlos A. Vega
- *A Generalized Buckley-Leverett System*
  by Nikolai Chemetov & Wladimir Neves
- *The Quasineutral Limit for the Navier-Stokes-Fourier-Poisson System*
  by Donatella & Pierangelo Marcati
- *On Strong Local Alignment in the Kinetic Cucker-Smale Model*
  by Trygve K. Karper, Antoine Mellet & Konstantina Trivisa

Also included are articles that bridge the gap between these two perspectives, seeking synergetic links between theory, analysis, and applications:

- *Entropy, Elasticity, and the Isometric Embedding Problem:* $M^3 \to \mathbb{R}^6$
  by Gui-Qiang G. Chen, Marshall Slemrod & Dehua Wang
- *An Overview of Piston Problems in Fluid Dynamics*
  by Min Ding & Yachun Li

These papers cover a wide range of topics, including shock reflection-diffraction, stability of nonlinear viscous/inviscid waves, free boundary problems, transonic flow, isometric embedding, formation and dynamics of singularities, well-posedness and regularity of entropy solutions, piston problems, kinetic models, weak convergence methods, singular limits, divergence-measure fields, semigroup approach, approximations, and numerical methods. They are at the forefront of current exciting developments.

The editors express their gratitude to the authors and the invited speakers for their invaluable contribution, to all of the participants and attendees for making the workshop successful, and to the referees for their constructive criticisms and suggestions. As the organisers, it is our great pleasure to acknowledge the ICMS leadership, the support in part from the Oxford Centre for Nonlinear PDE (OxPDE) through the UK EPSRC Science and Innovation award (EP/E035027/1), as well as effective assistance of the ICMS/OxPDE staff members including Somthawin Carter, Helene Frossing, Jane Waler, Dawn Wasley, and Jonathan Whyman. Our thanks go especially to the former and current ICMS Scientific Directors Professors John Toland and Keith Ball, as well as to the OxPDE Director Professor Sir John Ball. Finally, the editors are indebted to Springer-Verlag GmbH, Heidelberg, especially Catriona M. Byrne (Editorial Director, Mathematics), Marina Reizakis (Associate Editor, Mathematics), and Rainer Justke (Editorial Rights) for their professional assistance.

Oxford, UK — Gui-Qiang G. Chen
Trondheim, Norway — Helge Holden
Oslo, Norway — Kenneth H. Karlsen

Contents

The Semigroup Approach to Conservation Laws with Discontinuous Flux

Boris Andreianov

Abstract The model one-dimensional conservation law with discontinuous *spatially heterogeneous* flux is

$$u_t + \mathfrak{f}(x,u)_x = 0, \quad \mathfrak{f}(x,\cdot) = f^l(x,\cdot)\mathbb{1}_{x<0} + f^r(x,\cdot)\mathbb{1}_{x>0}. \qquad \text{(EvPb)}$$

We prove well-posedness for the Cauchy problem for (EvPb) in the framework of solutions satisfying the so-called adapted entropy inequalities.

Exploiting the notion of integral solution that comes from nonlinear semigroup theory, we propose a way to circumvent the use of strong interface traces for the evolution problem (EvPb) (in fact, proving the existence of such traces for the case of x-dependent $f^{l,r}$ would be a delicate technical issue). The difficulty is shifted to the study of the associated one-dimensional stationary problem $u + \mathfrak{f}(x,u)_x = g$, where the existence of strong interface traces of entropy solutions is an easy fact. We give a direct proof of this, avoiding the subtle arguments of the kinetic formulation (Kwon YS, Vasseur A (2007) Arch Ration Mech Anal 185(3):495–513) and of the H-measure approach (Panov EY (2007) J Hyperbolic Differ Equ 4(4):729–770).

2010 Mathematics Subject Classification Primary: 35L65, 35L04; Secondary: 47H06, 47H20

B. Andreianov (✉)
Laboratoire de Mathématiques CNRS UMR 6623, Université de Franche-Comté,
16 route de Gray, 25000 Besançon, France
e-mail: boris.andreianov@univ-fcomte.fr

G.-Q.G. Chen et al. (eds.), *Hyperbolic Conservation Laws and Related Analysis with Applications*, Springer Proceedings in Mathematics & Statistics 49,
DOI 10.1007/978-3-642-39007-4_1,

1 Introduction

Scalar conservation laws with space-discontinuous flux have been the a subject of intense study for 20 years. The goal of this note is to highlight the results that can be inferred from the nonlinear semigroup approach to such problems (see [13, 17]), specifically for the case of space dimension one.

We stick to the unifying framework for proving the existence, uniqueness, stability and convergence of numerical approximations that was proposed in the paper [9] of K.H. Karlsen, N.H. Risebro and the author. In [9], we studied the model problem

$$u_t + \mathfrak{f}(x,u)_x = 0, \quad \mathfrak{f}(x,\cdot) = f^l(x,\cdot)1\!\!1_{x<0} + f^r(x,\cdot)1\!\!1_{x>0} \qquad \text{(EvPb)}$$

under the *space homogeneity* assumption $f^{l,r}(x,\cdot) \equiv f^{l,r}(\cdot)$. This assumption appears as a technical one, nevertheless it was a cornerstone of the entropy formulation because of the explicit use of *strong interface traces* within the uniqueness technique of [9]. Presently, to the authors' knowledge there is no proof of existence of strong traces for the non-homogeneous case. And even though such a result is expected to be true under some weak assumptions on the dependence of $f^{l,r}$ on u and x, the proof (following well-established kinetic techniques [24, 30] or H-measure techniques [27, 28]) would be rather lengthy and highly technical. The semigroup approach exploited in the present note permits us to circumvent the difficulty, for the one-dimensional case. Actually, we will justify the existence of strong interface traces in a particularly simple setting, using the least technical ideas from [28]. Then we will conduct a brief study of the operator governing (EvPb) and apply general principles of nonlinear semigroup theory.

Let us recall the main features of the entropy formulation of Karlsen, Riesbro and the author [9] for the case $f^{l,r}(x,u) \equiv f^{l,r}(u)$. We postulated that a function $u \in L^\infty((0,T)\times\mathbb{R})$ is a $\mathcal{G}$-entropy solution of (EvPb) if:

(i) It is an entropy solution in the classical sense of Kruzhkov [23] away from the interface $\{x=0\}$, i.e., in the subdomains $\Omega^l := (0,T)\times\mathbb{R}^-$ and $\Omega^r := (0,T)\times R^+$; and moreover
(ii) The two solutions are coupled across the interface $\{x=0\}$ by the relation

$$\Big(\gamma^l u, \gamma^r u\Big)(t) \in \mathcal{G} \quad \text{for a.e. } t \in (0,T). \qquad (1)$$

Here $\gamma^l u, \gamma^r u$ are strong (in the L^1 sense) traces of local entropy solutions $u|_{\Omega^l}$ and $u|_{\Omega^r}$, respectively: see [28] (and also [24]) for the proof of existence of these traces in the homogeneous case.[1] Further, $\mathcal{G} \subset \mathbb{R}^2$ is an L^1*-dissipative germ*, that is, a set of pairs (u^l, u^r) encoding the Rankine-Hugoniot (conservativity) condition

[1] Actually, a non-degeneracy of $f^{l,r}$ on intervals is needed for existence of such traces, see assumption (H3). But if the degeneracy happens, one can reformulate (1) in terms of the traces of some "singular mapping functions" $Vf^{l,r}(u)$, see [9].

$$\forall (u^l, u^r) \in \mathcal{G} \qquad f^l(u^l) = f^r(u^r) \tag{2}$$

and the interface dissipation condition

$$\forall (u^l, u^r), (c^l, c^r) \in \mathcal{G} \qquad q^l(u^l, c^l) \geq q^r(u^r, c^r) \tag{3}$$

with $q^{l,r}$ the Kruzhkov entropy fluxes given by

$$q^{l,r}(\cdot, c) = \text{sign}(\cdot - c)\Big(f^{l,r}(\cdot) - f^{l,r}(c)\Big). \tag{4}$$

Further, [9] provides a *global entropy formulation* (see Definition 3 below) which is shown to be equivalent to (ii) whenever the one-sided traces $\gamma^{l,r}u$ on $\{x = 0\}$ do exist. Yet the global entropy formulation avoids the explicit use of interface traces (such as (1) above); for this reason, it is especially useful for proving the existence of solutions and convergence of various approximation procedures. Our goal is to provide a uniqueness proof that relies on this global entropy formulation. To this end, we combine two ideas.

Firstly, we observe that one can use the technique of the "comparison" proof of [9, Theorem 3.28] in the case where one works with solutions u and $\hat{u}$ such that *only one of them* (say, $\hat{u}$) has strong interface traces. In this paper, we will say that $\hat{u}$ is *trace-regular* if $\gamma^l \hat{u}$ and $\gamma^r \hat{u}$ exist in the sense of Definition 1 below.

Thus, we are able to "compare" a general solution and a trace-regular solution. Here the second ingredient comes into play. Indeed, the trace-regularity issue is particularly simple in the one-dimensional case for the so-called *stationary problem*:

$$u + \mathfrak{f}(x, u)_x = g \tag{StPb}$$

where $g \in L^1(\mathbb{R}) \cap L^\infty(\mathbb{R})$ and a $\mathcal{G}$-entropy solution of (StPb) is sought. In Lemma 1 we give a trace-regularity result based on elementary arguments. Now, problem (StPb) can be seen as the resolvent equation

$$u + A_{\mathcal{G}} u \ni g \tag{AbSt}$$

associated with the abstract evolution equation

$$\frac{d}{dt}u + A_{\mathcal{G}} u \ni h, \quad u(0) = u_0. \tag{AbEv}$$

Here $A_{\mathcal{G}}$ is the operator $u \mapsto \mathfrak{f}(x, u)_x$ defined on the appropriate domain $D(A_{\mathcal{G}}) \subset L^1(\mathbb{R})$ by its graph: $A_{\mathcal{G}} = \{(u, z) \in (L^1(\mathbb{R}))^2 \mid z \in A_{\mathcal{G}} u\}$. As a matter of fact, we will require that $u \in D(A_{\mathcal{G}})$ be trace-regular functions. Then the notion of integral solution can be exploited, following [13, 17], as it was done in [4, 6, 15] in various contexts. Indeed, u is an integral solution of (AbEv) if the comparison inequality in $\mathcal{D}'(0, T)$ holds:

$$\forall (\hat{u}, z) \in A_{\mathcal{G}} \qquad \frac{d}{dt}\|u(t) - \hat{u}\|_{L^1} \leq \Big[u(t) - \hat{u}, h - z\Big]_{L^1} \tag{5}$$

where the right-hand side is the so-called L^1 bracket (see Definition 5 below). Notice that within the semigroup approach, we limit our attention to $L^1 \cap L^\infty$ data (see Corollary 1 and Sect. 5 for a generalization to L^∞ data, which is not trivial).

Here is our point:

property (5) *(with $z = g - u$) can be established*
whenever u is a $\mathcal{G}$-entropy solution of (EvPb)
and $\hat{u}$ is a trace-regular $\mathcal{G}$-entropy solution of (StPb).

This observation closes the loop, because we deduce uniqueness of a $\mathcal{G}$-entropy solution to the evolution problem from the uniqueness of the integral solution. The latter uniqueness comes for free from the general principles of nonlinear semigroup theory as soon as we prove that $A_{\mathcal{G}}$ is a densely defined accretive operator on $L^1(\mathbb{R})$ with m-accretive closure.

The paper is organized as follows. In Sect. 2 we state the assumptions, definitions and the main result. In Sect. 3 we study the stationary problem (StPb) and establish the main properties of the operator $A_{\mathcal{G}}$ on $L^1(\mathbb{R})$ associated with the formal expression $u \mapsto \mathfrak{f}(x,u)_x$. In particular, we show that the domain of $A_{\mathcal{G}}$ can be restricted to trace-regular functions. Then in Sect. 4 we deduce the uniqueness in the setting of $\mathcal{G}$-entropy solutions for problem (EvPb) with $L^1 \cap L^\infty$ data. Finally, in Sect. 5 we discuss the application of the idea of this paper for the one-dimensional Dirichlet boundary-value problem for the conservation law; we also treat the case of merely L^∞ data for problem (EvPb). The Appendix of the paper contains a technical result on entropy solutions of a spatially non-homogeneous conservation law; this result has some interest on its own.

2 Assumptions, Definitions and Results

Let $\mathbb{R}^l := (-\infty, 0)$ and $\mathbb{R}^r := (0, +\infty)$, so that $\Omega^{l,r} = (0,T) \times \mathbb{R}^{l,r}$. For the sake of simplicity of presentation, let us assume

$$\forall x \in \mathbb{R}^{l,r} \text{ the functions } u \mapsto f^{l,r}(x,u) \text{ are supported in } [0,1]. \tag{H1}$$

This assumption is only used to ensure a uniform L^∞ bound on solutions and on approximate solutions.[2] For the sake of generality we will consider $\mathbb{R}$-valued bounded functions u_0 and g, although (H1) naturally appears in the case where solutions are $[0,1]$-valued (such solutions represent saturations in porous media, sedimentation or road traffic models; see, e.g., [1, 5, 18]).

[2]See [9] for more general assumptions that ensure L^∞ bounds, which have to be adapted to the inhomogeneous case.

Throughout this paper, we assume that $f^{l,r}$ satisfy the following:

$$\begin{array}{c} f^{l,r} \text{ are Lipschitz continuous in } (x,u) \in \mathbb{R}^{l,r} \times [0,1] \\ \text{and } f^{l,r}(0,\cdot) \text{ have a finite number of extrema on } [0,1]. \end{array} \tag{H2}$$

We will also require the genuine nonlinearity property:

$$\forall x \in \mathbb{R}^{l,r} \text{ the functions } u \mapsto (f^{l,r})_u(x,u) \text{ do not vanish on subintervals of } [0,1]. \tag{H3}$$

Notice that these assumptions can be relaxed but we stick to the above hypotheses for the sake of simplicity.

Let us give the main definitions. Firstly, we recall the notion of strong boundary trace in the case of the domain $(0,T) \times \mathbb{R}^l$ (the case of $(0,T) \times \mathbb{R}^r$ is analogous).[3] What is needed for our case is:

Definition 1. Let $u \in L^\infty((0,T) \times (-\infty,0))$. Then $\gamma^l u \in L^\infty(0,T)$ is the strong trace of u on the boundary $\{x = 0\} := \{(t,0) \mid t \in (0,T)\}$ if $u(\cdot,x)$ converges to $(\gamma^l u)(\cdot)$ essentially in $L^1(0,T)$ as $x \uparrow 0$.

Next, we define *germs* in terms of fluxes $f^{l,r}$ corresponding to the "frozen" value $x = 0$. Prescribing a *complete, maximal* L^1D-germ is a way to prescribe the interface coupling at $\{x = 0\}$ (see [9]).

Definition 2 (L^1-dissipative germs). A subset $\mathcal{G}$ of $\mathbb{R}^2$ is called an L^1D-germ (germ, for short) if it satisfies (2) and (3) with the fluxes $f^{l,r}$ evaluated at $x = 0$.

Such a germ is called maximal if it possesses no non-trivial extension; it is called definite if it possesses only one maximal extension, in which case the extension is denoted by $\mathcal{G}^*$. Finally, it is called complete if any Riemann problem for the auxiliary conservation law

$$u_t + \left(f^l(0,u)\mathbb{1}_{x<0} + f^r(0,u)\mathbb{1}_{x>0}\right)_x = 0 \tag{6}$$

admits a solution satisfying (i) and (ii) in the Introduction.

Completeness means that for any $(u_-, u_+) \in \mathbb{R}^2$ there exists a pair $(c^l, c^r) \in \mathcal{G}$ such that u_- can be joined to c^l by a Kruzhkov-admissible wave fan with negative speed for the flux $f^l(0,\cdot)$ and c^r can be joined to u_+ by a Kruzhkov-admissible wave fan with positive speed for the flux $f^r(0,\cdot)$. Notice that in this case, the so constructed function u is self-similar, therefore it possesses interface traces (in the strong sense of $L^1(0,T)$ convergence of $u(r,\cdot) \to (\gamma^r u)(\cdot)$ and of $u(-r,\cdot) \to (\gamma^l u)(\cdot)$ as $r \to 0^+$) that satisfy $\gamma^{l,r} u = c^{l,r}$.

The following definition (cf. [9–11, 18]), however, avoids explicit reference to point (ii) of the introduction.

---

[3] For the multi-dimensional domains treated in the Appendix, one uses an analogous definition based upon a parametrizaton of a neighbourhood of $\partial\Omega$ by $(\sigma, h) \in \partial\Omega \times (0,1)$.

Definition 3 ($\mathcal{G}$-entropy solution of the evolution problem). Assume we are given an L^1 D-germ $\mathcal{G}$. A function $u \in L^\infty((0,T)\times\mathbb{R})$ is called a $\mathcal{G}$-entropy solution of (EvPb) with an initial datum $u(0,\cdot) = u_0 \in L^\infty(\mathbb{R})$ if it satisfies the Kruzhkov entropy inequalities away from the interface $\{x = 0\}$:

$$\forall c \in \mathbb{R}\ \ |u-c|_t + \mathrm{sign}(u-c)\mathfrak{f}_x(x,c) + \mathfrak{q}(x;u,c)_x \le 0,\ \ |u-c|\Big|_{t=0} = |u_0 - c| \ \text{ in } \mathcal{D}'\Big([0,T)\times(\mathbb{R}\setminus\{0\})\Big) \tag{7}$$

and if, in addition, it satisfies the global adapted entropy inequalities

$$|u-c(x)|_t + \mathrm{sign}(u-c(x))\mathfrak{f}_x(x,c(x)) + \mathfrak{q}(x;u,c(x))_x \le 0 \ \text{ in } \mathcal{D}'((0,T)\times\mathbb{R}) \tag{8}$$

for every function $c(\cdot)$ of the form

$$c(x) = c^l 1\!\!1_{x<0} + c^r 1\!\!1_{x>0} \quad \text{with } (c^l,c^r) \in \mathcal{G}^*. \tag{9}$$

In the inequalities (7) and (8) the Kruzhkov entropy flux $\mathfrak{q} = q^l 1\!\!1_{x<0} + q^r 1\!\!1_{x>0}$ is computed with the help of (4), with the tacit x-dependency in $f^{l,r}$. Notice that with respect to the case of spatially homogeneous $f^{l,r}$, there is the additional term $\mathfrak{f}_x(x,c(x))$; the notation $\mathfrak{f}_x(x,c(x))$ ignores the discontinuity at zero, i.e.,

$$\mathfrak{f}_x(x,c(x)) := f^l_x(x,c^l)1\!\!1_{x<0} + f^r_x(x,c^r)1\!\!1_{x>0}.$$

Remark 1. Note that it can be assumed, without loss of restriction, that a $\mathcal{G}$-entropy solution u belongs to $C([0,T];L^1_{loc}(\mathbb{R}))$. This is a consequence of the Kruzhkov inequalities in domains $\Omega^{l,r}$; see, e.g., [8, 19, 27] and references therein. In the sequel, we will always select the time-continuous representative of u; in particular, the initial condition can be taken in the sense $u(0,\cdot) = u_0$.

The definition for the stationary problem (StPb) is analogous, cf. [14].

Definition 4 ($\mathcal{G}$-entropy solution of the stationary problem). Assume we are given an L^1 D-germ $\mathcal{G}$. A function $u \in L^\infty(\mathbb{R})$ is called a $\mathcal{G}$-entropy solution of (StPb) if it satisfies the Kruzhkov entropy inequalities

$$\forall c \in \mathbb{R}\ \ \mathrm{sign}(u-c)(u + \mathfrak{f}_x(x,c) - g) + \mathfrak{q}(x;u,c)_x \le 0 \ \text{ in } \mathcal{D}'\Big(\mathbb{R}\setminus\{0\}\Big) \tag{10}$$

and if for every function $c(\cdot)$ of the form (9) it satisfies the global adapted entropy inequalities:

$$\mathrm{sign}(u-c(x))(u + \mathfrak{f}_x(x,c(x)) - g) + \mathfrak{q}(x;u,c(x))_x \le 0 \ \text{ in } \mathcal{D}'(\mathbb{R}). \tag{11}$$

Remark 2. In the homogeneous case (see [9]) one can replace $\mathcal{G}^*$ by $\mathcal{G}$ in (9) for the evolution problem (EvPb). This weaker assumption leads to a smaller number of global adapted entropy inequalities to be checked. E.g., in the situation where the fluxes $f^{l,r}$ are "bell-shaped", only one global adapted entropy inequality is needed in (8), see [5, 8, 18].

In the present paper, one can replace $\mathcal{G}^*$ by $\mathcal{G}$ in the above definition for the stationary problem (StPb) but not for the evolution problem. At the present stage, this drawback appears to be the price to pay for the approach which does not rely upon the existence of strong interface traces for solutions of (EvPb) (see also [9, Sect. 3.4]).

Here is the main result of this paper.

Theorem 1 (Well-posedness for (EvPb)**).** *Assume $f^{l,r}$ satisfy* (H1)–(H3). *Let $\mathcal{G}$ be a definite maximal L^1D germ. Then for all $u_0 \in L^1(\mathbb{R}) \cap L^\infty(\mathbb{R})$ there exists a unique $\mathcal{G}$-entropy solution of* (EvPb) *with the initial datum u_0. It depends continuously on u_0, namely, if $u, \hat{u}$ are the $\mathcal{G}$-entropy solutions corresponding to $L^1(\mathbb{R}) \cap L^\infty(\mathbb{R})$ data $u_0, \hat{u}_0$, respectively, then $\|u(t,\cdot) - \hat{u}(t,\cdot)\|_{L^1} \le \|u_0 - \hat{u}_0\|_{L^1}$ for all $t \in [0,T]$.*

As stated in the introduction, the uniqueness claim is shown in an indirect way, with the help of abstract tools from nonlinear semigroup theory. The existence can also be obtained in an abstract way, as in [17]. However, here we prefer to justify the existence by constructing solutions with a well-chosen finite volume scheme. Alternatively, in the cases where $\mathcal{G}$ is compatible with some vanishing viscosity approach, the adapted viscosity approximation can be used.

Exploiting the property of finite speed of propagation and a continuation argument for entropy solutions in which we solve auxiliary Dirichlet problems, we can extend the result to general L^∞ data. Namely, we get the following:

Corollary 1. *Under the assumptions of Theorem 1, the existence and uniqueness of a $\mathcal{G}$-entropy solution still holds if $u_0 \in L^\infty(\mathbb{R})$. If u is the $\mathcal{G}$-entropy solution with $u(t,0) = u_0$, then for all $t \in [0,T]$ the function $u(t,\cdot)$ depends continuously on u_0 in the $L^1_{loc}(\mathbb{R})$ topology.*

In the opposite direction, starting from Theorems 2 and 1 we can drop the L^∞ assumption on the data. Indeed, Theorem 2 permits the definition of solutions of the abstract evolution problem (AbEv) for merely L^1 data. In the context of conservation laws of the form $u_t + \operatorname{div}_x f(u) = 0$, such solutions can be characterized intrinsically as its *renormalized solutions* (see [16]). We expect that for general L^1 data, the integral solutions of (AbEv) are renormalized solutions of (EvPb); but this issue is beyond the scope of this paper.

3 The Stationary Problem (StPb) and the Underlying *m*-Accretive Operator

Let us define the operator $A_\mathcal{G}$ on $L^1(\mathbb{R})$ by its graph:

$$(u,z) \in A_\mathcal{G} \ \text{ iff } u \text{ is a } \mathcal{G}\text{-entropy solution of (StPb) with } g = z + u. \qquad (12)$$

Thus, the domain $D(A_{\mathcal{G}})$ is defined implicitly. Let us show that it consists of trace-regular functions.

Lemma 1 (Trace-regularity). *If $u \in L^\infty(\mathbb{R})$ satisfies the away-from-the-boundary Kruzhkov entropy inequalities* (10) *and $f^{l,r}$ satisfies* (H2) *and* (H3), *then $\gamma^l u := \lim_{x\uparrow 0} u(x)$ and $\gamma^r u := \lim_{x\downarrow 0} u(x)$ exist.*

Proof. Consider, for instance, $u|_{\mathbb{R}^l}$. From the entropy inequalities (10) it follows that for all $c \in \mathbb{R}$ there exist non-negative Borel measures γ_c^+ on $\mathbb{R}^l = (-\infty, 0)$ such that

$$\text{sign}^+(u-c)(u+\mathfrak{f}_x(x,c)-g)+Q_c(x)_x = -\gamma_c^+ \tag{13}$$

where $Q_c(x) := \text{sign}^+(u(x)-c)(f^l(x,u(x)) - f^l(x,c))$. Because $Q_c(x) \in L^\infty(\mathbb{R}^l)$, it is easy to see that the variation of γ_c^+ is finite up to the boundary. Indeed, taking (by approximation) the test function

$$\xi_h(x) = (1-\min\{1, -x/h\})\min\{1,(1+x)^+\}$$

in the entropy formulation, we find

$$|\gamma_c^+|([-1,0)) = \lim_{h\to 0}\int_{[-1,0)} \xi_h\, d\gamma_c^+ \int_{-1}^0 |u-g| + \int_{-\infty}^0 \big|Q_c(x)\big|\,|(\xi_h)_x|\,dx.$$

The right-hand side is finite, since $\|(\xi_h)_x\|_1 \le 2$ uniformly in $h \in (0,1)$.

Now, let $M = \|u\|_\infty$ and $c_0, \dots, c_N$ be a partition of $[-M, M]$ such that f' keeps constant sign on each interval (c_{i-1}, c_i), $i = 1, \dots, N$ (this is possible due to (H2)). For instance, assume that this sign is "–" if i is odd and "+" if i is even. Then the variation function (Vf^l) on $[-M, M]$ can be represented as

$$(Vf^l)(x,u) := \int_{-M}^u |(f^l)_u(x,z)|\,dz = \int_{-M}^u \eta'(z)(f_u^l)(x,z)\,dz$$

where $\eta'|_{(c_{i-1},c_i)} = (-1)^i$. Then (Vf^l) is the entropy-flux corresponding to the (non-convex) entropy η with

$$\eta'(z) = sign^+(z-c_0) + 2\sum_{i=1}^{N-1}(-1)^i\,\text{sign}^+(z-c_i),$$

hence a linear combination of equalities (13) yields

$$(Vf^l)(x,u(x))_x = \gamma_{c_0}^+ - 2\sum_{i=1}^{N-1}(-1)^i\gamma_{c_i}^+ - \eta'(u)(u-g+\mathfrak{f}_x(x,c)) \ \text{ in } \mathcal{D}'(-\infty,0).$$

From the facts that $(u-g)+\mathfrak{f}_x(u,x) \in L^1(\mathbb{R}^l) + L^\infty(\mathbb{R}^l)$ and that γ_{c_i} are finite up to the boundary, it follows that $(Vf^l)(x,u(x)) \in C((-\infty,0])$. Now, notice that the map $W(\cdot) := (Vf^l)(0,\cdot)$ is non-decreasing, by construction; moreover, due to assumption (H3) the map W is strictly increasing (and furthermore, we can assume that it is bijective, upon modifying the definition of W outside $[-M, M]$). Therefore the map $x \mapsto W^{-1} \circ (Vf^l)(x, u(x))$ is continuous on $(-\infty, 0]$. Hence its limit at zero exists; let us denote it by $\gamma^l u$.

It remains to notice that $\gamma^l u = \lim_{x \uparrow 0} u(x)$. Indeed, because f^l is continuous in $(x, u) \in \mathbb{R}^l \times \mathbb{R}$, this is also the case for Vf^l. Moreover, $W^{-1} \circ (Vf^l)(0, \cdot)$ is the identity map. Hence

$$|u(x) - W^{-1} \circ (Vf^l)(x, u(x))| = |W^{-1} \circ (Vf^l)(0, u(x)) - W^{-1} \circ (Vf^l)(x, u(x))|$$

vanishes as $x \to 0$ (notice that $u(x)$ stays in a compact set on which W^{-1} is uniformly continuous). This concludes the proof. □

Now, we can reformulate Definition 4 as follows.

Lemma 2 (Interface coupling for (StPb)**).** *Assume $(H2)$ and $(H3)$. A function $u \in L^\infty(\mathbb{R})$ is a $\mathcal{G}$-entropy solution of* (StPb) *if and only if it satisfies* (10) *and, in addition, $(\gamma^l u, \gamma^r u) \in \mathcal{G}^*$.*

Note that by Lemma 1 the existence of $\gamma^{l,r} u$ is automatic in the above statement.

Proof. Let us prove that an entropy solution of (StPb) satisfies $(\gamma^l u, \gamma^r u) \in \mathcal{G}^*$. It is enough to take $\xi_h = 1 - \min\{|x|/h, 1\}$ as a test function in (11) and let $h \to 0$; one finds

$$\forall (c^l, c^r) \in \mathcal{G}^* \quad q^l(0, \gamma^l u, c^l) - q^r(0, \gamma^r u, c^r) \geq 0. \tag{14}$$

Because $\mathcal{G}^*$ is a maximal $L^1 D$ germ associated with the fluxes $f^{l,r}(0, \cdot)$, the claim follows.

Conversely, by the definition of an $L^1 D$ germ, the property $(\gamma^l u, \gamma^r u) \in \mathcal{G}^*$ implies (14). It remains to take $(1 - \xi_h)\xi$ as a test function in (10), where $\xi \in \mathcal{D}(\mathbb{R})$. One deduces (11). □

Now, let us study the operator $A_{\mathcal{G}}$. We refer to [13, 15, 17] for the definitions.

Proposition 1 (Accretivity). *Let $\mathcal{G}$ be a definite $L^1 D$ germ. Assume $f^{l,r}$ satisfy $(H2)$ and $(H3)$. Then the operator $A_{\mathcal{G}}$ is accretive on $L^1(\mathbb{R})$.*

Proof. One has to prove that for all $(u, z), (\hat{u}, \hat{z}) \in A_{\mathcal{G}}$ the following holds

$$\forall \lambda > 0 \quad \|u - \hat{u}\|_{L^1} \leq \|(u + \lambda z) - (\hat{u} + \lambda \hat{z})\|_{L^1}. \tag{15}$$

It is easily seen that u and $\hat{u}$ are $\mathcal{G}$-entropy solutions of the stationary problem (StPb) with the flux $\lambda \mathfrak{f}$ in place of $\mathfrak{f}$ and with the source terms $h = u + \lambda z$, $\hat{h} = \hat{u} + \lambda \hat{z}$, respectively. For instance, the entropy inequality (10) with $g = u + z$ can be rewritten as

$$\operatorname{sign}(u - c)\Big(u - (u + \lambda z - \lambda \mathfrak{f}_x(x, c))\Big) + \lambda \mathfrak{q}(x, u, c)_x \leq 0. \tag{16}$$

Based on (16) and its analogue written for $\hat{u}$, we can use the Kruzhkov doubling of variables to deduce the so-called *Kato inequality*:

$$|u - \hat{u}| + \lambda \mathfrak{q}(x, u, \hat{u})_x \leq |h - \hat{h}| \ \text{ in } \mathcal{D}'(\mathbb{R} \setminus \{0\}). \tag{17}$$

The argument we use to derive this inequality is essentially based on the fundamental work of Kruzhkov [23], but it is not entirely classical. Indeed, notice that we have the dependency of $\mathfrak{f}$ on x but we are able to drop the "$\mathfrak{f}_x(x, c)$" term that appears in [23]. Roughly speaking, we justify that a Kruzhkov entropy solution (even a local one!) is a vanishing viscosity limit; and we observe that the solution operator for $u + f(x, u)_x - \varepsilon u_{xx} = h$ leads to a Kato inequality the limit of which, as $\varepsilon \to 0$, yields (17). The details of the justification of (17) are postponed to the Appendix (see in particular Remark 5).

It then remains to take the test function $\xi_s(x) = \exp(-s|x|) \min\{1, |x|/s\}$ in (17); this can be done by approximation. Taking into account the fact that $|q(x, u, \hat{u})| \leq L|u - \hat{u}|$ where L is a uniform in x Lipschitz constant of $\mathfrak{f}(x, \cdot)$ (here we use (H2)), at the limit $s \to 0^+$ we infer

$$\|u - \hat{u}\|_{L^1} \leq \|h - \hat{h}\|_{L^1} - \Big(q^l(0, \gamma^l u, \gamma^l \hat{u}) - q^r(0, \gamma^r u, \gamma^r \hat{u})\Big) \leq \|h - \hat{h}\|_{L^1};$$

the latter inequality follows by Lemma 2 and the L^1-dissipativity of $\mathcal{G}^*$. In view of the definition of h and $\hat{h}$, this proves (15). □

Proposition 2 (m-accretivity of the closure of $A_{\mathcal{G}}$). *Let $\mathcal{G}$ be a complete maximal L^1D germ. Assume $f^{l,r}$ satisfy* (H1)–(H3).

(i) We have $L^\infty(\mathbb{R}) \cap L^1(\mathbb{R}) \cap BV(\mathbb{R}) \subset Im(I + \lambda A_{\mathcal{G}})$, for all $\lambda > 0$.
(ii) The domain $D(A_{\mathcal{G}})$ is dense in $L^1(\mathbb{R})$.

Proof. For the proof of (i), we construct solutions of $u + \lambda A_{\mathcal{G}} u = g$ for $g \in L^1(\mathbb{R}) \cap L^\infty(\mathbb{R}) \cap BV(\mathbb{R})$ using a monotone two-point finite volume scheme, in the vein of [9, Theorem 6.4]. See Remark 3 for an alternative construction. For the proof of (ii), we denote by u^λ the solution of the problem treated in the first part; letting $\lambda \to 0$, we will prove the convergence of u^λ to g for an L^1-dense set of source terms g. Now, let us give the details.

Let us approximate problem $u + \mathfrak{f}(x, u)_x = g$ (indeed, it is enough to consider $\lambda = 1$) by piecewise constant functions $u_h := \sum_{n=-\infty}^{+\infty} u_n \mathbb{1}_{((n-1)h, nh)}$ using a finite volume scheme. To this end, discretize $x \mapsto \mathfrak{f}(x, \cdot)$ by

$$f_n(z) = f^l(nh, z) \ \text{ if } n < 0 \quad \text{and} \quad f_n(z) = f^r(nh, z) \ \text{ if } n > 0.$$

For every $n \neq 0$, we take a monotone two-point flux F_n (see, e.g., [22]) consistent with f_n. Since $\mathcal{G}$ is a complete germ, for $i = 0$ we can take the Godunov flux F_0 associated with the Riemann solver for the auxiliary discontinuous-flux problem (6) associated with the fluxes $f^{l,r}(0,\cdot)$ (cf. [9, Sect. 6.3]). Now, the finite volume scheme to be solved reads

$$\forall n \in \mathbb{Z}\ \ u_n + h\Big(F_n(u_{n+1}, u_n) - F_{n-1}(u_n, u_{n-1})\Big) = g_n \tag{18}$$

where $g_h := \sum_{n=-\infty}^{+\infty} g_n \mathbb{1}_{((n-1)h, nh)}$ is an approximation of g in $L^1(\mathbb{R})$ such that $\|g_h\|_{L^1} \leq \|g\|_{L^1}$, $\|g_h\|_{L^\infty} \leq \|g\|_{L^\infty}$ and $\|g_h\|_{BV} \leq \|g\|_{BV}$.

Due to (H1) and (H2) we can choose F_n to be Lipshitz continuous in both variables, uniformly in n. Therefore, for h small enough, the scheme can be rewritten in the form

$$\forall n \in \mathbb{Z}\ \ H_n(u_{n-1}, u_n, u_{n+1}) = g_n \ \ \text{with } H_n \text{ monotone in each variable.}$$

From this property and assumption (H1) we get the uniform L^∞ a priori bound $\min\{0, m\} \leq u_h \leq \max\{1, M\}$ where m and M are such that $m \leq g_h \leq M$ a.e. on $\mathbb{R}$.

The existence of a solution to the scheme can be inferred from the topological degree theorem as follows. One first truncates the system at ranks $\pm N$, setting $u_{-N} = 0 = u_N$ and considering only the equations for $|n| < N$ with F_n, g_n substituted by θF_n, θg_n, respectively, where $\theta \in [0,1]$. For $\theta = 0$ the problem has the trivial zero solution. The a priori L^∞ estimate still holds for the truncated problem, and the topological degree theorem ensures the existence of a solution $U^N \in \mathbb{R}^{2N-1}$ (for $\theta = 1$) to the finite-dimensional system. We consider U^N as an element of $\mathbb{R}^{\mathbb{Z}}$, setting to zero the components with $|n| \geq N$. Then compactness (component per component) and diagonal extraction are used to obtain an accumulation point $U := \lim_{N_k \to \infty} U^{N_k}$ in the topology of component-wise convergence in $\mathbb{R}^{\mathbb{Z}}$. Then by passage to the limit in the truncated problem, it is easily seen that $U = (u_n)_{n \in \mathbb{Z}}$ solves problem (18).

Now we have to prove that, first, there exists a convergent subsequence $(u_h)_h$ (not labelled); and second, that $u := \lim_{h \downarrow 0} u_h$ is a $\mathcal{G}$-entropy solution of (StPb).

Let us assess the $BV_{loc}(\mathbb{R} \setminus \{0\})$ compactness of $(u_h)_h$. We can restrict our attention to $h \in \{2^{-j} \mid j \in \mathbb{N}\}$. Let us normalize u_h so that it is left-continuous for $x < 0$ and right-continuous for $x > 0$. Using the diagonal extraction argument we can ensure that $u^h(\pm 2^{-\ell})$ converge to some limits $u_\ell^\pm$ as $h \to 0$, for all $\ell \in \mathbb{N}$. Similarly, we can assume that $u^h(\pm 2^\ell \mp 0) \to U_\ell^\pm$ as $h \to 0$. Then we can consider that u^h approximate the Dirichlet boundary-value problems in $(-2^\ell, -2^{-\ell})$ (with the boundary values U_ℓ^- and u_ℓ^- at the extremities) and in $(2^{-\ell}, 2^\ell)$ (with the boundary values u_ℓ^+ and U_ℓ^+). By standard arguments (see in particular [22] and [9, 18]) using the monotonicity of H_n and the fact that $\sup_{a,b,c \in [m,M]} |H_n(a,b,c) - H_{n-1}(a,b,c)| \leq const\, h$ (this comes from (H1) and (H2)) we deduce a uniform BV bound on $(u^h)_h$ in $\{x \in \mathbb{R} \,|\, 2^{-\ell} < |x| < 2^\ell\}$. Another application of the diagonal extraction argument proves the BV_{loc} compactness in $\mathbb{R} \setminus \{0\}$.

It remains to pass to the limit in the scheme, as $h \to 0$. Thanks to the local variation bound, this is a standard issue (see [22] and the arguments of [9] for the discontinuous-flux context). One first gets approximate entropy inequalities and approximate adapted entropy inequalities for u^h; here, it is important that we use the Godunov flux at the interface. Then one sends h to zero using the L^1_{loc} compactness of $(u_h)_h$. In particular, consistency of the numerical fluxes and the continuity of $f^{l,r}$ in x permit to passage to the limit in the nonlinear terms. This concludes the proof of (i).

Now, we turn to the proof of (ii). Indeed, let g be a compactly supported, piecewise constant function. We will use λ-dependent test functions ψ_λ on each interval where g is constant. Namely, let $g = c_i$ on a finite or semi-infinite interval (a_{i-1}, a_i); without loss of generality we may assume that $0 \notin (a_{i-1}, a_i)$. From the Kruzhkov entropy inequalities for u^λ, which is a $\mathcal{G}$-entropy solution of $u + \lambda \mathfrak{f}(x,u)_x = g$, we have

$$\operatorname{sign}(u^\lambda - c_i)\Big(u^\lambda - c_i + \lambda \mathfrak{f}_x(x, u^\lambda, c_i)\Big) + \lambda \mathfrak{f}(x, u^\lambda)_x \le 0 \ \text{ in } \mathcal{D}'((a_{i-1}, a_i)).$$

Taking test functions ψ_λ in this inequality such that $\psi_\lambda \to 1\!\!1_{(a_{i-1},a_i)}$ with $\|\psi'_\lambda\|_\infty \le \lambda^{-1/2}$, we find

$$\lim_{\lambda \downarrow 0} \int_{a_{i-1}}^{a_i} |u^\lambda - g| = \lim_{\lambda \downarrow 0} \int_{a_{i-1}}^{a_i} |u^\lambda - c_i| \le 0.$$

Summing in i, we deduce that $u^\lambda \to g$ in $L^1(\mathbb{R})$ as $\lambda \to 0$. This ends the proof. □

Remark 3. Notice that in many cases, the existence of a $\mathcal{G}$-entropy solution can be shown using an *adapted vanishing viscosity* approximation.

For instance, in the case of bell-shaped fluxes, one looks at the definite germs of the form $\mathcal{G}_{(A,B)} = \{(A,B)\}$ where (A,B) are the so-called *connections* (see [2, 5, 18]). For each of these germs, there exists a choice of *adapted viscosity approximations* that take the form

$$u^\varepsilon + \mathfrak{f}(x, u^\varepsilon)_x = g + \varepsilon(\mathfrak{a}(x, u^\varepsilon))_{xx},$$

and for which $u = A 1\!\!1_{x<0} + B 1\!\!1_{x>0}$ is an obvious solution with $g = u + f^l_x(x,A) 1\!\!1_{x<0} + f^r_x(x,B) 1\!\!1_{x>0}$, for every $\varepsilon > 0$. As in [9, Theorem 6.3], one deduces the convergence of u^ε to a $\mathcal{G}$-entropy solution u of (StPb). Moreover, one can use viscosity approximations having the physical meaning of *vanishing capillarity*, see [5].

Recall the definition of an integral solution for an evolution equation governed by an accretive operator on L^1.

Definition 5 (Integral solution). A function $u \in C([0,T], L^1(\mathbb{R}))$ is an integral solution of $\frac{d}{dt}u + Au \ni h$ with A defined on $L^1(\mathbb{R})$ if $u(0) = u_0$ and (5) holds in $\mathcal{D}'(0,T)$, with the notation $\Big[u, f\Big]_{L^1} := \int \operatorname{sign} u\, f + \int |f|\, 1\!\!1_{u=0}$.

Now we can apply the key result of nonlinear semigroup theory.

Theorem 2 (Uniqueness of an integral solution). *Assume* (H1)–(H3). *For all $u_0 \in L^1$ there exists one and only one integral solution to the problem $\frac{d}{dt}u + \overline{A_{\mathcal{G}}}u \ni 0$ with the initial datum u_0. If $\hat{u}$ is the integral solution corresponding to $\hat{u}_0$, then $\|u(t) - \hat{u}(t)\|_{L^1} \leq \|u_0 - \hat{u}_0\|_{L^1}$.*

Proof. It is enough to apply [17, Theorem 6.6] to the closure of $A_{\mathcal{G}}$. Indeed, according to Propositions 1 and 2, $\overline{A_{\mathcal{G}}}$ is a densely defined m-accretive operator. Therefore there exists a mild solution to the abstract evolution problem governed by $\overline{A_{\mathcal{G}}}$; hence the mild solution is the unique integral solution of this problem. □

4 $\mathcal{G}$-Entropy Solutions of the Evolution Problem

In this section, the main issue is the uniqueness of a solution to (EvPb) in the sense of Definition 3. We first derive an equivalent form of this definition (note the difference with the stationary case: we do not ensure nor exploit the trace-regularity of the solution u of (EvPb)).

Lemma 3 (Interface coupling for (EvPb)**).** *Assume* (H2) *and* (H3). *A function $u \in L^\infty(\mathbb{R})$ is a $\mathcal{G}$-entropy solution of* (EvPb) *iff it satisfies* (7) *and, in addition,*

$$\forall (c^l, c^r) \in \mathcal{G}^* \;\; (\gamma_w^l q^l(\cdot, u(\cdot), c^l))(t) \geq (\gamma_w^r q^r(\cdot, u(\cdot), c^r))(t) \;\; \textit{for a.e. } t \in (0,T). \tag{19}$$

Here $\gamma_w^{l,r} q^{l,r}(u, c^{l,r})$ denote the weak interface traces *of the respective fluxes.*

Note that the existence of $\gamma_w^{l,r} q^{l,r}(\cdot, u(\cdot), c^{l,r})$ comes from the Kruzhkov entropy inequalities (7), the Schwartz lemma on non-negative distributions and the general result of [21]. At this point, it should be stressed that the left-hand side of (7) is a non-positive Radon measure that is, in addition, finite up to the interface $\{x = 0\}$ (cf. the corresponding argument of the proof of Lemma 1).

Proof. As in the proof of Lemma 2, we use $\xi_h = 1 - \min\{|x|/h, 1\}$. Taking $\xi_h(x)\theta(t)$ (with $\theta \in \mathcal{D}(0,T)$, $\theta \geq 0$) as test function in (8), using the existence of weak traces $\gamma_w^{l,r} q^{l,r}(u, c^{l,r})$ we find the $\mathcal{D}'$ formulation of (19). Since θ is arbitrary, we get (19) by localization at every point of $(0,T)$ that is a Lebesgue point of the weak trace functions $t \mapsto (\gamma^{l,r} q^{l,r}(u, c^{l,r}))(t)$. Conversely, in a similar way to Lemma 2, it can also be shown that (19) and (7) imply (8). □

As was the case for Lemmas 2 and 3 provides an equivalent definition of a $\mathcal{G}$-entropy solution.

Now, note the following elementary property.

Lemma 4. *Let u and $\hat{u}$ be two bounded functions for which we assume that*

$$\textit{the weak interface traces} \quad \gamma_w^{l,r} q^{l,r}(\cdot, u(\cdot), \hat{u}(\cdot)) \quad \textit{exist.}$$

If $\hat{u}$ is a trace-regular function (i.e., there exist strong interface traces $(\gamma^{l,r}\hat{u})(t)$ for a.e. $t \in (0,T)$), then

$$\gamma_w^{l,r} q^{l,r}(\cdot, u(\cdot), \hat{u}(\cdot)) = \gamma_w^{l,r} q^{l,r}(\cdot, u(\cdot), c^{l,r}) \;\; \textit{with } c^{l,r}(t) = (\gamma^{l,r}\hat{u})(t), \textit{for a.e. } t. \tag{20}$$

Proof. Property (20) stems from the definition of a weak trace in the L^∞ sense (actually, this is in the weak-* sense) and the fact that due to the continuity of $f^{l,r}$ and the existence of strong traces, one has for instance

$$\operatorname{ess}\lim_{x\uparrow 0} |q^l(x, u(t,x), \hat{u}(t,x)) - q^l(x, u(t,x), c^l(t))| = 0 \;\; \text{for a.e. } t$$

while $q^l(x, u, \hat{u})$ remains uniformly bounded. □

We are now in a position to deliver the key observation of our method:

Proposition 3. *Assume* (H2). *Let u be a $\mathcal{G}$-entropy solution u of* (EvPb) *with $u(0,\cdot) \in L^1(\mathbb{R}) \cap L^\infty(\mathbb{R})$. Then the map $t \mapsto u(t,\cdot)$ is an integral solution of the associated abstract evolution problem governed by the operator $\overline{A_{\mathcal{G}}}$ (with $h = 0$ and the initial datum $u(0,\cdot)$, cf. Remark 1).*

Proof. By a density argument and the upper semi-continuity in $L^1(\mathbb{R})$ of the bracket $\Big[\cdot,\cdot\Big]_{L^1}$, it is enough to prove (5) (i.e., we need only consider $(u,z) \in A_{\mathcal{G}}$ in place of $(u,z) \in \overline{A_{\mathcal{G}}}$). Recall that we have $h = 0$.

By the definition (12) of $A_{\mathcal{G}}$, we take $\hat{u}$, a $\mathcal{G}$-entropy solution of the stationary problem (StPb). Then we "compare" u and $\hat{u}$ using the Kruzhkov doubling of variables: more precisely, we use it *away from the interface*. Using the version of the Kruzhkov argument presented in Appendix, we deduce the local (in $\mathbb{R} \setminus \{0\}$) *Kato inequality*

$$|u-\hat{u}|_t + \mathfrak{q}(x,u,\hat{u})_x \le \operatorname{sign}(u-\hat{u})(u-g) + 1\!\!1_{[u=\hat{u}}|g-\hat{g}| = \Big[u(t)-\hat{u}, u-g\Big]_{L^1} \;\; \text{in } \mathcal{D}'(R\setminus\{0\}). \tag{21}$$

Letting the test function in (21) converge to 1 in the same way as in the proof of Proposition 1, we generate the weak interface traces:

$$\begin{aligned} &\frac{d}{dt}\|u(t)-\hat{u}\|_{L^1} \le \Big[u(t)-\hat{u}, u-g\Big]_{L^1} - \Delta, \\ &\text{with } \Delta := \int_0^T \Big((\gamma_w^l q^l(\cdot, u(\cdot), \hat{u}(\cdot)))(t) - (\gamma_w^r q^r(\cdot, u(\cdot), \hat{u}(\cdot)))(t) \Big)\, dt. \end{aligned} \tag{22}$$

It remains to combine Lemma 4 (note that $\hat{u}$ is trace-regular, by Lemma 1), Lemmas 2 and 3. One finds that the term Δ in (22) is non-negative, which leads to inequalities (5). Indeed, we have $c^{l,r} = \gamma^{l,r}\hat{u}$ that satisfy $(c^l, c^r) \in \mathcal{G}^*$; then Δ can be re-written using (20); finally, (19) guarantees that the integrand in Δ is non-negative. This ends the proof. □

Proof of Theorem 1. The uniqueness claim and the L^1-contraction property are straightforward from Proposition 3 and Theorem 2. In order to conclude the proof of Theorem 1, it remains to ensure the existence of an entropy solution. We refer to the existence arguments used for the stationary problem (see Proposition 2(i) and Remark 3). For the evolution problem, analogous approximation arguments apply: either approximation by a finite volume scheme or, in the case of bell-shaped fluxes, the use of adapted viscosity approximations. One should pay attention to heterogeneity, as in the proof of Proposition 2(i) and in Remark 3. The delicate point is the BV_{loc} estimate, the proof of which is more involved than the arguments used to justify Proposition 2(i); one has to argue in the same way as in [9, 18]. □

5 On the Dirichlet Problem for the One-Dimensional Conservation Law

5.1 Application of the Semigroup Method to the Dirichlet Problem

The fundamental reference for the Dirichlet problem

$$\begin{cases} u_t + f(x,u)_x = 0 \text{ in } (0,T)\times(0,+\infty) \\ u|_{x=0} = u^D \\ u|_{t=0} = u_0 \end{cases} \tag{23}$$

is the Bardos, LeRoux and Nédélec paper [12]. The setting of [12] is the space $L^\infty(0,T;BV((0,+\infty))$, thus $u_0 \in BV(0,+\infty)$ and $u^D \in BV(0,T)$; moreover, f should be BV in x. These restrictions are due to the fact that the formulation of [12] uses the strong boundary trace γu of u on $\{x = 0\}$. More recently, Vasseur [30] (see also [28] for the most general argument) proved the existence of such traces for the spatially homogeneous case and thus dropped the BV assumptions of [12]. Notice that the result of [12] is used[4] in our proof of Theorem 1 through the justification of Lemma 6 in the Appendix; thus we have kept the BV assumption on f.

For the non-homogeneous case $f = f(x,u)$, with the same method as in the present paper we can treat the particular case where u^D is a constant in t function (this restriction is inherent to the semigroup approach). To do so, we can exploit the notion of a solution for (23) based upon the up-to-the-boundary entropy inequalities introduced in [7]. The arguments of the well-posedness proof (see [3]) are almost

[4]To be specific, the Bardos-LeRoux-Nédélec formulation with a strong boundary trace (cf. [30]) is used not in Ω but in specially selected subdomains of Ω, so that the existence of strong boundary traces comes "for free"

identical to those developed for problem (EvPb); the use of a germ is replaced by the use of some maximal monotone graph which encodes a boundary dissipation property analogous to (3).

Yet let us stress that the method of weak boundary trace formulation (Otto [26] and Málek et al. [25]; see also the slightly different definition in [31]) gives the general well-posedness result for the Dirichlet problem (23); indeed, the case of a non-homogeneous flux function $f = f(t,x,u)$ has been treated in the work of Vallet [29]. In an opposite direction, we refer to [7] for a thorough treatment of conservation laws with different nonlinear boundary conditions, in the case of a homogeneous flux $f = f(u)$ and in the strong trace setting. Our argument can be used in the setting of [7] with $f = f(x,u)$, for various boundary conditions.

5.2 *Continuation of Local Entropy Solutions and Justification of Corollary 1*

Let us justify the extension to L^∞ data of the results obtained for $L^1 \cap L^\infty$ data. To this end, we exploit the Dirichlet problem (in its strong-trace formulation) for conservation laws with (x,u)-continuous flux.

Proof of Corollary 1 (sketched). The existence arguments for Theorem 1 do not require the L^1 assumption on the data, hence there is nothing to be generalized at this point.

In order to deduce the uniqueness and the continuous dependence on the data for (EvPb) with L^∞ data, we use the property of finite speed of propagation. Indeed, let u be a $\mathcal{G}$-entropy solution of (EvPb) with some L^∞ datum. Firstly, applying the result of [23] (for conservation laws in Ω^l and Ω^r) we readily see that the solution is uniquely defined by the datum outside the triangle $\mathcal{T} := \{(t,x) \mid t \in (0,T],\ |x| \le Lt\}$ where $L = L_0 + 1$ and L_0 is the uniform in x Lipschitz constant of the flux $\mathfrak{f}(x,\cdot)$. To prove the uniqueness of the solution in $\mathcal{T}$, we construct another $\mathcal{G}$-entropy solution $\tilde{u}$ that coincides with u in $\mathcal{T}$ but which corresponds to an $L^1 \cap L^\infty$ initial datum. Let us give the idea of the construction and sketch the details, which require some careful analysis of the Dirichlet problem for non-homogeneous conservation laws with a "space-like" boundary.[5]

For $h > 2LT$, consider the segments $S_h^\pm := \{x = \pm(h - Lt),\ t \in [0,T]\}$. A.e. $h > 0$ is a Lebesgue point of the maps $h \mapsto u|_{S_h^\pm}$ with values in L^1. Thus, we

[5]Consider a conservation law of the form $\mathrm{div}_{(t,x)}\, \phi(t,x,u) = h(t,x)$ set up in a space-time domain Q. We say that the boundary ∂Q is *space-like* if the map $u \mapsto \phi(t,x,u) \cdot n(t,x)$ is strictly decreasing for all points (t,x) of the boundary. In this case, the local change of variables $w(t,x) := \phi(t,x,u) \cdot n(t,x)$ (the field of exterior unit normal vectors $n(\cdot)$ on ∂Q should be lifted in a neighbourhood of ∂Q) reduces the situation to a standard conservation law with the time direction given by the vector field $n(\cdot)$.

can pick $h_0 > 2LT$ such that strong traces of u on both $S^+_{h_0}$ and $S^-_{h_0}$ exist. Then we set $\tilde{u} \equiv u$ for $t \in [0,T]$ and $|x| \leq h_0 - Lt$ (note that this domain contains $\mathcal{T}$, by the choice of h_0). We extend $\tilde{u}$ to the remaining part of the strip $[0,T] \times \mathbb{R}$ by solving two Cauchy-Dirichlet problems with fluxes $f^l(x,\cdot)$ (for $x < 0$) and $f^r(x,\cdot)$ (for, $x > 0$). For instance, in the domain where $x < -(h_0 + Lt)$ we take the flux $f^l(x,\cdot)$, use the zero initial datum and the boundary datum which is the strong trace γu on $S^-_{h_0}$. To construct the solution in the domain with slanted boundary, it is enough to change the variables. Setting $y = x - Lt + h_0$, in variables (t,y) we obtain a new conservation law in the domain $\Theta = (0,T) \times (-\infty, 0)$, moreover, its characteristics are outgoing on the boundary (this is due to the choice of L and to the change of variable we make). For instance, the result of [29] ensures that there exists a solution to such Cauchy-Dirichlet problem in the domain Θ. Moreover, because the boundary is space-like it can be shown that the solution assumes, in the strong sense, the Dirichlet datum that was prescribed on the boundary.[6] Consider the domain Ω^l; now $\tilde{u}|_{\Omega^l}$ is the juxtaposition of two Kruzhkov entropy solutions on the two sides from the segment $S^-_{h_0}$. It is a Kruzhkov entropy solution, due to the continuity of $\tilde{u}$ that we enforced across the segment $S^-_{h_0}$. In the same way, we see that $\tilde{u}$ is a Kruzhkov entropy solution in the domain Ω^r. Moreover, the trace property (19) for u is inherited by $\tilde{u}$. Thus, using the characterization of Lemma 3 we see that $\tilde{u}$ is indeed a $\mathcal{G}$-entropy solution of (EvPb) corresponding to the truncated initial datum $\tilde{u}_0 = u_0 \mathbb{1}_{[-h_0,h_0]}$. Further, by assumption (H1) it is easy to deduce that, whatever the L^1D germ $\mathcal{G}$ is, the pairs (r,r) with $r \notin (0,1)$ belong to $\mathcal{G}$. Then from the entropy formulation one readily gets the $L^\infty(0,T;L^1(\mathbb{R}))$ bound on $\tilde{u}$.

Now we are in a position to apply the result of Theorem 1. Given two solutions u and $\hat{u}$ with the same initial datum, we obtain $\tilde{u}$ and $\tilde{\hat{u}}$ to which the result of the theorem applies (notice that a common value of h_0 can be taken while constructing $\tilde{u}$ and $\tilde{\hat{u}}$). This ensures that u and $\hat{u}$ coincide between the segments $S^-_{h_0}$ and $S^+_{h_0}$, thus they coincide in the triangle $\mathcal{T}$. This ends the proof of uniqueness. Repeating back to the same arguments but using different initial data, we readily deduce an L^1_{loc} estimate of $u(t,\cdot) - \hat{u}(t,\cdot)$ in terms of the L^1_{loc} distance between u_0 and $\hat{u}_0$. □

Appendix

Throughout the Appendix, we assume that

$$f \text{ is a Lipschitz continuous function of } (t,x,u) \in (0,T) \times \Omega \times \mathbb{R}, \qquad \text{(HA)}$$

[6]To justify this claim, the arguments are the same as for the time-continuity of entropy solutions. Indeed, we have ensured that the normal component of the flux is a strictly increasing function: this makes the normal direction to the boundary *time-like*. Let us stress that the existence of a strong trace for this case is considerably easier to justify than in the general case: as a matter of fact, it follows from a local application of entropy inequalities. We refer to [19] and to [8, Lemma A4] for the arguments that can be used in this context.

where Ω is an open domain of $\mathbb{R}^N$. Our objective is to prove the following "sharp Kato inequality":

Theorem 3. *Assume* (HA). *Let u be a Kruzhkov entropy solution of a conservation law*

$$u_t + \operatorname{div}_x f(t,x,u) = g(t,x) \tag{24}$$

in $(0,T)\times\Omega$. Let $\hat{u}$ be another Kruzhkov entropy solution corresponding to a source term $\hat{g}$. Then one has in $\mathcal{D}'((0,T)\times\Omega)$ the inequality

$$|u-\hat{u}|_t + \operatorname{div}_x \operatorname{sign}(u-\hat{u})\Big(f(t,x,u)-f(t,x,\hat{u})\Big) \le \operatorname{sign}(u-\hat{u})(g-\hat{g}) + \mathbb{1}_{[u=\hat{u}]}|g-\hat{g}|. \tag{25}$$

Remark 4. Notice that the "rough Kato inequality" with the additional term $Const\,|u-\hat{u}|$ on the right-hand side of (25) can be deduced directly from the doubling of variables approach of Kruzhkov [23]. This additional term originates from a bound on $\Big|(\operatorname{div}_x f)(t,x,u) - (\operatorname{div}_x f)(t,x,\hat{u})\Big|$; although this latter term is absent from the formal computation, it appears in the proof whenever the regularity of u is not sufficient to write

$$\operatorname{div}_x \operatorname{sign}(u-k)\Big(f(x,u)-f(x,k)\Big) = \operatorname{sign}(u-k) f_u(x,u)\cdot\nabla u + \operatorname{sign}(u-k)\Big((\operatorname{div}_x f)(x,u)-(\operatorname{div}_x f)(x,k)\Big).$$

Therefore, we argue at the level of the more regular vanishing viscosity approximations, and then observe that locally, every entropy solution of (24) can be seen as a vanishing viscosity limit.

Remark 5. Notice that, considering solutions of the stationary problem $u + \operatorname{div}_x f(x,u) = g$ as time-independent solutions of the corresponding conservation law with the source term $h = g - u$, one deduces (17) from (25).

The proof of Theorem 3 is a straightforward combination of the two following lemmas.

Lemma 5. *Assume* (HA). *Assume that $u \in L^\infty((0,T)\times\Omega)$ is the L^1_{loc} limit, as $\varepsilon \to 0$, of functions u^ε that are solutions (in the variational sense: namely, $u \in V := L^2(0,T;H^1_{loc}(\Omega))$ with the equation satisfied in the dual space of V) of the viscosity approximated equation* (24)*:*

$$u^\varepsilon_t + \operatorname{div}_x f(t,x,u^\varepsilon) - \varepsilon\Delta u^\varepsilon = g(t,x). \tag{26}$$

Similarly, assume $\hat{u} \in L^\infty((0,T)\times\Omega)$ is the L^1_{loc} limit of functions $\hat{u}^\varepsilon$ that are the viscosity approximations of the corresponding equation with the source term $\hat{g}$. Then (25) *holds in $\mathcal{D}'((0,T)\times\Omega)$.*

Lemma 6. *Assume* (HA). *Let u be a Kruzhkov entropy solution of a conservation law* (24) *in* $(0,T)\times\Omega$. *Then there exists a sequence* $(\omega_n)_n$ *of open subdomains of* Ω *such that* $\Omega=\cup_{n=1}^{\infty}\omega_n$ *and in each domain* $(0,T)\times\omega_n$, *the function u is the a.e. limit, as* $\varepsilon\to 0$, *of some solutions* u_n^ε *of equations* (26) *in the domain* $(0,T)\times\omega_n$.

Proof of Lemma 5. The argument is a classical one. One takes $H_\alpha : z\mapsto \int_0^z \frac{1}{\alpha}\mathbb{1}_{[-\alpha,\alpha]})(s)\,ds$ (this is a Lipschitz approximation of the sign function). Set $I_\alpha : z\mapsto \int_0^z H_\alpha(s)\,ds$; we have $I_\alpha(\cdot)\to|\cdot|$ uniformly on $\mathbb{R}$.

Fix $\xi\in\mathcal{D}((0,T)\times\Omega)$. Take the difference of equations (26) written for u^ε and $\hat u^\varepsilon$ and take the test function $H_\alpha(u^\varepsilon-\hat u^\varepsilon)\xi\in L^2(0,T;H^1(\Omega))$ in the corresponding variational formulation. We get

$$\int_0^T\int_\Omega\Big\{-I_\alpha(u^\varepsilon-\hat u^\varepsilon)\,\xi_t-H_\alpha(u^\varepsilon-\hat u^\varepsilon)\Big(f(x,u^\varepsilon)-f(x,\hat u^\varepsilon)-\varepsilon(\nabla u^\varepsilon-\nabla\hat u^\varepsilon)\Big)\cdot\nabla\xi\Big\}$$

$$\le\int_0^T\int_\Omega H_\alpha(u^\varepsilon-\hat u^\varepsilon)(g-\hat g)\,\xi+\frac{1}{\alpha}\iint_{[0<|u^\varepsilon-\hat u^\varepsilon|<\alpha]}(f(x,u^\varepsilon)-f(x,\hat u^\varepsilon))\cdot\nabla(u^\varepsilon-\hat u^\varepsilon)\xi.$$

Here, we have used two chain rules (see in particular [20]) and the fact that for a.e. t, the gradient of the $H^1(\Omega)$ function $(u^\varepsilon-\hat u^\varepsilon)(t,\cdot)$ is zero a.e. on the set where $u^\varepsilon(t,\cdot)-\hat u^\varepsilon(t,\cdot)=const$. Due to the Lipschitz assumption (HA) the last term of the above inequality vanishes, as $\alpha\to 0$. Indeed, it is bounded by the integral of the L^1 function $Const|\nabla u^\varepsilon-\nabla\hat u^\varepsilon|\xi$ over the set $[0<|u^\varepsilon-\hat u^\varepsilon|<\alpha]:=\Big\{(t,x)\,|\,0<|u^\varepsilon(t,x)-\hat u^\varepsilon(t,x)|<\alpha\Big\}$ the measure of which vanishes as $\alpha\to 0$. Thus letting $\alpha\to 0$ then $\varepsilon\to 0$, we deduce (25) in $\mathcal{D}'((0,T)\times\Omega)$. □

Proof of Lemma 6. We will select ω_n in such a way that $u_{(0,T)\times\partial\omega_n}$ admit a strong trace $u^D:=\gamma_{\omega_n}u$ in the L^1 sense, and construct u^ε as solutions to the Cauchy-Dirichlet problem with a smooth boundary datum $u^{D,\delta}$ converging to u^D as $\delta\to 0$.

Indeed, one can represent any open domain Ω in $\mathbb{R}^N$ as a countable union of bounded subdomains Ω_k with C^2 boundary. In each of these subdomains, one considers the parametrization of a neighbourhood of $\partial\Omega_k$ by parameters $\sigma\in\partial\Omega_k$ and $h\in(0,h_{max})$, where $h=\mathrm{dist}(x,\partial\Omega_k)$. A.e. h is a Lebesgue point of the map $h\mapsto u|_{(0,T)\times\Sigma_k^h}$ where $\Sigma_k^h:=\{x\in\Omega_k\mid \mathrm{dist}(x,\partial\Omega_k)=h\}$. Thus for every k, one can pick a countable sequence $(\omega_{k,m})_m$ of Lipschitz subdomains of Ω_k such that $\Omega_k=\cup_m\omega_{k,m}$ and u has a *strong* trace (in the L^1 sense) on $(0,T)\times\partial\omega_{k,m}$. We can re-label $\omega_{k,m}$ by a subscript $n\in\mathbb{N}$. From now on, we fix n and write ω for ω_n.

To conclude the proof, combining classical techniques we will construct a vanishing viscosity limit $\tilde u$ which is a Kruzhkov entropy solution of the problem (24) in $(0,T)\times\omega$ with the initial condition $\tilde u(0,\cdot)=u(0,\cdot)$ (cf. Remark 1 for the issue of time-continuity of local entropy solutions) and the formal boundary condition $\tilde u|_{(0,T)\times\partial\omega}=u^D$, where u^D is the strong trace of u on $(0,T)\times\partial\omega$. Then we will justify the fact that u and $\tilde u$ coincide; notice that at this level, the "rough version" of the Kato inequality (25) (see Remark 4) is enough to "compare" u and $\tilde{u}$. Let us provide the details of these arguments.

First, one approximates u^D and $u_0 := u(0,\cdot)$ a.e. on their respective domains by BV functions $u^{D,\delta}$ and u_0^δ. Then one constructs the solutions $\tilde{u}^{\varepsilon,\delta}$ of (26) in $(0,T)\times\omega$ with the corresponding initial and boundary data u_0^δ and $u^{D,\delta}$ using the results of the classical work [12]. As shown in [12], $\tilde{u}^{\varepsilon,\delta}$ converge, as $\varepsilon \to 0$, to an entropy solution $\tilde{u}^\delta$ of the conservation law (24) with the same initial datum u_0^δ and with the same Dirichlet datum $u^{D,\delta}$ understood in the Bardos-LeRoux-Nédélec sense. It remains to obtain $\tilde{u} = \lim_{\delta\to 0} \tilde{u}^\delta$ and to prove that $\tilde{u}$ and u coincide. To this end, we exploit the "rough Kato inequality" of [23] (see Remark 4) with test functions of the form $\xi_s(x)\eta(t)$, where $\eta \in \mathcal{D}(0,T)$, $\eta \geq 0$, and $(\xi_s)_{s>0}$ is the sequence in $W_0^{1,\infty}(\omega)$ given by $\xi_s = \min\{1, \mathrm{dist}(x,\partial\omega)/s\}$. By a straightforward calculation, at the limit $s \to 0$ we find the inequality

$$-\int_0^T\!\!\int_\omega |u-\tilde{u}^\delta|\,\eta_t \leq Const \int_0^T\!\!\int_\omega |u-\tilde{u}^\delta|\eta - \int_0^T\!\!\int_{\partial\omega} \Big(\mathrm{sign}(u^D-\gamma\tilde{u}^\delta)\Big(f(t,x,u^D)-f(t,x,\gamma\tilde{u}^\delta)\Big)\cdot n_{\partial\omega}\Big)\,\eta, \tag{27}$$

where $n_{\partial\omega}$ is the exterior unit normal vector to $\partial\omega$ and $\gamma\tilde{u}^\delta$ is the strong trace of the $L^\infty(0,T;BV(\omega))$ function $\tilde{u}^\delta$. By the result of [12], one has for a.e. (t,x) (with respect to the Hausdorff measure on $(0,T)\times\partial\omega$) the property $(\gamma\tilde{u}^\delta)(t,x) \in I(t,x,u^{D,\delta}(t,x))$ where

$$I(t,x,v) = \{u \in \mathbb{R} \,|\, \forall k \in [\min\{v,u\},\max\{v,u\}]\ \ \mathrm{sign}(k-v)\Big(f(t,x,k)-f(t,x,v)\Big)\cdot n_{\partial\omega} \geq 0\}.$$

From the definition of $I(t,x,u^{D,\delta})$ and assumption (HA) it is easily seen that the last term in (27) is bounded from above by $Const|u^D - u^{D,\delta}|$, which vanishes as $\delta \to 0$. Letting $\delta \to 0$, using the Gronwall inequality one sees that $\tilde{u}^\delta \to u$ as $\delta \to 0$. Hence one can extract a family $\tilde{u}^{\varepsilon(\delta),\delta}$ of local solutions on (26) that converges to u, as $\delta \to 0$. This concludes the proof of the lemma. □

Remark 6. For the one-dimensional stationary problem (i.e., in the context of Proposition 1) a simpler construction can be used in place of the one exploited in the proof of Lemma 6. Indeed, it is enough to take, for example, the function $u|_{\mathbb{R}^l}$ and extend it to $\mathbb{R}$ by setting $\tilde{u}(x) \equiv \gamma^l u = const$ for $x > 0$. Then it is clear that the extension $\tilde{u}$ of u is an entropy solution on $\mathbb{R}$ of the stationary problem $\tilde{u} + \tilde{f}^l(x,\tilde{u}) = \tilde{h}$ with the flux $f^l(x,\cdot)$ extended by $f^l(0,\cdot)$ for $x \geq 0$; also the source term h has to be extended by $\tilde{h}(x) = \gamma^l u = const$ for $x > 0$. Then one can use the classical result of Kruzhkov [23] which guarantees uniqueness of entropy solutions and convergence of vanishing viscosity approximations for the conservation law in the whole space.

Acknowledgements The author thanks Kenneth H. Karlsen for turning his attention to the difficulty treated in the Appendix. The work on this paper was partially supported by the French ANR project CoToCoLa.

References

1. J.J. Adimurthi, G.D. Veerappa Gowda, Godunov-type methods for conservation laws with a flux function discontinuous in space. SIAM J. Numer. Anal. **42**(1), 179–208 (2004)
2. S.M. Adimurthi, G.D. Veerappa Gowda, Optimal entropy solutions for conservation laws with discontinuous flux-functions. J. Hyperbolic Differ. Equ. **2**(4), 783–837 (2005)
3. B. Andreianov, One-dimensional conservation law with boundary conditions: general results and spatially inhomogeneous case, in *Proceedings of HYP-2012 Conference*, Padua (accepted). Available as HAL preprint http://hal.archives-ouvertes.fr/hal-00761664
4. B. Andreianov, F. Bouhsiss, Uniqueness for an elliptic-parabolic problem with Neumann boundary condition. J. Evol. Equ. **4**(2), 273–295 (2004)
5. B. Andreianov, C. Cancès, Vanishing capillarity solutions of Buckley-Leverett equation with gravity in two-rocks' medium. Comput. Geosci. **17**(3), 551–572 (2013)
6. B. Andreianov, M.K. Gazibo, Entropy formulation of degenerate parabolic equation with zero-flux boundary condition. ZAMP – Zeitschr. Angew. Math. Phys. (2013). doi:10.1007/s00033-012-0297-6
7. B. Andreianov, K. Sbihi, Well-posedness of general boundary-value problems for scalar conservation laws. Trans. AMS (to appear). Available as HAL preprint http://hal.archives-ouvertes.fr/hal-00708973
8. B. Andreianov, M. Bendahmane, K.H. Karlsen, Discrete duality finite volume schemes for doubly nonlinear degenerate hyperbolic-parabolic equations. J. Hyperbolic Differ. Equ. **7**(1), 1–67 (2010)
9. B. Andreianov, K.H. Karlsen, N.H. Risebro, A theory of L^1-dissipative solvers for scalar conservation laws with discontinuous flux. Arch. Ration. Mech. Anal. **201**, 27–86 (2011)
10. E. Audusse, B. Perthame, Uniqueness for scalar conservation laws with discontinuous flux via adapted entropies. Proc. R. Soc. Edinb. A **135**(2), 253–265 (2005)
11. P. Baiti, H.K. Jenssen, Well-posedness for a class of 2×2 conservation laws with L^∞ data. J. Differ. Equ. **140**(1), 161–185 (1997)
12. C. Bardos, A.Y. Le Roux, J.-C. Nédélec, First order quasilinear equations with boundary conditions. Commun. Partial Differ. Equ. **4**(4), 1017–1034 (1979)
13. P. Bénilan, *Équations d'évolution dans un espace de Banach quelconque et applications* (Thèse d'état, Orsay, 1972)
14. P. Bénilan, S.N. Kruzhkov, Conservation laws with continuous flux functions. NoDEA Nonlinear Differ. Equ. Appl. **3**(4), 395–419 (1996)
15. P. Bénilan, P. Wittbold, On mild and weak solutions of elliptic-parabolic problems. Adv. Differ. Equ. **1**(6), 1053–1073 (1996)
16. P. Bénilan, J. Carrillo, P. Wittbold, Renormalized entropy solutions of scalar conservation laws. Ann. Scuola Norm. Sup. Pisa Cl. Sci. (4) **29**(2), 313–327 (2000)
17. P. Bénilan, M.G. Crandall, A. Pazy, *Nonlinear Evolution Equations in Banach Spaces* (preprint book)
18. R. Bürger, K.H. Karlsen, J. Towers, An Engquist-Osher type scheme for conservation laws with discontinuous flux adapted to flux connections. SIAM J. Numer. Anal. **47**, 1684–1712 (2009)
19. C. Cancès, T. Gallouët, On the time continuity of entropy solutions. J. Evol. Equ. **11**, 43–55 (2011)
20. J. Carrillo, P. Wittbold, Uniqueness of renormalized solutions of degenerate elliptic-parabolic problems. J. Differ. Equ. **156**(1), 93–121 (1999)
21. G.-Q. Chen, H. Frid, Divergence-Measure fields and hyperbolic conservation laws. Arch. Ration. Mech. Anal. **147**, 89–118 (1999)
22. R. Eymard, T. Gallouët, R. Herbin, in *Finite Volume Methods*, ed. by P. Ciarlet, J.-L. Lions. Handbook of Numerical Analysis, vol. VII (North-Holland, 2000)
23. S.N. Kruzhkov, First order quasilinear equations with several independent variables. Mat. Sb. (N.S) **81**(123), 228–255 (1970)

24. Y.-S. Kwon, A. Vasseur, Strong traces for solutions to scalar conservation laws with general flux. Arch. Ration. Mech. Anal. **185**(3), 495–513 (2007)
25. J. Málek, J. Nečas, M. Rokyta, M. Ružička, *Weak and Measure-Valued Solutions to Evolutionary PDEs* (Chapman & Hall, London, 1996)
26. F. Otto, Initial-boundary value problem for a scalar conservation law. C. R. Acad. Sci. Paris Sér. I **322**, 729–734 (1996)
27. E.Y. Panov, Existence of strong traces for generalized solutions of multidimensional scalar conservation laws. J. Hyperbolic Differ. Equ. **2**(4), 885–908 (2005)
28. E.Y. Panov, Existence of strong traces for quasi-solutions of multidimensional conservation laws. J. Hyperbolic Differ. Equ. **4**(4), 729–770 (2007)
29. G. Vallet, Dirichlet problem for a nonlinear conservation law. Rev. Math. Complut. **13**(1), 231–250 (2000)
30. A. Vasseur, Strong traces for weak solutions to multidimensional conservation laws. Arch. Ration. Mech. Anal. **160**(3), 181–193 (2001)
31. J. Vovelle, Convergence of finite volume monotone schemes for scalar conservation laws on bounded domains. Numer. Math. **90**(3), 563–596 (2002)

On Numerical Methods for Hyperbolic Conservation Laws and Related Equations Modelling Sedimentation of Solid-Liquid Suspensions

F. Betancourt, R. Bürger, R. Ruiz-Baier, H. Torres, and C.A. Vega

Abstract A classical kinematical model of sedimentation of small equal-sized particles dispersed in a viscous fluid leads to a scalar conservation law with a nonlinear flux. Several extensions of this model are reviewed, with a strong focus on recently developed numerical methods. These extensions include a one-dimensional clarifier-thickener model giving rise to a conservation law with discontinuous flux, a conservation law with nonlocal flux, systems of nonlinear conservation modelling the sedimentation of polydisperse suspensions, and sedimentation-flow models consisting of a conservation law coupled with the Stokes or Navier-Stokes system in two space dimensions. Numerical examples are presented.

F. Betancourt
Departamento de Ingeniería Metalúrgica, Facultad de Ingeniería, Universidad de Concepción, Casilla 160-C, Concepción, Chile
e-mail: fbetancourt@udec.cl

R. Bürger (✉)
CI^2MA and Departamento de Ingeniería Matemática, Facultad de Ciencias Físicas y Matemáticas, Universidad de Concepción, Casilla 160-C, Concepción, Chile
e-mail: rburger@ing-mat.udec.cl

R. Ruiz-Baier
Modeling and Scientific Computing, MATHICSE, Ecole Polytechnique Fédérale de Lausanne EPFL, Station 8, CH-1015, Lausanne, Switzerland
e-mail: ricardo.ruiz@epfl.ch

H. Torres
Departamento de Matemáticas, Facultad de Ciencias, Universidad de La Serena, Av. Cisternas 1200, La Serena, Chile
e-mail: htorres@ing-mat.udec.cl

C.A. Vega
Departamento de Matemáticas y Estadística, División de Ciencias Básicas, Universidad del Norte, Barranquilla, Colombia
e-mail: cvega@uninorte.edu.co

G.-Q.G. Chen et al. (eds.), *Hyperbolic Conservation Laws and Related Analysis with Applications*, Springer Proceedings in Mathematics & Statistics 49,
DOI 10.1007/978-3-642-39007-4_2,

2010 Mathematics Subject Classification 65M06, 65M08, 65M60, 76M20, 76T20

1 Introduction

1.1 Scope

The sedimentation of small particles dispersed in a viscous fluid under the influence of a (mostly gravitational) body force is a process of theoretical and practical interest that appears as a controlled unit operation in mineral processing, wastewater treatment, the pulp-and-paper and chemical industry, medicine, volcanology, and other areas where a suspension must be separated into a clarified liquid and concentrated sediment. The particles are small compared with typical length scales (diameter and depth) of the settling vessel. Moreover, sedimentation models for these applications should be able to predict the behaviour of a given unit on relatively large temporal and spatial scales, while microscopical information such as, for instance, the position of a given particle is of little interest. These considerations justify representing the liquid and the solid particles as superimposed continuous phases, namely a liquid phase and one or several solid phases.

The most widely used sedimentation model goes back to Kynch [64], who postulated that (under idealizing circumstances) the settling velocity v_s of a single particle in a batch column is a given function of the local solids volume fraction u only, $v_s = v_s(u)$. Inserting this assumption into the one-dimensional solids continuity equation, written in differential form as

$$u_t + (uv_s)_x = 0, \tag{1}$$

where t is time and x is depth, yields the first-order scalar conservation law

$$u_t + b(u)_x = 0, \quad b(u) := uv_s(u), \tag{2}$$

which is supplied with suitable initial and boundary conditions.

If we assume (for simplicity, but without loss of generality) that u varies between $u = 0$, the clear-liquid limit, and $u = u_{\max}$ with $u_{\max} = 1$ for a packed bed, then a common approach is

$$v_s(u) = v_{St}V(u), \tag{3}$$

where v_{St} is the Stokes velocity, that is, the settling velocity of a single particle in an unbounded fluid, and the so-called hindered settling factor $V = V(u)$ can, for instance, be the one given by Richardson and Zaki [75]

$$V(u) = (1-u)^{n_{RZ}}, \quad n_{RZ} \geq 1, \tag{4}$$

so that $b(u) = u(1-u)^{n_{\mathrm{RZ}}}$. For $n_{\mathrm{RZ}} > 1$, this function has an inflection point $u_{\mathrm{infl}} = 2/(1+n_{\mathrm{RZ}}) \in (0,1)$. Thus, the basic mathematical model is a nonlinear, scalar conservation law with non-convex flux. The precise algebraic form of the batch flux density function $b = b(u)$ is a specific property of the material under consideration.

As it stands, (2) only applies to batch settling of a suspension of small equal-sized (monodisperse) spherical particles. Extensions of (2) have been made, for instance, to include continuously operated so-called clarifier-thickener units, to handle suspensions of particles forming compressible sediments, and to describe polydisperse suspensions with particles having different sizes and densities. Moreover, the dependence of v_{s} on the spot value $u = u(x,t)$ has been replaced by a non-local one, and multi-dimensional versions of (2) have been formulated, which require the solution of additional equations for the motion of the mixture. These extensions give rise to conservation laws with a flux that depends discontinuously on x, strongly degenerate parabolic equations, strongly coupled systems of nonlinear, first-order conservation laws, conservation laws with non-local flux, and multi-dimensional conservation laws coupled with the Stokes or Navier-Stokes system. Thus, the mathematical framework for many sedimentation models relevant to applications includes the well-posedness and numerical analysis of nonlinear hyperbolic conservation laws and related equations. The resulting models have some intriguing non-standard properties that make them interesting objects of study for the well-posedness and numerical analysis of conservation laws and related equations. On the other hand, a thorough understanding of the properties of these models is necessary for the design of reliable numerical simulation tools. This is a particular challenge for clarifier-thickener units. It is the purpose of this contribution to review recent advances in this area.

1.2 Some Historical Remarks and Motivation

To put the original research problem into the proper historical perspective of the engineering application, we first mention that extensive historical accounts are provided in [17, 33]. The exploitation of the difference in density between solid particles and fluid for operations of washing ores can be traced back at least to the ancient Egyptians [94]. The use of settling tanks, operated in a batch or semi-continuous manner, for processes that can now be identified as classification, clarification and thickening, was described in detail in Georgius Agricola's book *De Re Metallica*, first published in 1556 [17, 33]. The most important technological invention that would rationalize the settling process is the continuous thickener, introduced by J.V.N. Dorr, a chemist, cyanide mill owner, consulting engineer and plant designer, in the early twentieth century [44]. A continuous thickener is essentially a cylindrical settling tank into which the feed suspension to be separated is fed continuously, the sediment forming by settling of particles is removed continuously, and the clear liquid produced is removed by a circumferential launder,

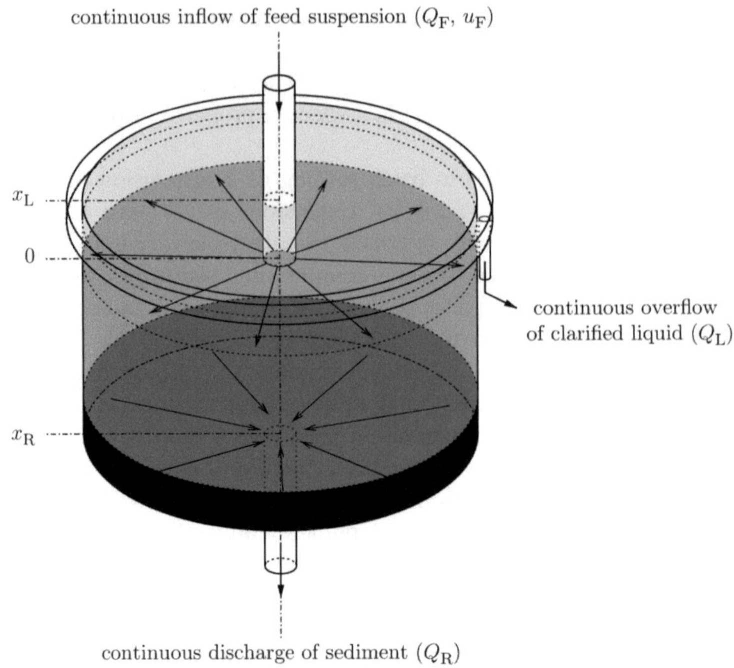

Fig. 1 Schematic view of a clarifier-thickener (CT). Technical details are omitted

see Fig. 1. This design is widely used today in mineral processing and in secondary settling tanks in wastewater treatment.

The invention of the clarifier-thickener was soon followed by efforts to mathematically model its operation. It was recognized early [35] that understanding the dynamics of the batch settling process of a suspension at different solids concentrations is fundamental for effective thickener design and control.

The starting point of the mathematical modelling of sedimentation is the well-known Stokes formula, which states that the settling velocity of a sphere of size (diameter) d and density ρ_s in an unbounded fluid of density ρ_f and viscosity μ_f is given by

$$v_{St} = \frac{gd^2(\rho_s - \rho_f)}{18\mu_f}, \tag{5}$$

where g denotes acceleration of gravity. The settling velocity of a particle in a concentrated suspension is, however, smaller than (5) due to the hindrance exerted by the presence of other particles. This effect can be expressed as an increase in viscosity of the suspension. Explicit formulas describing the phenomenon of hindered settling of the type (3), where the hindered settling factor $V = V(u)$ should satisfy $V(0) = 1$, $V(u_2) < V(u_1)$ for $u_1 < u_2$ and $V(u_{max}) = 0$, were derived in the dilute limit $u/u_{\max} \ll 1$ more than a century ago by A. Einstein [45], and in the 1940s for both dilute and concentrated suspensions (see, e.g., [55, 84, 89]). It was in Kynch's specific contribution [64] that he explicitly *solved* the governing equation (1) under the assumption $v_s = v_{St} V(u)$, for initially constant concentrations. In mathematical terms, if the function b has support on the interval $(0, u_{\max})$, then the settling of an initially homogeneous suspension of concentration $u_0 \in (0, u_{\max})$ in a column of depth L can be described by the initial-value problem for (2) defined by the piecewise constant initial datum

$$u(x,0) = \begin{cases} 0 & \text{for } x < 0, \\ u_0 & \text{for } 0 < x < L, \\ u_{\max} & \text{for } x > L \end{cases} \tag{6}$$

corresponding to two adjacent Riemann problems. Kynch [64] applied the method of characteristics and resolving cases of intersection by discontinuities based on physical principles that agree with theoretically motivated entropy conditions to be introduced much later. One piece of insight these constructions could provide is the explanation why fairly dilute and concentrated suspensions would settle with a sharp interface and a zone of continuous transition of concentration separating the growing sediment from the bulk suspension; namely, the former situation gives rise to a kinematic shock (in u) and the latter to a rarefaction.

Kynch's efforts were followed by systematic classifications of qualitatively different solutions to (2) and (6) [51, 90]. Based on work by Ballou [3], K.S. Cheng [34] and Liu [67] (see [33]), Bustos and Concha [32] and Diehl [40] appropiately embedded these constructions into the theory of entropy solutions of a scalar conservation law with non-convex flux. The interest Kynch's theory immediately caused in mineral processing, wastewater treatment (where it has become known as the *solids flux theory*) and other applicative areas has been widely discussed in some reviews (e.g., [17, 42]). Clearly, to make this theory applicable to the settling of a given suspension one must assume that the factor $V = V(u)$ is known. The reliable identification of this factor or equivalently, of the function $b = b(u)$, from experimental data is a current research problem in itself [37, 41, 50].

The model is very similar to the well-known Lighthill-Whitham-Richards (LWR) model for traffic flow. In fact, in textbooks on hyperbolic conservations, the LWR model forms the preferred example, since the typical flux $b(u) = u(1-u)$ arising in that model is convex and allows for simpler construction of solutions, and the initial value problem (Riemann problem) for such an equation is easier to handle, than for the problem (2) and (6) with b non-convex. The construction of solutions for the direct problem of (2) with piecewise constant initial data and constant u_0 (6) is in any case well understood and for decades has formed standard material for engineering textbooks including [74, 91]. The extensions mentioned in Sect. 1.1 do, however, give rise to research problems centering around the well-posedness and numerical analysis and efficient numerical simulation of the corresponding model.

1.3 Outline of This Contribution

The model for continuous sedimentation was later improved to the configuration of a so-called clarifier-thickener. The basic idea is to replace boundary conditions that would describe feed and discharge operations in a continuously operated unit by changes of the definition of the convective flux. This results in a flux with discontinuities with respect to spatial position, which reflect the injection of feed suspension at a certain level of height into an idealized unit, and the split of the feed flow into upward- and downward-directed bulk flows of the mixture. If sediment compressibility is ignored for the moment, then the resulting model can be expressed as a conservation law with a discontinuous flux:

$$u_t + f\big(\boldsymbol{\gamma}(x), u\big)_x = 0, \quad (x,t) \in \Pi_T := \mathbb{R}, \times (0,T], \tag{7}$$

$$u(x,0) = u_0(x), \quad x \in \mathbb{R}, \tag{8}$$

where $\boldsymbol{\gamma}(x)$ is a given vector of discontinuous parameters. The basic associated difficulty is that well-posedness for (7) is ensured [62] for smooth functions $\boldsymbol{\gamma} = \boldsymbol{\gamma}(x)$, but the theory for discontinuous $\boldsymbol{\gamma} = \boldsymbol{\gamma}(x)$ does *not* emerge as a "limit case" for smooth parameter vectors that approximate a discontinuous one. It turns out that one has to explicitly specify which discontinuities of the solution u are considered to be admissible across the jumps in $\boldsymbol{\gamma}$.

The model was later extended to include the effect of sediment compressibility; the governing equation can then be expressed as

$$u_t + f\big(\boldsymbol{\gamma}(x), u\big)_x = \big(\gamma_2(x) A(u)_x\big)_x, \tag{9}$$

where $\boldsymbol{\gamma} = (\gamma_1, \gamma_2)$ and γ_2 are now discontinuous vectorial and scalar functions, respectively, of x, and $A(\cdot)$ typically has the behaviour

$$A(u) := \int_0^u a(s)\,\mathrm{d}s, \quad a(u) \begin{cases} = 0 & \text{for } u \leq u_\mathrm{c}, \\ > 0 & \text{for } u > u_\mathrm{c}, \end{cases} \quad u_\mathrm{c} > 0, \tag{10}$$

where u_c is a critical concentration above which the solid particles touch each other.

The well-posedness analysis of the model (7) or (9), together with (8), has been a small part of the tremendous interest and activity conservation laws and related equations with dicontinuous flux have seen in recent years. Partial overviews are given in [16, 23], while a comprehensive and unifying treatment is provided by Andreianov, Karlsen, and Risebro [2]. While some of the previous existence results are based on the convergence of suitable monotone, and therefore first-order, finite difference schemes (cf., e.g., [19–21, 23, 25, 59, 88] and [60] for the underlying L^1 stability theory), it is desirable for practical purposes to construct higher order schemes, for examples analogues to second-order TVD schemes for standard conservation laws, for which one would be able to prove convergence at least to a weak solution. In Sect. 2, which summarizes results of [25], two different methodologies to construct a simple TVD scheme and a flux-TVD scheme, respectively, are illustrated, along with an outline of the convergence analysis for the flux-TVD scheme that is based on a nonlocal flux limiter algorithm.

In Sect. 3 we study the family of conservation laws with nonlocal flux

$$u_t + \big(u(1-u)^{\alpha} V(K_a * u)\big)_x = 0, \quad x \in \mathbb{R}, \quad t \in (0,T], \tag{11}$$

together with the initial datum

$$u(0,x) = u_0(x), \quad 0 \le u_0(x) \le 1, \quad x \in \mathbb{R}, \tag{12}$$

where either $\alpha = 0$ or $\alpha \ge 1$. Usually, one defines a kernel $K = K(x)$ with support on $[-2,2]$ and sets $K_a(x) := a^{-1}K(a^{-1}x)$ with support on $[-2a, 2a]$. The basic motivation of the nonlocal dependence (34) lies in the observation that Kynch's theory, despite being a useful approximation, sharply contrasts with the theoretical result that the velocity of each particle is determined by the size and position of all spheres and the nature of possible boundaries. The convolution of u with a weighting function, an assumption that eventually leads to (34) (see [12]), is a compromise.

In [12] the well-posedness of (11) and (12) is studied. The main results are the uniqueness and existence of entropy solutions. This is done by proving convergence of a difference-quadrature scheme based on the standard Lax-Friedrichs scheme. It turns out that for $\alpha = 0$, solutions are bounded by a constant that depends on the final time T, and are Lipschitz continuous if u_0 is Lipschitz continuous. In contrast, for $\alpha \ge 1$ solutions are in general discontinuous even if u_0 is smooth, but assume values within the interval $[0, 1]$ for all times. Some numerical examples illustrate the solution behaviour, in particular the so-called effect of layering in sedimenting suspensions and the differences between the cases $\alpha = 0$ and $\alpha \ge 1$. These results are summarized in Sect. 3.

Next, in Sect. 4, we will consider models of sedimentation of polydisperse suspensions. These mixtures consist of small solid particles that belong to a number N of species that may differ in size or density, and which are dispersed in a viscous fluid. Here we only consider particles of the same density. If ϕ_i denotes the volume fraction of particle species i having diameter D_i, where we assume that $D_1 > D_2 > \cdots > D_N$, and v_i is the phase velocity of species i, then the continuity equations of the N species are $\partial_t \phi_i + \partial_x(\phi_i v_i) = 0$, where t is time and x is depth. (In this section any statement involving a free index i is supposed to hold for $i = 1, \dots, N$.) The velocities v_i are assumed to be given functions of the vector $\Phi := \Phi(x,t) := (\phi_1(x,t), \dots, \phi_N(x,t))^{\mathrm{T}}$ of local concentrations. This yields nonlinear, strongly coupled systems of conservation laws of the type

$$\partial_t \Phi + \partial_x \boldsymbol{f}(\Phi) = 0, \quad \boldsymbol{f}(\Phi) := \big(f_1(\Phi), \dots, f_N(\Phi)\big)^{\mathrm{T}}, \quad f_i(\Phi) := \phi_i v_i(\Phi). \tag{13}$$

We seek solutions $\Phi = \Phi(x,t)$ that take values in the closure of the set

$$\mathcal{D}_{\phi_{\max}} := \left\{ \Phi \in \mathbb{R}^N \, : \, \phi_1 > 0, \dots, \phi_N > 0, \, \phi := \phi_1 + \cdots + \phi_N < \phi_{\max} \right\}.$$

The parameter $0 < \phi_{\max} \le 1$ is a given maximum solids concentration. For batch settling in a column of height L, (13) is defined on $\Omega_T := \{(x,t) \in \mathbb{R}^2 \,|\, 0 \le x \le L, \; 0 \le t \le T\}$ for a given final time $T > 0$ along with the initial condition

$$\Phi(x,0) = \Phi^0(x) = \left(\phi_1^0(x), \dots, \phi_N^0(x)\right)^{\mathrm{T}}, \quad \Phi^0(x) \in \bar{\mathcal{D}}_{\phi_{\max}}, \quad x \in [0, L]$$

and the zero-flux boundary conditions

$$\boldsymbol{f}\,|_{x=0} = \boldsymbol{f}\,|_{x=L} = 0. \tag{14}$$

Several choices of v_i ("models") as functions of Φ, and depending on the vector of normalized particle sizes $\boldsymbol{d} := (d_1, \dots, d_N)^{\mathrm{T}}$, where $d_i := D_i/D_1$, have been proposed [96]. We here discuss the models due to Masliyah [68] and Lockett and Bassoon [65] (the "MLB model") and Höfler and Schwarzer [56] (the "HS model"), respectively. Both models are strictly hyperbolic for all $\Phi \in \mathcal{D}_{\phi_{\max}}$, for arbitrary N, and under certain restrictions on model parameters and d_N [24]. We mention here that hyperbolicity for a large range of parameter values is a desirable property for polydisperse sedimentation models with equal-density particles, since such mixtures have been observed to always settle stably, i.e., under the formation of horizontal layers and interfaces. Instabilities, such as the formation of blobs and columns, have been observed with particles having different densities only [93], and their ocurrence is predicted by a criterion equivalent to loss of hyperbolicity [6, 18].

In Sect. 4 the main results of [26] are summarized. Specifically, the results in [24] provide a good estimate of the viscosity coefficient in a Lax-Friedrichs-type flux splitting. This allows one to construct high-resolution component-wise weighted essentially non-oscillatory (WENO) schemes (cf. [79] and its references) for the numerical solution of (13)–(14). In addition, the full spectral decomposition of $\mathcal{J}_f(\Phi)$, which can now be computed numerically, can be used to obtain *characteristic-based* WENO schemes, for which the WENO reconstruction procedure is applied to the local characteristic variables and fluxes at each cell-interface. When combined with a strong stability preserving (SSP) Runge-Kutta-type time discretization (see [49]), the resulting SSP-WENO-SPEC schemes turn out to be extremely robust. Here we summarize results related to the hyperbolicity analysis and the construction of the aforementioned schemes, and present some numerical examples.

In Sect. 5 we are concerned with the simulation of sedimentation of monodisperse suspensions in several space dimensions. In fact, for the realistic description of the sedimentation of suspensions in two- or three-dimensional (2D, 3D) domains the governing system of PDEs is a (possibly degenerate) convection-diffusion equation coupled with a version of the Stokes or Navier-Stokes system, supplied with suitable initial and boundary conditions.

A prototype model of this kind is given by the following system, where the local solids concentration u, the mixture velocity $\boldsymbol{v}$ and the pressure p are sought:

$$u_t + \nabla \cdot (u\boldsymbol{v} + f(u)\boldsymbol{k}) = \Delta A(u), \quad \boldsymbol{x} \in \Omega \subset \mathbb{R}^d, \quad t \in (0,T], \tag{15}$$

$$\begin{aligned} \nu(\rho_s u + \rho_f(1-u))(\boldsymbol{v}_t + \boldsymbol{v}\cdot\nabla\boldsymbol{v}) - \nabla\cdot\big(\mu(u)\nabla\boldsymbol{v}\big) + \lambda\nabla p &= \zeta u\boldsymbol{k}, \\ \nabla\cdot\boldsymbol{v} &= 0, \end{aligned} \tag{16}$$

where $d = 2$ or 3, $f(u) = uV(u)$, $\boldsymbol{k}$ is the upwards-pointing unit vector, the term $\Delta A(u)$ accounts for sediment compressibility where the integrated diffusion coefficient $A(\cdot)$ has the behaviour (10), $\mu(u)$ is a viscosity function, and $\nu \geq 0$, $\zeta > 0$ and $\lambda > 0$ are constants. Note that the convection-diffusion equation (15) involves the linear transport term $u\boldsymbol{v}$, while $\boldsymbol{v}$ (and p) are determined by the Navier-Stokes or Stokes (for $\nu > 0$ and $\nu = 0$, respectively) system (16). This strong coupling of (15) and (16) is the main challenge for solving this sedimentation-flow model. The equations (16) do not have to be solved in a 1D setting, since then $v_x = 0$, so in absence of sources or sinks, $v = v(t)$ becomes controllable. We present numerical results for two-dimensional subcases of (15) and (16) discretized either by finite volume schemes combinded with an adaptive multiresolution technique or by a finite volume element scheme.

Some open research problems and alternate treatments are discussed in Sect. 6.

2 TVD and Flux-TVD Schemes for Clarifier-Thickener Models

2.1 Clarifier-Thickener Models

The basic principle of operation of a clarifier-thickener can be inferred from Fig. 1. The feed suspension, which is to be separated into a concentrated sediment and a clarified liquid, is fed into a cylindrical vessel at depth level $x = 0$, at a volume rate $Q_F \geq 0$ and with a feed solids volume fraction $u_F \geq 0$. The feed flow immediately spreads over the whole cross section, and is separated into upward- and downward-directed bulk flows forming the so-called clarification and thickening zones $x_L < x < 0$ and $0 < x < x_R$, respectively. The solid particles settle downward, forming a concentrated sediment at the bottom which is continuously removed at a controllable discharge volume rate $Q_R \geq 0$, while the overflowing supernatant liquid is collected in a circumferential launder. The (signed) liquid overflow rate is $Q_L \leq 0$, such that $Q_F = Q_R - Q_L$. We assume that solid-liquid separation takes place within the unit only, but not in the overflow and discharge flows, where both phases move with the

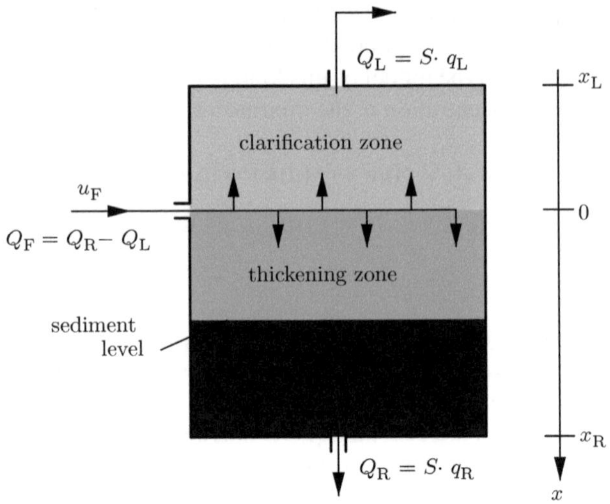

Fig. 2 One-dimensional idealized clarifier-thickener model

same speed. In applications, real-world units usually have a gently sloped bottom; however in this review we assume that the cross-sectional area S is constant.

If we assume that all flow variables are horizontally constant and wall effects are negligible, then the conceptual model reduces to the setup shown in Fig. 2. To derive the final mathematical model, we replace the solids and fluid phase velocities v_s and v_f by the volume average velocity of the mixture, $q := uv_s + (1-u)v_f$ and the solid-fluid relative velocity $v_r = v_s - v_f$. One then always has $q_x = 0$, i.e. $q = q(t)$ in the absence of sources and sinks, and $v_s = q + (1-u)v_r$. In particular, $q = 0$ for settling in a closed column. For the clarifier-thickener model of Fig. 2, the velocities q_R, q_L and q_F are related to the signed volume bulk flows by $q_R = Q_R/S$, $q_L = Q_L/S$ and $q_F = Q_F/S$. Moreover, stating the constitutive assumption as

$$v_r(u) = \frac{b(u)}{u(1-u)},$$

we obtain the governing equation (7), where

$$f\big(\boldsymbol{\gamma}(x), u\big) = \gamma_1(x)b(u) + \gamma_2(x)(u - u_F).$$

The parameters γ_1 and γ_2 are defined as follows, and discriminate between the interior and exterior of the unit and the directions of the bulk flows, respectively:

$$\gamma_1(x) := \begin{cases} 1 & \text{for } x \in (x_L, x_R), \\ 0 & \text{for } x \notin (x_L, x_R), \end{cases} \qquad \gamma_2(x) := \begin{cases} q_L & \text{for } x < 0, \\ q_R & \text{for } x > 0. \end{cases} \tag{17}$$

If we include the effect of sediment compressibility, then the governing equation is given by (9), where γ_1 and γ_2 are still given by (17).

By a solution to the hyperbolic problem (7) and (8) we understand the following, where BV_t denotes the space of locally integrable functions on Π_T for which u_t (but not u_x) is a locally bounded measure, which is a superset of BV.

Definition 1 (**BV_t weak solution**). A measurable function $u : \Pi_T \to \mathbb{R}$ is a BV_t weak solution of (7) and (8) if $u \in (L^\infty \cap BV_t)(\Pi_T)$, and if for all test functions $\phi \in \mathcal{D}(\mathbb{R} \times [0, T))$,

$$\iint_{\Pi_T} \Big(u\phi_t + f\big(\boldsymbol{\gamma}(x), u\big)\phi_x \Big)\, dx\, dt + \int_{\mathbb{R}} u_0 \phi(x, 0)\, dx = 0.$$

2.2 TVD and Flux-TVD (FTVD) Schemes

We start with a description of the scheme under study in general form, and identify terms that ensure that the resulting scheme has second order accuracy. To this end we consider the case $A \equiv 0$ and select $\Delta x > 0$ and set $x_j := j\Delta x$, $\boldsymbol{\gamma}_{j+1/2} := \boldsymbol{\gamma}(x_{j+1/2}+)$ and $U_j^0 := u_0(x_j+)$ for $j \in \mathbb{Z}$. Here $x_{j+1/2} := x_j + \Delta x/2$. Let $t_n := n\Delta t$ and let χ^n denote the characteristic function of $[t_n, t_{n+1})$, χ_j the characteristic function of $[x_{j-1/2}, x_{j+1/2})$, and $\chi_{j+1/2}$ the characteristic function of the interval $[x_j, x_{j+1})$. Our difference algorithm will produce an approximation U_j^n associated with (x_j, t_n). We then define

$$u^\Delta(x,t) := \sum_{n \geq 0} \sum_{j \in \mathbb{Z}} U_j^n \chi_j(x) \chi^n(t), \quad \boldsymbol{\gamma}^\Delta(x) := \sum_{j \in \mathbb{Z}} \boldsymbol{\gamma}_{j+1/2} \chi_{j+1/2}(x). \tag{18}$$

We recall the definition of the standard difference operators $\Delta_- V_j := V_j - V_{j-1}$ and $\Delta_+ V_j := V_{j+1} - V_j$. Then our algorithm is defined by

$$U_j^{n+1} = U_j^n - \lambda \Delta_-\big(h_{j+1/2}^n + \hat{F}_{j+1/2}^n\big), \quad \lambda = \frac{\Delta t}{\Delta x}, \quad j \in \mathbb{Z}, \quad n = 0, 1, 2, \dots. \tag{19}$$

Here $h_{j+1/2}^n := h(\boldsymbol{\gamma}_{j+1/2}, U_{j+1}^n, U_j^n)$, where h is the Engquist-Osher (EO) flux [46]:

$$h(\boldsymbol{\gamma}, v, u) := \frac{1}{2}\big(f(\boldsymbol{\gamma}, u) + f(\boldsymbol{\gamma}, v)\big) - \frac{1}{2}\int_u^v \big|f_u(\boldsymbol{\gamma}, w)\big|\, dw, \tag{20}$$

and $\hat{F}_{j+1/2}^n$ is a correction term that is required in order to achieve second-order accuracy. Without those terms, (19) is the first-order scheme analyzed in [20]. Finally, we keep λ constant as we refine the mesh.

Focusing on the difference scheme (19) for (7), we now define second-order correction terms $d^n_{j+1/2}$, $e^n_{j+1/2}$ that are appropriate if $\boldsymbol{\gamma}$ is piecewise constant. We are seeking formal second-order accuracy at points (x,t) where the solution u is smooth. At jumps in $\boldsymbol{\gamma}$, u will generally be discontinuous, so for the purpose of defining correction terms, we concentrate on points located away from the jumps in $\boldsymbol{\gamma}$. In light of our (temporary) assumption that $\boldsymbol{\gamma}$ is piecewise constant we obtain the following Lax-Wendroff type correction terms that are well known to provide for formal second-order accuracy in both space and time (see e.g. [86]):

$$d^n_{j+1/2} = \frac{\alpha^+_{j+1/2}}{2}\bigl(1-\lambda\alpha^+_{j+1/2}\bigr)\Delta_+ U^n_j, \quad e^n_{j+1/2} = \frac{\alpha^-_{j+1/2}}{2}\bigl(1+\lambda\alpha^-_{j+1/2}\bigr)\Delta_+ U^n_j. \tag{21}$$

Here the quantities $\alpha^{\pm}_{j+1/2}$ are the positive and negative wave speeds associated with the cell boundary located at $x_{j+1/2}$:

$$\alpha^+_{j+1/2} := \frac{1}{\Delta_+ U^n_j}\int_{U^n_j}^{U^n_{j+1}} \max\bigl(0, f_u(\boldsymbol{\gamma}_{j+1/2}, w)\bigr)\,\mathrm{d}w = \frac{f(\boldsymbol{\gamma}_{j+1/2}, U^n_{j+1}) - h^n_{j+1/2}}{\Delta_+ U^n_j} \geq 0,$$

$$\alpha^-_{j+1/2} := \frac{1}{\Delta_+ U^n_j}\int_{U^n_j}^{U^n_{j+1}} \min\bigl(0, f_u(\boldsymbol{\gamma}_{j+1/2}, w)\bigr)\,\mathrm{d}w = \frac{h^n_{j+1/2} - f(\boldsymbol{\gamma}_{j+1/2}, U^n_j)}{\Delta_+ U^n_j} \leq 0.$$

The scheme defined by (19) and (20), and with the flux correction terms not in effect, i.e., $\hat{F}^n_{j+1/2} = 0$ for all j and n, is only first-order accurate. We now set out to find second-order correction terms that are required when $x \mapsto \boldsymbol{\gamma}(x)$ is piecewise C^2, and start by identifying the truncation error of the first-order scheme. For the case $f_u(\boldsymbol{\gamma}, u) \geq 0$ the first-order version of the scheme (19) simplifies to

$$U^{n+1}_j - U^n_j + \lambda\Delta_- f\bigl(\boldsymbol{\gamma}_{j+1/2}, U^n_j\bigr) = 0.$$

Inserting a smooth solution $u(x,t)$ into this scheme, using u^n_j to denote $u(x_j,t^n)$, substituting $u_t = -f(\boldsymbol{\gamma}, u)_x$ into the resulting expression (as well as differentiated versions of this identity) and applying Taylor expansions, we get (see [25] for details)

$$TE^+ = -\Delta x^2\lambda\left[\frac{1}{2}f_u(1-\lambda f_u)u_x - \frac{1}{2}\lambda f_u f_{\boldsymbol{\gamma}}\boldsymbol{\gamma}_x\right]_x + \mathcal{O}(\Delta^3).$$

Similarly, when $f_u \leq 0$, we arrive at the following formula for the truncation error:

$$TE^- = \Delta x^2\lambda\left[\frac{1}{2}f_u(1+\lambda f_u)u_x + \frac{1}{2}\lambda f_u f_{\boldsymbol{\gamma}}\boldsymbol{\gamma}_x\right]_x + \mathcal{O}(\Delta^3).$$

So, when $\boldsymbol{\gamma}$ is piecewise smooth (not piecewise constant), we see from these expressions that appropriate second-order correction terms are given by the following modified versions of (21):

$$\begin{aligned} F^n_{j+1/2} &:= D^n_{j+1/2} - E^n_{j+1/2}, \\ D^n_{j+1/2} &:= d^n_{j+1/2} - \frac{1}{2}\lambda\alpha^+_{j+1/2} f_{\boldsymbol{\gamma}}\big(\boldsymbol{\gamma}_{j+1/2}, U^n_{j+1/2}\big)\Delta_+\boldsymbol{\gamma}_j, \\ E^n_{j+1/2} &:= e^n_{j+1/2} + \frac{1}{2}\lambda\alpha^-_{j+1/2} f_{\boldsymbol{\gamma}}\big(\boldsymbol{\gamma}_{j+1/2}, U^n_{j+1/2}\big)\Delta_+\boldsymbol{\gamma}_j. \end{aligned} \tag{22}$$

For the values $f_{\boldsymbol{\gamma}}(\boldsymbol{\gamma}_{j+1/2}, U^n_{j+1/2})$ appearing in (22), we use the approximation

$$f_{\boldsymbol{\gamma}}(\boldsymbol{\gamma}_{j+1/2}, U^n_{j+1/2}) \approx \frac{1}{2}\big(f_{\boldsymbol{\gamma}}(\boldsymbol{\gamma}_{j+1/2}, U^n_j) + f_{\boldsymbol{\gamma}}(\boldsymbol{\gamma}_{j+1/2}, U^n_{j+1})\big). \tag{23}$$

Even without the jumps in $\boldsymbol{\gamma}$, the solution will generally develop discontinuities. If we use the correction terms above without further processing, the solution will develop spurious oscillations near these discontinuities. To damp out the oscillations, we apply so-called flux limiters, resulting in the flux-limited quantities $\hat{F}_{j+1/2}$.

A simple limiter that enforces the TVD property when $\boldsymbol{\gamma}$ is constant is

$$\begin{aligned} \hat{F}^n_{j+1/2} &= \hat{D}^n_{j+1/2} - \hat{E}^n_{j+1/2}, \\ \hat{D}^n_{j+1/2} &= \operatorname{minmod}\big(D^n_{j+1/2}, 2D^n_{j-1/2}\big), \\ \hat{E}^n_{j+1/2} &= \operatorname{minmod}\big(E^n_{j+1/2}, 2E^n_{j+3/2}\big), \end{aligned} \tag{24}$$

where we recall that the m-variable minmod function is defined by

$$\operatorname{minmod}(p_1, \dots, p_m) = \begin{cases} \min\{p_1, \dots, p_m\} & \text{if } p_1 \geq 0, \dots, p_m \geq 0, \\ \max\{p_1, \dots, p_m\} & \text{if } p_1 \leq 0, \dots, p_m \leq 0, \\ 0 & \text{otherwise.} \end{cases}$$

When $\boldsymbol{\gamma}$ is not constant, the actual solution u is not TVD, but numerical experiments [25] indicate that (24) is an effective method of damping oscillations even in the variable-$\boldsymbol{\gamma}$ context considered here. The only negative practical aspect that we have observed is a small amount of overshoot in certain cases when a shock collides with a stationary discontinuity at a jump in $\boldsymbol{\gamma}$, see Fig. 4.

Next, we wish to eliminate the non-physical overshoot observed with the simple TVD limiter (24), and also put the resulting difference scheme on a firm theoretical basis. For a conservation law having a flux with a discontinuous spatial dependency, it is natural to expect not the conserved variable, but the flux, to be TVD [88]. Consequently, we require that

$$\sum_{j\in\mathbb{Z}} \left|\Delta_+ h^{n+1}_{j-1/2}\right| \leq \sum_{j\in\mathbb{Z}} \left|\Delta_+ h^{n}_{j-1/2}\right|, \quad n = 0, 1, \ldots.$$

We call this property flux-TVD, or FTVD. We will see that under an appropriate CFL condition, the FTVD property (along with a bound on the solution) holds if

$$\left|\Delta_+ \hat{F}^n_{j+1/2}\right| \leq \left|\Delta_+ h^n_{j+1/2}\right|, \quad j \in \mathbb{Z}, \quad n = 0, 1, 2, \ldots. \tag{25}$$

It is reasonable to also impose the condition

$$0 \leq \hat{F}^n_{j+1/2} / F^n_{j+1/2} \leq 1, \quad j \in \mathbb{Z}, \quad n = 0, 1, 2, \ldots. \tag{26}$$

in addition to (25), so that after we have applied the correction terms, the numerical flux lies somewhere between the first-order flux and the pre-limiter version of the second-order flux.

We can view (25) and (26) as a system of inequalities, and ask if it is possible to find a solution that keeps the ratio $\hat{F}^n_{j+1/2}/F^n_{j+1/2}$ appearing in (26) close enough to unity that we still have formal second-order accuracy. This leads us to propose the nonlocal limiter algorithm that we describe in Algorithm 1.

For the case of piecewise constant $\boldsymbol{\gamma}$, the results produced by the two algorithms (namely the "simple TVD scheme" (STVD) and the "flux-TVD scheme" (FTVD)) usually differ by only a small amount. However, we have observed one situation where there is a discernable difference—the case of a shock impinging on a discontinuity in $\boldsymbol{\gamma}$. As mentioned above, the STVD limiter sometimes allows overshoots by a small amount in this situation. We have not observed any such overshoot with the FTVD limiter, see Example 2 in Sect. 2.3.

Finally, we mention that at a steady sonic rarefaction, both the Engquist-Osher (EO) scheme and the Godunov scheme are slightly overcompressive, leading to a so-called dogleg feature in the solution. This feature vanishes as the mesh size tends to zero, but it is distracting. This dogleg artifact is present in certain situations with both the STVD and the FTVD versions of our second-order schemes. It turns out that if the corrections (21) are replaced by

$$d^n_{j+1/2} = \frac{1}{2}\alpha^+_{j+1/2}\left(\frac{\alpha^+_{j+1/2}}{\alpha^+_{j+1/2} - \alpha^-_{j+1/2}} - \lambda\alpha^+_{j+1/2}\right)\Delta_+ U^n_j,$$

$$e^n_{j+1/2} = \frac{1}{2}\alpha^-_{j+1/2}\left(-\frac{\alpha^-_{j+1/2}}{\alpha^+_{j+1/2} - \alpha^-_{j+1/2}} + \lambda\alpha^-_{j+1/2}\right)\Delta_+ U^n_j,$$

the scheme only changes near sonic points, but the dogleg feature diminishes noticeably. We have implemented this refinement in Examples 1–3.

Next, we describe a method for solving the system of inequalities (25) and (26) while trying to maximize $\hat{F}^n_{j+1/2}/F^n_{j+1/2}$ to maintain formal second-order accuracy wherever possible. We set $z_i := F^n_{i+1/2}$, $\theta_i := |\Delta_+ h^n_{i+1/2}|$ and $\hat{z}_i := \hat{F}^n_{i+1/2}$, and then restate the system of inequalities (25) and (26) in the form

$$|\hat{z}_{i+1} - \hat{z}_i| \le \theta_i, \quad 0 \le \hat{z}_i/z_i \le 1. \tag{27}$$

The unknowns are $\hat{z}_i$, and the data are z_i, $\theta_i \ge 0$. Moreover, there are indices i_*, i^* such that $z_i = 0$ for $i \le i_*$ and $i \ge i^*$ since u_0 has compact support. Thus we may always assume that U_j^n and $F_{j+1/2}^n$ vanish for sufficiently; large j.

Algorithm 1 (Nonlocal limiter algorithm).
Input: data $z_i \ge 0$, $\theta_i \ge 0$, $i = i_*, \dots, i^*$.
Output: a vector $\hat{Z} = \{\hat{z}_{i_*}, \dots, \hat{z}_{i^*}\}$ *such that* (27) *is satisfied, where* z_i *denotes the data before application of the algorithm.*
Initialization: The sequence $\zeta_i \ge 0$, $\theta_i \ge 0$, $i = i_*, \dots, i^*$ *is initialized to the input data* $z_i \ge 0$, $\theta_i \ge 0$, $i = i_*, \dots, i^*$.

1. *Preprocessor step:*

    do $i = i_*, i_* + 1, \dots, i^* - 1$

    if $\zeta_{i+1}\zeta_i < 0$ ***and*** $|\zeta_{i+1} - \zeta_i| > \theta_i$ ***then***

    $\zeta_i \leftarrow \text{sgn}(\zeta_i) \min\{|\zeta_i|, \theta_i/2\}$
    $\zeta_{i+1} \leftarrow \text{sgn}(\zeta_{i+1}) \min\{|\zeta_{i+1}|, \theta_i/2\}$

    endif

    enddo

2. *Forward sweep:*

    do $i = i_*, i_* + 1, \dots, i^* - 1$

    if $|\zeta_{i+1}| > |\zeta_i|$ ***then***

    $\zeta_{i+1} \leftarrow \zeta_i + \text{sgn}(\zeta_{i+1} - \zeta_i) \min\{|\zeta_{i+1} - \zeta_i|, \theta_i\}$

    endif

    enddo

3. *Backward sweep:*

    do $i = i^*, i^* - 1, \dots, i_* + 1$

    if $|\zeta_{i-1}| > |\zeta_i|$ ***then***

    $\zeta_{i-1} \leftarrow \zeta_i + \text{sgn}(\zeta_{i-1} - \zeta_i) \min\{|\zeta_{i-1} - \zeta_i|, \theta_{i-1}\}$

    endif

    enddo

Generate output:

do $i = i_*, i_* + 1, \dots, i^*$

$\hat{z}_i \leftarrow \zeta_i$

enddo

Here the left arrow $\leftarrow$ is the replacement operator. Algorithm 1 can be written compactly as $\hat{Z} = \Phi(Z,\Theta) = \Phi^-(\Phi^+(\tilde{Z},\Theta),\Theta)$, where $\tilde{Z} = \mathrm{Pre}(Z,\Theta)$. Here Φ^+ and Φ^- represent the forward and backward sweeps, Pre represents the preprocessor step, and $\hat{Z} = \{\hat{z}_i\}$, $\tilde{Z} = \{\tilde{z}_i\}$, $Z = \{z_i\}$ and $\Theta = \{\theta_i\}$. In [25] it is shown that the output of Algorithm 1 solves the system of inequalities (27), and that the limiter Φ is consistent with formal second-order accuracy in the following sense.

Lemma 1. *Let u and $\boldsymbol{\gamma}$ be C^2 in a neighborhood of the point $\bar{x}$ where*

$$f\big(\boldsymbol{\gamma}(\bar{x}), u(\bar{x})\big)_x \neq 0. \tag{28}$$

Assume that $u(\pm x) = u_{\pm\infty}$ for x sufficiently large, so that the limiter Φ is well-defined on the flux corrections $F^{\Delta}_{j+1/2} = F_{j+1/2}$. Let

$$\hat{F}^{\Delta} = \Phi\left(\left\{F^{\Delta}_{j+1/2}\right\}_{j\in\mathbb{Z}}, \left\{\left|\Delta_+ h_{j+1/2}\right|\right\}_{j\in\mathbb{Z}}\right). \tag{29}$$

Then there is a mesh size $\Delta_0 = \Delta_0(\bar{x}) > 0$ and a $\delta(\bar{x}) > 0$ such that for $\Delta \leq \Delta_0$, we have

$$\hat{F}^{\Delta}_{j+1/2} = F^{\Delta}_{j+1/2} \quad \textit{for all } x_j \in \{x : |x - \bar{x}| < \delta\}.$$

Consequently, the scheme defined by (18)–(23), including the flux corrections $\hat{F}^n_{j+1/2}$ produced by (29) will have formal second-order accuracy at any point where u andγ are smooth, and where (28) is satisfied. Thus, the resulting FTVD scheme is given by $U^{n+1}_j = U^n_j - \lambda\Delta_-(h^n_{j+1/2} + \hat{F}^n_{j+1/2})$. In [20] the first-order version of this scheme, $U^{n+1}_j = U^n_j - \lambda\Delta_- h^n_{j+1/2}$, was analyzed. Clearly, this scheme results by setting $\hat{F}^n_{j+1/2} = 0$ for all j and n. Moreover in [20] we assumed that $\boldsymbol{\gamma}$ is piecewise constant, while in [19] we dealt with a piecewise smooth coefficient function $\boldsymbol{\gamma}$. The convergence analysis for the FTVD scheme strongly relies on results from [19] and [20]. We assume that the following CFL condition is satisfied:

$$\lambda\Big(\max\{-q_{\mathrm{L}}, q_{\mathrm{R}}\} + \|\gamma_1 b'\|\Big) \leq \frac{1}{4}, \tag{30}$$

where $\|\gamma_1 b'\| := \max\{|\gamma_1(x) b'(u)| : x \in [x_{\mathrm{L}}, x_{\mathrm{R}}],\ u \in [0, u_{\max}]\}$.

Our theorem concerning convergence is the following.

Theorem 1 (Convergence of the FTVD scheme). *Let u^{Δ} be defined by* (18)–(23). *Assume that the flux corrections $\hat{F}^n_{j+1/2}$ are produced by applying Algorithm 1 to the non-limited flux corrections $F^n_{j+1/2}$. Let $\Delta \to 0$ with λ constant and the CFL condition* (30) *be satisfied. Then u^{Δ} converges along a subsequence in $L^1_{\mathrm{loc}}(\Pi_T)$ and boundedly a.e. in Π_T to a BV_t weak solution of the CT model* (7) *and* (8).

The proof of Theorem 1 amounts to checking that Lemmas 1–7, along with the relevant portion of Theorem 1, of [19] remain valid in the present context. See [25] for details. We resume the essential steps of the proof.

One first shows that under the CFL condition (30) we get a uniform bound on U_j^n, specifically $U_j^n \in [0,1]$, and that the flux-TVD property is satisfied, i.e.,

$$\sum_{j\in\mathbb{Z}} \left|h_{j+1/2}^{n+1} - h_{j-1/2}^{n+1}\right| \leq \sum_{j\in\mathbb{Z}} \left|h_{j+1/2}^{n} - h_{j-1/2}^{n}\right|, \quad n = 0,1,2,\ldots.$$

The proof of these properties follows that of [19, Lemma 1].

The flux-TVD property is the ingredient that allows us to maintain time continuity even though the present scheme, as a second-order scheme, is no longer monotone. Thus, there exists a constant C, independent of Δ and n, such that

$$\Delta x \sum_{j\in\mathbb{Z}} \left|U_j^{n+1} - U_j^n\right| \leq \Delta x \sum_{j\in\mathbb{Z}} \left|U_j^1 - U_j^0\right| \leq C\Delta t.$$

As in [19], to prove that the difference scheme converges, one needs to establish compactness for the transformed quantity z^Δ that emerges from the numerical solution by a singular mapping Ψ also known as the Temple functional [87]. The critical ingredient is a bound on its total variation. We then derive compactness for u^Δ by appealing to the monotonicity and continuity of the mapping $u \mapsto \Psi(\boldsymbol{\gamma}, u)$. To show that z^Δ has bounded variation it then suffices to invoke Lemmas 2–7 of [19], making modifications where necessary to account for the addition of the second-order correction terms. See [25].

We now use the notation $\mathcal{O}(\Delta\boldsymbol{\gamma}_j)$ to denote terms which sum (over j) to $\mathcal{O}(|\boldsymbol{\gamma}|_{BV})$, and employ the Kružkov entropy-entropy flux pair indexed by c, i.e. $q(u) := |u-c|$ and $\eta(\boldsymbol{\gamma}, u) := \operatorname{sgn}(u-c)(f(\boldsymbol{\gamma}, u) - f(\boldsymbol{\gamma}, c))$. One then obtains that for each $c \in \mathbb{R}$, the following inequality holds:

$$\begin{aligned} q(U_j^{n+1}) \leq{} & q(U_j^n) - \lambda\Big[H(\boldsymbol{\gamma}_{j+1/h}, U_{j+1}^n, U_j^n) - H(\boldsymbol{\gamma}_{j+1/h}, U_j^n, U_{j-1}^n)\Big] \\ & + \lambda\left|\Delta_+ h_{j-1/2}^n\right| + \lambda\mathcal{O}(\Delta\boldsymbol{\gamma}_j), \quad j \in \mathbb{Z}, \quad n = 0,1,2\ldots, \end{aligned} \tag{31}$$

where the EO numerical entropy flux is given by

$$H(\gamma, v, u) = \frac{1}{2}\big(\eta(\boldsymbol{\gamma}, u) + \eta(\boldsymbol{\gamma}, v)\big) - \frac{1}{2}\int_u^v \operatorname{sgn}(w-c)\left|f_u(\boldsymbol{\gamma}, w)\right| \mathrm{d}w.$$

It is now possible to repeat the proofs of Lemmas 3–7 of [19], the only change being the contribution of the term $\lambda|\Delta_- h_{j+1/2}^n|$ appearing in (31).

2.3 *Numerical Examples (Examples 1 and 2)*

Consider a suspension characterized by $b(u) = v_{\mathrm{St}} u V(u)$, where $v_{\mathrm{St}} = 10^{-4}\,\mathrm{m/s}$ and $V(u)$ is given by (4) with $n_{\mathrm{RZ}} = 5$ and $u_{\max} = 1$. We assume that $A \equiv 0$ and consider a cylindrical CT with $x_{\mathrm{L}} = -1\,\mathrm{m}$ and $x_{\mathrm{R}} = 1\,\mathrm{m}$ with (nominal) interior cross-sectional area $S = 1\,\mathrm{m}^2$. The CT is assumed to initially contain no

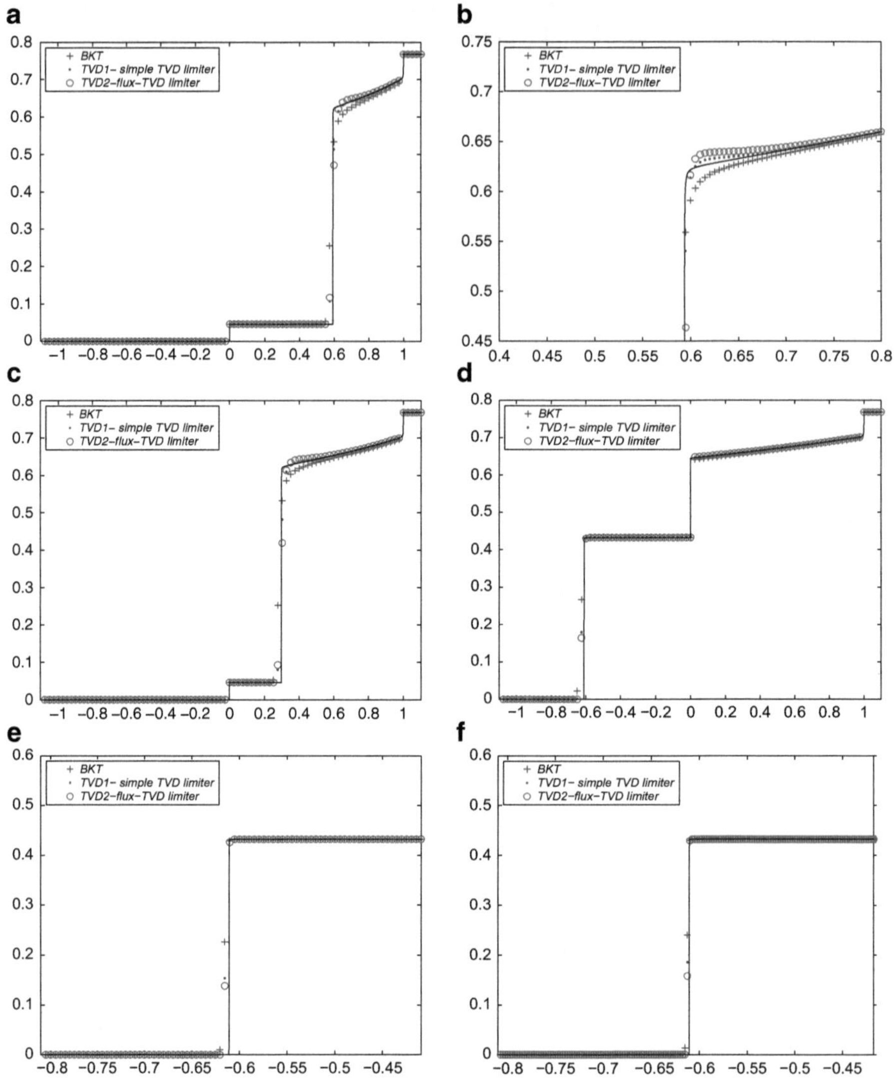

Fig. 3 Example 1: numerical solution at (**a**, **b**) $t = 150{,}000\,\mathrm{s}$ with (**a**) $J = 40$, (**b**) $J = 200$ (enlarged view around $x = 0.6$), at (**c**) $t = 250{,}000\,\mathrm{s}$ with $J = 40$, and at (**d–f**) $t = 500{,}000\,\mathrm{s}$ with (**d**) $J = 40$, (**e**) $J = 200$, (**f**) $J = 400$ ((**e**, **f**): enlarged view around $x = -0.61$). The *solid line* is the reference solution

solids ($u_0 \equiv 0$), is operated with a feed concentration $u_F = 0.3$ in Example 1 and $u_F = 0.5$ in Example 2, and the flow velocities are $q_L = -1.0 \times 10^{-5}\,\mathrm{m/s}$ and $q_R = 2.5 \times 10^{-6}\,\mathrm{m/s}$. In these examples, the solution is clearly not TVD, since $\mathrm{TV}(u_0) = 0$. Figure 3 shows the numerical solution for Example 1 calculated by the first-order scheme of [21] (BKT), the scheme described herein that uses the

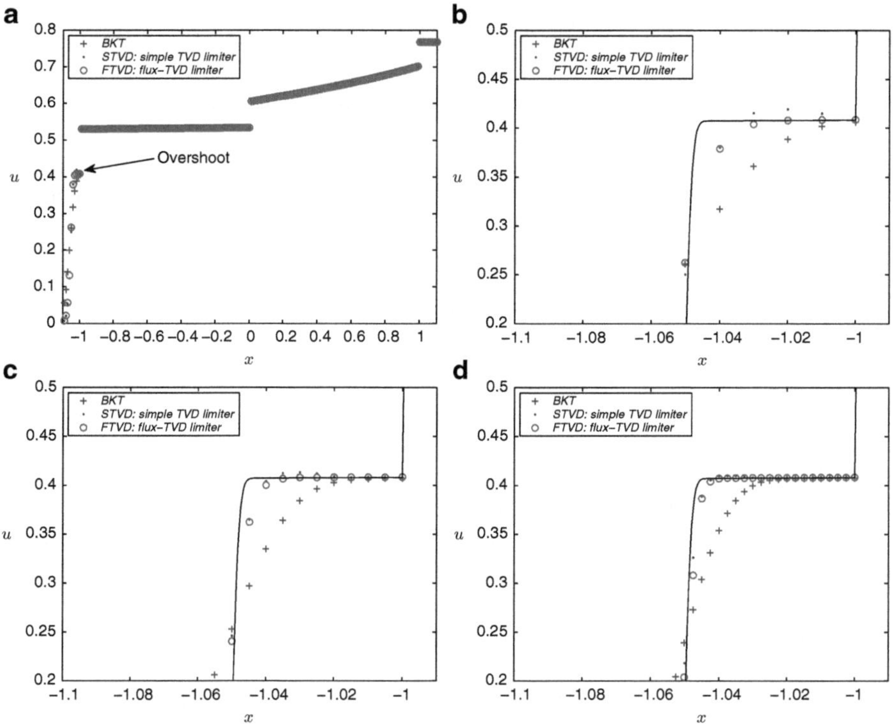

Fig. 4 Example 2: numerical solution at $t = 272{,}760$ s with (**a**, **b**) $J = 100$, (**c**) $J = 200$ and (**d**) $J = 400$ ((**b**–**d**): enlarged views around $x = -1$). The *solid line* is the reference solution

simple TVD (STVD) limiter (in short, STVD scheme), and the FTVD scheme. All calculations were performed with $\lambda = 2{,}000$ s/m, and results are compared against a reference solution calculated by the first-order scheme of [22] with $J = 100{,}00$, where $J = 1/\Delta x$ (in meters). Example 2 illustrates the overshoot mentioned in Sect. 2.2, see Fig. 4. We observe that Fig. 4 illustrates how the "overshoot" phenomenon diminishes as $\Delta x \to 0$.

The numerical solutions of Examples 1 and 2 indicate that the STVD and FTVD schemes are significantly more accurate than their first-order counterpart. It seems that both schemes STVD and FTVD, have comparable accuracy. A significant difference in solution behaviour between both schemes becomes visible in Fig. 4.

2.4 A Note on Second-Order Degenerate Parabolic Equations (Example 3)

The model (9) with a degenerate diffusion term can be handled by a Strang-type operator splitting scheme [85]. To describe it, let U^n denote the approximate solution at time level n, and write the scheme (19) in operator notation via $U^{n+1} = \mathcal{H}(\Delta t)U^n$. Then the proposed operator splitting scheme for (9) is

$$U^{n+1} = \big[\mathcal{H}(\Delta t/2) \circ \mathcal{P}(\Delta t) \circ \mathcal{H}(\Delta t/2)\big]U^n, \quad n = 0, 1, 2, \dots. \tag{32}$$

Here $\mathcal{P}(\Delta t)$ represents a second-order scheme for $u_t = (\gamma_1(x)A(u)_x)_x$ written as $U^{n+1} = \mathcal{P}(\Delta t)U^n$. If we employ the Crank-Nicolson (CN) scheme, which has second-order accuracy in space and time, then $\mathcal{P}(\Delta t)$ is defined by

$$U_j^{n+1} = U_j^n + \frac{\Delta t}{2\Delta x^2}\Big[\Delta_+\Big(s_{j-1/2}\Delta_- A_j^n\Big) + \Delta_+\Big(s_{j-1/2}\Delta_- A_j^{n+1}\Big)\Big]. \tag{33}$$

Here $s_{j-1/2}$ denotes our discretization of the parameter $\gamma_1(x)$. The CN scheme is stable with linear stability analysis. For our nonlinear problem, we generally need a very strong type of stability, both from a practical and theoretical point of view. It seems that it is impossible to get this type of strong stability for implicit schemes of accuracy greater than one [49]. On the other hand, the solution u is continuous in the regions where the parabolic operator is in effect (cf., e.g., [21]), which seems to stabilize the numerical approximation. The CN scheme leads to a nonlinear system of equations, which are solved here iteratively; each step of iteration requires solving a tridiagonal linear system (see [25]). These iterations have turned out to converge rather quickly.

Since each of the parabolic and hyperbolic operators has formal second-order accuracy in both space and time, we will maintain overall second order accuracy with the Strang splitting [85]. This is a well-known result, see, e.g., [48].

Next, we include the strongly degenerate diffusion term (10) with

$$a(u) = \frac{b(u)\sigma_e'(u)}{(\rho_s - \rho_f)gu},$$

where the so-called effective solid stress function $\sigma_e(u)$ is given by

$$\sigma_e(u) = \begin{cases} 0 & \text{for } u \le u_c, \\ \sigma_0\big((u/u_c)^k - 1\big) & \text{for } u > u_c, \end{cases}$$

where we use $\sigma_0 = 1\,\text{Pa}$, $u_c = 0.1$ and $k = 6$ along with $\Delta\rho = 1{,}500\,\text{kg/m}^3$ and $g = 9.81\,\text{m/s}^2$ [21]. The vessel and control variables are the same as in Example 1, and we again set $u_0 \equiv 0$. Figure 5 shows the numerical solution calculated by the semi-implicit scheme described in [21] (BKT-SI), the operator splitting scheme described herein (BKT-OS), the operator splitting scheme (32) and (33) including the simple TVD limiter (STVD-OS), and the analogue scheme involving the non-local limiter (FTVD-OS). All calculations were performed with $\lambda = 2{,}000\,\text{s/m}$.

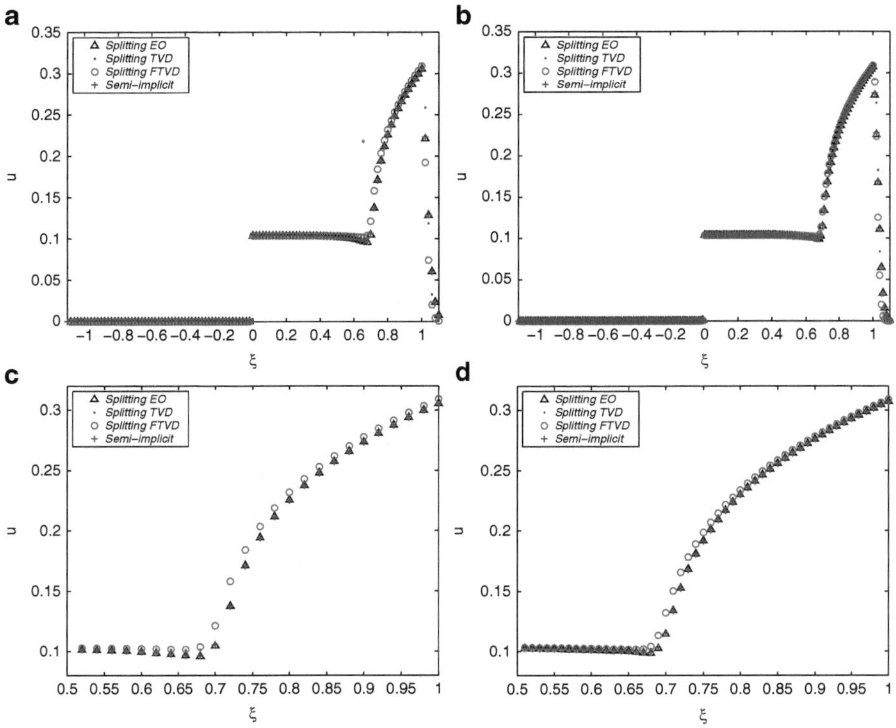

Fig. 5 Example 3: numerical solution at $t = 25{,}000$ s with (**a**) $J = 50$, (**b**) $J = 100$, (**c**) at $t = 25{,}000$ s with $J = 50$, (**d**) at $t = 100{,}000$ s with $J = 50$

3 A Conservation Law with Nonlocal Flux Modeling Sedimentation

When diffusion is negligible, the one-dimensional continuity equation is (1), and the solids phase velocity v_{s} is given by (3) and (5). Assume now that V is given by (4) but depends on u in the nonlocal form

$$V = V(K_a * u), \quad (K_a * u)(x,t) = \int_{-2a}^{2a} K_a(y)u(x+y,t)\,\mathrm{d}y, \tag{34}$$

where K_a is a symmetric, non-negative piecewise smooth kernel with support on $[-2a, 2a]$ for a parameter $a > 0$ and $\int_{\mathbb{R}} K_a(x)\,\mathrm{d}x = 1$. Then (1) takes the form

$$u_t + v_{\mathrm{St}}\big(u(1 - K_a * u)^{n_{\mathrm{RZ}}}\big)_x = 0. \tag{35}$$

On the other hand, starting from the relation $v_{\mathrm{s}} = (1-u)v_{\mathrm{r}}$ valid for batch settling, we obtain the alternative governing equation $u_t + (u(1-u)v_{\mathrm{r}})_x = 0$.

If v_r (instead of v_s) has a nonlocal behaviour and the local versions based on specifying either v_s or v_r should coincide, then the constitutive assumption for v_r becomes $v_r = V(K_a * u)/(1-u)$. For instance, (4) leads to the conservation law

$$u_t + v_{St}\big(u(1-u)(1-K*u)^{n_{RZ}-1}\big)_x = 0. \tag{36}$$

Both (35) and (36) are special cases of (11).

3.1 Properties of the Nonlocal Equation

Insight into properties of (11) can be gained by analyzing an approximate local PDE (the "effective" local PDE [99]) obtained from the Taylor expansion of $K_a * u$. If $2M_2$ denotes the second moment of K_a, then we obtain the approximate diffusive-dispersive local PDE

$$u_t + \big(u(1-u)^{\alpha}V(u)\big)_x = -a^2M_2\big(V'(u)u(1-u)^{\alpha}u_{xx}\big)_x \tag{37}$$

(see [12] for details). For $\alpha \geq 1$ the factor $u(1-u)$ in the right-hand side and in the flux has a "saturating" effect; it prevents solution values from leaving $[0,1]$. Thus, we should expect that the nonlocal PDE (11) also satisfies an invariant region principle for $\alpha \geq 1$. This is indeed the case, as will be shown below.

We mention that Zumbrun [99] studied an equation equivalent to (11) in the case $\alpha = 0$ and $V(w) = v_{St}(1-\beta w)$, namely

$$u_t + \big(uK_a * u\big)_x = 0, \tag{38}$$

where $K_a(x) := a^{-1}K(a^{-1}x)$ and K is the truncated parabola given by

$$K(x) = \frac{3}{8}\left(1-\frac{x^2}{4}\right) \text{ for } |x|<2; \quad K(x) = 0 \text{ otherwise.} \tag{39}$$

He showed global existence of weak solutions for (12) and (38) in L^∞ and uniqueness in the class BV, and derived the effective local, dispersive, KdV-like PDE

$$u_t + (u^2)_x = -M_2a^2(uu_{xx})_x. \tag{40}$$

He showed by analyzing (40) that (38) supports travelling waves, but not viscous shocks. This result is based on the symmetry of K, which makes (38) completely dispersive. Moreover, an L^2 stability argument is invoked to conclude that smooth solutions of the Burgers-like first-order conservation law $u_t + (u^2)_x = 0$ arise from smooth solutions of (38) as $a \to 0$. Zumbrun [99] also studied the effect of artificial diffusion added to (38), and showed that for the corresponding effective local PDE, solutions of shock initial data converge to a stable, oscillatory travelling wave.

For $\alpha = 0$, the notion of weak solution is sufficient for uniqueness and stability (at least in the Wasserstein distance, see [11, 66]), since the convolution introduces sufficient regularization to ensure that the advective velocity is Lipschitz continuous. This is true even with discontinuous data. For the case $\alpha = 0$, the analysis of [12] based on a quadrature-difference scheme comes to a corresponding Lipschitz continuity result for Lipschitz continuous initial data, as will be discussed below.

3.2 Numerical Scheme and Well-Posedness Analysis

We discretize (11) on a fixed grid given by $x_j = j\Delta x$ for $j \in \mathbb{Z}$ and $t_n = n\Delta t$ for $n \le N := T/\Delta t$, where T is the finite final time. As usual, u_j^n approximates the cell average of $u(\cdot, t_n)$ on $(x_{j-1/2}, x_{j+1/2})$, and we define $U^n := (\dots, u_{j-1}^n, u_j^n, u_{j+1}^n, \dots)^{\mathrm{T}}$. The initial datum u_0 is discretized accordingly. We define the second spatial difference operator $\Delta^2 u_j^n := \Delta_+\Delta_- u_j^n$.

We assume that K_a is a positive symmetric kernel, has compact support on $[-2a, 2a]$, $K_a \in C^{0,1}(\mathbb{R}) \cap C^2([-2a, 2a])$ and $\int_{-2a}^{2a} K_a(y)\,\mathrm{d}y = 1$. (The same analysis remains valid for more general kernels [12].) The integral in (34) is approximated by the quadrature formula

$$(K_a * u)_j^n \approx \tilde{u}_{a,j}^n := \sum_{i=-l}^{l} \gamma_i u_{j-i}^n, \text{ where } \gamma_i = \int_{x_{i-1/2}}^{x_{i+1/2}} K_a(y)\,\mathrm{d}y,\ l = \left\lceil \frac{2a}{\Delta x} \right\rceil + 1.$$

Due to the properties of K_a, $\gamma_{-l} + \dots + \gamma_l = 1$. Furthermore, we require that u_0 has compact support, $u_0(x) \ge 0$ for $x \in \mathbb{R}$ and $u_0 \in BV(\mathbb{R})$. The function $u \mapsto V(u)$ and its derivatives are locally Lipschitz continuous for $u \ge 0$ (which occurs, for example, if $V(\cdot)$ is a polynomial). When we send $\Delta x, \Delta t \downarrow 0$ then it is understood that $\lambda := \Delta t/\Delta x$ is kept constant. Moreover, for the case $\alpha \ge 1$ we suppose that $u_0(x) \le 1$ for all $x \in \mathbb{R}$.

From now on we let the function u^Δ be defined by

$$u^\Delta(x,t) = U_j^n \quad \text{for } (x,t) \in [j\Delta x, (j+1)\Delta x) \times [n\Delta t, (n+1)\Delta t).$$

Definition 2. A measurable, non-negative function u is an *entropy solution* of the initial value problem (11) and (12) if it satisfies the following conditions:

1. We have $u \in L^\infty(\Pi_T) \cap L^1(\Pi_T) \cap BV(\Pi_T)$.
2. The initial condition (12) is satisfied in the following sense:

$$\lim_{t \downarrow 0} \int_{\mathbb{R}} \left| u(x,t) - u_0(x) \right| \mathrm{d}x = 0.$$

3. For all non-negative test functions $\varphi \in C_0^\infty(\Pi_T)$, the following Kružkov-type [62] entropy inequality is satisfied, where we define $f(u) := u(1-u)^\alpha$:

$$\forall k \in \mathbb{R}: \quad \iint_{\Pi_T} \Big\{ |u-k|\varphi_t + \operatorname{sgn}(u-k)\big(f(u)-f(k)\big)V(K_a * u)\varphi_x \\ - \operatorname{sgn}(u-k) f(k) V'(K_a * u)(\partial_x K_a * u)\varphi \Big\} \, \mathrm{d}x \, \mathrm{d}t \geq 0. \tag{41}$$

An entropy solution is, in particular, a weak solution of (11) and (12), which is defined by (1) and (2) of Definition 2, and the following equality, which must hold for all $\varphi \in C_0^\infty(\Pi_T)$:

$$\iint_{\Pi_T} \Big\{ u\varphi_t + f(u)V(K_a * u)\varphi_x - f(u)V'(K_a * u)(\partial_x K_a * u)\varphi \Big\} \, \mathrm{d}x \, \mathrm{d}t = 0.$$

Suitable Rankine-Hugoniot and entropy jump conditions can be derived from (41).

The uniqueness of entropy solutions follows from a result proved in [58] regarding continuous dependence of entropy solutions with respect to the flux function:

Theorem 2. *If u and v are entropy solutions of* (11) *and* (12) *with initial data u_0 and v_0, respectively, then for $T > 0$ there exists a constant C_1 such that*

$$\big\| u(\cdot,t) - v(\cdot,t) \big\|_{L^1(\mathbb{R})} \leq C_1 \, \| u_0 - v_0 \|_{L^1(\mathbb{R})} \quad \forall t \in (0,T].$$

In particular, an entropy solution of (11) *and* (12) *is unique.*

Finally, let us briefly address the convergence analysis and the related result of existence of entropy solutions. To this end, let $V_j^n := V(\tilde{u}_{a,j}^n)$. Then the marching formula for the approximation of solutions of (11) and (12) reads

$$u_j^{n+1} = \frac{u_{j-1}^n + u_{j+1}^n}{2} - \frac{\lambda}{2} u_{j+1}^n \big(1 - u_{j+1}^n\big)^\alpha V_{j+1}^n + \frac{\lambda}{2} u_{j-1}^n \big(1 - u_{j-1}^n\big)^\alpha V_{j-1}^n. \tag{42}$$

We assume that $\lambda = \Delta t / \Delta x$ satisfies the following CFL condition:

$$\lambda \max_{u \leq u^*} \big| V(u) \big| < 1 \text{ for } \alpha = 0, \; u^* := \|K_a\|_\infty \|u_0\|_1;$$

$$\lambda \max_{0 \leq u \leq 1} \big| V(u) \big| < 1 \text{ for } \alpha \geq 1.$$

The convergence proof of the numerical scheme is based on the usual L^∞, BV and L^1 Lipschitz continuity in time bounds, where the latter two depend on T and adversely on a. The L^∞ bound is as follows:

$$0 \le u_j^n \le \begin{cases} C_3 & \text{if } \alpha = 0, \\ 1 & \text{if } \alpha \ge 1, \end{cases} \quad \text{for } j \in \mathbb{Z} \text{ and } 0 \le n \le N, \tag{43}$$

where the constant C_3 is independent of Δx and Δt but depends on T. This bound represents the most important estimate of the convergence analysis [12, Lemma 5.3]. In view of (37), one should expect an "invariant region" principle to hold for (11), (12) with $\alpha \ge 1$. The estimate (43) shows that this property indeed holds.

Invoking the bounds established so far and applying a Lax-Wendroff-type argument to the discrete entropy inequality

$$|u_j^{n+1} - k| - |u_j^n - k| + \mathcal{G}_{j+}^n - \mathcal{G}_{j-}^n + \operatorname{sgn}(u_j^{n+1} - k)\frac{\lambda}{2} f(k)(V_{j+1}^n - V_{j-1}^n) \le 0$$

satisfied by the scheme, where we define

$$\mathcal{G}_{j\pm}^n := \frac{\lambda}{2}\left[\left(f\left(u_{j\pm1}^n \vee k\right) - f\left(u_{j\pm1}^n \wedge k\right)\right) V_{j\pm1}^n - \frac{1}{\lambda}\Delta_{\pm}\left(\left|u_j^n - k\right|\right)\right],$$

we can conclude by Helly's theorem that u^{Δ} converges to a function $u \in L^{\infty}(\Pi_T) \cap L^1(\Pi_T) \cap BV(\Pi_T)$ as Δx, $\Delta t \to 0$, and prove the following theorem.

Theorem 3. *The numerical solution generated by* (42) *converges to the unique entropy solution of* (11) *and* (12)*.*

As an additional regularity result for $\alpha = 0$, it can be shown that for $T > 0$, u^{Δ} converges to a Lipschitz continuous function u provided u_0 is also Lipschitz continuous. This result is as expected since in the simplest case, V constant, (11) becomes a linear advection equation, whose solution has a regularity that is the same as that of u_0. Moreover, as a Lipschitz continuous weak solution of (11) and (12), u will automatically be an entropy solution.

3.3 Numerical Examples

We illustrate in Example 4 how the value of a affects the numerical solution of (11) and (12) for $\alpha = 0$ and $\alpha = 1$. We use (4) with $n_{\mathrm{RZ}} = 5$ for $\alpha = 0$ and, correspondingly, (4) with $n_{\mathrm{RZ}} = 4$ for $\alpha = 1$. In both cases, K is given by (39) with $a = 0.4, 0.2, 0.1$, and 0.01. The initial datum is

$$u_0(x) = \begin{cases} 0.0 & \text{for } x \le 0.2, \\ 0.6 & \text{for } x > 0.2, \end{cases} \quad \text{and} \quad u_0(x) = \begin{cases} 0.0 & \text{for } x \le 0.2, \\ 0.01 & \text{for } x > 0.2, \end{cases}$$

for the two cases of a concentrated and a dilute suspension with $\Delta x = 0.0005$ and $\lambda = 0.2$. Figure 6 shows the numerical results. The case $a = 0.01$ was calculated

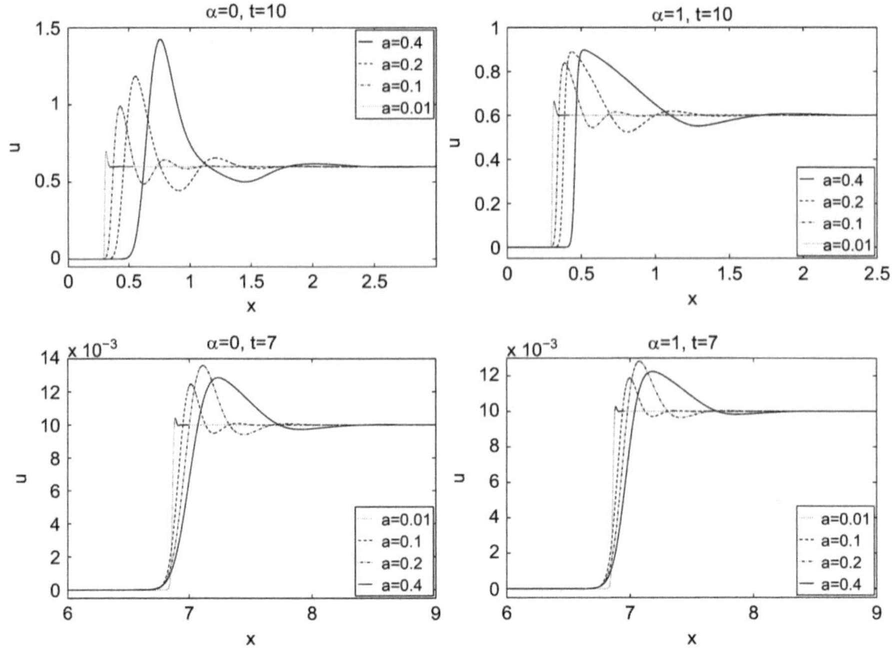

Fig. 6 Example 4: numerical solutions of (11) and (12) (*top*) for an initially concentrated suspension at $t = 10$ and (*bottom*) for an initially dilute suspension at $t = 7$

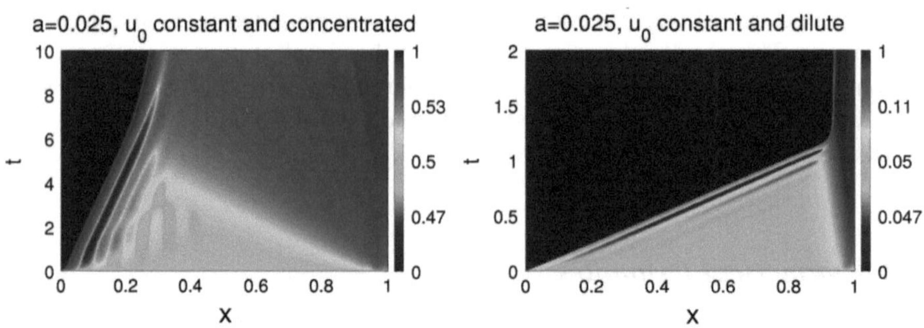

Fig. 7 Example 5: numerical solution of (11) and (12) with $\alpha = 1$ and initial data (44)

with $\Delta x = 0.0002$ since otherwise the stencil of the convolution includes just a few points. We observe a more strongly oscillatory behaviour with $a = 0.4, 0.2$ and 0.1 than with $a = 0.01$, and that the period of the oscillation is proportional to a.

In Example 5 we attempt to reproduce the layering phenomenon observed by Siano [81] for batch settling. In Fig. 7 we show the numerical results for $\alpha = 1$, with $V(u) = (1-u)^4$, K as in (39), $a = 0.025$, $\Delta x = 0.00025$, $\lambda = 0.5$ and the initial datum for the respective concentrated and dilute case

$$u_0(x) = \begin{cases} 0 & \text{for } x < 0, \\ 0.5 & \text{for } 0 \leq x < 1, \\ 1 & \text{for } x \geq 1 \end{cases} \quad \text{and} \quad u_0(x) = \begin{cases} 0 & \text{for } x < 0, \\ 0.05 & \text{for } 0 \leq x < 1, \\ 1 & \text{for } x \geq 1. \end{cases} \tag{44}$$

We observe layers of concentrations smaller or larger than the initial value u_0. These "stripes" are initially close to parallel to the supernate-suspension interface. However, stripes are obliterated as soon as interaction with concentration information travelling upwards from the vessel bottom takes place.

4 Kinematic Models of Polydisperse Sedimentation

Polydisperse sedimentation models belong to the wider class of multi-species kinematic flow models given by (13) with explicit velocity functions v_i, including the multi-class Lighthill-Whitham-Richards (MCLWR) kinematic traffic model [8,95]. The basic phenomenon of interest in these models, the segregation of species, is usually associated with the formation of discontinuities in Φ, so-called kinematic shocks. Other multi-species kinematic flow models also include the settling of oil-in-water dispersions [76] and of emulsions (cf., e.g., [22,47]).

For many multi-species kinematic flow models, the velocities v_i do not depend on each of the N components of Φ in an individual way, but are functions of $m \ll N$ ($m \leq 4$ for all models of interest) scalar functions of Φ, i.e.,

$$v_i = v_i(p_1, \dots, p_m), \quad p_l = p_l(\Phi), \quad l = 1, \dots, m. \tag{45}$$

Thus, $\mathcal{J}_f(\Phi)$ is a rank-m perturbation of $\boldsymbol{D} := \operatorname{diag}(v_1, \dots, v_N)$ of the form

$$\mathcal{J}_f = \boldsymbol{D} + \boldsymbol{B}\boldsymbol{A}^{\mathrm{T}}, \quad \begin{cases} \boldsymbol{B} := (B_{il}) = (\phi_i \partial v_i / \partial p_l), \\ \boldsymbol{A} := (A_{jl}) = (\partial p_l / \partial \phi_j), \end{cases} \quad 1 \leq i, j \leq N, \ 1 \leq l \leq m. \tag{46}$$

The analysis in [24] also provides sharp bounds of the eigenvalues of $\mathcal{J}_f(\Phi)$. This information permits to numerically calculate the eigenvalues and corresponding eigenvectors of $\mathcal{J}_f(\Phi)$ with acceptable effort. This characteristic (or spectral) information can be exploited for the implementation of high-resolution schemes.

High-resolution shock capturing schemes can be applied to systems of conservation laws either in a component-wise or in a characteristic-wise (spectral) fashion. The latter requires a detailed knowledge of the spectral decomposition of the Jacobian matrix of the system. For multi-species kinematic flow models, however, eigenvalues are not available in closed form. Nevertheless, it has been possible to prove strict hyperbolicity of some of these models by an explicit representation of the characteristic polynomial [10,76,97], as well as to obtain an interlacing property of the (unknown) eigenvalues λ_i of the Jacobian with the (known) velocities v_i, which provide excellent starting values for a root finder. For the MCLWR model, these results can be found in [97, 98] and in references cited in these papers.

Donat and Mulet [43] showed that the hyperbolicity calculus of multi-species kinematic flow models satisfying (45) can be greatly simplified by using the so-called secular equation [1], which provides a systematic algebraic framework to determine the eigenvalues, and eventually the eigenvectors, but avoids the explicit representation of the characteristic polynomial. The hyperbolicity analysis for the MCLWR model becomes very simple. Via the secular approach, hyperbolicity of the MLB model for equal-density spheres (a case of $m = 2$) can be proved in a few lines [43], which contrasts with several pages of computation necessary to exhibit the characteristic polynomial in [10]. In [24] the secular approach was used to estimate the region of hyperbolicity of the HS model, for which $m = 3$ or $m = 4$. In [26] the results of [24] are employed to implement characteristic-wise WENO schemes for the polydisperse sedimentation model. On the other hand, there are also other polydisperse sedimentation models (besides the MLB and HS models) for which the flux Jacobian is a rank-m perturbation of a diagonal, and to which a version of the present numerical technique can be applied [27, 38, 72].

4.1 Hyperbolicity Analysis

The hyperbolicity analysis of (13) under the assumption (45) is then based on the following theorem.

Theorem 4 (The secular equation, [1, 43]). *Assume that $v_i > v_j$ for $i < j$, and that $\boldsymbol{A}$ and $\boldsymbol{B}$ have the formats specified in* (46)*. We denote by S_r^p the set of all (ordered) subsets of r elements taken from a set of p elements. If $\boldsymbol{X}$ is an $m \times N$ matrix, $I := \{i_1 < \cdots < i_k\} \in S_k^N$ and $J := \{j_1 < \cdots < j_l\} \in S_l^m$, then we denote by $\boldsymbol{X}^{I,J}$ the $k \times l$ submatrix of $\boldsymbol{X}$ given by $(\boldsymbol{X}^{I,J})_{p,q} = X_{i_p,j_q}$. Let $\lambda \neq v_i$ for $i = 1, \dots, N$. Then λ is an eigenvalue of $\boldsymbol{D} + \boldsymbol{B}\boldsymbol{A}^{\mathrm{T}}$ if and only if*

$$R(\lambda) := \det\bigl(\boldsymbol{I} + \boldsymbol{A}^{\mathrm{T}}(\boldsymbol{D} - \lambda \boldsymbol{I})^{-1}\boldsymbol{B}\bigr) = 1 + \sum_{i=1}^{N} \frac{\gamma_i}{v_i - \lambda} = 0, \tag{47}$$

$$\textit{where} \quad \gamma_i := \sum_{r=1}^{\min\{N,m\}} \sum_{i \in I \in S_r^N, J \in S_r^m} \frac{\det \boldsymbol{A}^{I,J} \det \boldsymbol{B}^{I,J}}{\prod_{l \in I, l \neq i}(v_l - v_i)}.$$

The relation $R(\lambda) = 0$, (47)*, is known as the* secular equation *[1].*

Assuming that $m < N$, with $\boldsymbol{A}$ and $\boldsymbol{B}$ defined in (46) we can write

$$\gamma_i = \phi_i \sum_{r=1}^{m} \gamma_{r,i}, \quad \gamma_{r,i} = \sum_{i \in I \in S_r^N} \prod_{l \in I, l \neq i} \frac{\phi_l}{v_l - v_i} \sum_{J \in S_r^m} \det\left(\frac{\partial v_I}{\partial p_J}\right) \det\left(\frac{\partial p_J}{\partial \phi_I}\right).$$

When $m \leq 2$, these quantities can be easily computed and the hyperbolicity analysis via Theorem 4 is much less involved than explicitly deriving and discussing $\det(\mathcal{J}_f(\Phi) - \lambda \boldsymbol{I})$. For $m = 3$ or $m = 4$, the computations are more involved [24,27], but provide at least partial results concerning hyperbolicity, where the theoretical analysis of $\det(\mathcal{J}_f(\Phi) - \lambda \boldsymbol{I})$ is essentially out of reach.

The following corollary follows from Theorem 4 by a discussion of the poles of $R(\lambda)$ and its asymptotic behaviour as $\lambda \to \pm\infty$.

Corollary 1 ([24]). *If $\gamma_i \cdot \gamma_j > 0$ for $i, j = 1, \dots, N$, then $\boldsymbol{D} + \boldsymbol{B}\boldsymbol{A}^{\mathrm{T}}$ is diagonalizable with real eigenvalues λ_i. If $\gamma_1, \dots, \gamma_N < 0$, then the* interlacing property

$$M_1 := v_N + \gamma_1 + \dots + \gamma_N < \lambda_N < v_N < \lambda_{N-1} < \dots < \lambda_1 < v_1$$

holds, while for $\gamma_1, \dots, \gamma_N > 0$, the following analogous property holds:

$$v_N < \lambda_N < v_{N-1} < \lambda_{N-1} < \dots < v_1 < \lambda_1 < M_2 := v_1 + \gamma_1 + \dots + \gamma_N.$$

The analysis of (47) also leads to an explicit spectral decomposition of $\mathcal{J}_f$ required for spectral schemes. Assume λ is an eigenvalue of $\mathcal{J}_f$ that satisfies $\lambda \neq v_i$ for all $i = 1, \dots, N$. Then $\boldsymbol{\xi} = \boldsymbol{A}^{\mathrm{T}}\boldsymbol{x}$ is a solution of $\boldsymbol{M}_\lambda \boldsymbol{\xi} = \boldsymbol{0}$, where the $m \times m$ matrix $\boldsymbol{M}_\lambda := \boldsymbol{I} + \boldsymbol{A}^{\mathrm{T}}(\boldsymbol{D} - \lambda \boldsymbol{I})^{-1}\boldsymbol{B}$ can easily be computed. In fact, given $\boldsymbol{g}, \boldsymbol{h} \in \mathbb{R}^N$, if we use the notation

$$[\boldsymbol{g}, \boldsymbol{h}] := [\boldsymbol{g}, \boldsymbol{h}]_\lambda := \boldsymbol{g}^{\mathrm{T}}(\boldsymbol{D} - \lambda \boldsymbol{I})^{-1}\boldsymbol{h} = \sum_{k=1}^{N} \frac{g_k h_k}{v_k - \lambda},$$

then $\boldsymbol{M}_\lambda = \boldsymbol{I} + ([\boldsymbol{a}_i, \boldsymbol{b}_j])_{1 \leq i,j \leq m}$, where $\boldsymbol{a}_i$ and $\boldsymbol{b}_j$ are the columns of $\boldsymbol{A}$ and $\boldsymbol{B}$, respectively. If $\boldsymbol{\xi} \neq \boldsymbol{0}$ solves $\boldsymbol{M}_\lambda \boldsymbol{\xi} = \boldsymbol{0}$, then we can use $\boldsymbol{x} + (\boldsymbol{D} - \lambda \boldsymbol{I})^{-1}\boldsymbol{B}(\boldsymbol{A}^{\mathrm{T}}\boldsymbol{x}) = \boldsymbol{0}$ to compute a right eigenvector of $\mathcal{J}_f$ as $\boldsymbol{x} = -(\boldsymbol{D} - \lambda \boldsymbol{I})^{-1}\boldsymbol{B}\boldsymbol{\xi}$. The same procedure may be employed to calculate the left eigenvectors of $\mathcal{J}_f$.

The MLB model arises from the mass and linear momentum balance equations for the solid species and the fluid [10]. For equal-density particles, its final form is

$$v_i(\Phi) = v_i^{\mathrm{MLB}}(\Phi) := \frac{(\rho_{\mathrm{s}} - \rho_{\mathrm{f}})gD_1^2}{18\mu_{\mathrm{f}}}(1 - \phi)\mathcal{V}(\phi)\left(d_i^2 - \sum_{m=1}^{N} \phi_m d_m^2\right),$$

where μ_{f} is the fluid viscosity, and $\phi = \phi_1 + \dots + \phi_N$ is the total solids volume fraction. Here $\mathcal{V}(\phi)$ is assumed to satisfy $\mathcal{V}(0) = 1$, $\mathcal{V}(\phi_{\max}) = 0$ and $\mathcal{V}'(\phi) \leq 0$

for $\phi \in [0, \phi_{\max}]$, where the maximum total solids concentration is assumed to be given by the constant $\phi_{\max}$. A standard choice for $\mathcal{V}(\phi)$ is the equation

$$\mathcal{V}(\phi) = \begin{cases} (1-\phi)^{n_{\mathrm{RZ}}-2} & \text{if } \Phi \in \mathcal{D}_{\phi_{\max}} \\ 0 & \text{otherwise,} \end{cases} \quad n_{\mathrm{RZ}} > 2. \tag{48}$$

(This formula is consistent with (4) for $N = 1$, i.e., $V(\phi) = (1-\phi)^2\mathcal{V}(\phi)$.) We may write the components of the flux vector $\boldsymbol{f}(\Phi)$ of the MLB model as

$$f_i(\Phi) = f_i^{\mathrm{MLB}}(\Phi) := v_1^{\mathrm{MLB}}(\mathbf{0})\phi_i(1-\phi)\mathcal{V}(\phi)\left(d_i^2 - \sum_{m=1}^N \phi_m d_m^2\right). \tag{49}$$

The present version of the MLB model corresponds to $m = 2$, where $p_1 = \phi$ and $p_2 = \mathcal{V}(\phi)(d_1^2\phi_1 + \cdots + d_N^2\phi_N)$. For this model, we have:

Lemma 2 ([24]). *The MLB model* (13) *and* (49) *is strictly hyperbolic on* $\mathcal{D}_{\phi_{\max}}$. *The eigenvalues* $\lambda_i = \lambda_i(\Phi)$ *of* $\mathcal{J}_f(\Phi) = \mathcal{J}_{f^{\mathrm{MLB}}}(\Phi)$ *satisfy the interlacing property*

$$M_1(\Phi) < \lambda_N(\Phi) < v_N(\Phi) < \lambda_{N-1}(\Phi) < v_{N-1}(\Phi) < \cdots < \lambda_1(\Phi) < v_1(\Phi), \tag{50}$$

$$M_1(\Phi) := v_1^{\mathrm{MLB}}(\mathbf{0})\left(d_N^2 V(\Phi) + \big((1-\phi)V'(\phi) - 2V(\phi)\big)\sum_{m=1}^N \phi_m d_m^2\right).$$

Furthermore, if $\lambda \notin \{v_1, \ldots, v_N\}$ *is an eigenvalue of* $\mathcal{J}_f(\Phi)$, *then the discussion following Corollary 1 allows us to express the corresponding eigenvector in closed algebraic form (not detailed here).*

The Höfler and Schwarzer (HS) model is motivated by the following expression for v_i by Batchelor and Wen [5,7], valid for a dilute suspension (i.e., $\phi \ll \phi_{\max}$):

$$v_i(\Phi) = \frac{(\rho_{\mathrm{s}} - \rho_{\mathrm{f}})gD_1^2}{18\mu_{\mathrm{f}}} d_i^2(1 + \boldsymbol{s}_i^{\mathrm{T}}\Phi). \tag{51}$$

Here, $\boldsymbol{s}_i^{\mathrm{T}} := (S_{i1}, \ldots, S_{iN})$ is the i-th row of the matrix $\boldsymbol{S} = (S_{ij})_{1 \le i,j \le N}$ of dimensionless sedimentation coefficients S_{ij}, which are negative functions of $\lambda_{ij} := d_j/d_i$ and depend on certain other parameters. They can be reasonably approximated by

$$S_{ij} = \sum_{l=0}^3 \beta_l \left(\frac{d_j}{d_i}\right)^l, \quad 1 \le i, j \le N \quad \text{with coefficients } \beta_0, \ldots, \beta_3 \le 0. \tag{52}$$

Some authors set $\beta_3 = 0$ a priori; for example, Höfler and Schwarzer [56] obtained

$$\boldsymbol{\beta}^{\mathrm{T}} = (\beta_0, \dots, \beta_3) = (-3.52, -1.04, -1.03, 0) \tag{53}$$

by fitting data from [7] to a second-order polynomial. For simplicity, we also consider $\beta_3 = 0$ in this work.

To overcome the limitation of (51) to dilute suspensions, Höfler and Schwarzer [56] extended (51) to the whole range of concentrations by the formula

$$v_i(\Phi) = v_i^{\mathrm{HS}}(\Phi) := \frac{(\rho_{\mathrm{s}} - \rho_{\mathrm{f}}) g D_1^2}{18\mu_{\mathrm{f}}} d_i^2 \exp(\boldsymbol{s}_i^{\mathrm{T}} \Phi + n\phi)(1-\phi)^n, \quad n \geq 0.$$

The corresponding flux vector of the HS model is given by

$$f_i(\Phi) = f_i^{\mathrm{HS}}(\Phi) := v_1^{\mathrm{HS}}(\mathbf{0}) \phi_i d_i^2 \exp(\boldsymbol{s}_i^{\mathrm{T}} \Phi + n\phi)(1-\phi)^n.$$

For the HS model it is straightforward to verify strict hyperbolicity on $\mathcal{D}_1$ for $N = 2$, arbitrary non-positive Batchelor matrices $\boldsymbol{S}$ and arbitrarily small values of d_2. The analysis of [24] ensures hyperbolicity for arbitrary N and in the case of the coefficients (53) under the fairly mild restriction $d_N > 0.0078595$.

For the hyperbolicity analysis of the HS model, we define

$$\boldsymbol{a}_\nu := \boldsymbol{d}_{\nu-1}^{\mathrm{T}} := \left(d_1^{\nu-1}, d_2^{\nu-1}, \dots, d_N^{\nu-1}\right), \quad p_\nu := \boldsymbol{a}_\nu^{\mathrm{T}} \Phi, \quad \nu = 1, \dots, 4,$$

and taking into account that $\beta_3 = 0$, we obtain from (51) and (52)

$$v_i(\Phi) = v_1^{\mathrm{HS}}(\mathbf{0}) d_i^2 \exp\left((\beta_0 + n) p_1 + \frac{\beta_1}{d_i} p_2 + \frac{\beta_2}{d_i^2} p_3 \right) (1 - p_1)^n.$$

Thus, the hyperbolicity of the HS model can be analyzed by Theorem 4, where $m = 3$ if $\beta_3 = 0$ and $m = 4$ if $\beta_3 \neq 0$. The calculations become involved, but still lead to estimates of the hyperbolicity region. A typical result is the following.

Lemma 3. *Assume that $\boldsymbol{\beta}$, $\phi_{\max}$, and the width of the particle size distribution characterized by the value of $d_N \in (0, 1]$ satisfy*

$$H(\phi, \boldsymbol{\beta}, d_N) := -\tilde{\beta}_0 \big(\beta_1 d_N + \beta_2 (1 + d_N)^2\big) - \beta_2 \beta_1 d_N - \phi (1 - d_N)^2 \tilde{\beta}_0 \beta_1 \beta_2 < 0$$

for all $\phi \in (0, \phi_{\max})$. Then the HS model is strictly hyperbolic for $\Phi \in \mathcal{D}_{\phi_{\max}}$. The eigenvalues satisfy the interlacing property (50). *(The fairly involved algebraic expression for γ_i for this model is not written out here for brevity. We refer to [24] and [27] for the respective cases $\beta_3 = 0$ and $\beta_3 < 0$.)*

For $\boldsymbol{\beta}$ given by (53) the region of hyperbolicity for the HS model ensured by Lemma 3 is illustrated in Fig. 8. The spectral decomposition of $\mathcal{J}_f(\Phi)$, i.e., the eigenvectors corresponding to the eigenvalues $\lambda_i(\Phi)$, is easy to obtain from Theorem 4, see [26] for details. Similar estimates of the hyperbolicity region for

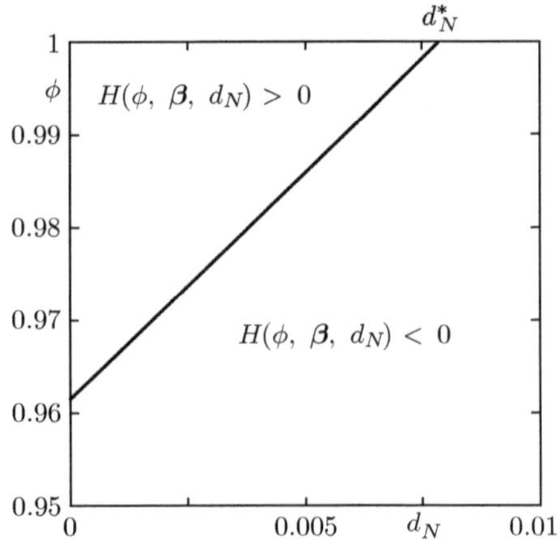

Fig. 8 Region of hyperbolicity ($H(\phi, \boldsymbol{\beta}, d_N) < 0$) for the HS model for the coefficients (53) [24]

the original model by Batchelor and Wen [5, 7], which is not discussed in this contribution, and for the HS model can be obtained by the same method for the case $\beta_3 < 0$, which gives rise to a perturbation rank of $m = 4$ [27].

4.2 *Spectral and Component-Wise Numerical Schemes*

For grid points $x_j = j\Delta x$, $t_n = n\Delta t$, a conservative scheme for $\Phi_i^n \approx \Phi(x_j, t_n)$ is given by

$$\Phi_j^{n+1} = \Phi_j^n - \frac{\Delta t}{\Delta x}(\hat{\boldsymbol{f}}_{j+1/2} - \hat{\boldsymbol{f}}_{j-1/2}), \quad \hat{\boldsymbol{f}}_{j+1/2} = \hat{\boldsymbol{f}}\left(\Phi_{j-s+1}^n, \dots, \Phi_{j+s}^n\right), \quad j \in \mathbb{Z}.$$

The resulting scheme should be (at least second-order) accurate and stable. The most common design of numerical fluxes $\hat{\boldsymbol{f}}_{j+1/2}$ is to solve Riemann problems, either exactly (as in the original Godunov scheme, which is very costly), or approximately (e.g., as in the Roe scheme). For polydisperse sedimentation, exact Riemann solvers are out of reach, since the eigenstructure of $\mathcal{J}_f$ is hard to compute.

In [26] Shu-Osher's technique [80] is used along with the information provided by the secular equation to get efficient schemes for the MLB and HS models. This scheme is based on the method of lines, that is, on applying an ODE solver to spatially semi-discretized equations. For the discretization of the flux derivative we use local characteristic projections. Local characteristic information to compute $\hat{\boldsymbol{f}}_{j+1/2}$ is provided by the eigenstructure of $\mathcal{J}_f(\Phi_{j+1/2})$, where $\Phi_{j+1/2} = \frac{1}{2}(\Phi_j + \Phi_{j+1})$, given by the right and left eigenvectors that form the respective matrices

$$\boldsymbol{R}_{j+1/2} = \left[\boldsymbol{r}_{j+1/2,1} \dots \boldsymbol{r}_{j+1/2,N}\right], \quad \left(\boldsymbol{R}_{j+1/2}^{-1}\right)^{\mathrm{T}} = \left[\boldsymbol{l}_{j+1/2,1} \dots \boldsymbol{l}_{j+1/2,N}\right].$$

From a local flux-splitting $\boldsymbol{f}^{\pm,k}$ (we omit its dependency on $j + 1/2$) given by $\boldsymbol{f}^{-,k} + \boldsymbol{f}^{+,k} = \boldsymbol{f}$, where $\pm\lambda_k(\mathcal{J}_{\boldsymbol{f}^{\pm,k}}(\Phi)) \geq 0$, $\Phi \approx \Phi_{i+1/2}$ and λ_k is the k-th eigenvalue, $k = 1, \dots, N$, we can define the k-th characteristic flux as

$$g_j^{\pm,k} = \boldsymbol{l}_{j+1/2,k}^{\mathrm{T}} \cdot \boldsymbol{f}^{\pm,k}(\Phi_j).$$

If $\mathcal{R}^+$ and $\mathcal{R}^-$ denote upwind-based reconstructions, then

$$\hat{g}_{j+1/2,k} = \mathcal{R}^+\big(g_{j-s+1}^{+,k}, \dots, g_{j+s-1}^{+,k}; x_{j+1/2}\big) + \mathcal{R}^-\big(g_{j-s+2}^{-,k}, \dots, g_{j+s}^{-,k}; x_{j+1/2}\big),$$

$$\hat{\boldsymbol{f}}_{j+1/2} = \boldsymbol{R}_{j+1/2}\hat{\boldsymbol{g}}_{j+1/2} = \sum_{k=1}^{n} \hat{g}_{j+1/2,k}\boldsymbol{r}_{j+1/2,k}.$$

If we do not want to use local characteristic information, we can use the previous formula with $\boldsymbol{R}_{j+1/2} = \boldsymbol{I}_N$, where $\boldsymbol{I}_N$ denotes the $N \times N$ identity matrix, and a global flux splitting $\boldsymbol{f}^- + \boldsymbol{f}^+ = \boldsymbol{f}$, where $\pm\lambda_k(\mathcal{J}_{\boldsymbol{f}^\pm}(\Phi)') \geq 0$ for all k. With this choice, and denoting by $\boldsymbol{e}_k$ the kth unit vector, we get $g_j^{\pm,k} = \boldsymbol{e}_k^{\mathrm{T}}\boldsymbol{f}^\pm(\Phi_j) = f_k^\pm(\Phi_j)$, i.e., $g_j^{\pm,k}$ are the components of the split fluxes, and the numerical flux is computed component by component by reconstructing the split fluxes component by component, i.e., $\hat{\boldsymbol{f}}_{j+1/2} = (\hat{f}_{j+1/2,1}, \dots, \hat{f}_{j+1/2,N})^{\mathrm{T}}$, where

$$\begin{aligned}\hat{f}_{j+1/2,k} &= \mathcal{R}^+\big(g_{j-s+1}^{+,k}, \dots, g_{j+s-1}^{+,k}; x_{j+1/2}\big)\\ &\quad + \mathcal{R}^-\big(g_{j-s+2}^{-,k}, \dots, g_{j+s}^{-,k}; x_{j+1/2}\big), \quad k = 1, \dots, N.\end{aligned}$$

This scheme will be referred to as COMP-GLF and it is a high-order extension of the Lax-Friedrichs scheme.

We now explain the SPEC-INT scheme. If $\lambda_k(\mathcal{J}_f(\Phi)) > 0$ (respectively, < 0) for all $\Phi \in [\Phi_j, \Phi_{j+1}]$, where $[\Phi_j, \Phi_{j+1}] \subset \mathbb{R}^N$ denotes the segment joining both states, then we upwind (since then there is no need for flux splitting):

$$\boldsymbol{f}^{+,k} = \boldsymbol{f},\ \boldsymbol{f}^{-,k} = \boldsymbol{0} \quad \text{if } \lambda_k(\mathcal{J}_f(\Phi)) > 0, \quad \boldsymbol{f}^{+,k} = \boldsymbol{0},\ \boldsymbol{f}^{-,k} = \boldsymbol{f} \quad \text{if } \lambda_k(\mathcal{J}_f(\Phi)) < 0.$$

On the other hand, if $\lambda_k(\mathcal{J}_f(\Phi))$ changes sign on $[\Phi_j, \Phi_{j+1}]$, then we use a Local Lax-Friedrichs flux splitting given by $\boldsymbol{f}^{\pm,k}(\Phi) = \boldsymbol{f}(\Phi) \pm \alpha_k\Phi$, where the numerical viscosity parameter α_k should satisfy

$$\alpha_k \geq \max_{\Phi \in [\Phi_j, \Phi_{j+1}]} \big|\lambda_k\big(\mathcal{J}_f(\Phi)\big)\big|. \tag{54}$$

The following usual choice of α_k produces oscillations in the numerical solution indicating that the amount of numerical viscosity is insufficient:

$$\alpha_k = \max\big\{\big|\lambda_k\big(\mathcal{J}_f(\Phi_j)\big)\big|, \big|\lambda_k\big(\mathcal{J}_f(\Phi_{j+1})\big)\big|\big\}.$$

The right-hand side of (54) can usually cannot be evaluated exactly in closed form. However, for the present class of models, Corollary 1 generates a fairly sharp bound for that expression. In the case of the MLB model, we have $\gamma_k < 0$ and the interlacing property leads to the efficiently computable bounds [26]

$$\max_{\Phi \in [\Phi_j, \Phi_{j+1}]} |\lambda_k(\Phi)| \leq \alpha_k := \max \left\{ \max_{\Phi \in [\Phi_j, \Phi_{j+1}]} \left| \nu_k(\Phi) \right|, \max_{\Phi \in [\Phi_j, \Phi_{j+1}]} \left| \nu_{k+1}(\Phi) \right| \right\},$$
$$k = 1, \dots, N. \tag{55}$$

(This property also holds for other models, under appropriate circumstances [24, 27].) "SPEC-INT" denotes the scheme for which $\alpha_1, \dots, \alpha_N$ are defined by (55).

4.3 Numerical Examples

The zero-flux boundary conditions (14) are implemented by setting $\hat{f}_{-1/2} = \hat{f}_{M-1/2} = 0$. We recall that a WENO5 scheme requires the consideration of two additional ghost cells on each boundary of the computational domain. To guarantee that all the interpolatory stencils remain inside the computational domain we set large values for the concentrations in the ghost cells, which produce large variations, so that the WENO procedure avoids the use of any stencil involving the ghost cells. The time discretization employs the well-known optimal third-order, three-stage Runge-Kutta method named SSPRK(3,3). SSP time discretization methods are widely used for hyperbolic PDE because they preserve the nonlinear stability properties which are necessary for problems with non-smooth solutions. To satisfy the CFL condition, the value of Δt is computed adaptively for each step ν. More precisely, the solution $\mathbf{\Phi}^{\nu+1}$ at $t_{\nu+1} = t_\nu + \Delta t$ is calculated from $\mathbf{\Phi}^\nu$ by using the time step $\Delta t = \mathrm{CFL} * \Delta x / \rho^\nu_{\max}$, where $\rho^\nu_{\max}$ is an estimate of the maximal characteristic velocity for $\mathbf{\Phi}^\nu$.

From [26] we select the case $N = 4$ for the MLB and HS models (Examples 6 and 7, respectively). We consider $d_1 = 1$, $d_2 = 0.8$, $d_3 = 0.6$ and $d_4 = 0.4$, $\phi_{\max} = 0.6$, and $\phi_i^0 = 0.05$ for $i = 1, \dots, 4$. We furthermore choose $D_1 = 4.96 \times 10^{-4}$ m, a settling vessel of (unnormalized) depth $L = 0.3$ m and $\phi_{\max} = 0.68$. We employ (48) with $n_{\mathrm{RZ}} = 4.7$. The remaining parameters are $g = 9.81\,\mathrm{m/s^2}$, $\mu_{\mathrm{f}} = 0.02416\,\mathrm{Pa\,s}$ and $\rho_{\mathrm{f}} = 1{,}208\,\mathrm{kg/m^3}$. Moreover, the spatial coordinate $x \in [0, 1]$ refers to normalized depth. In this section, we take CFL $= 0.5$ throughout.

Figures 9a and 10a display the reference solution obtained with SPEC-INT and $M_{\mathrm{ref}} = 6{,}400$ for $t = 50$ s and $t = 300$ s respectively, while plots (b–d) of both figures are enlarged views of the corresponding numerical solutions obtained with SPEC-INT and COMP-GLF with $M = 400$. Figure 11 shows the corresponding results for Example 7. Both series of plots show that at $M = 400$ the quality of approximation of piecewise constant portions of the solution and the resolution of kinematic shocks by SPEC-INT is superior to that of COMP-GLF. For the times

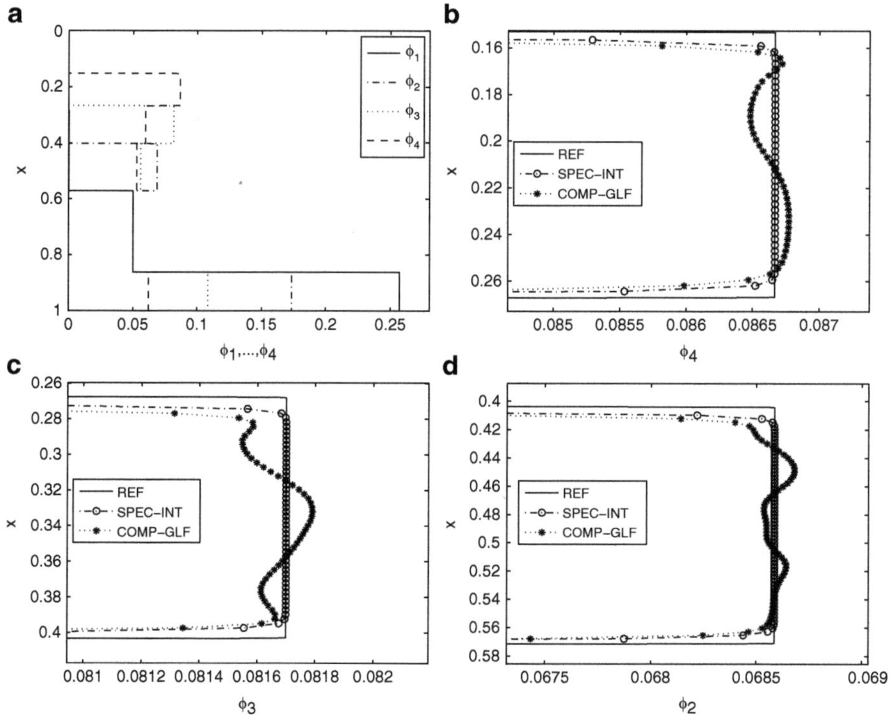

Fig. 9 Example 6: reference solution for $\phi_1, \ldots, \phi_4$ computed by SPEC-INT with $M_{\mathrm{ref}} = 6{,}400$ (**a**), and details of numerical solutions with $M = 400$ (**b**–**d**) at $t = 50\,\mathrm{s}$

considered the average convergence rate using SPEC-INT is close to one. On the other hand, as time increases, the errors increase considerably. Of course, for a given value of M, COMP-GLF is faster than SPEC-INT. Nevertheless, if we seek a fixed level of resolution in the numerical simulation, then SPEC-INT turns out to be computationally more efficient, see [26].

As in the case of the MCLWR kinematic traffic models, the characteristic-based schemes, which use the full spectral decomposition of $\mathcal{J}_f$ at each cell-interface, are more robust and lead to numerical solutions which are essentially oscillation free. This situation is similar to what is observed for the Euler equations for gas dynamics, where the superiority of characteristic-based schemes is well known. For gas dynamics, the spectral decomposition of the Jacobian matrix is given in closed form, hence characteristic-based schemes pose no special difficulties. For polydisperse models, the spectral decomposition can only be computed numerically. In addition, the characteristic fields are neither genuinely nonlinear nor linearly degenerate, hence the determination of the viscosity coefficients in flux-vector splitting schemes becomes a non-trivial task. In any case we have shown that SPEC-INT gives a good resolution on the numerical approximation with a relative small number of mesh points, hence it is competitive with respect to the simpler component-wise

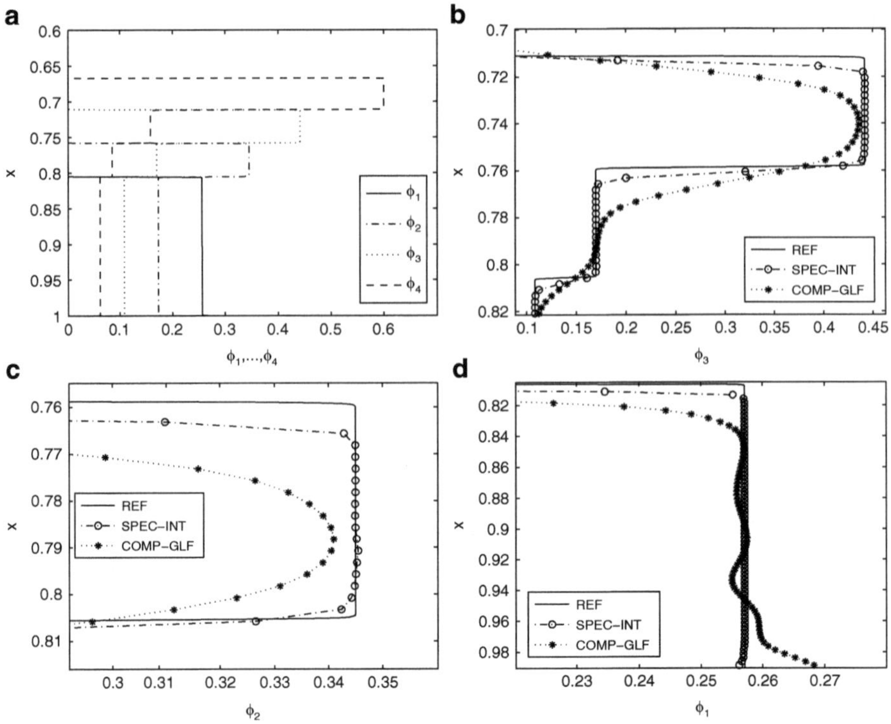

Fig. 10 Example 6: reference solution for $\phi_1, \ldots, \phi_4$ and ϕ computed by SPEC-INT with $M_{\mathrm{ref}} =$ 6,400 (**a**, **b**), and details of numerical solutions with $M = 400$ (**c**–**f**), at $t = 300$ s.

schemes. In recent work [30] it is shown that SPEC-INT is even more competitive than cheaper component-wise schemes, such as COMP-GLF, in an Adaptive Mesh Refinement (AMR) framework, since its non-oscillatory properties will help to avoid unnecessary refinement in regions of constant concentration.

5 Multidimensional Models

5.1 Adaptive Multiresolution (MR) Techniques

Adaptive multiresolution (MR) techniques are naturally fitted for FV schemes [13, 54, 69, 77]. They are based on representing the numerical solution on a fine grid by values on a much coarser grid plus a series of differences at different levels of nested dyadic grids. These differences are small in regions where the solution is smooth. Therefore, by discarding small details (the so-called "thresholding" operation), data compression can be achieved [13]. This automatic grid

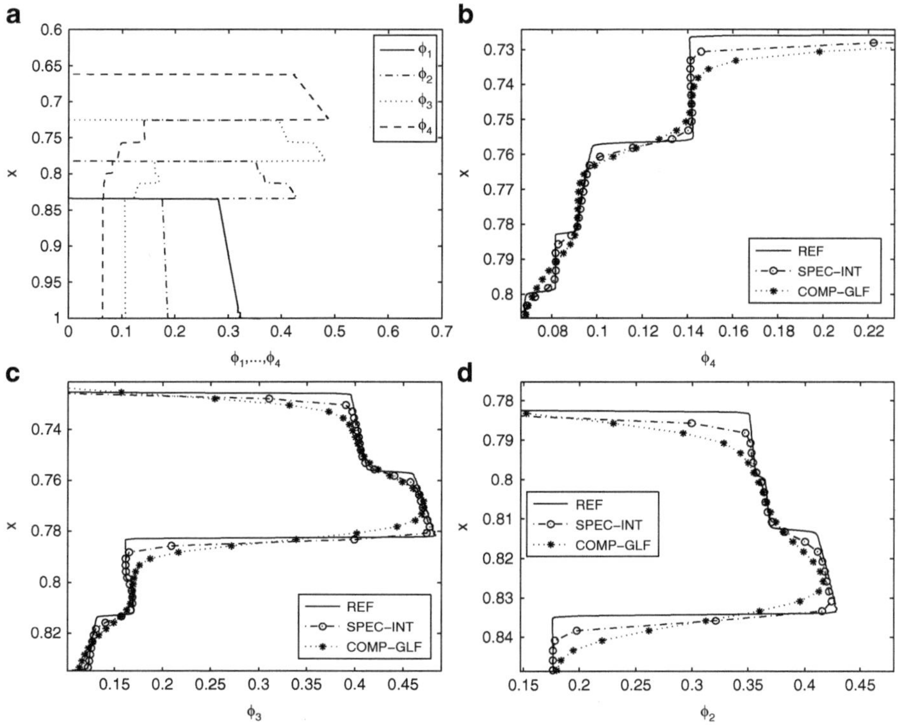

Fig. 11 Example 7: numerical solution for $\phi_1, \ldots, \phi_4$ with $M = 400$: at $t = 250$ s (**a**) and enlarged views (**b–d**), where the reference solution is computed using SPEC-INT with $M_{\mathrm{ref}} = 6{,}400$

refinement allows for memory and CPU time reductions while the approximation error remains controlled. The governing equations, in the present case (15) and (16), are discretized with a classical FV discretization. This approach has been implemented in [28] for (15) and (16) with $A \equiv 0$ and $\nu = 0$ to simulate the settling of a monodisperse suspension in a tilted narrow channel, which gives rise to the so-called "Boycott effect" [14], namely an increase of settling rates compared with a vertical channel. This effect is related to the formation of discontinuities in u and a boundary layer beneath a downward-facing inclined wall, occurs in vessels of simple geometry, and is therefore suitable for testing the capability of adaptive methods to concentrate computational effort on zones of strong variation such as discontinuities in u and boundary layers. In [28] the MR technique indeed produced a significant gain in efficiency.

Figure 12 (Example 8) shows an example from [28] with $L = 8$ resolution levels in total, corresponding to a finest grid of 256×256 cells on which (16) (with $\nu = 0$, $\lambda = 1, \boldsymbol{k} = (\cos\theta, \sin\theta)$, $\zeta = 0.67$ and $\mu(u) = (1-u)^{-2}$ and pressure stabilization [15]) is solved by a finite volume scheme, while (16) (with $f(u) = u(1-u)^2$ and $A \equiv 0$) is solved on an adaptive grid by the first-order Godunov scheme.

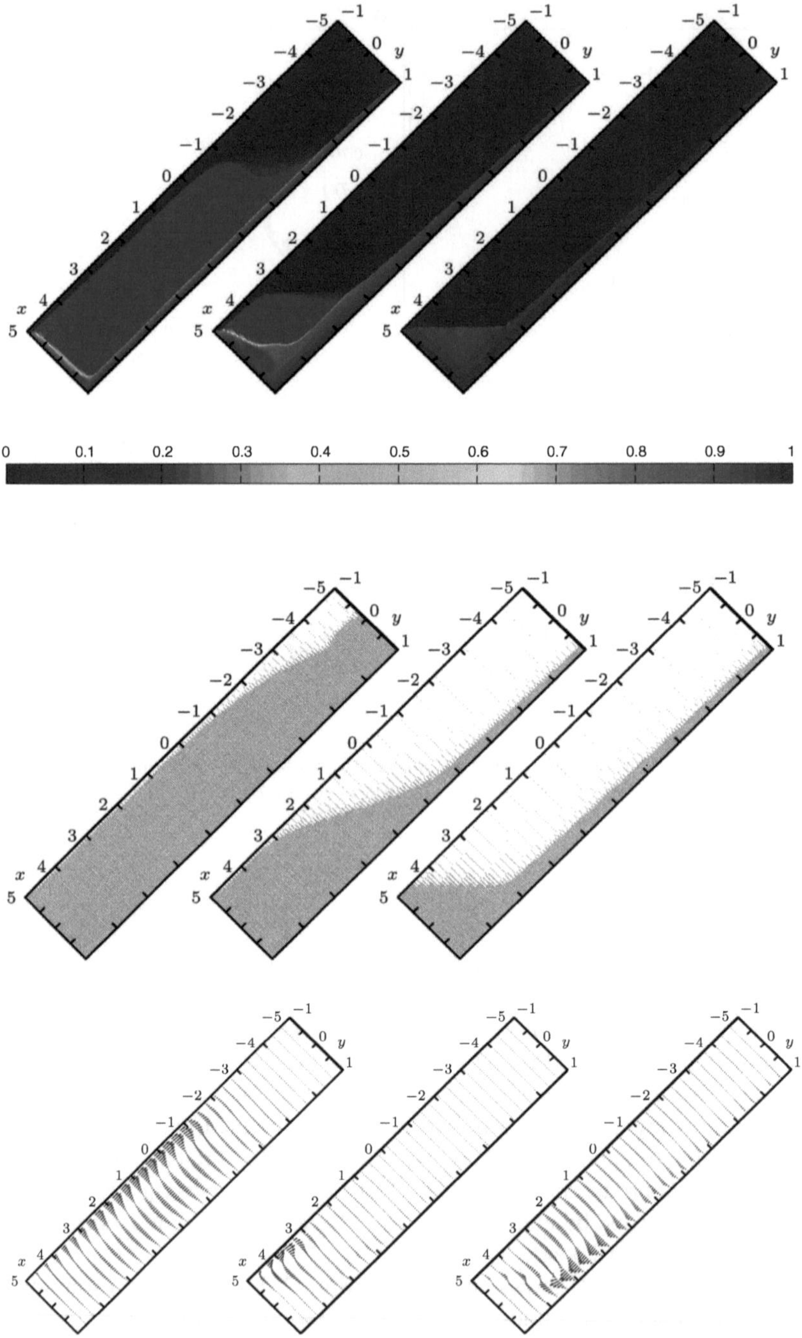

Fig. 12 Example 8: simulation of the settling of a suspension of constant initial concentration $u_0 = 0.2$ in a channel inclined by $\theta = 45^\circ$ [28]. *Top*: concentration u, *middle*: leaves of the adaptive tree, *bottom*: velocity $\boldsymbol{v}$, at times $t = 1.5$ (*left*), $t = 3.75$ (*middle*), and $t = 11.25$ (*right*). We have $\|\boldsymbol{v}(1.5)\| = 11.84$, $\|\boldsymbol{v}(3.75)\| = 3.72$ and $\|\boldsymbol{v}(11.25)\| = 2.7 \times 10^{-2}$

5.2 *Finite Volume Element (FVE) Methods*

If $A \not\equiv 0$ and $A(\cdot)$ has the behaviour (10), then (15) becomes strongly degenerate parabolic. Usually the type-change interface $u = u_c$ is associated with a discontinuity in the solution. An open problem of interest in applications is the development of numerical methods for (15) and (16) under the assumption of strong degeneracy. While FV methods are the best choice to discretize (15) (due to its convection-dominated nature along with the strong gradients in the solution), they are outperformed by finite element (FE) methods for what concerns the discretizations of the momentum and continuity equations forming the Stokes equations [31]. This observation motivated the FVE method (cf. [73] and the references cited in that paper) as a "hybrid" methodology, which is intermediate between FV and FE methods: the method is locally conservative (like a classical FV method) while it allows for L^2 estimates in a rather natural way (as in classical FE methods). The basic idea is to reformulate the FE scheme as a FV scheme on a dual mesh (see [4, 73] for details). The FVE methodology permits treating the full system (15) and (16) by a unified approach.

This method is implemented in [29] for a 2D section of an axisymmetric vessel (which requires cylindrical coordinates, cf., e.g., [9,52]), $\nu = 0$ (the Stokes system), and pointwise degeneracy ($a(u) = 0$ at $u = 0$ and $u = u_{\max}$ only). The last restriction was found necessary since the particular Galerkin discretization used in [29] relies on formulas like $\Delta A(u) = \nabla \cdot (a(u)\nabla u)$ which are not valid in general in the strongly degenerate case. However, numerical solutions behave reasonably in both the pointwise and strongly degenerate cases.

As a numerical example, we consider the fill-up of a cylindrical settling tank with a so-called skirt baffle and circumferential suction lifts introduced in [61, 92]. The essential parameters are $\rho_s - \rho_f = 1{,}562\,\mathrm{kg/m^3}$, $f(u) = 2.2 \times 10^{-3} u(1 - u/0.9)^2\,\mathrm{m/s}$, $u_{\mathrm{in}} = 0.1$, $g = 9.81\,\mathrm{m/s^2}$, $v_{z,\mathrm{out}} = \nu v_{z,\mathrm{in}}$, $v_{r,\mathrm{ofl}} = \frac{9-\nu}{52} v_{z,\mathrm{in}}$, $v_{r,\mathrm{in}} = 0.019\,\mathrm{m/s}$ and $\Delta t = 5\,\mathrm{s}$. The primal mesh $\mathcal{T}$ is composed of 7,410 elements and 4,206 interior nodes. The boundary conditions for velocity at the suction lifts are given by $\boldsymbol{v} = (0, -u_{z,\mathrm{out}}/4)$, where $v_{z,\mathrm{out}} = \nu v_{r,\mathrm{in}}$. See Figs. 13 and 14 for numerical results.

6 Alternate Treatments and Some Open Problems

Concerning the analysis of TVD and FTVD schemes of Sect. 2, we mention that in [20] an entropy inequality similar to (31) was used to prove that the first-order version of our scheme converges to a unique entropy solution of the conservation law. Although our numerical experiments indicate that the second-order schemes STVD and FTVD also converge to the unique entropy solution, the entropy inequality (31) is not quite in a form that allows us to repeat the uniqueness argument in [20]. We leave this as an open problem.

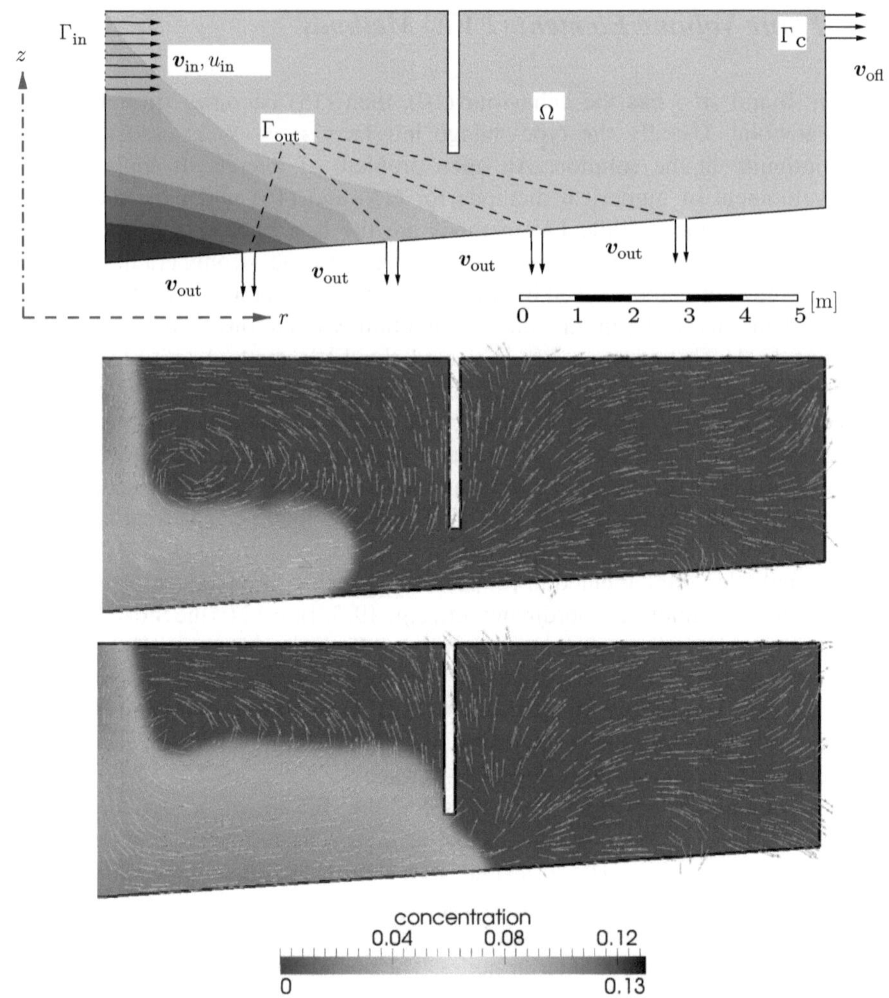

Fig. 13 Numerical simulation of the fill-up of a settling basin with skirt baffle (*top*) showing the solids concentration u at $t = 500$ s (*middle*) and 1,000 s (*bottom*)

Let us mention some of the works that analyze problems related to the conservation law with discontinuous flux (11) analyzed in Sect. 3. Another spatially one-dimensional, nonlocal sedimentation model was studied by Sjögreen et al. [82], who consider a hyperbolic-elliptic model problem given by (1) coupled with $-\eta(v_s)_{xx} + v_s = u$, where $\eta > 0$ is a viscosity parameter. Clearly, at any fixed position x_0, $v_s(x_0, t)$ will depend on $u(\cdot, t)$ as a whole; the nonlocal dependence is not limited to a neighborhood, as in [99] and herein. They prove that their model has a smooth solution, and present numerical solutions obtained by a high-order difference scheme. Furthermore, the (local) kinematic model of sedimentation (2) is similar to the well-known Lighthill-Whitham-Richards (LWR) model of

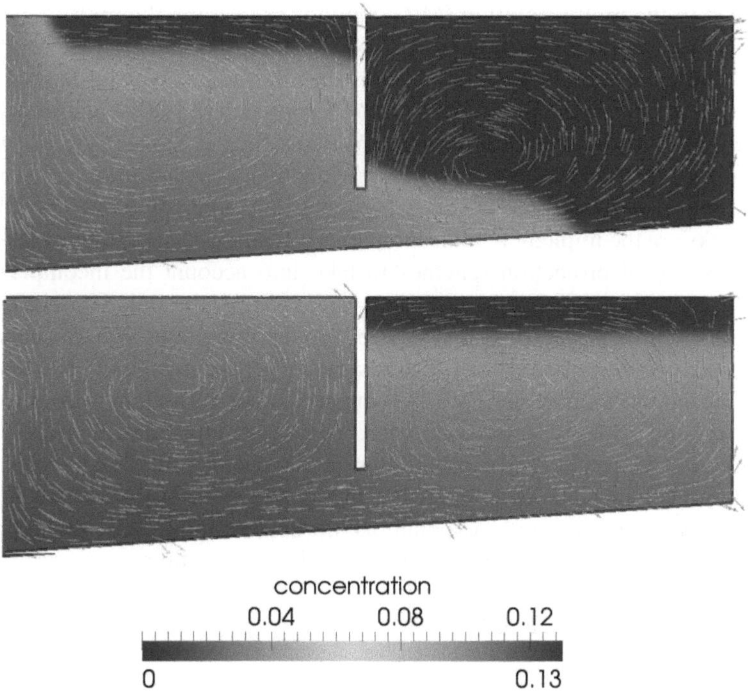

Fig. 14 Continuation of Fig. 13 showing the solids concentration u at $t = 2{,}000$ s (*top*) and 7,500 s (*bottom*)

vehicular traffic. Sopasakis and Katsoulakis [83] extended the LWR model to a nonlocal version by a "look-ahead" rule, i.e. drivers choose their velocity taking account the density on a stretch of road ahead of them. Kurganov and Polizzi [63] showed that an extension of the well-known Nesshayu-Tadmor (NT) central nonoscillatory scheme [71] is suitable for the nonlocal model of [83], which can be written as (11) for $\alpha = 1$ and $V(w) = \exp(-w)$, and if we replace K_a by a particular non-symmetric kernel function with compact support. Related models with a nonlocal convective flux that have been analyzed within an entropy solution framework (as done herein and in [12]) include the continuum model for the flow of pedestrians by Hughes [57], which gives rise to a multi-dimensional conservation law with a nonlocal flux; see also [36, 39]. See [12] for further applications.

As another open research problem, a systematic travelling wave analysis of (11), which would extend the results of [99], is still lacking. Such an analysis could explain whether new phenomena, e.g. nonclassical shocks, should be expected when one considers the formal limit $a \to 0$ of entropy solutions of (11), especially in the case $\alpha \geq 1$. Unfortunately, most of the constants appearing in the compactness estimates of [12] are not uniform, i.e. they blow up when $a \to 0$. It is therefore not clear whether a sequence of entropy solutions converges to a meaningful limit as $a \to 0$. This problem should at first be explored numerically.

Related to the multiresolution (MR) method for tackling the multi-dimensional system (15) and (16) outlined in Sect. 5, we mention that in [28] MR was applied to the solution of (15) (with $A \equiv 0$) only, but the more involved Stokes system was always solved on the finest grid. The MR approach of [28] should be extended to a method that solves both (15) and (16) (first, for the Stokes system ($\nu = 0$), and then for the Navier-Stokes case) on an adaptively refined grid. Further speed-up of adaptive methods is achieved using local time stepping strategies [70]. The central tasks are the implementation and numerical analysis of pressure stabilization techniques and of projection schemes to take into account the incompressibility of $\boldsymbol{v}$ (cf., e.g., [53, 78]). Further research will concern the polydisperse case, for which (15) will be replaced by a system of conservation laws. Finally, concerning the FVE method described in Sect. 5.2, besides incorporating the full Navier-Stokes terms, one should modify the FVE scheme so that its formulation from the onset also covers the strongly degenerate case. Thus, discretizations alternative to the Discontinuous Galerkin (DG) formulation employed in [29] should be tested by adequately choosing the numerical flux associated with (15), and we intend to investigate whether the choice of a diamond mesh (one of the dual meshes) made in [29] will in general capture the hyperbolic-parabolic transition.

Acknowledgements RB acknowledges support by Fondecyt project 1090456 and BASAL project CMM, Universidad de Chile and Centro de Investigación en Ingeniería Matemática (CI^{2}MA), Universidad de Concepción. RR is supported by ERC advanced grant 227058, project *Mathcard, Mathematical Modelling and Simulation of the Cardiovascular System.* HT is supported by Fondecyt project 11110264, and RB and HT are funded by CONICYT project Anillo ACT1118 (ANANUM). CV is supported by Universidad del Norte, project 2012-002.

References

1. J. Anderson, A secular equation for the eigenvalues of a diagonal matrix perturbation. Linear Algebra Appl. **246**, 49–70 (1996)
2. B. Andreianov, K.H. Karlsen, N.H. Risebro, A theory of L^1-dissipative solvers for scalar conservation laws with discontinuous flux. Arch. Ration. Mech. Anal. **201**, 27–86 (2011)
3. D.P. Ballou, Solutions to non-linear hyperbolic Cauchy problems without convexity conditions. Trans. Am. Math. Soc. **152**, 441–460 (1970)
4. R.E. Bank, D.J. Rose, Some error estimates for the box method. SIAM J. Numer. Anal. **24**, 777–787 (1987)
5. G.K. Batchelor, Sedimentation in a dilute polydisperse system of interacting spheres. Part 1. General theory. J. Fluid Mech. **119**, 379–408 (1982)
6. G.K. Batchelor, R.W.J. van Rensburg, Structure formation in bidisperse sedimentation. J. Fluid Mech. **166**, 379–407 (1986)
7. G.K. Batchelor, C.S. Wen, Sedimentation in a dilute polydisperse system of interacting spheres. Part 2. Numerical results. J. Fluid Mech. **124**, 495–528 (1982)
8. S. Benzoni-Gavage, R.M. Colombo, An n-populations model for traffic flow. Eur. J. Appl. Math. **14**, 587–612 (2003)
9. C. Bernardi, M. Dauge, Y. Maday, *Spectral Methods for Axisymmetric Domains* (Gauthier-Villars, Paris, 1999)

10. S. Berres, R. Bürger, K.H. Karlsen, E.M. Tory, Strongly degenerate parabolic-hyperbolic systems modeling polydisperse sedimentation with compression. SIAM J. Appl. Math. **64**, 41–80 (2003)
11. A.L. Bertozzi, T. Laurent, J. Rosado, L^p theory for the multidimensional aggregation equation. Commun. Pure Appl. Math. **64**, 45–83 (2011)
12. F. Betancourt, R. Bürger, K.H. Karlsen, E.M. Tory, On nonlocal conservation laws modelling sedimentation. Nonlinearity **24**, 855–885 (2011)
13. B.L. Bihari, A. Harten, Multiresolution schemes for the numerical solution of 2-D conservation laws I. SIAM J. Sci. Comput. **18**, 315–354 (1997)
14. A.E. Boycott, Sedimentation of blood corpuscles. Nature **104**, 532 (1920)
15. F. Brezzi, J. Pitkäranta, On the stabilization of finite element approximations of the Stokes equations, in *Efficient Solutions of Elliptic Systems*, Kiel. Notes on Numerical Fluid Mechanics, vol. 10 (Vieweg, Braunschweig, 1984), pp. 11–19
16. R. Bürger, K.H. Karlsen, Conservation laws with discontinuous flux: a short introduction. J. Eng. Math. **60**, 241–247 (2008)
17. R. Bürger, W.L. Wendland, Sedimentation and suspension flows: historical perspective and some recent developments. J. Eng. Math. **41**, 101–116 (2001)
18. R. Bürger, K.H. Karlsen, E.M. Tory, W.L. Wendland, Model equations and instability regions for the sedimentation of polydisperse suspensions of spheres. ZAMM Z. Angew. Math. Mech. **82**, 699–722 (2002)
19. R. Bürger, K.H. Karlsen, N.H. Risebro, J.D. Towers, Monotone difference approximations for the simulation of clarifier-thickener units. Comput. Vis. Sci. **6**, 83–91 (2004)
20. R. Bürger, K.H. Karlsen, N.H. Risebro, J.D. Towers, Well-posedness in BV_t and convergence of a difference scheme for continuous sedimentation in ideal clarifier-thickener units. Numer. Math. **97**, 25–65 (2004)
21. R. Bürger, K.H. Karlsen, J.D. Towers, A mathematical model of continuous sedimentation of flocculated suspensions in clarifier-thickener units. SIAM J. Appl. Math. **65**, 882–940 (2005)
22. R. Bürger, A. García, K.H. Karlsen, J.D. Towers, A family of numerical schemes for kinematic flows with discontinuous flux. J. Eng. Math. **60**, 387–425 (2008)
23. R. Bürger, K.H. Karlsen, J.D. Towers, An Engquist-Osher type scheme for conservation laws with discontinuous flux adapted to flux connections. SIAM J. Numer. Anal. **47**, 1684–1712 (2009)
24. R. Bürger, R. Donat, P. Mulet, C.A. Vega, Hyperbolicity analysis of polydisperse sedimentation models via a secular equation for the flux Jacobian. SIAM J. Appl. Math. **70**, 2186–2213 (2010)
25. R. Bürger, K.H. Karlsen, H. Torres, J.D. Towers, Second-order schemes for conservation laws with discontinuous flux modelling clarifier-thickener units. Numer. Math. **116**, 579–617 (2010)
26. R. Bürger, R. Donat, P. Mulet, C.A. Vega, On the implementation of WENO schemes for a class of polydisperse sedimentation models. J. Comput. Phys. **230**, 2322–2344 (2011)
27. R. Bürger, R. Donat, P. Mulet, C.A. Vega, On the hyperbolicity of certain models of polydisperse sedimentation. Math. Methods Appl. Sci. **35**, 723–744 (2012)
28. R. Bürger, R. Ruiz-Baier, K. Schneider, H. Torres, A multiresolution method for the simulation of sedimentation in inclined channels. Int. J. Numer. Anal. Model. **9**, 479–504 (2012)
29. R. Bürger, R. Ruiz-Baier, H. Torres, A stabilized finite volume element formulation for sedimentation-consolidation processes. SIAM J. Sci. Comput. **34**, B265–B289 (2012)
30. R. Bürger, P. Mulet, L.M. Villada, Spectral WENO schemes with adaptive mesh refinement for models of polydisperse sedimentation. ZAMM Z. Angew. Math. Mech. **93**, 373–386 (2013)
31. E. Burman, A. Quarteroni, B. Stamm, Interior penalty continuous and discontinuous finite element approximations of hyperbolic equations. J. Sci. Comput. **43**, 293–312 (2010)
32. M.C. Bustos, F. Concha, On the construction of global weak solutions in the Kynch theory of sedimentation. Math. Method Appl. Sci. **10**, 245–264 (1988)
33. M.C. Bustos, F. Concha, R. Bürger, E.M. Tory, *Sedimentation and Thickening. Phenomenological Foundation and Mathematical Theory* (Kluwer, Dordrecht, 1999)
34. K.S. Cheng, Constructing solutions of a single conservation law. J. Differ. Equ. **49**, 344–358 (1983)

35. H.S. Coe, G.H. Clevenger, Methods for determining the capacity of slime settling tanks. Trans. AIME **55**, 356–385 (1916)
36. R.M. Colombo, M. Herty, M. Mercier, Control of the continuity equation with a non local flow. ESAIM Control Opt. Calc. Var. **17**, 353–379 (2011)
37. A. Coronel, F. James, M. Sepúlveda, Numerical identification of parameters for a model of sedimentation processes. Inverse Probl. **19**, 951–972 (2003)
38. R.H. Davis, H. Gecol, Hindered settling function with no empirical parameters for polydisperse suspensions. AIChE J. **40**, 570–575 (1994)
39. M. Di Francesco, P.A. Markowich, J.-F. Pietschmann, M.-T. Wolfram, On the Hughes' model for pedestrian flow: the one-dimensional case. J. Differ. Equ. **250**, 1334–1362 (2011)
40. S. Diehl, Shock behaviour of sedimentation in wastewater treatment. M. Sc. thesis, University of Lund, Lund (1988)
41. S. Diehl, Estimation of the batch-settling flux function for an ideal suspension from only two experiments. Chem. Eng. Sci. **62**, 4589–4601 (2007)
42. S. Diehl, Shock-wave behaviour of sedimentation in wastewater treatment: a rich problem, in *Analysis for Science, Engineering and Beyond*, ed. by K. Åström et al. (Springer, Berlin, 2012), pp. 175–214
43. R. Donat, P. Mulet, A secular equation for the Jacobian matrix of certain multi-species kinematic flow models. Numer. Methods Partial Differ. Equ. **26**, 159–175 (2010)
44. J.V.N. Dorr, The use of hydrometallurgical apparatus in chemical engineering. J. Ind. Eng. Chem. **7**, 119–130 (1915)
45. A. Einstein, Eine neue Bestimmung der Moleküldimensionen. Ann. Phys. **19**, 289–306 (1906); A. Einstein, Berichtigung zu meiner Arbeit: "Eine neue Bestimmung ...". Ann. Phys. **34**, 591–592 (1911)
46. B. Engquist, S. Osher, One-sided difference approximations for nonlinear conservation laws. Math. Comput. **36**, 321–351 (1981)
47. T. Frising, C. Noïk, C. Dalmazzone, The liquid/liquid sedimentation process: from droplet coalescence to technologically enhanced water/oil emulsion gravity separators: a review. J. Dispers. Sci. Technol. **27**, 1035–1057 (2006)
48. E. Godlewski, P. Raviart, *Numerical Approximation of Hyperbolic Systems of Conservation Laws* (Springer, New York, 1996)
49. S. Gottlieb, C.-W. Shu, E. Tadmor, Strong stability-preserving high-order time discretization methods. SIAM Rev. **43**, 89–112 (2001)
50. P. Grassia, S.P. Usher, P.J. Scales, Closed-form solutions for batch settling height from model settling flux functions. Chem. Eng. Sci. **66**, 964–972 (2011)
51. P. Grassmann, R. Straumann, Entstehen und Wandern von Unstetigkeiten der Feststoffkozentration in Suspensionen. Chem. Ing. Tech. **35**, 477–482 (1963)
52. A. Guardone, L. Vigevano, Finite element/volume solution to axisymmetric conservation laws. J. Comput. Phys. **224**, 489–518 (2007)
53. J.L. Guermond, P. Minev, J. Shen, An overview of projection methods for incompressible flows. Comput. Method Appl. Eng. **195**, 6011–6045 (2006)
54. A. Harten, Multiresolution algorithms for the numerical solution of hyperbolic conservation laws. Commun. Pure Appl. Math. **48**, 1305–1342 (1995)
55. P.G.W. Hawksley, The effect of concentration on the settling of suspensions and flow through porous media, in *Some Aspects of Fluid Flow* (E. Arnold, London, 1951), pp. 114–135
56. K. Höfler, S. Schwarzer, The structure of bidisperse suspensions al low Reynolds numbers, in *Multifield Problems: State of the Art*, ed. by A.M. Sändig, W. Schiehlen, W.L. Wendland (Springer, Berlin, 2000), pp. 42–49
57. R.L. Hughes, A continuum theory for the flow of pedestrians. Transp. Res. B **36**, 507–535 (2002)
58. K.H. Karlsen, N.H. Risebro, On the uniqueness and stability of entropy solutions for nonlinear degenerate parabolic equations with rough coefficients. Discret. Contin. Dyn. Syst. **9**, 1081–1104 (2003)

59. K.H. Karlsen, N.H. Risebro, J.D. Towers, Upwind difference approximations for degenerate parabolic convection-diffusion equations with a discontinuous coefficient. IMA J. Numer. Anal. **22**, 623–664 (2002)
60. K.H. Karlsen, N.H. Risebro, J.D. Towers, L^1 stability for entropy solutions of nonlinear degenerate parabolic convection-diffusion equations with discontinuous coefficients. Skr. K. Nor. Vid. Selsk (3), 1–49 (2003)
61. D. Kleine, B.D. Reddy, Finite element analysis of flows in secondary settling tanks. Int. J. Numer. Methods Eng. **64**, 849–876 (2005)
62. S.N. Kružkov, First order quasilinear equations in several independent variables. Math. USSR Sb. **10**, 217–243 (1970)
63. A. Kurganov, A. Polizzi, Non-oscillatory central schemes for traffic flow models with Arrhenius look-ahead dynamics. Netw. Heterog. Media **4**, 431–451 (2009)
64. G.J. Kynch, A theory of sedimentation. Trans. Farad. Soc. **48**, 166–176 (1952)
65. M.J. Lockett, K.S. Bassoon, Sedimentation of binary particle mixtures. Powder Technol. **24**, 1–7 (1979)
66. G. Loeper, Uniqueness of the solution of the Vlasov-Poisson system with bounded density. J. Math. Pures Appl. **86**, 68–79 (2006)
67. T.P. Liu, Invariants and asymptotic behavior of solutions of a conservation law. Proc. Am. Math. Soc. **71**, 227–231 (1978)
68. J.H. Masliyah, Hindered settling in a multiple-species particle system. Chem. Eng. Sci. **34**, 1166–1168 (1979)
69. S. Müller, *Adaptive Multiscale Schemes for Conservation Laws* (Springer, Berlin, 2003)
70. S. Müller, Y. Stiriba, Fully adaptive multiscale schemes for conservation laws employing locally varying time stepping. J. Sci. Comput. **30**, 493–531 (2007)
71. H. Nessyahu, E. Tadmor, Non-oscillatory central differencing for hyperbolic conservation laws. J. Comput. Phys. **87**, 408–463 (1990)
72. V.S. Patwardhan, C. Tien, Sedimentation and liquid fluidization of solid particles of different sizes and densities. Chem. Eng. Sci. **40**, 1051–1060 (1985)
73. A. Quarteroni, R. Ruiz-Baier, Analysis of a finite volume element method for the Stokes problem. Numer. Math. **118**, 737–764 (2011)
74. H.-K. Rhee, R. Aris, N.R. Amundson, *First-Order Partial Differential Equations*. Volume 1: Theory and Applications of Single Equations (Prentice Hall, Englewood Cliffs, 1986)
75. J.F. Richardson, W.N. Zaki, Sedimentation and fluidization: part I. Trans. Instn. Chem. Eng. (Lond.) **32**, 35–53 (1954)
76. F. Rosso, G. Sona, Gravity-driven separation of oil-water dispersions. Adv. Math. Sci. Appl. **11**, 127–151 (2001)
77. K. Schneider, O. Vasilyev, Wavelet methods in computational fluid dynamics. Ann. Rev. Fluid Mech. **42**, 473–503 (2010)
78. L. Shen, Z. Chen, Analysis of a stabilized finite volume method for the transient Stokes equations. Int. J. Numer. Anal. Model. **6**, 505–519 (2009)
79. C.-W. Shu, High order weighted essentially nonoscillatory schemes for convection dominated problems. SIAM Rev. **51**, 82–126 (2009)
80. C.-W. Shu, S. Osher, Efficient implementation of essentially non-oscilatory shock capturing schemes. J. Comput. Phys. **77**, 439–471 (1988)
81. D.B. Siano, Layered sedimentation in suspensions of monodisperse spherical colloidal particles. J. Colloid Interface Sci. **68**, 111–127 (1979)
82. B. Sjögreen, K. Gustavsson, R. Gudmundsson, A model for peak formation in the two-phase equations. Math. Comput. **76**, 1925–1940 (2007)
83. A. Sopasakis, M.A. Katsoulakis, Stochastic modelling and simulation of traffic flow: asymmetric single exclusion process with Arrhenius look-ahead dynamics. SIAM J. Appl. Math. **66**, 921–944 (2006)
84. H.H. Steinour, Rate of sedimentation. Nonflocculated suspensions of uniform spheres. Ind. Eng. Chem. **36**, 618–624 (1944)

85. G. Strang, On the construction and comparison of difference schemes. SIAM J. Numer. Anal. **5**, 506–517 (1968)
86. P.K. Sweby, High resolution schemes using flux limiters for hyperbolic conservation laws. SIAM J. Numer. Anal. **21**, 995–1011 (1984)
87. B. Temple, Global solution of the Cauchy problem for a class of 2×2 nonstrictly hyperbolic conservation laws. Adv. Appl. Math. **3**, 335–375 (1982)
88. J.D. Towers, A difference scheme for conservation laws with a discontinuous flux: the nonconvex case. SIAM J. Numer. Anal. **39**, 1197–1218 (2001)
89. V. Vand, Viscosity of solutions and suspensions. I. Theory. J. Phys. Chem. **52**, 277–299 (1948)
90. G.B. Wallis, A simplified one-dimensional representation of two-component vertical flow and its application to batch sedimentation, in *Proceedings of the Symposium on the Interaction Between Fluids and Particles*, London, 20–22 June 1962 (Institution of Chemical Engineers, London, 1962), pp. 9–16
91. G.B. Wallis, *One-Dimensional Two-Phase Flow* (McGraw-Hill, New York, 1969)
92. R.W. Watts, S.A. Svoronos, B. Koopman, One-dimensional modeling of secondary clarifiers using a concentration and feed velocity-dependent dispersion coefficient. Water Res. **30**, 2112–2124 (1996)
93. R.H. Weiland, Y.P. Fessas, B.V. Ramarao, On instabilities arising during sedimentation of two-component mixtures of solids. J. Fluid Mech. **142**, 383–389 (1984)
94. A.J. Wilson, *The Living Rock* (Woodhead Publishing, Cambridge, 1994)
95. G.C.K. Wong, S.C.K. Wong, A multi-class traffic flow model—an extension of LWR model with heterogeneous drivers. Transp. Res. A **36**, 827–841 (2002)
96. A. Zeidan, S. Rohani, A. Bassi, P. Whiting, Review and comparison of solids settling velocity models. Rev. Chem. Eng. **19**, 473–530 (2003)
97. M. Zhang, C.-W. Shu, G.C.K. Wong, S.C. Wong, A weighted essentially non-oscillatory numerical scheme for a multi-class Lighthill-Whitham-Richards traffic flow model. J. Comput. Phys. **191**, 639–659 (2003)
98. P. Zhang, R.X. Liu, S.C. Wong, S.Q. Dai, Hyperbolicity and kinematic waves of a class of multi-population partial differential equations. Eur. J. Appl. Math. **17**, 171–200 (2006)
99. K. Zumbrun, On a nonlocal dispersive equation modeling particle suspensions. Q. Appl. Math. **57**, 573–600 (1999)

SBV Regularity Results for Solutions to 1D Conservation Laws

Laura Caravenna

Abstract A well-posedness theory has been established for entropy solutions to strictly hyperbolic systems of conservation laws, in one space variable, with small total variation. We give in this note an introduction to SBV-regularity results: when the characteristic fields are genuinely nonlinear, the derivative of an entropy solution consists only of the absolutely continuous part and of the jump part, while a fractal behavior (the Cantor part) is ruled out. We first review the scalar uniformly convex case, related to the Hopf-Lax formula. We then turn to the case of systems: one has a decay estimate for both positive and negative waves, obtained considering the interaction-cancellation measures and balance measures for the jump part of the waves. When the Cantor part of the time restriction of the entropy solution does not vanish, either the Glimm functional has a downward jump, or there is a cancellation of waves or this wave balance measure is positive, and this can occur at most at countably many times. We then remove the assumption of genuine nonlinearity. The Cantor part is in general present. There are however interesting nonlinear functions of the entropy solution which still enjoy this regularity.

2010 Mathematics Subject Classification 35-02/35-06, 35L65, 35B65

1 Introduction

We give an informal introduction to an issue of regularity for entropy solutions to strictly hyperbolic systems of conservation laws, mainly based on [1,5,9,11,16,21] and in particular on the joint work [6] of the author with Stefano Bianchini.

L. Caravenna (✉)
Mathematical Institute, University of Oxford, Oxford OX1 2LB, UK
e-mail: laura.caravenna@maths.ox.ac.uk

G.-Q.G. Chen et al. (eds.), *Hyperbolic Conservation Laws and Related Analysis with Applications*, Springer Proceedings in Mathematics & Statistics 49,
DOI 10.1007/978-3-642-39007-4_3,

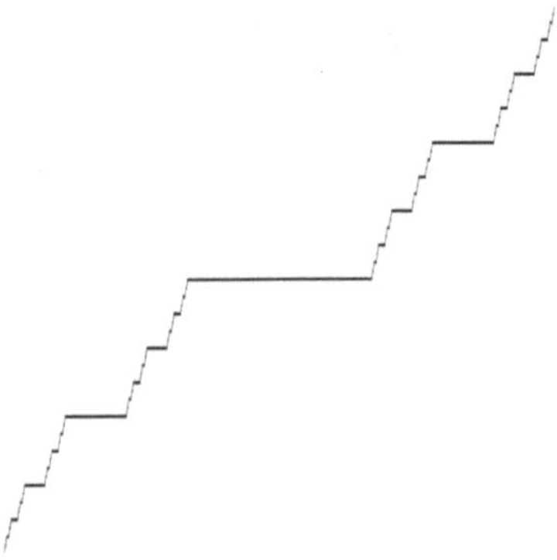

Fig. 1 The Cantor-Vitali function (or Devil's Staircase)

In the first section we present the case of a prototype equation, Burgers' equation

$$D_t u(t,x) + D_x \frac{u^2(t,x)}{2} = 0, \qquad u : \mathbb{R}^+ \times \mathbb{R} \to \mathbb{R}. \tag{1}$$

As an effect of nonlinearity, entropy solutions corresponding to bounded initial data satisfy the Oleinik decay estimate, and they have bounded variation at any positive time. Although BV-functions have a fine structure, a fractal behavior of the derivative as in the Cantor-Vitali function is still allowed (Fig. 1). It was proved by Ambrosio-De Lellis that this indeed is not the case for this conservation law.

Theorem 1 ([1]). *Let u be an $L^\infty(\Omega)$-entropy solution to* (1)*, for an open $\Omega \subset \mathbb{R}^+ \times \mathbb{R}$. Except at most at countably many times $\bar{t}$, the derivative $D_x u(t = \bar{t}, x)$ is the sum of an absolutely continuous measure and of a purely atomic measure, corresponding to the jumps of u at time $t = \bar{t}$.*

We first sketch the construction given by Ambrosio-De Lellis. It relies, by the Hopf-Lax formula, on a decomposition of the domain into jump points of u and into backward characteristics through continuity points, which allows the analysis of the push forward of the Lebesgue measure along characteristics. The Cantor part may be created when a positive set of characteristics cannot be extended beyond that time. We then conclude the first section motivating the same result with a different argument, with the aim of introducing in a simpler setting the strategy employed for systems. The presence of the Cantor part is quantitatively related to the formation of jumps.

In Sect. 3 we review the generalization to strictly hyperbolic, genuinely nonlinear systems

$$D_t u + D_x f(u) = 0 \qquad f \in C^2(\mathbb{R}^N; \mathbb{R}^N). \tag{2}$$

We recall that the assumption of strict hyperbolicity means that $Jf(z)$ has real eigenvalues $\lambda_1(z) < \cdots < \lambda_n(z)$, while genuine nonlinearity can be expressed as $\nabla\lambda_i(z) \cdot r_i(z) \geq k > 0$, where $r_1(z), \ldots, r_n(z)$ are corresponding right eigenvectors.

In the context of the well-posedness theory for entropy solutions with bounded variation (see [16]), the same regularity holds as for the scalar uniformly convex case.

Theorem 2 ([6]). *Let $u : \Omega \to \mathbb{R}^n$ be an entropy solution to* (2) *with sufficiently small total variation. Except at most at countably many times $\bar{t}$, the derivative $\mathrm{D}_x u(t = \bar{t}, x)$ is the sum of an absolutely continuous measure and of a purely atomic measure, corresponding to the jumps of u at time $t = \bar{t}$.*

The first step is to reduce the problem from the vector measure $\mathrm{D}_x u$ to scalar quantities, by a decomposition of $\mathrm{D}_x u$ along r_i. The strategy we then adopt differs from the scalar case: computations are not performed directly on the entropy solution but on the wave front-tracking approximations. For these, we manage to give bounds for the wave balance measure and for the jump wave balance measures, first defined as distributions. As Oleinik's estimate was first generalized for this class of systems in [12], with these new measures it is possible to extend it to a complementary decay estimate for negative waves [6]. The full decay estimate yields the answer to the question that we reformulated for the scalar quantities.

Section 4 treats the optimality and extensions of this regularity. The genuine nonlinearity assumption is crucial for ruling out the Cantor part of the derivative of the entropy solution, even for one equation. There are however interesting nonlinear functions, reducing to $f'(u)$ for the scalar case $\mathrm{D}_t u + \mathrm{D}_x f(u) = 0$, that always enjoy this regularity. They are called i-th components of the space derivative of λ_i [9]. In particular SBV-regularity holds also for fluxes whose second derivative vanishes on a Lebesgue negligible set [9, 21]. An SBV regularity result has been obtained in [16] out of the context of entropy solutions, still under the assumption of genuine nonlinearity, but restricted to BV-solutions which are self-similar.

Although it may have been a natural and interesting addition to the article, we have not included a review of the literature on the decay estimates of waves.

2 The Scalar Case for Burgers' Equation

We firstly outline the idea behind two proofs of the SBV regularity result in the scalar case: first the one originally given in [1], then the approach by front-tracking that we followed for the case of systems.

2.1 Statement and Notation

The Cauchy problem for Burgers' equation $\mathrm{D}_t u + \mathrm{D}_x \frac{u^2}{2} = 0$ on the half-plane $(t, x) \in \mathbb{R}^+ \times \mathbb{R}$, starting from a 'generic' smooth data $u(t = 0, x)$, develops at most finitely many discontinuity curves, and it remains smooth elsewhere [22]. It may however happen that some particular smooth data develop infinitely many

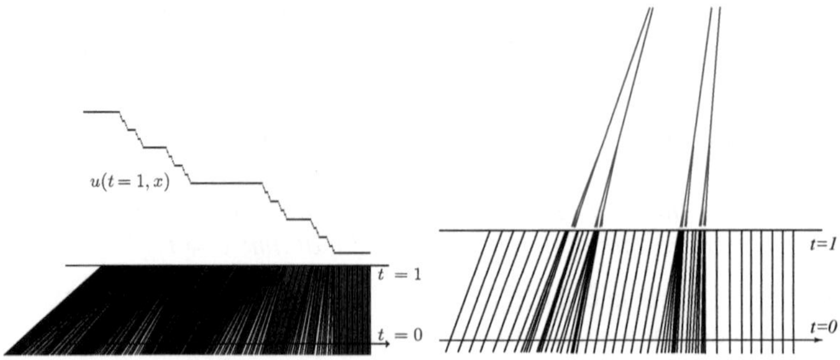

Fig. 2 The solution to Burges' equation with a smooth data developing a singular part. In the *left figure* one has the profile of the solution at a given time $t = 1$, and below some of the characteristics arriving at that time. Before time $t = 1$ the solution is absolutely continuous, and may be assumed smooth. On the *right* one has a sketch of the shock set that develops immediately after $t = 1$, with a fractal behavior. This example is taken from Remark 3.3 of [1], emphasizing the presence of the Cantor part at $t = 1$. Li Bang-He also gives examples where the origin points of jumps have positive measure

discontinuities (Fig. 2) and, as well as for initial data which are bounded, the solution is found in general in the space of functions of locally bounded variation [18, 20]. The space of functions of bounded variation (BV) also contains functions which are continuous but whose derivative has a fractal behavior, like the Cantor-Vitali function, which is monotone, almost everywhere differentiable with vanishing derivative, but runs continuously from 0 to 1 (Fig. 1). The special functions of bounded variation (SBV functions) better generalize the concept of piecewise Sobolev functions and rule out this fractal behavior of the derivative.

The central regularity result of this note states that this fractal behavior of the derivative is essentially not present for entropy solutions of strictly hyperbolic systems of conservation laws whose fields are genuinely nonlinear [1, 6]. In particular, in the scalar case this means for uniformly convex fluxes.

Definition 1. An integrable function $v : \mathbb{R} \to \mathbb{R}^N$ is a special function of bounded variation if its distributional derivative is the sum of a finite, absolutely continuous Borel measure and of a purely atomic finite measure.

Theorem 3 (Ambrosio–De Lellis). *Consider an entropy solution* $u : \mathbb{R}^+ \times \mathbb{R} \to \mathbb{R}$ *with bounded variation of the scalar conservation law* $\mathrm{D}_t u + \mathrm{D}_x \frac{u^2}{2} = 0$. *Then there exists an at most countable set of times* N *such that for* $\bar{t} \notin N$ *the restriction* $u(t = \bar{t}, x)$ *is a special function of bounded variation.*

2.1.1 The Statement for Two Variables

Instead of looking at the time sections $u(t = \bar{t}, \cdot)$ of u, one can state the consequent result for the function of two variables $u(t, x)$. We give now a rough (hopefully intuitive) presentation and we refer to [2] and [15] for precise notations and proofs.

We recall that $u : \mathbb{R}^+ \times \mathbb{R} \to \mathbb{R}$ has locally bounded variation if the distributional derivatives $\partial_t u$, $\partial_x u$ are (locally) finite Borel measures. This implies a particular structure (see Sects. 3.8 and 3.9 in [2]). One of the features is that the set of approximate discontinuity points can be covered by finitely or countably many Lipschitz curves, plus a Lebesgue negligible subset of the plane, while elsewhere the function u is approximatively continuous.

When u is an entropy solution of Burgers' equation this structure improves further. Due to the finite speed of propagation and to a fine analysis of u (see Theorem 10.4 in [11] for a stronger result), the set of jump points can be written as the union of countably many Lipschitz graphs $\{(t, \gamma_i(t))\}_{i\in I, t\in[a_i,b_i]}$, $I \subset \mathbb{N}$, except for at most countably many points. We mention however that in choosing $\{(t, \gamma_i(t))\}_{i\in I, t\in[a_i,b_i]}$ which only cover the jump set, and possibly also include some points where u is continuous, we would not encounter a contradiction in the constructions below: in writing the jump part of u below we would simply have at those points the coefficients $u^+ - u^-$, $f(u^+) - f(u^-)$ which would vanish.

Remark 1. This countable covering does not contradict the uncountability of 'discontinuity lines' underlined by Li Bang-He [19]; our notations differ. Focusing on a non-increasing initial data, Li Bang-He counts the set of maximal Lipschitz curves $\{(t, \gamma(t)) : (t_\gamma, +\infty) \to \mathbb{R}^+ \times \mathbb{R}\}$ which have image discontinuity points of u, maximal in the sense that he rules out the restrictions to subintervals. These curves are in bijective correspondence with the 'origin points' of discontinuities $\{(t_\gamma, \gamma(t_\gamma))\}$.

In order to better visualize the difference, we focus on a ramified discontinuity pattern like the one in Fig. 2, even though it is not accurate. The number of discontinuities at a fixed time decreases in time. Counting them from the future, there are $1, 2, 4 \ldots, 2^k$. In particular, it is clear that $1 + 2 + \cdots + 2^k = 2^{k+1} - 1$ curves, each living on different time intervals, cover the jump set from ∞ up to a certain time greater than 1, and countably many curves cover the entire jump set. Each maximal Lipschitz curve γ defined by Li Bang-He can instead be associated to a sequence $.1011001\ldots$, where the k-th digit is 0 if at the k-th bifurcation, from the future, the curve γ proceeds on the left side, and is equal to 1 if it proceeds on the right. In particular, as he claims they are uncountable, as well as the origin points of discontinuities at time $t = 1$.

If a continuity point (t, x) of u is an origin point of a shock, then the backward characteristic from (t, x) lives precisely up to time t.

If $u^\pm(t, x)$ denote the approximate left and right limits at approximate jump points, the distributional derivative $\mathrm{D}\,u$ is a locally finite vector Borel measure on $\mathbb{R}^+ \times \mathbb{R}$ of the form

$$\mathrm{D}\,u = \binom{\mathrm{D}_t\,u}{\mathrm{D}_x\,u} = \mathrm{D^a}\,u + \mathrm{D^j}\,u + \mathrm{D^C}\,u,$$

with the absolutely continuous part D^{a}, the jump part D^{j} and the Cantor part D^{C} defined by:

$\mathrm{D}^{\mathrm{a}}\, u = \nabla u \mathscr{L}^n$, with ∇u denoting the approximate differential;
$\mathrm{D}^{\mathrm{j}}\, u(B) = \sum_{i\in\mathbb{N}} \int \left(\frac{-[f(u^+)-f(u^-)]}{u^+-u^-} \right) (t, \gamma_i(t))\, dt$;
$\mathrm{D}^{\mathrm{C}}\, u$ the remaining part, which in particular vanishes on sets which are finite w.r.t. $\mathscr{H}^1$.

Notice that the Cantor part D^{C} is the only part not recoverable by a blow-up procedure. One also calls $\overline{\mathrm{D}} = \mathrm{D}^{\mathrm{a}} + \mathrm{D}^{\mathrm{C}}$ the diffuse part of the derivative.

Definition 2. The function of (locally) bounded variation $u : \mathbb{R}^+ \times \mathbb{R} \to \mathbb{R}$ is a *special function of (locally) bounded variation* if its distributional derivative has a null Cantor part.

Corollary 1 (Ambrosio–De Lellis). *Let $\Omega \subset \mathbb{R}^+ \times \mathbb{R}$ be an open set. Entropy solutions $u \in L^\infty(\Omega)$ of the scalar conservation law $\mathrm{D}_t\, u + \mathrm{D}_x \frac{u^2}{2} = 0$ are special functions of locally bounded variation on Ω.*

The fact that here u is assumed to be bounded instead of requiring bounded variation is mainly due to a local argument and to the well-known regularizing effect of the uniformly convex fluxes, where bounded data to the Cauchy problem instantaneously acquire locally bounded variation. The fact that $\mathrm{D}_x\, u$ does not have a Cantor part derives from the slicing theory of BV functions and by the above stated regularity result for slices at fixed time. By the PDE $\mathrm{D}_t\, u = -\mathrm{D}_x \frac{u^2}{2}$ together with the Vol'pert chain rule this implies in turn that $\mathrm{D}_t\, u$ is also free of a Cantor part, details can be found in [1]. There is no loss of generality in focusing below on $\Omega = \mathbb{R}^+ \times \mathbb{R}$, by the finite speed of propagation.

2.2 A Sketch of the Proof

In order to prove that the time sections of $u(t,x)$ are special functions of bounded variation, the most important point is understanding what happens at those times $\bar{t}$ when a Cantor part is present in the space derivative of the restriction $u(\bar{t}, x)$. In this section we first outline the original argument given by Ambrosio and De Lellis in the case of a single equation, and then we introduce a different approach which will be used for the case of systems.

2.2.1 The Original Argument

In the proof given by Ambrosio and De Lellis, the authors exploit a natural partition of the domain into different lines depending upon the entropy solution u that now we describe. Being in the scalar, uniformly convex case, they can solve the Cauchy problem for the conservation law

$$\begin{cases} \mathrm{D}_t u(t,x) + \mathrm{D}_x \frac{u^2(t,x)}{2} = 0 & \text{in } \mathbb{R}^+ \times \mathbb{R} \\ u(0,x) = u_0(x) & \text{with } u_0(x) \in L^1(\mathbb{R}) \cap L^\infty(\mathbb{R}) \end{cases}$$

by the Hopf-Lax formula. As well known, positive waves decay according to the Oleinik E-condition

$$u(t, x+y) - u(t,x) \leq \frac{y}{t}.$$

Notice that the Oleinik E-condition implies the absolute continuity of the positive part of the measure $\mathrm{D}\,u(t,\cdot)$, therefore in the following the focus will be on the negative part in order to show that, apart from the atoms resulting from the jumps, it is absolutely continuous. At each time t the suitable representative of u is continuous except at most countably many jump points, where one has left and right limits. Excluding the jump points of u, one can define, here by the Hopf-Lax formula, backward characteristic curves through each continuity point (t,x): they are straight lines with slope $u(t,x)$ which do not intersect each other and which go back to the line $\{t = 0\}$. At jump points, characteristic lines which originated at time 0 from different points intersect each other. At jump points it is therefore no longer possible to define a unique backward characteristic. Nevertheless, one can define the minimal and maximal characteristics, which have slope respectively $u(t,x^-)$, $u(t,x^+)$ and which delimitate a backward cone with vertex (t,x) and basis on $\{t = 0\}$. If one moves points forward in time following characteristics, and follows the jump curve whenever they collapse, then the basis of each cone is mapped into a discontinuity point. More formally, one can move points along generalized characteristics, which here are the unique Filippov solution to the differential inclusion $\dot{x} \in [u(t,x^+), u(t,x^-)]$ (Fig. 3).

The main ingredient of the proof by Ambrosio and De Lellis consists in estimating the measure $F(t)$ of the set of initial points $\{(x,0) \;:\; x \in \mathbb{R}\}$ which in the way described above are mapped at time t into a discontinuity point of $u(t,\cdot)$. This is the measure of the union of the bases of backward cones emanating from the discontinuity points of $u(t,\cdot)$, cones whose sides have slope $u(x^-,\cdot)$ and $u(x^+,\cdot)$. Therefore $F(t)$ is precisely t times the variation of $\mathrm{D}_x^{\mathrm{j}}\,u(t,\cdot)$. By the non-crossing property of characteristics, moreover, one can see that $F(t)$ is nondecreasing, as intersecting cones are included one into the other.

One can then conclude observing that the monotone function $F(t)$ has a jump precisely at those times t where a Cantor part is present in $u(t,\cdot)$, a situation which can therefore occur at most countably many times. The estimate which could be derived collecting the computations in [1] is the lower decay estimate

$$F(t^+) - F(t^-) \geq -t \cdot \mathrm{D}^{\mathrm{C}}\,u(t,\cdot),$$

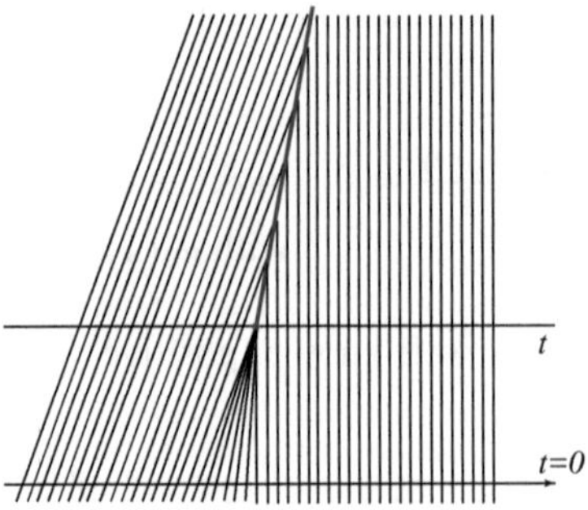

Fig. 3 A simple generalized characteristic consists, in the picture, of a classical characteristic until it merges into a shock curve, at which point it follows the shock curve. While at continuity points there is a unique backward classical characteristic, at jump points one considers minimal and maximal characteristics, drawing a backward cone

which is strictly positive when the Cantor part of $u(t,\cdot)$ does not vanish. Notice that $F(t^+) - F(t^-)$ gives precisely the measure of the set of initial points of characteristics which arrive at time t, but which cannot be prolonged beyond that time because between time t and time $t+\varepsilon$ they would cross other characteristics, for every $\varepsilon > 0$ (Fig. 2).

In view of the generalization to the case systems, we adopt in [1] another viewpoint for observing this phenomenon. Here we present it in a simplified setting where it simply provides an alternative description, and we outline the extension to the case of systems in the next section.

2.2.2 The Jump Wave Balance Measure for Burgers' Equation

In the previous argument one estimates the formation of the Cantor part in $\mathrm{D}_x u(t,\cdot)$ associating it to the initial points of characteristics which live precisely up to time t: times where the Cantor part is present correspond to disjoint non-negligible subsets of $\mathbb{R}$, the starting point of characteristics living precisely up to time t, therefore by σ-additivity there can be at most countably many such sets, and consequently at most countably many such times. Here instead we try to give a more local argument introducing a new measure, the wave balance measure. Here instead we try to give a more local argument introducing a new measure, the wave balance measure.

Given an entropy solution u with bounded variation, one can consider the wave measure $\nu = \mathrm{D}_x u$. Denoting by ∂_x, ∂_t distributional derivatives, $\nu = \mathrm{D}_x u$ satisfies the transport equation

$$0 = \partial_t \nu + \partial_x (\tilde{u}\nu) = \partial_x \left[\mathrm{D}_t u + \mathrm{D}_x \frac{u^2}{2} \right]$$

where $\tilde{u}(t,x)$ is defined as $u(t,x)$ at continuity points of $u(t,\cdot)$ and as $\frac{u(t,x^+)+u(t,x^-)}{2}$ at jump points.

One can distinguish the jump wave measure $\nu_{\text{jump}} = D_x^{\text{j}} u$. Define, for the case of Burgers' equation, the *jump wave balance measure* as the source term in a transport equation for the jump wave measure:

$$\mu_{\text{jump}} = \partial_t \nu_{\text{jump}} + \partial_x \left(\tilde{u}\nu_{\text{jump}}\right).$$

By the BV-structure mentioned in Sect. 2.1.1, one can easily compute that

$$\begin{aligned}
-\int \varphi(t,x)\mu_{\text{jump}} &= \sum_{i\in\mathbb{N}} \int [\varphi_t(t,\gamma_i(t)) + \lambda(t,\gamma_i(t))\varphi_x(t,\gamma_i(t))] \\
&\quad \cdot \left[u(t,\gamma_i(t)^+) - u(t,\gamma_i(t)^-)\right] dt \\
&= \sum_{i\in\mathbb{N}} \int \frac{d}{dt}\varphi(t,\gamma_i(t)) \cdot \left[u(t,\gamma_i(t)^+) - u(t,\gamma_i(t)^-)\right] dt.
\end{aligned}$$

This gives a precise idea of what is meant by this measure. Notice however that it is not a priori clear that the distribution defined above is actually a measure, as presently we do not know the BV-regularity in time of the left/right values on the jumps $u(t,\gamma_i(t)^+) - u(t,\gamma_i(t)^-)$. Indeed, part of our proof is devoted to estimating the bound for μ_{jump}. In particular, notice that the measure vanishes for nondecreasing initial data, as such data do not develop jumps. The measure is non-positive for nonincreasing initial data, as there are no cancellations and the size of jumps may only increase in time. In general, μ_{jump} is a signed measure.

Example 1. It is clear that μ_{jump} (here) measures the variation in time of the jump part of $D_x^{\text{j}} u$. When a single jump is formed at a point $(\bar{t},\bar{x})$ and later remains constant, we have that $\mu_{\text{jump}} = \left(u^+ - u^-\right)\delta_{(\bar{t},\bar{x})}$, where $u^+ \leq u^-$ are the right and left values of u at the jump (Fig. 3).

Example 2. The measure μ_{jump} of a horizontal strip delimitated by times $t_1 < t_2$ is

$$\mu_{\text{jump}}([t_1,t_2]\times\mathbb{R}) = D_x^{\text{j}} u(t_2,\cdot) - D_x^{\text{j}} u(t_1,\cdot).$$

We introduced the measure above in order to obtain the following lower estimate for the diffuse part of $D_x u(t,\cdot)$. Denote by $\left[\mu_{\text{jump}}\right]^-$ the negative part of μ_{jump}. While Oleinik's estimate gives the upper bound

$$\left[D_x u(t,\cdot)\right](B) \leq \frac{\mathscr{L}^1(B)}{t} \qquad \text{for all Borel sets } B,$$

one can obtain the following complementary lower estimate for $\overline{D} u(t,\cdot) = D^{\text{a}} u(t,\cdot) + D^{\text{C}} u(t,\cdot)$:

$$\left[\overline{D} u(t,\cdot)\right](B) \geq -\frac{\mathscr{L}^1(B)}{\tau - t} - \left[\mu_{\text{jump}}\right]^-([t,\tau]\times\mathbb{R}) \qquad \tau > t > 0,\ B \text{ Borel set.} \tag{3}$$

The new measure gives the same quantitative lower bound as that given by the strategy of Ambrosio-De Lellis. This alternative point of view will be useful for the generalization.

We now give an heuristic idea for the above estimate. Consider a space interval $B = [a, b]$ where at time t there is no jump of u. The first case which may happen is that the characteristics starting from a, b reach time τ. If so, since $u(t, a)$ and $u(t, b)$ are the constant speeds of the characteristics starting from the points a and b respectively, clearly one has

$$u(t, b) - u(t, a) = \left[\overline{\mathrm{D}} u(t, \cdot)\right]([a, b]) \geq -\frac{b-a}{\tau - t}.$$

In particular, the estimate (3) holds with the first addend. If this does not occur, consider the simplified situation where the characteristics cross each other before τ developing a jump at that moment (Fig. 3): at that point $\mu_{\text{jump}}(\bar{t}, \bar{x}) = u(t, b) - u(t, a)$ and the estimate (3) holds with the second addend. The general situation is more complex, but the two cases give a good idea of what happens.

3 The Case of Genuinely Nonlinear Systems

In this section we outline how the argument introduced in the scalar uniformly convex case works for a genuinely nonlinear strictly hyperbolic system

$$\mathrm{D}_t u + \mathrm{D}_x f(u) = 0 \qquad f \in C^2(\mathbb{R}^N; \mathbb{R}^N).$$

We try to describe qualitatively the new behaviors that one faces, and why the argument of the scalar case is still applicable, skipping all the technical details which can be found in [6] and [11]. We just recall that, if $\lambda_1(z) < \cdots < \lambda_N(z)$ are the eigenvalues of the Jacobian matrix of $f(z)$, then the assumption of genuine nonlinearity is a uniform convexity assumption of the primitives of each $\lambda_i(z_i(t))$, where $z_i(t)$ is an integral curve of the vector field given by the (right) eigenvectors of the Jacobian matrix of f.

This is a different approach than that taken by Ancona-Nguyen, exploiting Riemann invariants, used for genuinely nonlinear Temple systems of balance laws [4].

3.1 *Dimensional Reduction*

If we had N uncoupled equations, each one would have his own characteristics and we would study them independently, repeating for each the scalar argument. This is clearly not the case, but since we are studying solutions with small total variation we can get information from the linearization of the system. If $\lambda_1 < \cdots < \lambda_N$ are the eigenvalues of the Jacobian matrix of f, one can define i-characteristics by the differential inclusion

$$\dot{x}_i(t) \in [\lambda_i(u(t,x^+)), \lambda_i(u(t,x^-))].$$

As one decomposes a vector into its components, one can also decompose the vector measure $\mathrm{D}_x u$ along a basis of (right) eigenvectors of the Jacobian matrix of f, into N-scalar measures ν_i, each of which is called the *i-th wave measure*. In the case of Burgers' equation, these measures reduce to the previous definition, while for a single equation with a different uniformly convex flux function f they reduce to $\nu = \mathrm{D}_x f'(u)$, which by a normalization choice is not equal to $\mathrm{D}_x u$ (Example 5.2 in [6]).

This projection reduces the regularity problem for $\mathrm{D} u$ to a regularity problem for these scalar measures: if one can show that each of them comprises only an absolutely continuous part and a jump part, then the conditions of our thesis are satisfied. One can now see the analogy with the scalar case, because we will exclude the presence of the Cantor part by the decay estimates: for $\tau > t > \theta > 0$ and B a Borel set

$$-C\left\{\frac{\mathscr{L}^1(B)}{\tau - t} + \mu^{ICJ}((t,\tau)\times\mathbb{R})\right\} \leq [\bar{\nu}_i(t)](B) \leq C\left\{\frac{\mathscr{L}^1(B)}{t-\theta} + Q(\theta) - Q(t)\right\},$$

where $\bar{\nu}_i(t)$ is the non-atomic (also called diffuse) part of $\nu_i(t)$, the Glimm functional Q is a nonincreasing functional and the interaction-cancellation-jump measure μ^{ICJ} is a positive, finite measure.

The upper bound, due to Bressan and Colombo [11, 12, 14], generalizes Oleinik's estimate, while we derived instead the lower bound [6]. Notice that this estimate yields the thesis for ν_i: if there is a Lebesgue negligible set B where we have a positive Cantor part, then the monotone functional Q has a jump. If we have a negative Cantor part, then there is an atom of the time marginal of the interaction-cancellation-jump measure. Each of the two situation can occur at most countably many times.

We give below an idea of the derivation of the above decay estimates, sufficient for the thesis.

3.2 A Proof by Approximation

For the case of systems, i-characteristics are no longer straight lines, even in the absence of the complication of jumps. Clearly u is non-constant on them. While in the scalar case the thesis is derived by a direct approach, here it is more difficult. On one hand there are new behaviors arising from the presence of more characteristic fields which interact. On the other hand, there is the technical difficulty that the restriction of u to i-characteristics is much more difficult to analyze, and a direct approach would pass through this. The technical difficulty is presently overcome by proving the estimate with a limiting procedure by front-tracking approximations. As well known, these are piecewise constant approximations of the solution to the Cauchy problem which are obtained by piecing together (approximate) solutions to Riemann problems. We refer to the monograph [11] for more details.

An essential intermediate step toward the estimate is the global structure of solution established by Bressan and LeFloch [13]. They give an algorithm for distinguishing in the front-tracking approximations a jump part—roughly made by jumps of size over a threshold—and the remaining continuous part, and they prove a fine convergence for the two parts.

In particular, for each characteristic field i, we would like to study, as for the scalar case, the wave balance measure and the jump wave balance measure

$$\mu_i := \partial_t \nu_i + \partial_x \left(\tilde{\lambda}_i \nu_i\right)$$

$$\mu_{i,\text{jump}} = \partial_t \nu_{i,\text{jump}} + \partial_x \left(\tilde{u}\nu_{i,\text{jump}}\right),$$

where ν_i is the i-th wave measure introduced in Sect. 3.1 and $\nu_{i,\text{jump}}$ is its jump part. However, it is much easier to define and study the same quantity in the front tracking approximations because in each approximation the measures will consist of finitely many atoms at interaction points. The proof that the above defined distributions are actually measures, as we claimed, relies on uniform estimates given for the corresponding measures in the front-tracking approximations. These estimates rely on the known interaction-cancellation measures [11]. However, to control the negative part of $\mu_{i,\text{jump}}$ we need a nonlocal argument. Indeed, $\mu_{i,\text{jump}}$ is not absolutely continuous w.r.t. the interaction-cancellation measure. After providing uniform estimates for the front tracking approximations, due to the above mentioned theorem on the global structure of the solution, passing to non-sharp bounds for the limit is just a technicality.

3.3 *Estimates for Wave Balance Measures and Jump Wave Balance Measures*

In the following, we assume without mention that we are working with the wave front-tracking approximations and not directly on the solution. Furthermore, we will focus throughout on a single fixed characteristic field.

Suppose two fronts interact at one point P, which is the unique interaction at that time. The first estimate is natural: by the definition of the wave balance measure one can see that

$$\mu_i(P) = \nu_i(\{t\} \times \mathbb{R}) - \lim_{\varepsilon \downarrow 0} \nu_i(\{t - \varepsilon\} \times \mathbb{R}).$$

By the classical interaction estimates (see for example Lemma 7.2 in [11]), this quantity is controlled by the measure of interaction μ^I, in this case equal to a delta measure in P with size the product of the strengths of the incoming waves.

Similarly, one can estimate

$$\mu_{i,\mathrm{jump}}(P) = \nu_{i,\mathrm{jump}}(\{t\} \times \mathbb{R}) - \lim_{\varepsilon \downarrow 0} \nu_{i,\mathrm{jump}}(\{t - \varepsilon\} \times \mathbb{R}).$$

Estimating this quantity is considerably more challenging. If two i-jumps merge, then one can still apply the classical interaction estimate at the point of interaction, obtaining that $\mu_{i,jump}(P)$ is controlled by the interaction-cancellation measure. If instead, for example, a shock gets cancelled, then the estimate is no longer at a single point but one should look backward at the value of the interaction-cancellation measure along the shock front γ, before it was cancelled. The result is that $\mu_{i,\mathrm{jump}}$ is no longer absolutely continuous w.r.t. the interaction-cancellation measure, otherwise it would not be capable of capturing the formation of the Cantor part. It is indeed a new measure, and only the value of its total variation on the half-plane, $|\mu_{i,\mathrm{jump}}|(\mathbb{R}^+ \times \mathbb{R})$, not $\mu_{i,\mathrm{jump}}$ itself, can be controlled by the total interaction cancellation of $\mathbb{R}^+ \times \mathbb{R}$ (see Lemma 5.4 in [6]).

Very roughly, the interaction cancellation measures were meant to give balances for the total amount of i-th waves entering/exiting a region of the plane and, separately, also for their positive and negative parts. The jump wave balance measure permits instead balances regarding only the jump part of the waves, and consequently balances also for only the continuous part. In particular, if one considers a trapezoidal region T with basis $J(t_{t_0})$, $J(t)$ at times $t_0 < t$ and sides minimal i-th characteristics without interaction points, the front-tracking approximations satisfy a balance of the kind

$$\nu_i(t)(J(t)) - \nu_i(t_0)(J_{t_0}) \leq \mu_{i,\mathrm{jump}}(T).$$

3.4 The Decay Estimate

As mentioned, there is a wide literature on the decay of positive waves, with different important contributions, not reported. We try to justify here the decay estimate for the negative part of the wave measures μ_i relative to an interval $B = [a, b]$, as made for the scalar case. It follows the guideline of the proof which has been given for the positive part, but it clearly needs the new estimates mentioned above involving the wave jump balance measure.

We want to prove that the diffuse part $\bar{\nu}_i$ of ν_i satisfies

$$[\bar{\nu}_i(t)]([a,b]) \geq -C \left\{ \frac{b-a}{t} + \mu^{ICJ}\left(\{(x,s) : a(s) \leq x \leq b(s),\ t \leq s \leq \tau\}\right) \right\} \tag{4}$$

where $a(t), b(t)$ are i-characteristics starting from a, b at time t and the interaction-cancellation-jump measure is the sum of the variations of the measures above μ_i, $\mu_{i.\text{jump}}$ for $i = 1, \dots, N$.

We focus our attention only on the i-th fronts, because the further interference of the others that we hide can be controlled. Suppose moreover that they are negative, since we are considering the lower estimate. The genuine nonlinearity implies

$$\frac{d}{ds}(b(s) - a(s)) \leq \nu_i(s),$$

corresponding to approaching i-characteristics. Two cases can occur.
In the first case two characteristics may keep on approaching at a uniform positive rate, namely

$$\frac{d}{ds}(b(s) - a(s)) \leq \frac{[\bar{\nu}_i(t)]([a,b])}{4} \qquad \forall s \in [t, \tau].$$

In this case integrating in s one has immediately (4) with the first addend.
If this is not the case, then at some time $\bar{s}$ we will have the reverse inequality

$$\frac{[\bar{\nu}_i(t)]([a,b])}{4} < \frac{d}{ds}(b(\bar{s}) - a(\bar{s})) \leq [\bar{\nu}_i(s)]([a(s), b(s)]),$$

where the last inequality follows by genuine nonlinearity. By the balances for the jump part mentioned above, however, the waves exiting the region at time s are controlled by the ones entering the region at time t and by those interaction-cancellation-jump measure of the region:

$$[\bar{\nu}_i(s)]([a(s), b(s)]) \leq [\bar{\nu}_i(t)]([a,b]) + C\mu_{ICJ}(\{(x,r) :\ a(r) \leq x \leq b(r),\ t \leq r \leq\}).$$

The last two inequalities give the estimate (4) with the second addend.

4 Optimality, Extensions Without Nonlinearity and Counterexamples

It is worth mentioning that the issue of SBV-regularity presented for Burgers' equation reads also as a regularity result for the associated Hamilton-Jacobi equation. In that direction, the result has been extended first in the work of Bianchini-De Lellis-Robyr to the case of multidi-D uniformly convex Hamiltonians depending only on the gradient of the viscosity solution [10], then by Bianchini-Tonon in the case firstly of a uniformly convex Hamiltonian also dependent on (t, x) [7], and secondly relaxing the assumption of uniform convexity [8].

We mention here instead the extensions still related to $1D$-conservation laws.

4.1 A Counterexample Concerning Convex Fluxes

We illustrated above the regularity result for Burgers' equation as a prototype of the scalar conservation law with uniformly convex flux. It is sharp in the sense that it is possible to have a Cantor part at countably many times (variation of Fig. 2), but not more than countably many. We first want to mention that even in the scalar case the hypothesis of uniform convexity cannot be relaxed to strict convexity, the regularity present is really an effect of the nonlinearity, as is Olenik's estimate.

Example 3. Consider a flux $f \in C^2(\mathbb{R})$ with $f'' \geq 0$ vanishing on a set which has positive Lebesgue measure. A particular flux is constructed as follows. Let $w(x) = v(x) + x$, where $v : [0, 1] \to [0, 1]$ is the (non-decreasing) Cantor-Vitali function (Fig. 1). The function w is continuous, strictly increasing from 0 (at 0) to 2 (at 1). Let $f' = w^{-1}$ be its inverse, which is then a monotone Lipschitz function. The primitive function $f(z)$ is therefore continuously differentiable ($W^{2,\infty}_{\text{loc}}(\mathbb{R})$)—it could also be assumed to be twice continuously differentiable by minor modifications of v—and it has a strictly convex epigraph. However, if one considers the Riemann problem with left value 0 and right value 2, the self-similar solution is fixed by $u(t, x) = w(x/t)$ (see e.g. Theorem 4 in [17]). It therefore has a Cantorian part at all positive times.

It is important to observe in this example that the nonlinear function of u defined by $f'(u(t, x)) = x/t$ belongs to $\mathrm{SBV}_{\text{loc}}(\mathbb{R})$, and that this quantity is particularly interesting because it is the slope of the characteristics. Indeed, f'' vanishes on the image, by $u(t, \cdot)$, of the Cantor set where the Cantor part of $\mathrm{D}_x u(t, x)$ is concentrated, and this yields a null Cantor part for the composition:

$$\mathrm{D}^{\mathrm{C}} f'(u(t, \cdot)) = f''(u(t, \cdot))\, \mathrm{D}^{\mathrm{C}} u(t, \cdot) = 0.$$

4.2 Extensions Without Convexity

The first extension of the SBV-regularity result, by Robyr, was for a single balance law, where the second derivative of the function may vanish on at most countably many points and the source term, depending on space, time and on the solution, must be continuously differentiable. The lack of convexity is overcome by using an appropriate covering of the domain and working locally in order to reduce the problem to the convex or concave case. The presence of a source term, which is treated basically for the convex flux case, implies that characteristics are no longer straight lines, but in general Lipschitz curves. However, the author is able to overcome this difficulty by using the non-crossing property between genuine characteristics, which reduces the problem to that where the source term is absent.

A generalization for the single conservation law has now been given by Bianchini-Yu [9]. They show that the SBV-regularity result holds for $f'(u(t, x))$. In particular, if $f''(z)$ vanishes on a Lebesgue-negligible set, the SBV-regularity also holds for $u(t,x)$. The result for $f'(u(t,x))$ is clearly of independent interest because it is a physically relevant quantity.

The proof still relies on a local argument where one applies the result in the convex or concave case, by a more careful partition of the space-time domain. Roughly, in the domain of influence of small intervals where one has a uniform bound on the second derivative of the flux, the SBV-regularity is known for u, and therefore, by the Vol'per chain rule, also for $f'(u)$. The thesis then amounts to showing that there is a countable covering by such domains of influence, and in the residual part of the domain, excluding at most countably many time (coordinate) lines, is made by points where either f'' vanishes, or one has jumps.

An extension holds for strictly hyperbolic systems [9]. When allowing linear degeneracy the SBV-regularity does not in general hold for u. It does not hold even for the eigenvalue maps $t \mapsto \lambda_i(u(t,x))$, $i = 1, \dots, N$, as one might suspect at first glance. The objects which gain regularity are what are called i-th components of the space derivative of these eigenvalues: the i-th component of the space derivative is defined as the derivative of λ_i in the direction of the i-th vector which is obtained in the decomposition of $D_x u$ along a basis of (right) eigenvectors of the Jacobian matrix of f, suitably specified at jump points. If $D_x u = (v_i^{\text{cont}} + v_i^{\text{jump}})\tilde{r}_i$, then the i-th component is

$$(\nabla\lambda_i \cdot \tilde{r}_i)\, v_i^{\text{cont}} + \big(\lambda_i(u^+) - \lambda_i(u^-)\big)\frac{|v_i^{\text{jump}}|}{\sum_k |v_k^{\text{jump}}|}.$$

When suitably specified at jump points, one can insert this decomposition into $D_x(\lambda_i \circ u) = \nabla\lambda_i \cdot (D_x^{\text{a}} u + D_x^{\text{C}} u) +$, and the i-th component in this decomposition is what gains SBV regularity [9]. Technically, as the proof for the genuinely nonlinear case relies on the classical version of the front-tracking algorithm [11], this extension relies on the extension of the front-tracking algorithm given by Ancona-Marson [3], which is much more complicated but with similar interaction estimates. Although it does not differ substantially from the genuinely nonlinear case, the proof shows that the quantities above are really the ones which come into play.

We conclude by drawing attention to an earlier interesting extension by Dafermos [16], who establishes the SBV-regularity for self-similar solutions of genuinely nonlinear strictly hyperbolic systems of conservation laws. This is achieved regardless of the entropy condition, but the analysis is different, reducing, by the self-similarity assumption $u(t,x) = v\left(\frac{x}{t}\right)$, the problem to ODEs along the straight rays exiting the origin. As mentioned in Sect. 4.1, the genuine nonlinearity assumption cannot be substantially weakened, even in the self-similar case.

Acknowledgements The author wishes to thank Gui-Qiang G. Chen and the anonymous referee for interesting comments. She has been supported by the UK EPSRC Science and Innovation award to the Oxford Centre for Nonlinear PDE (EP/E035027/1).

References

1. L. Ambrosio, C. De Lellis, A note on admissible solutions of 1d scalar conservation laws and 2d Hamilton-Jacobi equations. J. Hyperbolic Differ. Equ. **31**(4), 813–826 (2004)
2. L. Ambrosio, N. Fusco, D. Pallara, *Functions of Bounded Variation and Free Discontinuity Problems* (Clarendon, Oxford, 2000)
3. F. Ancona, A. Marson, Existence theory by front tracking for general nonlinear hyperbolic systems. Arch. Ration. Mech. Anal. **185**(2), 287–340 (2007)
4. F. Ancona, K.T. Nguyen, SBV regularity of L^∞ solutions to genuinely nonlinear temple systems of balance laws (in preparation)
5. S. Bianchini, SBV regularity of genuinely nonlinear hyperbolic systems of conservation laws in one space dimension. Acta Math. Sci. **32B**(1), 380–388 (2012)
6. S. Bianchini, L. Caravenna, SBV regularity for genuinely nonlinear, strictly hyperbolic systems of conservation laws in one space dimension. Commun. Math. Phys. **313**(1), 1–33 (2012)
7. S. Bianchini, D. Tonon, SBV regularity for Hamilton-Jacobi equations with Hamiltonian depending on (t, x). Commun. Pure Appl. Anal. **10**(6), 1549–1566 (2011)
8. S. Bianchini, D. Tonon, SBV-like regularity for Hamilton-Jacobi equations with a convex Hamiltonian. J. Math. Anal. Appl. **391**(1), 190–208 (2012)
9. S. Bianchini, L. Yu, SBV-like regularity for general hyperbolic systems of conservation laws in one space dimension. arXiv:1202.2680
10. S. Bianchini, C. De Lellis, R. Robyr, SBV regularity for Hamilton-Jacobi equations in $\mathbb{R}^n$. Arch. Ration. Mech. Anal. **200**(3), 1003–1021 (2011)
11. A. Bressan, *Hyperbolic Systems of Conservation Laws. The One-Dimensional Cauchy Problem* (Oxford University Press, Oxford, 2000)
12. A. Bressan, R. Colombo, Decay of positive waves in nonlinear systems of conservation laws. Ann. Scuola Norm. Sup. Pisa Cl. Sci. **26**(2), 133–160 (1998)
13. A. Bressan, P.G. LeFloch, Structural stability and regularity of entropy solutions to hyperbolic systems of conservation laws. Indiana Univ. Math. J. **48**(1), 43–84 (1999)
14. A. Bressan, T. Yang, A sharp decay estimate for nonlinear positive waves. SIAM J. Math. Anal. **36**, 659–677 (2004)
15. C.M. Dafermos, *Hyperbolic Conservation Laws in Continuous Physics* (Springer, New York, 2000)
16. C.M. Dafermos, Wave fans are special. Acta Math. Appl. Sin. English Ser. **24**(3), 369–374 (2008)
17. L.C. Evans, *Partial Differential Equations*. Graduate Studies in Mathematics, vol. 19 (American Mathematical Society, Providence, 1998)
18. J. Glimm, P.D. Lax, *Decay of Solutions of Systems of Nonlinear Hyperbolic Conservation Laws*. Memoirs of the American Mathematical Society, vol. 101 (American Mathematical Society, Providence, 1970), pp. 95–172
19. B.H. Li, Global structure of shock waves. Sci. Sin. **22**, 979–990 (1979)
20. O. Oleinik, Discontinuous solutions of nonlinear differential equations. Am. Math. Soc. Transl. **26**, 95–172 (1963). Translation of Uspekhi Mat. Nauk vol. 12, **3**(75), 3–73 (1957)
21. R. Robyr, SBV regularity of entropy solutions for a class of genuinely nonlinear scalar balance laws with non-convex flux function. J. Hyperbolic Differ. Equ. **5**(2), 449–475 (2008)
22. D. Schaeffer, A regularity theorem for conservation laws. Adv. Math. **11**, 368–386 (1973)

A Generalized Buckley-Leverett System

Nikolai Chemetov and Wladimir Neves

Abstract We show the existence of a solution to a new mathematical model of the Buckley-Leverett system, describing two-phase flows in porous media. To prove the solvability result, we consider an approximate parabolic-elliptic system, the approximate solutions of which do not have **any type of standard BV estimates**. Therefore, we justify the limit transition using a kinetic method. More precisely, we use the transport property of the derived linear (kinetic) transport equation, and the strong trace results proved for the kinetic function.

2010 Mathematics Subject Classification 35D30, 35L65, 35L60

1 Introduction

The simultaneous motion of two immiscible incompressible liquids (e.g. water and oil) in a porous medium can be described by the famous Buckley-Leverett system

$$\partial_t u + \operatorname{div}\big(\mathbf{v}\, g(u)\big) = 0, \qquad \operatorname{div}(\mathbf{v}) = 0, \tag{1}$$

$$h(u)\mathbf{v} = -\nabla p, \tag{2}$$

N. Chemetov
CMAF/Universidade de Lisboa, Av. Prof. Gama Pinto 2, 1649-003 Lisbon, Portugal

W. Neves (✉)
Instituto de Matemática, Universidade Federal do Rio de Janeiro C.P. 68530, Cidade Universitaria 21945-970, Rio de Janeiro, Brazil
e-mail: wladimir@im.ufrj.br

G.-Q.G. Chen et al. (eds.), *Hyperbolic Conservation Laws and Related Analysis with Applications*, Springer Proceedings in Mathematics & Statistics 49,
DOI 10.1007/978-3-642-39007-4_4,

where u, p and $\mathbf{v}$ are the saturation, the pressure and the total velocity of the two-phase flow respectively. The saturation dependent functions $h(u)$ and $g(u)$ describe physical properties of the porous media. Equation (2) is Darcy's Law, being an empirical equation. The study of system (1)–(2) has practical interest in connection with the planning and operation of oil wells, but also brings some challenging mathematical questions.

The solvability of this system has only recently been established. In order to pass this difficulty, the Buckley-Leverett system has been significantly simplified in many works, for instance see Cordoba, Gancedo and Orive [8] and Perepelitsa and Shelukhin [13]. Many authors have proposed interesting ideas, but most of them have focused on the saturation equation (1), reducing the Buckley-Leverett system to an elliptic-parabolic partial differential system. Some of the important works on this subject include Antontsev, Kazikhov and Monakhov [1], Chen [5] and Lenzinger and Schweizer [11] and further references cited therein.

In the present work we instead focus our attention on the equation of velocity. So we propose a generalized Darcy's law equation, which is physically no longer than the standard equation. One observes that, for very short time scales or high frequency oscillations, a time derivative of flux may be added to Darcy's law, which results in valid solutions at very small times

$$\tau\, \partial_t \mathbf{v} + h(u)\mathbf{v} = -\nabla p, \tag{3}$$

where $\tau > 0$ is a very small time constant. Another extension to the traditional form of Darcy's law is Brinkman's term, which is used to account for transitional flow between boundaries

$$-\,\nu\, \Delta \mathbf{v} + h(u)\mathbf{v} = -\nabla p, \tag{4}$$

where $\nu > 0$ is an effective viscosity. This correction term accounts for flow through a medium where the grains of the media are porous themselves. In the porous media literature [15] the combination of (3) and (4) is known as Brinkman-Forchheimer's law

$$\tau\, \partial_t \mathbf{v} - \nu\, \Delta \mathbf{v} + h(u)\mathbf{v} = -\nabla p. \tag{5}$$

It is important to observe that generalized Darcy's laws such as (3)–(5) have also been deduced via homogenization theory [10].

2 A Generalized Buckley-Leverett Model

Let $\Omega \subset \mathbb{R}^d$ (with $d = 1, 2$ or 3) be a bounded domain having a C^2-smooth boundary Γ. In this section we will study the generalized Buckley-Leverett model (1), (5) for given $\nu, \tau > 0$:

$$\partial_t u + \operatorname{div}\big(\mathbf{v}\, g(u)\big) = 0 \qquad \operatorname{div}(\mathbf{v}) = 0, \tag{6}$$

$$\tau\, \partial_t \mathbf{v} - \nu \Delta \mathbf{v} + h(u)\mathbf{v} = -\nabla p, \qquad \text{in } \Omega_T := \Omega \times (0,T), \tag{7}$$

satisfying the boundary-initial conditions

$$(u,\mathbf{v}) = (u_b, \mathbf{b}) \quad \text{on } \Gamma_T := \Gamma\times(0,T) \quad \text{and} \quad (u,\mathbf{v})|_{t=0} = (u_0, \mathbf{v}_0) \quad \text{in } \Omega. \tag{8}$$

Before the formulation of the main result let us introduce the following spaces

$$\mathbf{V}^s(\Omega) : = \{\mathbf{u} \in H^s(\Omega) : \quad \operatorname{div}(\mathbf{u}) = 0 \quad \text{in} \quad \mathscr{D}'(\Omega), \quad \int_\Gamma \mathbf{u}\cdot\mathbf{n}\, d\mathbf{x} = 0\},$$

$$\mathbf{V}^s(\Gamma) : = \{\mathbf{u} \in H^s(\Gamma) : \quad \int_\Gamma \mathbf{u}\cdot\mathbf{n}\, d\mathbf{x} = 0\}, \qquad \mathbf{V}^{-s}(\Gamma) := (\mathbf{V}^s(\Gamma))',$$

$$\mathbf{G}(\Gamma_T) : = \big\{\mathbf{u} \in L^2(0,T;\mathbf{V}^{1/2}(\Gamma)) : \quad \partial_t \mathbf{u} \in L^2(0,T;\mathbf{V}^{-1/2}(\Gamma)\big\},$$

where $\mathbf{n} = \mathbf{n}(\mathbf{x})$ is the outside normal to Ω at $\mathbf{x} \in \Gamma$. We assume that our data satisfy the following regularity properties

$$\begin{aligned} & g,\ h \in W^{1,\infty}_{\text{loc}}(\mathbb{R}) \qquad \text{with} \quad 0 < h_0 \le h(u), \\ & 0 \leqslant u_b \leqslant 1 \quad \text{on } \Gamma_T, \qquad 0 \leqslant u_0 \leqslant 1 \quad \text{on } \Omega, \\ & \mathbf{v}_0 \in \mathbf{V}^0(\Omega), \qquad \mathbf{b} \in \mathbf{G}(\Gamma_T) \qquad \text{and} \end{aligned} \tag{9}$$

$$\mathbf{b}(0)\cdot\mathbf{n} = \mathbf{v}_0\cdot\mathbf{n} \qquad \text{in} \quad H^{-1/2}(\Gamma). \tag{10}$$

Now, since the former equation in (6) is a hyperbolic scalar conservation law, the saturation function u may admit shocks. Therefore, in order to select a correct physical solution, we need the entropy concept of solution, as given in the following.

Definition 1. A pair of functions $u \in L^\infty(\Omega_T)$, $\mathbf{v} \in L^2(0,T;\mathbf{V}^1(\Omega))$ is called a weak solution of system (6)–(8) if this pair satisfies:

(1) The integral inequality

$$\iint_{\Omega_T} \big(|u-v|\, \phi_t + \operatorname{sgn}(u-v)\big(g(u)-g(v)\big)\, \mathbf{v}\cdot\nabla\phi\big)\, d\mathbf{x}\, dt$$

$$+ K\int_{\Gamma_T} |\mathbf{b}\cdot\mathbf{n}|\ |u_b - v|\phi\, d\mathbf{x}\, dt + \int_\Omega |u_0 - v|\phi(0,x)\, d\mathbf{x} \geqslant 0 \tag{11}$$

for all $x \in \mathbb{R}$ and for any positive $\phi \in C_0^\infty((-\infty,T)\times\mathbb{R}^d)$. Here $K{:=}||g'||_{L^\infty(\mathbb{R})}$.

(2) Equation (7) in the usual distributional sense where $\mathbf{v}|_{\Gamma_T} = \mathbf{b}$.

Theorem 1. *If the data* $g, h, u_b, u_0, \mathbf{v}_0, \mathbf{b}$ *have the regularity properties* (9)–(10), *then system* (6)–(8) *has a weak solution such that* $\mathbf{v} \in H^1(0,T;\mathbf{V}^{-1}(\Omega))$,

$$0 \leqslant u \leqslant 1 \qquad a.e. \ in \quad \Omega_T$$

$$\|\sqrt{\tau}\mathbf{v}\|_{C([0,T];\mathbf{V}^0(\Omega))} + \|\mathbf{v}\|_{L^2(0,T;\mathbf{V}^1(\Omega))} + \tau\|\mathbf{v}\|_{H^1(0,T;\mathbf{V}^{-1}(\Omega))} \leqslant C, \tag{12}$$

where C is a positive constant independent of τ.

The generalized Buckley-Leverett model (6)–(8) poses **specific difficulties** compared to *the usual theory of quasilinear scalar conservation laws*:

(1) It is not possible to obtain standard a priori compactness for approximate solutions (no BV-bounds or L^1-Kruzkov continuous compactness).
(2) Since we are dealing with the initial-boundary problem in the class of L^∞-bounded solutions, such solutions do not have trace values in the sense of Sobolev functions.

We stress that, to overcome these two difficulties, we use the kinetic theory [14] and the concept of trace values for non-regular functions developed for divergence type equations [6], see also [12].

2.1 An Approximate System

In order to show the solvability of system (6)–(8), we first study the following approximate parabolic system. For a fixed $\varepsilon > 0$, we consider

$$\partial_t u^\varepsilon + \operatorname{div}\big(\mathbf{v}^\varepsilon\, g(u^\varepsilon)\big) = \varepsilon\, \Delta u^\varepsilon \qquad \text{in } \Omega_T, \tag{13}$$

$$\tau\, \partial_t \mathbf{v}^\varepsilon - \nu\Delta\mathbf{v}^\varepsilon + h(u^\varepsilon)\mathbf{v}^\varepsilon = -\nabla p^\varepsilon, \quad \operatorname{div}(\mathbf{v}^\varepsilon) = 0 \quad \text{in } \Omega_T \tag{14}$$

jointly with the boundary-initial conditions

$$\varepsilon\frac{\partial u^\varepsilon}{\partial \mathbf{n}} + M(u^\varepsilon - u_b^\varepsilon) = 0 \qquad \text{and} \qquad \mathbf{v}^\varepsilon = \mathbf{b} \quad \text{on} \quad \Gamma_T$$

$$(u^\varepsilon, \mathbf{v}^\varepsilon)|_{t=0} = (u_0^\varepsilon, \mathbf{v}_0) \quad \text{in } \Omega, \tag{15}$$

where u_b^ε, u_0^ε are regularized boundary-initial data satisfying suitable compatibility conditions. Using the results of parabolic-elliptic theory, we obtain the solvability of system (13)–(15).

Proposition 1. *For each $\varepsilon > 0$, there exists a unique solution $(u^\varepsilon, \mathbf{v}^\varepsilon)$ of system* (13)–(15)*, which has the following regularity properties: $u^\varepsilon \in L^\infty(0,T;H^1(\Omega)) \cap L^2(0,T;H^2(\Omega))$ and $\mathbf{v}^\varepsilon \in L^2(0,T;\mathbf{V}^1(\Omega)) \cap H^1(0,T;\mathbf{V}^{-1}(\Omega))$ satisfying*

$$\varepsilon\|\nabla u^\varepsilon\|^2_{L^2(\Omega_T)} \leqslant C \qquad and \qquad 0 \leqslant u^\varepsilon \leqslant 1 \quad a.e.\ on\ \Omega_T, \text{and} \tag{16}$$

$$\|\sqrt{\tau}\mathbf{v}^\varepsilon\|_{C([0,T];\mathbf{V}^0(\Omega))} + \|\mathbf{v}^\varepsilon\|_{L^2(0,T;\mathbf{V}^1(\Omega))} + \|\tau\mathbf{v}^\varepsilon\|_{H^1(0,T;\mathbf{V}^{-1}(\Omega))} \leqslant C, \tag{17}$$

where C is a positive constant independent of ε (and τ).

2.2 *Sketch of the Proof (of Theorem 1)*

Let $(\eta(u), q(u))$ be an entropy pair for equation (6), i.e. $\eta = \eta(u)$ is a Lipschitz continuous convex function and $q'(u) = \eta'(u)g'(u)$ for $u \in \mathbb{R}$. Then from (13), we have in the distributional sense

$$\partial_t \eta(u^\varepsilon) + \operatorname{div}(\mathbf{v}^\varepsilon q(u^\varepsilon)) - \varepsilon\, \Delta\eta(u^\varepsilon) = -\varepsilon\, \eta''(u)\, |\nabla\eta(u)|^2 \leqslant 0. \tag{18}$$

For instance, we can take the entropy pair defined by

$$\eta(u) := |u-v|^+, \qquad q(u) := \operatorname{sgn}^+(u-v)\,(g(u)-g(v)) \quad \text{for each } v \in \mathbb{R}.$$

Then, we have in the distributional sense

$$\partial_t |u^\varepsilon - v|^+ + \operatorname{div}\left[\mathbf{v}^\varepsilon \operatorname{sgn}^+(u^\varepsilon - v)\,(g(u^\varepsilon) - g(v))\right] - \varepsilon\, \Delta |u^\varepsilon - v|^+ = -m^\varepsilon. \tag{19}$$

Here $|v|^+ := \max\{v,\ 0\}$, $\operatorname{sgn}^+(v)$ is equal to1 if $v > 0$ and 0 if $v \le 0$, and m^ε is a real nonnegative Radon measure. Differentiation of (19) with respect to the variable v gives that the function $f^\varepsilon(t,\mathbf{x},v) := \operatorname{sgn}^+(u^\varepsilon(t,\mathbf{x}) - v)$ satisfies

$$\partial_t f^\varepsilon + g'(v)\,\mathbf{v}^\varepsilon \cdot \nabla f^\varepsilon - \varepsilon\, \Delta f^\varepsilon = \partial_v m^\varepsilon \qquad \text{in } \mathscr{D}'(\Omega_T \times \mathbb{R}). \tag{20}$$

Let us point out that $0 \leqslant f^\varepsilon(t,\mathbf{x},v) \leqslant 1$ in $\Omega_T \times \mathbb{R}$. It is possible to show that m^ε is uniformly bounded with respect to ε. Hence, due to Proposition 1, there exist subsequences of f^ε, $\mathbf{v}^\varepsilon$, m^ε and the functions

$$f \in L^\infty(\Omega_T \times \mathbb{R}), \qquad \mathbf{v} \in L^2(0,T;\mathbf{V}^1(\Omega)) \tag{21}$$

and a real nonnegative Radon measure $m = m(t,\mathbf{x},v)$, such that

$$\begin{aligned}
&f^\varepsilon \to f && *\text{-weakly in } L^\infty(\Omega_T \times \mathbb{R}),\\
&\mathbf{v}^\varepsilon \to \mathbf{v}, && \varepsilon\, \nabla u^\varepsilon \to 0 \qquad \text{strongly in } L^2(\Omega_T),\\
&m^\varepsilon \to m && \text{weakly in } \mathscr{M}^+_{loc}(\overline{\Omega}_T \times \mathbb{R}).
\end{aligned}$$

Since (20) is linear, it follows that

$$\partial_t f + g'(v)\,\mathbf{v} \cdot \nabla f = \partial_v m \qquad \text{in } \mathscr{D}'(\Omega_T \times \mathbb{R}). \tag{22}$$

The $L^\infty(\Omega_T \times \mathbb{R})$-boundedness of f does not guarantee the existence of trace values, e.g. f at $t = 0$ and on Γ, which raises one of the major difficulties for the studied problem. Nevertheless, the divergence form of equation (22)

$$\operatorname{div}_{t,\mathbf{x},v}(\mathbf{F}) = 0 \qquad \text{with} \qquad \mathbf{F} = (f, g'(v)\mathbf{v} f, -m)$$

permits the introduction of a concept of trace values for f (for a theoretical discussion of this see the article [6]). Hence, taking into account the initial-boundary conditions for f^{ε}, we can show that the trace values of f exist, which is to say

$$\begin{aligned} f &= \operatorname{sgn}^{+}(u_0 - v) \qquad \text{for } t = 0, \\ f &= \operatorname{sgn}^{+}(u_b - v) \qquad \text{on} \quad \Gamma_T \times \mathbb{R}, \quad \text{where } g'(v)\mathbf{b}\cdot\mathbf{n} < 0. \end{aligned} \tag{23}$$

Since $\mathbf{v} \in L^2(0,T;\mathbf{V}^1(\Omega))$, we can apply DiPerna-Lions's theory for transport equations [9], and deduce that the solution f of (22)–(23) takes values equal only to 0 and 1 in $\Omega_T \times \mathbb{R}$. Since $f(\cdot,\cdot,v)$ is a monotone function on v, as a limit of monotone functions $f^{\varepsilon}(\cdot,\cdot,v)$, there exists a function $z = z(t,\mathbf{x})$ such that:

$$f = \operatorname{sgn}^{+}(z(t,\mathbf{x}) - v).$$

Therefore we derive the *-weak convergence in $L^{\infty}(\Omega_T)$

$$G(u^{\varepsilon}) = \int_0^1 G'(v) f^{\varepsilon}(\cdot,\cdot,v)\, dv \rightharpoonup \int_0^1 G'(v) f(\cdot,\cdot,v)\, dv = G(z)$$

for any $G \in C^1([0,1])$, $G(0) = 0$. This implies $z = u$ and

$$u^{\varepsilon} \to u \qquad \text{strongly in } L^p(\Omega_T) \quad \text{for all } p < \infty.$$

Hence the function $\mathbf{v}$ satisfies equality (7). Moreover, if we take Kružkov's entropy pair

$$\eta(u) := |u - v|, \qquad q(u) := \operatorname{sgn}(u - v)\,(g(u) - g(v)) \quad \text{for all } v \in \mathbb{R} \tag{24}$$

in inequality (18) and pass to the limit as $\varepsilon \to 0$, we deduce that u satisfies (11), which concludes the proof of Theorem 1.

3 A Quasi-stationary Buckley-Leverett Model

In this section we formulate a solvability result for the quasi-stationary Buckley-Leverett model (1), (5) with $\tau = 0$ and a given viscous parameter $\nu > 0$:

$$\begin{aligned} \partial_t u + \operatorname{div}\big(\mathbf{v}\, g(u)\big) &= 0 \qquad \operatorname{div}(\mathbf{v}) = 0, \\ -\nu\Delta\mathbf{v} + h(u)\mathbf{v} &= -\nabla p, \qquad \text{in } \Omega_T \end{aligned} \tag{25}$$

satisfying the boundary-initial conditions

$$(u,\mathbf{v}) = (u_b,\mathbf{b}) \quad \text{on } \Gamma_T \quad \text{and} \quad u = u_0 \quad \text{in } \Omega. \tag{26}$$

Theorem 2. *Let the data* g, h, u_b, u_0, $\mathbf{b}$ *satisfy the regularity properties* (9) *and* $\mathbf{b} \in \mathbf{G}(\Gamma_T)$. *Then system* (25)–(26) *has a weak solution* $(u, \mathbf{v})$, *which is understood in the sense of Definition* (1) *for* $\tau = 0$, *such that*

$$0 \leqslant u \leqslant 1 \qquad a.e. \text{ in } \quad \Omega_T,$$
$$\mathbf{v},\, \partial_t \mathbf{v} \in L^2(0,T;\mathbf{V}^1(\Omega)).$$

To prove the above theorem, we can use Theorem 1. By the last mentioned theorem, system (6)–(8) admits a solution $(u^\tau, \mathbf{v}^\tau)$ satisfying (12) for a fixed $\tau > 0$. Now the issue is to pass to the limit as $\tau \to 0$. Of course, the estimates (12) are not sufficient for the limit transition as $\tau \to 0$ in system (6)–(8), since we need the strong convergence of a subsequence for $\{u^\tau\}_{\tau>0}$. To get this strong convergence, we can apply the kinetic approach developed in Sect. 2.2, proving Theorem 2.

For complete details of the present exposition, we refer to our article [4], see also [2, 3]. Moreover, we wish to thank the anonymous referee, who brought our attention to the recent article related to the Buckley-Leverett System: Coclite, Karlsen, Mishra and Risebro [7].

References

1. S.N. Antontsev, A.V. Kazikhov, V.N. Monakhov, *Boundary-Value Problems in Mechanics of Non-homogeneous Fluids*. Studies in Mathematics and its Applications, vol. 22 (North-Holland, Amsterdam, 1990)
2. N.V. Chemetov, Nonlinear Hyperbolic-Elliptic systems in the bounded domain. Commun. Pure Appl. Anal. **10**(4), 1079–1096 (2011)
3. N.V. Chemetov, L. Arruda, L_p-Solvability of a full superconductive model. Nonlinear Anal. Real World Appl. **12**(4), 2118–2129 (2011)
4. N.V. Chemetov, W. Neves, The generalized Buckley-Leverett system. Solvability. To be published in: Arch. Ration. Mech. Anal. **208**, 1–24 (2013). http://arxiv.org/abs/1011.5461
5. Z. Chen, Degenerate two-phase incompressible flow: I. existence, uniqueness and regularity of a weak solution. J. Differ. Equ. **171**(2), 203–232 (2001)
6. G.-Q. Chen, H. Frid, Divergence measure fields and hyperbolic conservation laws. Arch. Ration. Mech. Anal. **147**, 89–118 (1999)
7. G.M. Coclite, K.H. Karlsen, S. Mishra, N.H. Risebro, A hyperbolic-elliptic model of two-phase flow in porous media—existence of entropy solutions. Int. J. Numer. Anal. Model. **9**(3), 562-583 (2012)
8. D. Cordoba, F. Gancedo, R. Orive, Analytical behavior of two-dimensional incompressible flow in porous media. J. Math. Phys. **48**(6), 1–19 (2007)
9. R.J. DiPerna, P.L. Lions, Ordinary differential equations, transport theory and Sobolev spaces. Invent. Math. **98**, 511–547 (1989)
10. U. Hornung, *Homogenization and Porous Media*. Interdisciplinary Applied Mathematics, vol. 6 (Springer, New York, 1996)
11. M. Lenzinger, B. Schweizer, Two-phase flow equations with outflow boundary conditions in the hydrophobic hydrophilic case. Nonlinear Anal. Theory Methods Appl. **73**(4), 840–853 (2010)

12. W. Neves, Scalar multidimensional conservation laws IBVP in noncylindrical Lipschitz domains. J. Differ. Equ. **192**, 360–395 (2003)
13. I. Perepetlitsa, V. Shelukhin, On Global solutions of a boundary-value problem for the one-dimensional Buckley-Leverett equations. Appl. Anal. **73**(3–4), 325–343 (1999)
14. B. Perthame, *Kinetic Formulation of Conservation Laws* (Oxford University Press, New York, 2002)
15. B. Straughan, *Stability and Wave Motion in Porous Media*. Applied Mathematical Sciences, vol. 165 (Springer, New York, 2008)

Entropy, Elasticity, and the Isometric Embedding Problem: $\mathbb{M}^3 \to \mathbb{R}^6$

Gui-Qiang G. Chen, Marshall Slemrod, and Dehua Wang

Abstract The balance laws for isometric embedding of a 3-dimensional Riemannian manifold into the 6-dimensional Euclidean space are exposited. It is shown that a necessary and sufficient condition for solving the equations of these balance laws is a system reminiscent of the equations of elastostatics. In turn, this elastostatic system possesses an elementary entropy equation in the sense of Lax. In addition, we use an entropy equality for the linearized system (linearized about a known embedding) to develop L^2 estimates for the dependent variables.

Keywords Balance laws • Isometric embedding • Gauss-Codazzi-Ricci equations • Entropy • Elasticity

2010 Mathematics Subject Classification Primary: 53C42, 53C21, 53C45, 35L65, 35M10; Secondary: 53C24, 57R40, 57R42, 58J32

G.-Q.G. Chen (✉)
Mathematical Institute, University of Oxford, Oxford, OX1 3LB, UK
e-mail: chengq@maths.ox.ac.uk

M. Slemrod
Department of Mathematics, University of Wisconsin, Madison, WI 53706, USA
e-mail: slemrod@math.wisc.edu

D. Wang
Department of Mathematics, University of Pittsburgh, Pittsburgh, PA 15260, USA
e-mail: dwang@math.pitt.edu

G.-Q.G. Chen et al. (eds.), *Hyperbolic Conservation Laws and Related Analysis with Applications*, Springer Proceedings in Mathematics & Statistics 49,
DOI 10.1007/978-3-642-39007-4_5,

1 Introduction

The purpose of this paper is to continue our investigation initiated in Chen-Slemrod-Wang [6] on the formulation of the balance laws describing the classical problem of isometric embedding of Riemann manifolds into the Euclidean spaces within the usual framework of continuum mechanics. In [6], we considered the problem of embedding a 2-dimensional Riemannian manifold $\mathbb{M}^2$ into the 3-dimensional Euclidean space $\mathbb{R}^3$ and showed that a natural analogy can be given by the equations of 2-dimensional, irrotational, inviscid gas dynamics, with an appropriate Bernoulli equation. These equations in turn imply additional balance laws or "entropy" equations. Here we continue this program now for embedding $\mathbb{M}^3 \to \mathbb{R}^6$ (cf. [1, 5, 7, 13–15]). In this case, we show that a natural continuum mechanical analogy is given by the equations of elastostatics with a special prescribed stress-energy functional. Once again, an "entropy" equality for smooth solutions is derived.

The rest of the paper is organized as follows. Section 2 provides a self-contained discussion of the isometric embedding problem and the Gauss-Codazzi-Ricci equations, while Sect. 3 provides the continuum mechanical analogy. Section 4 illustrates the issues for the linearized equations (linearized about a given smooth embedding). In Sect. 5, we show how a basic "entropy" equality will yield L^2 estimates for the perturbed dependent variables which are crucial for the existence theory (cf. [8, 9, 11, 12]).

2 The Gauss-Codazzi-Ricci Equations

In this section, we present a self-contained discussion of the isometric embedding problem and the Gauss-Codazzi-Ricci equations. We start with some basic lemmas.

2.1 Lemmas

The standard existence and uniqueness theorem of ordinary differential equations implies

Lemma 1. *Let $X = X' \times I \subset \mathbb{R}^n$, where $X' \subset \mathbb{R}^{n-1}$ is an open domain and I is a connected open interval. Given smooth functions $f : X \times \mathbb{R}^m \to \mathbb{R}^m$ and $A_0 : X' \to \mathbb{R}^m$, and $t \in I$, there exists a unique solution $A : X \to \mathbb{R}^m$ to*

$$\partial_n A = f(x', x_n, A), \quad A|_{x_n=t} = A_0(x') \qquad \text{for } x' \in X',$$

where $\partial_n := \partial_{x_n}$.

The following is a nonlinear Poincaré lemma, which has the same proof as the standard Poincaré lemma, except that the existence and uniqueness theorem of ordinary differential equations is used instead of the fundamental theorem of calculus.

Lemma 2. *Let $X \subset \mathbb{R}^n$ be an open contractible domain and let $f_i : X \times \mathbb{R}^m \to \mathbb{R}^m$ satisfy*

$$\frac{\partial f_i^a}{\partial x_j} + \frac{\partial f_i^a}{\partial A_b} f_j^b = \frac{\partial f_j^a}{\partial x_i} + \frac{\partial f_j^a}{\partial A_b} f_i^b \qquad \text{for each } (x, A) \in X \times \mathbb{R}^m,$$

here and hereafter we always use the Einstein summation convection unless specified. Then, given $x_0 \in X$ and $A_0 \in \mathbb{R}^m$, there exists a unique solution $A : X \to \mathbb{R}^m$ to

$$\partial_i A = f_i(x, A), \qquad A(x_0) = A_0,$$

where $\partial_i := \partial_{x_i}$ and $x := (x_1, \cdots, x_n)$.

2.2 Riemannian Structure in Local Coordinates

Let (X, g) be an n-dimensional connected Riemannian manifold. The Riemannian metric g uniquely determines a torsion free and metric-compatible connection, called the Levi-Civita or Riemannian connection ∇. In a local coordinate patch $(x_1, \ldots, x_n)$, its connection symbols Γ_{ij}^k, the Christoffel symbols, are calculated as

$$\Gamma_{ij}^k = \frac{1}{2} g^{kl} \big(\partial_i g_{lj} + \partial_j g_{il} - \partial_l g_{ij} \big),$$

where $g_{ij} = g(\partial_i, \partial_j)$ and (g^{pq}) is the inverse matrix of (g_{ij}), and the Riemann curvature tensor of the connection is calculated as

$$R_{ijk}^l = \partial_j \Gamma_{ki}^l - \partial_k \Gamma_{ji}^l + \Gamma_{jp}^l \Gamma_{ki}^p - \Gamma_{kp}^l \Gamma_{ji}^p.$$

As usual, we have

$$\begin{aligned}
\Gamma_{ij}^k &= \Gamma_{ji}^k, \\
\partial_k g_{ij} &= g_{ip} \Gamma_{kj}^p + g_{jp} \Gamma_{ik}^p, \\
\nabla_i \partial_j &= \Gamma_{ij}^l \partial_l, \\
R^l{}_{ijk} \partial_l &= -\nabla_j \nabla_k \partial_i + \nabla_k \nabla_j \partial_i,
\end{aligned}$$

where the Levi-Civita connection is denoted by $\nabla = (\nabla_1, \cdots, \nabla_n)$ with $\nabla_i = \nabla_{\partial_i}$.

2.3 Isometric Immersion

We use $\cdot$ to denote the canonical metric in a coordinate patch $(y^1, \cdots, y^m)$ in $\mathbb{R}^m$. An $\mathbb{R}^m$-valued function $y : (X, g) \to (\mathbb{R}^m, \cdot)$ is called an *isometric immersion* of X into $\mathbb{R}^m$ if the induced metric is the same as the original, that is, written locally using the coordinates $(x_1, \cdots, x_n)$,

$$\partial_i y \cdot \partial_j y = g_{ij} \qquad \text{for each } 1 \le i, j \le n. \tag{1}$$

Consider $y(X)$, the image of X under the map y. If y is injective and the quotient topology induced from X through y and the subset topology from $\mathbb{R}^m$ coincides on $y(X)$, then y is called a Riemannian embedding. At every point $y(x)$ of the image $y(X)$,

$$\{\partial_1 y(x), \ldots, \partial_n y(x)\}$$

span a linear subspace of $T_{y(x)}\mathbb{R}^m$ which is identical to $\mathbb{R}^m$. Let $T_x X$ denote this subspace and $N_x X$ denote the $(m-n)$-dimensional subspace orthogonal and complementary to $T_x X$. Fix an orthogonal basis $\{N_{n+1}(x), \ldots, N_m(x)\}$ of $N_x X$ for each x, and assume further that they depend smoothly on x.

2.4 Second Derivative of Immersion

For each x, the vectors $\{\partial_1 y(x), \ldots, \partial_n y(x),\ N_{n+1}(x), \ldots, N_m(x)\}$ comprise a basis of $\mathbb{R}^m$. Therefore, for each pair of indices $1 \le i, j \le n$, the vector $\partial^2_{ij} y(x)$ can be written as a linear combination of these vectors. In other words, there exist unique coefficients Γ^k_{ij}, $1 \le k \le n$, and H^μ_{ij}, $n+1 \le \mu \le m$, such that

$$\partial^2_{ij} y(x) = \Gamma^k_{ij}(x) \partial_k y(x) + H^\mu_{ij} N_\mu(x). \tag{2}$$

It can be checked that the functions Γ^k_{ij}, $1 \le i, j, k \le n$, are the Christoffel symbols for the metric g with respect to the coordinates $(x_1, \ldots, x_n)$. On the other hand, the coefficients H^μ_{ij} comprise what is called the second fundamental form.

2.5 First and Second Derivatives of Normal Vectors

Lemma 3. *There exist functions* $A^\nu_{\mu i} = -A^\mu_{\nu i}$ *such that*

$$\partial_i N_\mu = -g^{jk} H^\mu_{ik} \partial_j y + A^\nu_{\mu i} N_\nu. \tag{3}$$

Proof. First we know that there exist functions $A^{\nu}_{\mu i}$ and $B^{j}_{i\mu}$ such that

$$\partial_i N_\mu = B^j_{i\mu}\partial_j y + A^{\nu}_{\mu i} N_\nu. \tag{4}$$

Since

$$0 = \partial_i (N_\nu \cdot N_\mu) = N_\nu \cdot \partial_i N_\mu + N_\mu \cdot \partial_i N_\nu,$$

it follows from (4) that

$$A^{\nu}_{\mu i} = -A^{\mu}_{\nu i}.$$

On the other hand, we have

$$\begin{aligned}
0 &= g^{jk}\partial_i (N_\mu \cdot \partial_k y) = g^{jk}(\partial_k y \cdot \partial_i N_\mu + N_\mu \cdot \partial^2_{ik} y) = g^{jk}(\partial_k y \cdot \partial_i N_\mu + H^{\mu}_{ik}) \\
&= g^{jk}(\partial_k y \cdot \partial_p y B^p_{i\mu} + H^{\mu}_{ik}) = g^{jk}(g_{kp} B^p_{i\mu} + H^{\mu}_{ik}) \\
&= B^j_{i\mu} + g^{jk} H^{\mu}_{ik}.
\end{aligned}$$

Therefore, $B^j_{i\mu} = -g^{jk} H^{\mu}_{ik}$, which, together with (4), yields (3). This completes the proof.

Differentiating (3) one more time, we obtain

$$\begin{aligned}
\partial^2_{ij} N_\mu &= -\partial_j (g^{pq} H^{\mu}_{ip})\partial_q y - g^{pq} H^{\mu}_{ip}(\Gamma^k_{jq}\partial_k y + H^{\nu}_{jq} N_\nu) + \partial_j A^{\nu}_{\mu i} N_\nu \\
&\quad + A^{\nu}_{\mu i}(-g^{pq} H^{\nu}_{pj}\partial_q y + A^{l}_{\nu j} N_l) \\
&= -\big(\partial_j (g^{pq} H^{\mu}_{ip}) + g^{pk}\Gamma^q_{jk} H^{\mu}_{ip} + g^{pq} A^{\nu}_{\mu i} H^{\nu}_{pj}\big)\partial_q y \\
&\quad + \big(\partial_j A^{\nu}_{\mu i} - g^{pq} H^{\mu}_{ip} H^{\nu}_{jq} + A^{l}_{\mu i} A^{\nu}_{lj}\big) N_\nu.
\end{aligned}$$

The commutation of the partials implies two sets of equations: one using the tangential components of the right side, and the other using the normal components. The tangential component can be shown to be equivalent to the Codazzi equations (6). The normal components are given by the Ricci equations (7) below.

2.6 Gauss and Codazzi Equations

Now we return to equation (2). Differentiation of (2) and commutation of the partials give

$$
\begin{aligned}
0 &= \partial_k(\partial^2_{ij}y) - \partial_j(\partial^2_{ik}y) \\
&= \big(\partial_k \Gamma^l_{ij} - \partial_j \Gamma^l_{ik}\big)\partial_l y + \Gamma^l_{ij}\partial^2_{kl}y - \Gamma^l_{ik}\partial^2_{lj}y \\
&\quad + \big(\partial_k H^\mu_{ij} - \partial_j H^\mu_{ik}\big)N_\mu + H^\mu_{ij}\partial_k N_\mu - H^\mu_{ik}\partial_j N_\mu \\
&= \big(\partial_k \Gamma^l_{ij} - \partial_j \Gamma^l_{ik} + \Gamma^p_{ij}\Gamma^l_{kp} - \Gamma^p_{ik}\Gamma^l_{pj} + g^{lp}(-H_{ij}\cdot H_{kp} + H_{ik}\cdot H_{jp})\big)\partial_l y \\
&\quad + \big(\partial_k H^\mu_{ij} + H^\nu_{ij}A^\mu_{\nu k} - \Gamma^p_{ik}H^\mu_{jp} - \partial_j H^\mu_{ik} - H^\nu_{ik}A^\mu_{\nu j} + \Gamma^p_{ij}H^\mu_{kp}\big)N_\mu \\
&= g^{pl}\big(R_{pijk} + H_{ij}\cdot H_{pk} - H_{ik}\cdot H_{jp}\big)\partial_l y \\
&\quad + \big(\partial_k H^\mu_{ij} + H^\nu_{ij}A^\mu_{\nu k} - \Gamma^p_{ik}H^\mu_{pj} - \partial_j H^\mu_{ik} - H^\nu_{ik}A^\mu_{\nu j} + \Gamma^p_{ij}H^\mu_{kp}\big)N_\mu \\
&= g^{pl}\big(R_{pijk} + H_{ij}\cdot H_{pk} - H_{ik}\cdot H_{jp}\big)\partial_l y \\
&\quad + \big(\partial_k H^\mu_{ij} + H^\nu_{ij}A^\mu_{\nu k} - \Gamma^p_{ik}H^\mu_{pj} - \Gamma^p_{jk}H^\mu_{ip} - \partial_j H^\mu_{ik} - H^\nu_{ik}A^\mu_{\nu j} + \Gamma^p_{ij}H^\mu_{pk} + \Gamma^p_{jk}H^\mu_{ip}\big)N_\mu.
\end{aligned}
$$

Since $\{\partial_1 y, \dots, \partial_n y, N_{n+1}, \dots, N_m\}$ form a basis of $\mathbb{R}^m$, this implies the Gauss equations (5) and Codazzi equation (6) below.

2.7 Reconstructing an Isometric Embedding

Theorem 1. *Given a connected and simply connected Riemannian manifold X with coordinates $(x_1, \dots, x_n)$ and Riemannian metric $g = (g_{ij})$, if there exist functions $H^\mu_{ij} = H^\mu_{ji}$ and $A^\nu_{\mu i} = -A^\mu_{\nu i}$, $1 \le i, j \le n$, $n+1 \le \mu, \nu \le m$, and such that*

$$\sum_{\mu=n+1}^{m} H^\mu_{ik}H^\mu_{jl} - H^\mu_{il}H^\mu_{jk} = R_{ijkl}, \tag{5}$$

$$\partial_k H^\mu_{ij} + A^\mu_{\nu k}H^\nu_{ij} - \Gamma^p_{ki}H^\mu_{pj} - \Gamma^p_{kj}H^\mu_{ip} = \partial_j H^\mu_{ik} + A^\mu_{\nu j}H^\nu_{ik} - \Gamma^p_{ji}H^\mu_{pk} - \Gamma^p_{jk}H^\mu_{ip}, \tag{6}$$

$$\partial_i A^\nu_{\mu j} - \partial_j A^\nu_{\mu i} + A^\nu_{\eta i}A^\eta_{\mu j} - A^\nu_{\eta j}A^\eta_{\mu j} = g^{pq}H^\mu_{ip}H^\nu_{jq} - g^{pq}H^\mu_{jp}H^\nu_{iq}, \tag{7}$$

then there exist functions $N_{n+1}, \dots, N_m : X \to \mathbb{R}^m$ and a function $y : X \to \mathbb{R}^m$ such that the following hold:

$$N_\mu \cdot N_\nu = \delta_{\mu\nu}, \tag{8}$$

$$N_\mu \cdot \partial_i y = 0, \tag{9}$$

$$\partial_i y \cdot \partial_j y = g_{ij}, \tag{10}$$

and

$$\partial^2_{ij}y = \Gamma^k_{ij}\partial_k y + H^\mu_{ij}N_\mu, \tag{11}$$

$$\partial_i N_\mu = -g^{jk}H^\mu_{ik}\partial_j y + A^\nu_{\mu i}N_\nu. \tag{12}$$

Sketch of the proof. Let $\{e_1, \ldots, e_m\}$ denote the standard basis of $\mathbb{R}^m$. Fix a point $x_0 \in X$. Set $\{\partial_1 y(x_0), \ldots, \partial_n y(x_0), N_{n+1}(x_0), \ldots, N_m(x_0)\}$ so that equations (8)–(10) hold. One possibility is to set $N_\mu(x_0) = e_\mu$ and $y(x_0) = 0$, and choose $\{\partial_1 y(x_0), \ldots, \partial_n y(x_0)\}$ to be linear combinations of $\{e_1, \ldots, e_n\}$ such that (10) holds at x_0.

If we let $\varphi_i = \partial_i y$, then (11)–(12) form a total differential system for the unknown $\mathbb{R}^m$-valued functions $\{\varphi_1, \ldots, \varphi_n, N_{n+1}, \ldots, N_m\}$. We check by differentiating these equations that the compatibility conditions obtained by commuting partial derivatives are a consequence of the Gauss equations (5), Codazzi equations (6), Ricci equations (7), as well as the original equations (11)–(12). Therefore, by Lemma 2, there exists a unique solution extending the initial data specified at x_0.

Also, the differentials of equations (8)–(10) are consequences of (11)–(12). Therefore, they hold not only at x_0 but also on all of X.

Finally, (11) implies that $\partial_i \varphi_j = \partial_j \varphi_i$, because the right side is symmetric in i and j. Therefore, by Lemma 2, there exists a unique $\mathbb{R}^m$-valued function y on X such that $y(x_0) = 0$ and $\partial_i y = \varphi_i$, $1 \le i \le n$.

3 The Elasticity Formulation

Theorem 1 provides a possible but inconvenient method for proving the existence of a local embedding. In this section, we provide a new approach.

First we set

$$\varphi_j^p = \partial_j y^p, \tag{13}$$

for $i, j = 1, 2, 3$ and $p = 1, \cdots, 6$. Then (11) becomes

$$\nabla_j \varphi_i^p := \partial_j \varphi_i^p - \Gamma_{ij}^k \varphi_k^p = H_{ij}^\mu N_\mu^p, \tag{14}$$

where ∇_j denotes the covariant derivative. In order to avoid dealing with the Ricci system (12), we define

$$h_{ij}^p = H_{ij}^\mu N_\mu^p. \tag{15}$$

Lemma 4. *The Gauss relations* (5) *and* (15) *imply the Gauss relations for* h_{ij}^p*:*

$$h_{ik}^p h_{jl}^p - h_{il}^p h_{jk}^p = R_{ijkl}. \tag{16}$$

This can be seen by the following direct calculation:

$$\begin{aligned} h^p_{ik}h^p_{jl} - h^p_{il}h^p_{jk} &= \left(H^\mu_{ik}H^\nu_{jl} - H^\mu_{il}H^\nu_{jk}\right) N^p_\mu N^p_\nu \\ &= \left(H^\mu_{ik}H^\nu_{jl} - H^\mu_{il}H^\nu_{jk}\right) \delta_{\mu\nu} \\ &= R_{ijkl}. \end{aligned}$$

From (15), we now see that (14) is simply

$$\nabla_j \varphi^p_i = h^p_{ij}. \tag{17}$$

Furthermore, by definition of the curvature tensor: $R^l_{ijk}\, g_{lq} = R_{qijk}$, we have

$$\left(\nabla_j \nabla_k - \nabla_k \nabla_j\right) \varphi^p_i = -R^l_{ijk}\varphi^p_l, \tag{18}$$

and substitution of (17) into (18) yields the following system:

$$\nabla_k h^p_{ji} - \nabla_j h^p_{ki} = R^l_{ijk}\varphi^p_l. \tag{19}$$

Finally, since φ_k and N_μ are orthogonal, we see that h_{ij} and φ_k must be orthogonal:

$$h^p_{ij}\varphi^p_k = H^\mu_{ij} N^p_\mu \varphi^p_k = 0,$$

since $N^p_\mu \varphi^p_k = 0$. We write this as

Lemma 5. *A necessary condition for an isometric embedding is*

$$h^p_{ij}\varphi^p_k = 0. \tag{20}$$

Furthermore, we have

Lemma 6. *A necessary condition for an isometric embedding is that the Gauss equations* (16) *are satisfied.*

Proof. From the previously derived necessary conditions (19) (the Codazzi equations) and (20) (orthogonality), we see

$$\varphi^p_q \left(\nabla_k h^p_{ji} - \nabla_j h^p_{ki} - R^l_{ijk}\varphi^p_l\right) = 0,$$

which, by (20), implies

$$-\nabla_k \varphi^p_q h^p_{ji} + \nabla_j \varphi^p_q h^p_{ki} = R^l_{ijk}\varphi^p_l \varphi^p_q.$$

By (17) and the embedding assumption ($\varphi_l^p \varphi_q^p = g_{lq}$), we have

$$-h_{qk}^p h_{ji}^p + h_{qj}^p h_{ki}^p = R_{ijk}^l g_{lq} = R_{qijk}.$$

Lemma 7. *A necessary condition for an isometric embedding is $h_{ij} = h_{ji}$.*

This lemma follows from Theorem 1.

Lemma 8. *Relation* (17) *implies*

$$\nabla_l \operatorname{cof} h_{il}^p - \varepsilon_{ijk}\varepsilon_{mnl}\varphi_q^p R_{knl}^q h_{jm}^p = 0, \qquad (\textit{no sum on } p) \tag{21}$$

where ε_{ijk} is the Levi-Civita symbol (also called the permutation symbol, antisymmetric symbol, or alternating symbol).

Proof. Recall

$$\operatorname{cof} h_{il}^p = \varepsilon_{ijk}\varepsilon_{lmn} h_{kn}^p h_{jm}^p, \qquad (\text{no sum on } p)$$

and hence we see

$$\nabla_l \operatorname{cof} h_{il}^p = \varepsilon_{ijk}\varepsilon_{lmn}(\nabla_l h_{kn}^p) h_{jm}^p + \varepsilon_{ijk}\varepsilon_{lmn} h_{kn}^p (\nabla_l h_{jm}^p). \qquad (\text{no sum on } p) \tag{22}$$

Next note

$$\varepsilon_{lmn}\nabla_l h_{kn}^p = \varepsilon_{mnl}\nabla_l h_{kn}^p = \frac{1}{2}\varepsilon_{mnl} R_{knl}^q \varphi_q^p,$$

where we recall that (19) is equivalent to

$$\varepsilon_{ljk}\nabla_k h_{ij}^p = \frac{1}{2}\varepsilon_{ljk} R_{ijk}^q \varphi_q^p.$$

Similarly, we note

$$\varepsilon_{lmn}\nabla_l h_{jm}^p = \varepsilon_{mnl}\nabla_l h_{jm}^p = -\varepsilon_{nml}\nabla_l h_{jm}^p = -\frac{1}{2}\varepsilon_{nml} R_{jml}^q \varphi_q^p.$$

Inserting the above two relations into (22), we recover (21).

In summary, any isometric embedding y of $\mathbb{M}^3 \to \mathbb{R}^6$ must satisfy

$$\partial_i y^p = \varphi_i^p, \tag{23}$$

$$\varphi_i \cdot \varphi_j = g_{ij}, \tag{24}$$

$$\nabla_j \varphi_i^p = h_{ij}^p, \tag{25}$$

$$\nabla_k h_{ij}^p - \nabla_j h_{ik}^p = R_{ijk}^l \varphi_l^p, \tag{26}$$

$$\nabla_l \Big(\frac{\partial W}{\partial h_{il}^p}\Big) - \varepsilon_{ijk}\varepsilon_{mnl}\varphi_q^p R_{knl}^q h_{jm}^p = 0, \qquad \text{(no sum on } p) \tag{27}$$

$$h_{ik}^p h_{jl}^p - h_{il}^p h_{jk}^p = R_{ijkl}, \tag{28}$$

$$h_{ij}^p = h_{ji}^p, \tag{29}$$

where

$$W = \sum_{p=1}^{6} \det h^p,$$

and we have used

$$\operatorname{cof} h_{il}^p = \frac{\partial W}{\partial h_{il}^p}.$$

Equations (25)–(27) are reminiscent of the equations of elastostatics with W representing the strain energy. In elasticity, $\frac{\partial W}{\partial h_{il}^p}$ would represent the Piola-Kirchoff stress tensor and the nature of the equations as a system would be determined via the standard computation

$$\nabla_l \frac{\partial W}{\partial h_{il}^p} = \frac{\partial^2 W}{\partial h_{il}^p \partial h_{jk}^p} \nabla_l h_{jk}^p = C_{iljk}^p \nabla_l \nabla_j \varphi_k.$$

Of course, C_{ijkl}^p is linear in h_{jk}^p in the case of $\mathbb{M}^3 \to \mathbb{R}^6$. Equations (28)–(29) may be regarded as "constitutive" relations for (25)–(27).

Equation (24) comes from the definition of isometric embedding, and the existence of y^p in (23) may be regarded as a corollary of (25) and (29).

For the linearized theory in Sect. 4, we replace (23)–(24) by the orthogonality relation from Lemma 5:

$$\varphi_k \cdot h_{ij} = h_{ij}^p \varphi_k^p = 0, \tag{30}$$

which may be regarded as another "constitutive" relation, in addition to (28)–(29), for (25)–(27).

In the next section we continue our discussion in terms of linearized theory which is analogous to linear elasticity.

4 The Linearized System

We linearize (25)–(30) as follows: allow the system to depend on an artificial parameter t and then differentiate with respect to the parameter t. Denote the derivative with respect to t by an upper imposed dot "$\cdot$". We then find the linearized equation about the un-dotted base embedding:

$$\nabla_j \dot{\varphi}_i^p - \dot{h}_{ij}^p + Q_{ij}^p = 0, \tag{31}$$

$$\nabla_k \dot{h}_{ij}^p - \nabla_j \dot{h}_{ik}^p = R_{ijk}^l \dot{\varphi}_l^p + \dot{R}_{ijk}^l \varphi_l^p + P_{ijk}^p, \tag{32}$$

$$\nabla_l \left(\frac{\partial^2 W}{\partial h_{il}^p \partial h_{jk}^p} \dot{h}_{jk}^p \right) - \varepsilon_{ijk}\varepsilon_{mnl} \left(\dot{\varphi}_q^p R_{knl}^q h_{jm}^p + \varphi_q^p \dot{R}_{knl}^q h_{jm}^p + \varphi_q^p R_{knl}^q \dot{h}_{jm}^p \right)$$

$$+ S_i^p = 0, \qquad \text{(no sum on } p) \tag{33}$$

$$\dot{h}_{ik}^p h_{jl}^p + h_{ik}^p \dot{h}_{jl}^p - \dot{h}_{il}^p h_{jk}^p - h_{il}^p \dot{h}_{jk}^p = \dot{R}_{ijkl}, \tag{34}$$

$$\dot{h}_{ij}^p = \dot{h}_{ji}^p, \tag{35}$$

$$\dot{h}_{ij}^p \varphi_k^p + h_{ij}^p \dot{\varphi}_k^p = 0. \tag{36}$$

where the terms $Q_{ij}^p = Q_{ji}^p$, P_{ijk}^p, and S_i^p are linear in $\dot{\Gamma}_{ij}^q$.

Just as in classical linear elasticity, the nature of the equations is determined by Legendre-Hadamard quadratic form

$$\frac{\partial^2 W}{\partial h_{il}^p \partial h_{jk}^p} \dot{h}_{jk}^p \dot{h}_{il}^p, \tag{37}$$

where $\dot{h}_{ij}^p$ must satisfy the linear closure (constitutive) relations (34) provided by the Gauss relations.

An elementary example is seen from the classical case of surface theory, where we wish to embed $\mathbb{M}^2 \to \mathbb{R}^3$. There we have

$$W = \det h, \quad h = \begin{bmatrix} h_{11} & h_{12} \\ h_{12} & h_{22} \end{bmatrix}, \tag{38}$$

and (34) is just

$$(\dot{\det} h) = \dot{\kappa}, \tag{39}$$

where $\kappa = \det h$ is the Gauss curvature of the base embedding and $\dot{\kappa}$ is its first variation. Substitution of (38)–(39) into the Legendre-Hadamard quadratic form yields

$$\frac{1}{2h_{12}^2}\left[\dot{h}_{11}\ \dot{h}_{22}\right]\begin{bmatrix} h_{22}^2 & h_{11}h_{22}-2h_{12}^2 \\ h_{11}h_{22}-2h_{12}^2 & h_{11}^2 \end{bmatrix}\begin{bmatrix}\dot{h}_{11}\\ \dot{h}_{22}\end{bmatrix}.$$

The positive definiteness will be achieved when the determinant is positive, *i.e.* when

$$h_{22}^2h_{11}^2 - (h_{11}h_{22} - 2h_{12}^2)^2 = 4\kappa h_{12}^2 > 0.$$

Then $\kappa > 0$ yields ellipticity of linear elasticity system, while $\kappa < 0$ yields hyperbolicity.

For the case $\mathbb{M}^3 \to \mathbb{R}^6$, a result of Chern-Lewy (see [1]) says that there are no points of ellipticity and hence the Legendre-Hadamard quadratic form can never be positive definite.

5 The "Entropy" Equality and L^2 Estimates

In this section, we illustrate how the L^2 estimates for the dependent variables may be developed from the entropy equality for the linearized system.

Rewrite (33) as

$$\nabla_l\Big(\frac{\partial^2 W}{\partial h_{il}^p \partial h_{jk}^p}\dot{h}_{jk}^p\Big) + L_i^p = 0, \tag{40}$$

where

$$L_i^p = -\varepsilon_{ijk}\varepsilon_{mnl}\left(\dot{\varphi}_q^p R_{knl}^q h_{jm}^p + \varphi_q^p \dot{R}_{knl}^q + \varphi_q^p R_{knl}^q \dot{h}_{jm}^p\right) + S_i^p.$$

A subset of the Codazzi relations (32) is given by

$$\nabla_1 \dot{h}_{il}^p - \nabla_l \dot{h}_{i1}^p = R_{il1}^q \dot{\varphi}_q^p + \dot{R}_{il1}^q \varphi_q^p + P_{il1}^p. \tag{41}$$

Multiply (41) by $-\frac{\partial^2 W}{\partial h_{il}^p \partial h_{jk}^p}$ to obtain

$$-\frac{\partial^2 W}{\partial h_{il}^p \partial h_{jk}^p}\nabla_1 \dot{h}_{il}^p + \frac{\partial^2 W}{\partial h_{il}^p \partial h_{jk}^p}\nabla_l \dot{h}_{i1}^p = -\frac{\partial^2 W}{\partial h_{il}^p \partial h_{jk}^p}\left(R_{il1}^q \dot{\varphi}_q^p + \dot{R}_{il1}^q \varphi_q^p + P_{il1}^p\right). \tag{42}$$

Now (40) and (42) form a symmetrized system. For $p = 4, 5, 6$, (40) possesses 9 equations, (41) only has the cases $i = 1, 2, 3$ and $l = 2, 3$ (since, for $l = 1$, it just yields $0 = 0$) which are 18 equations. Thus, (40) and (41) together encompass all 27 Codazzi equations for $p = 4, 5, 6$. Furthermore, by incorporation of the linearized Gauss equations, they will provide the necessary and sufficient conditions to solve the linearized isometric embedding problem. This follows by a theorem of Blum [2–4] and also presented in Goenner [10]. In terms of our earlier Theorem 1, it just means that the Ricci equations will be automatically satisfied. This should come as no surprise since Lemma 11 will show that the additional $\dot{h}^p_{ij}$ terms can be eliminated. In mechanical terms, Lemma 11 will provide another "constitutive relation".

Multiply (40) by $\dot{h}^p_{i1}$ and (42) by $\dot{h}^p_{jk}$ and then add them to obtain

Lemma 9. *The following equality holds:*

$$\begin{aligned} &-\frac{1}{2}\nabla_1\Big(\frac{\partial^2 W}{\partial h^p_{il}\partial h^p_{jk}}\dot{h}^p_{jk}\dot{h}^p_{il}\Big) + \frac{1}{2}\nabla_1\Big(\frac{\partial^2 W}{\partial h^p_{il}\partial h^p_{jk}}\Big)\dot{h}^p_{jk}\dot{h}^p_{il} + \nabla_l\Big(\frac{\partial^2 W}{\partial h^p_{il}\partial h^p_{jk}}\dot{h}^p_{jk}\dot{h}^p_{i1}\Big) \\ &+ \dot{h}^p_{i1}L^p_i + \frac{\partial^2 W}{\partial h^p_{il}\partial h^p_{jk}}\dot{h}^p_{jk}\left(R^q_{il1}\dot{\varphi}^p_q + \dot{R}^q_{il1}\varphi^p_q + P^p_{il1}\right) = 0. \end{aligned} \tag{43}$$

Equation (43) might appear to be able to provide the L^2 estimates if

$$\nabla_1\left(\frac{\partial^2 W}{\partial h^p_{il}\partial h^p_{jk}}\right)\dot{h}^p_{jk}\dot{h}^p_{il}$$

was a positive (or negative) definite quadratic form in $\dot{h}^p_{jk}$. However, this is impossible: Since

$$\nabla_1\left(\frac{\partial^2 W}{\partial h^p_{il}\partial h^p_{jk}}\right) = \varepsilon_{ijm}\varepsilon_{nlk}\nabla_1 h^p_{mn},$$

we cannot produce terms of the form $(\dot{h}^p_{11})^2$, $(\dot{h}^p_{22})^2$, and $(\dot{h}^p_{33})^2$. Hence we must find a way to eliminate the diagonal terms from our quadratic form.

A direct way to eliminate the diagonal terms will now be given. The first step is to introduce the differentiated (or prolonged) system. Simply differentiate the existing linearized system (31)–(35) with respect to x_r. Hence, (40) and (41) become

$$\nabla_l\left(\frac{\partial^2 W}{\partial h^p_{il}\partial h^p_{jk}}\partial_r\dot{h}^p_{jk}\right) + \hat{L}^p_{ir} = 0, \tag{44}$$

$$\nabla_1\left(\partial_r\dot{h}^p_{il}\right) - \nabla_l\left(\partial_r\dot{h}^p_{i1}\right) = \partial_r(R^q_{il1}\dot{\varphi}^p_q + \dot{R}^q_{il1}\varphi^p_q) + \hat{P}^p_{il1r} = 0, \tag{45}$$

where $\hat{L}^p_{ir}$ and $\hat{R}_{il1r}$ are the lower order terms of the prolonged system. Hence, the same computation as used to provide Lemma 9 now gives us a new lemma.

Lemma 10. *The following equality holds:*

$$
\begin{aligned}
&-\frac{1}{2}\nabla_1\left(\frac{\partial^2 W}{\partial h_{il}^p \partial h_{jk}^p}\partial_r \dot{h}_{jk}^p \partial_r \dot{h}_{il}^p\right) + \frac{1}{2}\nabla_1\left(\frac{\partial^2 W}{\partial h_{il}^p \partial h_{jk}^p}\right)\partial_r \dot{h}_{jk}^p \partial_r \dot{h}_{il}^p \\
&+ \nabla_l\left(\frac{\partial^2 W}{\partial h_{il}^p \partial h_{jk}^p}\partial_r \dot{h}_{jk}^p \partial_r \dot{h}_{i1}^p\right) + \partial_r \dot{h}_{i1}^p \hat{L}_{ir}^p \qquad (46) \\
&+ \frac{\partial^2 W}{\partial h_{il}^p \partial h_{jk}^p}\partial_r \dot{h}_{jk}^p\, \partial_r\left(R_{il1}^q \dot{\varphi}_q^p + \dot{R}_{il1}^q + \hat{P}_{il1r}^p\right) = 0. \qquad \textit{(no sum on p)}
\end{aligned}
$$

To keep matters simple, let us focus on our critical term in (46):

$$
\frac{1}{2}\nabla_1\left(\frac{\partial^2 W}{\partial h_{il}^p \partial h_{jk}^p}\right)\partial_r \dot{h}_{jk}^p \partial_r \dot{h}_{il}^p. \qquad (47)
$$

We break this term into two pieces: a piece containing the diagonal terms and a piece with no diagonal terms, *i.e.*

$$
\begin{aligned}
\frac{1}{2}\nabla_1\left(\frac{\partial^2 W}{\partial h_{il}^p \partial h_{jk}^p}\right)\partial_r \dot{h}_{jk}^p \partial_r \dot{h}_{il}^p &= \frac{1}{2}\nabla_1\left(\frac{\partial^2 W}{\partial h_{il}^p \partial h_{jk}^p}\right)\partial_r \dot{h}_{jk}^p \partial_r \dot{h}_{il}^p\Big|_{\substack{j\neq k\\ i\neq l}} \\
&+ \frac{1}{2}\nabla_1\left(\frac{\partial^2 W}{\partial h_{il}^p \partial h_{jk}^p}\right)\partial_r \dot{h}_{jk}^p \partial_r \dot{h}_{il}^p\Big|_{\substack{j=k\\ i\neq l}} \qquad (48) \\
&+ \frac{1}{2}\nabla_1\left(\frac{\partial^2 W}{\partial h_{il}^p \partial h_{jk}^p}\right)\partial_r \dot{h}_{jk}^p \partial_r \dot{h}_{il}^p\Big|_{\substack{j\neq k\\ i=l}}.
\end{aligned}
$$

Of course, by symmetry, the second and third terms on the right-hand side of (48) are identical, so we now examine just the second term. Break this second term into two pieces:

$$
\begin{aligned}
&\frac{1}{2}\nabla_1\left(\frac{\partial^2 W}{\partial h_{il}^p \partial h_{jk}^p}\right)\partial_r \dot{h}_{jk}^p \partial_r \dot{h}_{il}^p\Big|_{\substack{j=k\\ i\neq l}} \\
&= \frac{1}{2}\nabla_1\left(\frac{\partial^2 W}{\partial h_{il}^p \partial h_{jk}^p}\right)\partial_r \dot{h}_{jk}^p \partial_r \dot{h}_{il}^p\Big|_{\substack{j=k\\ i\neq l\\ r\neq j}} + \frac{1}{2}\nabla_1\left(\frac{\partial^2 W}{\partial h_{il}^p \partial h_{jk}^p}\right)\partial_r \dot{h}_{jk}^p \partial_r \dot{h}_{il}^p\Big|_{\substack{j=k\\ i\neq l\\ r=j}} \\
&= \frac{1}{2}\nabla_1\left(\frac{\partial^2 W}{\partial h_{il}^p \partial h_{jk}^p}\right)\partial_k \dot{h}_{rj}^p \partial_r \dot{h}_{il}^p\Big|_{\substack{j=k\\ i\neq l\\ r\neq j}} + \frac{1}{2}\nabla_1\left(\frac{\partial^2 W}{\partial h_{il}^p \partial h_{jk}^p}\right)\partial_r \dot{h}_{jk}^p \partial_r \dot{h}_{il}^p\Big|_{\substack{j=k\\ i\neq l\\ r=j}} + l.o.t.
\end{aligned}
\qquad (49)
$$

Notice that all we have done is to use the Codazzi relation (32) to write

$$\partial_r \dot{h}^p_{jk} = \partial_k \dot{h}^p_{rj} + l.o.t.$$

Let us summarize what we have done so far:

(i) We started with our Codazzi relations (32)–(33).
(ii) We differentiated with respect to x_r and obtained the "entropy" equality given in Lemma 10.
(iii) We examined the crucial term (47) which will provide the L^2 estimates. As written in (47), this quadratic form still has no terms of the form $(\partial_r \dot{h}^p_{11})^2$, $(\partial_r \dot{h}^p_{22})^2$, and $(\partial_r \dot{h}^p_{33})^2$.
(iv) We again used the linearized Codazzi relations to write (47) as (49), where the only diagonal terms that appear are given by the last term in (49), *i.e.* the terms $\partial_1 \dot{h}^p_{11}$, $\partial_2 \dot{h}^p_{22}$, and $\partial_3 \dot{h}^p_{33}$.

Thus, the Codazzi relations on their own have yielded our relevant quadratic form (49). However, the consistency of the fundamental theorem of isometric embedding now presents itself: We know (cf. Goenner [10] and Blum [2–4]) that a necessary and sufficient condition for isometric embedding is that the Codazzi equations ($p = 4,5,6$) and the Gauss equations are satisfied. In terms of our computations, this means that the fact we have as of yet not written the terms $\partial_1 \dot{h}^p_{11}$, $\partial_2 \dot{h}^p_{22}$, and $\partial_3 \dot{h}^p_{33}$ as the derivatives of the off-diagonal terms $\dot{h}^p_{ij}$, $i \neq j$, is no surprise: We must use the linearized Gauss equation (34). The derivation is given in a sequence of elementary lemmas.

To keep things simple, as usual we assume that $x = (x_1, x_2, x_3) = 0$ is the origin of a system of the normal coordinates and, at $x = 0$, $\varphi^p_k = \delta^p_k$ for $p,k = 1,2,3$, $\varphi^p_k = 0$ for $p = 4,5,6$, and $h^p_{ij} = 0$ for $p,i,j = 1,2,3$.

Lemma 11. *The following identities hold:*

$$\sum_{p=1}^{6} \Big((\nabla_r \dot{h}^p_{ij})\varphi^p_k + h^p_{ij} \nabla_r \dot{\varphi}^p_q + (\nabla_r h^p_{ij})\dot{\varphi}^p_k + h^p_{ij} \nabla_r \dot{\varphi}^p_k \Big) = 0.$$

Hence, locally near $x = 0$, we can solve for $\nabla_r \dot{h}^p_{ij}, i,j,p = 1,2,3$, in terms of $\nabla_r \dot{h}^p_{ij}, p = 4,5,6, i,j = 1,2,3$; $\dot{h}^p_{ij}, p = 1,\ldots,6$; and $\dot{\varphi}^p_k, p = 1,\ldots,6$. Furthermore, the dependence on $\nabla_r \dot{h}^p_{ij}$, $p = 4,5,6$, vanishes as $x \to 0$.

Proof. Differentiate (36) with respect to x_r:

$$\sum_{p=1}^{6} \Big((\partial_r \dot{h}^p_{ij})\varphi^p_k + \dot{h}^p_{ij} \partial_r \varphi^p_k + (\partial_r h^p_{ij})\dot{\varphi}^p_k + h^p_{ij} \partial_r \dot{\varphi}^p_k \Big) = 0.$$

At $x = 0$, this equation reads

$$\partial_r \dot{h}_{ij}^k + \sum_{p=4}^{6} \dot{h}_{ij}^p h_{kr}^p + \sum_{p=1}^{6} (\partial_r h_{ij}^p)\dot{\varphi}_k^p + \sum_{p=4}^{6} h_{ij}^p \dot{h}_{rk}^p = 0.$$

Hence, by the inverse function theorem, we can solve for $\partial_r \dot{h}_{ij}^k, i, j, k = 1, 2, 3$. □

Finally, as noted earlier, we must use the Gauss equation (34) with respect to x_r:

$$\sum_{p=1}^{6} \Big((\partial_r \dot{h}_{ik}^p) h_{jl}^p + \dot{h}_{ik}^p (\partial_r h_{jl}^p) + (\partial_r h_{ik}^p) \dot{h}_{jl}^p + h_{ik}^p (\partial_r \dot{h}_{jk}^p) \\ - (\partial_r \dot{h}_{il}^p) h_{jk}^p - \dot{h}_{ik}^p (\partial_r h_{jk}^p) - (\partial_r h_{il}^p) \dot{h}_{jk}^p - h_{il}^p (\partial_r \dot{h}_{jk}^p) \Big) = \partial_r \dot{R}_{ijkl}. \tag{50}$$

In particular, enumerating the case $i = k = r = 1, 2, 3$, we have

$$\begin{aligned}
\sum_{p=4}^{6} (\partial_1 \dot{h}_{11}^p) h_{jl}^p &= - \sum_{p=1}^{3} (\partial_1 \dot{h}_{11}^p) h_{jl}^p + \sum_{p=1}^{6} h_{11}^p (\partial_1 \dot{h}_{jl}^p) - \sum_{p=1}^{6} h_{1l}^p (\partial_1 \dot{h}_{j1}^p) + l.o.t. + \partial_1 \dot{R}_{1j1l}, \\
\sum_{p=4}^{6} (\partial_2 \dot{h}_{22}^p) h_{jl}^p &= - \sum_{p=1}^{3} (\partial_2 \dot{h}_{22}^p) h_{jl}^p + \sum_{p=1}^{6} h_{22}^p (\partial_2 \dot{h}_{jl}^p) - \sum_{p=1}^{6} h_{2l}^p (\partial_2 \dot{h}_{j2}^p) + l.o.t. + \partial_2 \dot{R}_{2j2l}, \\
\sum_{p=4}^{6} (\partial_3 \dot{h}_{33}^p) h_{jl}^p &= - \sum_{p=1}^{3} (\partial_3 \dot{h}_{33}^p) h_{jl}^p + \sum_{p=1}^{6} h_{33}^p (\partial_3 \dot{h}_{jl}^p) - \sum_{p=1}^{6} h_{3l}^p (\partial_3 \dot{h}_{j3}^p) + l.o.t. + \partial_3 \dot{R}_{3j3l}.
\end{aligned} \tag{51}$$

In the first equation of (51), by the properties of the Riemann curvature tensor, we must have $l \neq 1, j \neq 1$. Thus, we have $j = 2, l = 3; j = 2, l = 2; j = 3, l = 3$, that is, three equations. Similarly, each of the other two equations in (51) provides three equations. Therefore, we may solve them under the assumption that $h_{23}^p, h_{22}^p, h_{33}^p$, $p = 4, 5, 6$, are linearly independent at $x = 0$ for $\partial_1 \dot{h}_{11}^p, p = 4, 5, 6$. Linear independence of $h_{11}^p, h_{32}^p, h_{13}^p$ at $x = 0$ allows us to solve for $\partial_2 \dot{h}_{22}^p$, $p = 4, 5, 6$, and linear independence of $h_{11}^p, h_{22}^p, h_{12}^p$, $p = 4, 5, 6$, allows us to solve for $\partial_3 \dot{h}_{33}^p$, $p = 4, 5, 6$. Notice from (51) that these solutions may depend on the terms of the form $\partial_r \dot{h}_{ii}^p, r \neq i$, but we know these terms can be changed into the derivatives of the off-diagonal terms $\partial_i \dot{h}_{ir}^p$ by the Codazzi equations. Finally, the dependence on $\partial_r h_{ij}^p, p = 1, 2, 3$, is removed by Lemma 11. We summarize as follows.

Lemma 12. *Linear independence of* $\{h_{11}^p, h_{33}^p, h_{13}^p\}$, $\{h_{11}^p, h_{22}^p, h_{12}^p\}$, $\{h_{22}^p, h_{33}^p, h_{23}^p\}$, $p = 4, 5, 6$, *at* $x = 0$ *allows the solvability of* $\partial_i \dot{h}_{ii}^p$, $p = 4, 5, 6$, *in terms of the derivatives of the off-diagonal terms and the lower order terms, i.e. in terms of* $\partial_r \dot{h}_{ij}^p, i \neq j, p = 4, 5, 6$, *and the lower order terms.*

We have thus eliminated all the derivatives of the diagonal terms $\partial_r \dot{h}^p_{ii}$ (no sum on i), $p = 4,5,6$, in favor of the derivatives of the off-diagonal terms, $\partial_r \dot{h}^p_{ij}$, $i \neq j, p = 4,5,6$. Hence, it appears at first glance that there are three off-diagonal $\dot{h}^p_{ij}$ for each p and $r = 1,2,3$, we have $3 \times 3 \times 3 = 27$ terms. However, the derivatives of the off-diagonal terms must satisfy the Codazzi relations (32). In particular, we have

$$\begin{aligned}
\nabla_2 \dot{h}^p_{13} - \nabla_3 \dot{h}^p_{12} &= R^l_{132} \dot{\varphi}^p_l + \dot{R}^l_{132} \varphi^p_l + P^p_{132}, \\
\nabla_3 \dot{h}^p_{12} - \nabla_1 \dot{h}^p_{23} &= R^l_{123} \dot{\varphi}^p_l + \dot{R}^l_{123} \varphi^p_l + P^p_{123}, \\
\nabla_2 \dot{h}^p_{31} - \nabla_1 \dot{h}^p_{32} &= R^l_{312} \dot{\varphi}^p_l + \dot{R}^l_{322} \varphi^p_l + P^p_{312},
\end{aligned} \tag{52}$$

$p = 4,5,6$. As the second equation of (52) is implied by the other two equations of (52), we see that the Codazzi equation allows us to lower the number of derivatives of the off-diagonal terms by 2 for each $p = 4,5,6$, *i.e.* by $2 \times 3 = 6$, and we have $27 - 6 = 21$ derivatives of the off-diagonal terms in our quadratic form.

Now we once more exploit the differentiated Gauss relation (50) which has $3 \times 6 = 18$ equations and, in solving (51), we have used 9 of them. An additional 6 may be used to eliminate a further 6 derivatives of the off-diagonal terms and $21 - 6 = 15$ derivatives of the off-diagonal terms in our "state" vector that enters the quadratic form (47), $p = 4,5,6$. Of course, linear independence of sets of these 3 vectors in $\mathbb{R}^3$ similar to those given in Lemma 12 will be needed. Also, since the set of equations has a non-trivial null space for the homogeneous equations, a solvability condition must be satisfied, which is guaranteed by the second Bianchi identity. Thus, only 6 terms can be eliminated.

We have seen that the quadratic form (47) can be written in terms of 15 terms. Incorporation of the quadratic form obtained from the terms $\partial_r \dot{h}^p_{il} \hat{L}^p_{ir}$ in (46) will have the same feature. Finally, the original energy estimate was done for fixed p. If we take a linear combination for $p = 4,5,6$, we have the final coefficient matrix which we wish to be positive (or negative) definite (and repeat when we take ∇_2, ∇_3 as well). Of course, this remains to be checked in examples. Finally, let us make the standard assumption of Bryant-Griffiths-Yang [5] and Poole [15] that the perturbations $\dot{g}_{ij}$ are compactly supported in a neighborhood of the origin. Then $\dot{\varphi}^p_i$ and $\dot{h}^p_{ij}$ are also compactly supported. Integrate the entropy equality (43) over the domain Ω. We see that our assumed definite quadratic form yields the $L^2(\Omega)$ estimates.

Acknowledgements The authors would like to thank the American Institute of Mathematics (Palo Alto, CA), especially Estelle Basor and Brian Conrey, for their generous support and encouragement of our research as part of the SQuaREs program *"Isometric Embedding of Higher Dimensional Riemannian Manifolds"*. Our special thanks go to Jeanne Clelland and Deane Yang for their constant and valuable discussions, generous sharing of their insights into the embedding problems, and their contributions to Sect. 2. G.-Q.G. Chen was supported in part by the National Science Foundation under Grant DMS-0807551, the UK EPSRC Science and Innovation award to the Oxford Centre for Nonlinear PDE (EP/E035027/1), the NSFC under a joint project

Grant 10728101, and the Royal Society–Wolfson Research Merit Award (UK); M. Slemrod was supported in part by the Simons Foundation Collaborative Research Grant 232531 and the Korean Mathematics Research Station at KAIST (Daejeon, S. Korea); D. Wang was supported in part by the National Science Foundation under Grant DMS-0906160 and ONR Grant N00014-07-1-0668.

References

1. E. Berger, R. Bryant, P. Griffiths, Some isometric embedding and rigidity results for Riemannian manifolds. Proc. Natl. Acad. Sci. U.S.A. **78**, 4657–4660 (1981)
2. R. Blum, Ueber die Bedingungsgleichungen einer Riemann'schen Mannigfaltigkeit, die in einer Euklidischen Mannigfaltigkeit eingebettet ist (in German). Bull. Math. Soc. Roum. Sci. **47**, 144–201 (1946)
3. R. Blum, Sur les tenseors dérivés de Gauss et Codazzi. C. R. Acad. Sci. Paris **244**, 708–709 (1947)
4. R. Blum, Subspaces of Riemannian spaces. Can. J. Math. **7**, 445–452 (1955)
5. R.L. Bryant, P.A. Griffiths, D. Yang, Characteristics and existence of isometric embeddings. Duke Math. J. **50**, 893–994 (1983)
6. G.-Q. Chen, M. Slemrod, D. Wang, Isometric immersions and compensated compactness. Commun. Math. Phys. **294**, 411–437 (2010)
7. G.-Q. Chen, M. Slemrod, D. Wang, Weak continuity of the Gauss-Codazzi-Ricci system for isometric embedding. Proc. Am. Math. Soc. **138**, 1843–1852 (2010)
8. C.M. Dafermos, *Hyperbolic Conservation Laws in Continuum Physics*, 3rd edn. (Springer, Berlin, 2010)
9. K.O. Friedrichs, Symmetric positive linear differential equations. Comm. Pure. Appl. Math. **11**, 333–418 (1956)
10. H.F. Goenner, On the interdependency of the Gauss-Codazzi-Ricci equations of local isometric embedding. Gen. Relativ. Gravitation **8**, 139–145 (1977)
11. Q. Han, J.-X. Hong, *Isometric Embedding of Riemannian Manifolds in Euclidean Spaces* (AMS, Providence, 2006)
12. C.-S. Lin, The local isometric embedding in $\mathbb{R}^3$ of 2-dimensional Riemannian manifolds with Gaussian curvature changing sign cleanly. Comm. Pure Appl. Math. **39**, 867–887 (1986)
13. G. Nakamura, Y. Maeda, Local isometric embedding problem of Riemannian 3-manifold into R^6. Proc. Jpn. Acad. Ser. A Math. Sci. **62**, 257–259 (1986)
14. G. Nakamura, Y. Maeda, Local smooth isometric embeddings of low-dimensional Riemannian manifolds into Euclidean spaces. Trans. Am. Math. Soc. **313**, 1–51 (1989)
15. T.E. Poole, The local isometric embedding problem for 3-dimensional Riemannian manifolds with cleanly vanishing curvature. Commun. Partial Differ. Equ. **35**, 1802–1826 (2010)

Existence and Stability of Global Solutions of Shock Diffraction by Wedges for Potential Flow

Gui-Qiang G. Chen and Wei Xiang

Abstract We present our recent results on the mathematical analysis of shock diffraction by two-dimensional convex cornered wedges in compressible fluid flow governed by the potential flow equation. The shock diffraction problem can be formulated as an initial-boundary value problem, which is invariant under self-similar scaling. Then, by employing its self-similar invariance, the problem is reduced to a boundary value problem for a first-order nonlinear system of partial differential equations of mixed elliptic-hyperbolic type in an unbounded domain. It is further reformulated as a free boundary problem for a nonlinear degenerate elliptic system of first-order in a bounded domain with a boundary corner whose angle is bigger than π. A first global theory of existence and regularity has been established for this shock diffraction problem for the potential flow equation.

Keywords Compressible flow • Potential flow equation • Shock diffraction • Mixed elliptic-hyperbolic type • Free boundary

2010 Mathematics Subject Classification Primary: 35M10, 35M12, 35B65, 35L65, 35L70, 35J70, 76H05, 35L67, 35R35; Secondary: 35L15, 35L20, 35J67, 76N10, 76L05

G.-Q.G. Chen (✉)
Mathematical Institute, University of Oxford, Oxford, OX1 3LB, UK
e-mail: chengq@maths.ox.ac.uk

W. Xiang
School of Mathematical Sciences, Fudan University, Shanghai 200433, China

Mathematical Institute, University of Oxford, Oxford, OX1 3LB, UK
e-mail: 071018004@fudan.edu.cn; xiang@maths.ox.ac.uk

G.-Q.G. Chen et al. (eds.), *Hyperbolic Conservation Laws and Related Analysis with Applications*, Springer Proceedings in Mathematics & Statistics 49,
DOI 10.1007/978-3-642-39007-4_6,

1 Introduction

We are concerned with shock diffraction by a two-dimensional convex cornered wedge, which is not only a longstanding open problem in fluid mechanics, but also fundamental in the mathematical theory of multidimensional conservation laws. When a vertical shock propagates to the right along the convex cornered wedge, the incident shock interacts with the wedge, and the shock diffraction occurs. The study of the shock diffraction problem dates back to the 1950s, in the work of Bargman [3], Lighthill [18, 19], Fletcher-Weimer-Bleakney [12], and Fletcher-Taub-Bleakney [13] via asymptotic or experimental analysis. See also Courant-Friedrichs [10] and Whitham [20].

One of the main challenges of this problem is that the expected elliptic domain of the solution is concave, since the angle exterior to the wedge at the origin is bigger than π, besides the other mathematical difficulties including free boundary problems without unform oblique derivative conditions and optimal regularity estimates along the degenerate elliptic curves that meets the free boundary. In general, the expected regularity of solutions at the corner in this domain, even for Laplace's equation, is only C^α with $\alpha \in (0, 1)$; however, the coefficients in (6) depend on the derivatives of ψ, due to the Bernoulli law (7). To overcome the difficulty, the physical boundary conditions must be exploited to force a finer regularity of solutions at the corner.

To date, all efforts to mathematically analyze the shock diffraction problem have focused on simplified models. For one of these models, the nonlinear wave system, Kim [15] studied this problem for the right-angle wedge with an additional physical assumption that the transonic shock will not collide with the sonic circle of the right-state. Recently, in Chen-Deng-Xiang [9], this assumption was removed, and the existence and optimal regularity of shock diffraction configurations were established for *all* angles of the convex wedge via a different approach, which has been further developed in Chen-Xiang [8] to deal with the problem for the potential flow equation.

The purpose of this paper is to present the recent results we have obtained in [8] on the mathematical analysis of this shock diffraction problem for the potential flow equation, which can be formulated as an initial-boundary value problem. By employing its self-similar invariance, this initial-boundary value problem is reduced to a boundary value problem for a first-order nonlinear system of partial differential equations of mixed elliptic-hyperbolic type in an unbounded domain. It is further reformulated as a free boundary problem for nonlinear degenerate elliptic systems of first-order in a bounded domain with a boundary corner whose angle is bigger than π. A first global theory of existence and regularity has been established for this shock diffraction problem for the potential flow equation. To achieve this, we develop several mathematical ideas and techniques, which are also useful for other related problems involving similar analytical difficulties.

The organization of this paper is as follows. In Sect. 2, we first formulate the shock diffraction problem as an initial-boundary value problem for the potential flow equation, and then reduce it into the boundary value problem (Problem 1) for a first-order nonlinear system of partial differential equations of mixed elliptic-hyperbolic type, and finally present the main theorem (Theorem 1). In Sect. 3, we first introduce some notions of admissible solutions and weighted Hölder norms and then present some *a priori* estimates of admissible solutions in the Hölder norms. In Sect. 4, based on the *a priori* estimates in Sect. 2, we then prove the existence of the shock diffraction configuration by a topological argument and establish Theorem 1.

Finally, we remark in passing that a closely related problem, shock reflection-diffraction by a concave cornered wedges for potential flow, has been analyzed in Chen-Feldman [6, 7] and Bae-Chen-Feldman [1], where the existence of regular shock reflection-diffraction configurations has been established up to the detached wedge-angle. The Prandtl-Meyer reflection for supersonic potential flow impinging onto a solid wedge has also been analyzed first in Elling-Liu [11] and recently in Bae-Chen-Feldman [2]. For other related references, we refer the reader to Canic-Keyfitz-Kim [4, 5] for the unsteady transonic small disturbance equation and the nonlinear wave system and Zheng [21] for the pressure-gradient system, and the references cited therein.

2 The Potential Flow Equation and the Shock Diffraction Problem

In this section, we first formulate the shock diffraction problem as an initial-boundary value problem for the potential flow equation, then reduce it to the boundary value problem (Problem 1) for a first-order nonlinear system of partial differential equations of mixed elliptic-hyperbolic type, and finally present the main theorem (Theorem 1).

2.1 The Potential Flow Equation and the Rankine-Hugoniot Conditions

The Euler equations for potential flow consist of the conservation law of mass and the Bernoulli law for the density ρ and velocity potential Φ with the velocity $(u, v) = \nabla_{\mathbf{x}}\Phi$:

$$\partial_t \rho + \nabla_{\mathbf{x}} \cdot (\rho \nabla \Phi) = 0, \tag{1}$$

$$\partial_t \Phi + \frac{1}{2}|\nabla_{\mathbf{x}}\Phi|^2 + i(\rho) = B_0, \tag{2}$$

where $i(\rho) = \frac{\rho^{\gamma-1}-1}{\gamma-1}$ for $\gamma > 1$ and $i(\rho) = ln\rho$ for $\gamma = 1$, and B_0 is the Bernoulli constant determined by the incoming flow and/or boundary conditions.

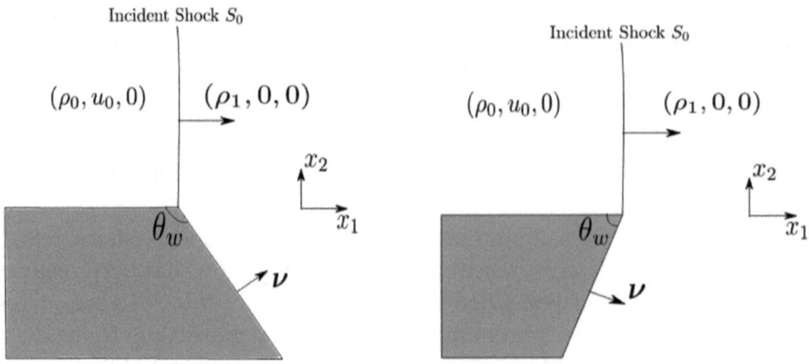

Fig. 1 Initial-boundary value problem

The shock diffraction can be formulated as an initial-boundary value problem: We seek a solution of system (1) and (2) with the initial condition at $t = 0$:

$$(\rho, \Phi)|_{t=0} = \begin{cases} (\rho_0, u_0 x_1) & \text{in } \{x_1 < 0, x_2 > 0\}, \\ (\rho_1, 0) & \text{in } \{-\pi + \theta_w \leq \arctan\left(\frac{x_2}{x_1}\right) \leq \frac{\pi}{2}\}, \end{cases} \tag{3}$$

and the slip boundary condition along the wedge boundary $\partial \mathscr{W}$:

$$\nabla_{\mathbf{x}} \Phi \cdot \boldsymbol{\nu}|_{\partial \mathscr{W} \times \mathbb{R}_+} = 0, \tag{4}$$

where $\boldsymbol{\nu}$ is the exterior unit normal to $\partial \mathscr{W}$ (see Fig. 1).

Notice that the initial-boundary value problem is invariant under the self-similar scaling:

$$(\mathbf{x}, t) \to (\alpha \mathbf{x}, \alpha t), \quad (\rho, \Phi) \to (\rho, \frac{\Phi}{\alpha}) \qquad \text{for } \alpha \neq 0.$$

Thus we seek self-similar solutions with the form:

$$(\rho, \Phi)(\mathbf{x}, t) = (\rho(\xi, \eta), \, t(\psi(\xi, \eta) + \frac{1}{2}(\xi^2 + \eta^2))) \qquad \text{for } (\xi, \eta) = \frac{\mathbf{x}}{t}, \tag{5}$$

where ψ is the pseudo-velocity potential, that is,

$$(\psi_\xi, \psi_\eta) = (u - \xi, v - \eta) =: (U, V),$$

which is called a pseudo-velocity. Then the pseudo–potential function ψ is governed by the following Euler equations:

$$\mathrm{div}(\rho D\psi) + 2\rho = 0, \tag{6}$$

$$\frac{1}{2}|D\psi|^2 + \psi + \frac{\rho^{\gamma-1}}{\gamma-1} = 0, \tag{7}$$

where the divergence div and gradient D are with respect to the self-similar variables (ξ, η). Here we have replaced ψ by $\psi - \frac{\rho_1^{\gamma-1}}{\gamma-1}$ to make the right-hand side of (7) zero.

Then (6) and (7) can be reduced to the following potential flow equation of second-order for the potential function ψ:

$$\operatorname{div}\big(\rho(|D\psi|^2, \psi)D\psi\big) + 2\rho(|D\psi|^2, \psi) = 0, \tag{8}$$

with

$$\rho(|D\psi|^2, \psi) = \Big(-(\gamma-1)(\psi + \frac{1}{2}|D\psi|^2)\Big)^{\frac{1}{\gamma-1}}. \tag{9}$$

Equation (8) is a second-order equation of mixed hyperbolic-elliptic type: It is elliptic if and only if $|D\psi| < c(|D\psi|^2, \psi) := \sqrt{-(\gamma-1)(\psi + \frac{1}{2}|D\psi|^2)}$, which is equivalent to

$$|D\psi| < c_\star(\psi, \gamma) := \sqrt{-\frac{2(\gamma-1)}{\gamma+1}\psi}. \tag{10}$$

Since one of the corners on the boundary of the pseudo-elliptic domain, *i.e.* the origin, is bigger than π, it is not clear in general whether we could obtain the C^1-regularity of ψ to ensure the ellipticity of (8) and (9) near the point, in comparison with [6]. In fact, there is a counterexample even for Laplace's equation for the general case so that the solution is only in $C^\alpha, \alpha \in (0, 1)$, at the corner. One of the key new ingredients here is to exploit the physical boundary conditions to ensure an additional regularity for the ellipticity. To achieve this, instead of studying (8) and (9) for ψ directly as in [6], we consider the corresponding system for $(\rho, U, V) = (\rho, u - \xi, v - \eta)$ to obtain the C^α-estimates of the solutions by exploiting the boundary conditions:

$$\begin{cases} \big(\rho(U, V, \psi)U\big)_\xi + \big(\rho(U, V, \psi)V\big)_\eta + 2\rho(U, V, \psi) = 0, \\ U_\eta = V_\xi, \\ \frac{\rho^{\gamma-1}(U,V,\psi)}{\gamma-1} + \frac{U^2+V^2}{2} = -\psi, \\ (\psi_\xi, \psi_\eta) = (U, V). \end{cases} \tag{11}$$

Since our global solutions involve shock waves in the problem, the solutions of (11) have to be considered as weak solutions in the distributional sense.

Definition 1. The vector function (U, V) is called a weak solution of (11) if there exists a function $\psi \in W^{1,1}_{loc}(\Omega)$ in a self-similar domain Ω such that:

(i) $\psi_\xi = U, \psi_\eta = V$ *a.e.* in Ω;
(ii) $-\psi - \frac{1}{2}(U^2 + V^2) \geq 0$ *a.e.* in Ω;
(iii) $(\rho(U, V, \psi), \rho(U, V, \psi)U, \rho(U, V, \psi)V) \in (L^1_{loc}(\Omega))^3$;
(iv) For every $\zeta \in C_c^\infty(\Omega)$,

$$\int_\Omega \rho(U, V, \psi)\big((U, V) \cdot D\zeta - 2\zeta\big)\, d\xi d\eta = 0,$$

and

$$\int_\Omega (V, -U) \cdot D\zeta\, d\xi d\eta = 0.$$

A piecewise smooth solution separated by a shock wave satisfies the conditions in Definition 1 if and only if it is a classical solution of (11) in each smooth subregion and satisfies the following Rankine-Hugoniot conditions across the shock wave:

$$\big[(\rho(U, V, \psi)U, \rho(U, V, \psi)V) \cdot \boldsymbol{\nu}\big]_S = 0, \tag{12}$$

$$[\psi]_S = 0, \tag{13}$$

where the bracket $[w]$ denotes the jump of the quantity w across the shock wave S, that is,

$$[w] = \lim_{\substack{(\xi,\eta)\to S+ \\ (\xi,\eta)\cdot\nu > 0}} w(\xi, \eta) - \lim_{\substack{(\xi,\eta)\to S- \\ (\xi,\eta)\cdot\nu < 0}} w(\xi, \eta).$$

Condition (12) follows from the conservation of mass, while (13) follows from irrotationality.

2.2 The Shock Diffraction Problem

If the initial left-state $(\rho_0, u_0, 0)$ is subsonic, *i.e.* $u_0 < c_0 := c(\rho_0)$, the degenerate boundary is the sonic circle centered at $(u_0, 0)$ with radius c_0, and the center rarefaction wave does not occur near the origin. In this paper, our focus is on this case where we consider system (11) in the pseudo-subsonic region.

A discontinuity of $D\psi$ satisfying the Rankine-Hugoniot conditions (12) and (13) is called a shock if it satisfies the following physical entropy condition: *The density ρ increases across a shock in the pseudo-flow direction.* From (12), the entropy condition indicates that the normal derivative function ψ_ν on a shock always decreases across the shock in the pseudo-flow direction, which implies that $\rho_0 > \rho_1$ and $u_0 > u_1 = 0$.

On the other hand, (13) implies

$$[v - \eta]d\eta = -[u - \xi]d\xi. \tag{14}$$

Then, as a direct corollary of (14), the Rankine-Hugoniot conditions are equivalent to:

$$u\big(\rho(u-\xi)+\rho_1\xi\big)+v\big(\rho(v-\eta)+\rho_1\eta\big)=0, \tag{15}$$

and

$$\psi=\psi_1 \tag{16}$$

along the incident shock. Let $\xi = \xi_1$ be the location of the incident shock. By a straightforward calculation, the incident shock position is

$$\xi_1=\sqrt{\frac{2\rho_0^2(c_0^2-c_1^2)}{(\gamma-1)(\rho_0^2-\rho_1^2)}}>0 \tag{17}$$

with the property:

$$0<u_0=\frac{\rho_0-\rho_1}{\rho_0}\xi_1<\xi_1. \tag{18}$$

Furthermore, we can show that the incident shock hits the sonic circle of the left-state, *i.e.* state (0), by the following relation:

$$0<\xi_1-u_0<c_0. \tag{19}$$

In the self-similar plane, the domain outside the wedge is

$$\Lambda:=\{(\xi,\eta)\;:\;-\pi+\theta_w\le\arctan\big(\frac{\eta}{\xi}\big)\le\pi\}.$$

Then the shock diffraction problem can be formulated as the following boundary value problem in the self-similar coordinates (ξ,η).

Problem 1 (Boundary Value Problem). (See Fig. 2). We seek a solution (U,V) of Eqs. (11) in the self-similar domain Λ with the slip boundary condition on the wedge boundary $\partial\Lambda$:

$$(U,V)\cdot\boldsymbol{\nu}|_{\partial\Lambda}=0$$

and the asymptotic boundary condition at infinity:

$$(\rho,U,V)\to(\bar\rho,\bar U,\bar V)\qquad\text{when } \xi^2+\eta^2\to\infty,$$

in the sense that

$$\lim_{R\to\infty}\|(\rho,U,V)-(\bar\rho,\bar U,\bar V)\|_{C(\Lambda\setminus B_R(0))}=0,$$

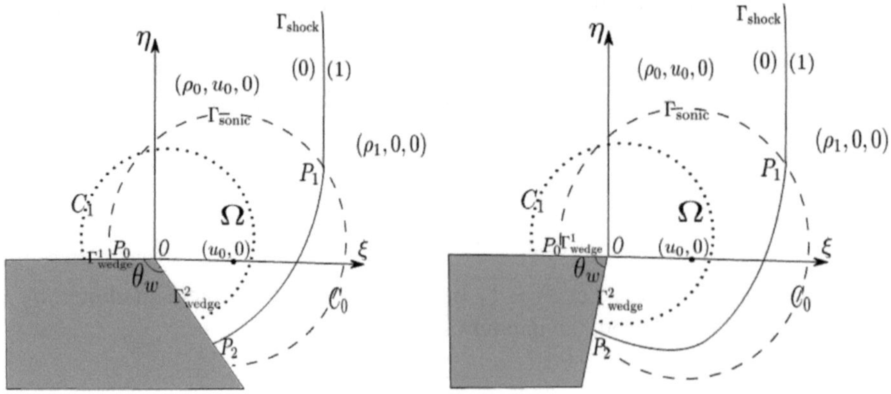

Fig. 2 Boundary value problem

where

$$(\bar{\rho}, \bar{U}, \bar{V}) = \begin{cases} (\rho_0, u_0 - \xi, -\eta) & \text{for } \{\xi < \xi_1, \eta > 0\}, \\ (\rho_1, -\xi, -\eta) & \text{for } \{\xi > \xi_1, \eta > 0\} \cup \{-\pi + \theta_w \leq \arctan\left(\frac{\eta}{\xi}\right) \leq 0\}. \end{cases}$$

Since $(u_0 - \xi, -\eta)$ does not satisfy the slip boundary condition for $\xi \geq 0$, the solution must differ from state (0) in $\{\xi < \xi_1\} \cap \Lambda$ near the wedge corner, which forces the shock to be diffracted by the wedge. In the domain Ω bounded by the pseudo-sonic circle of the left-state, *i.e.* state (0) with center $(u_0, 0)$ and radius $c_0 > 0$, and the shock wave, the solution is expected to be pseudo-subsonic and smooth, to satisfy the slip boundary condition along the wedge, and to be at least continuous across the pseudo-sonic circle to be pseudo-supersonic. The main theorem we have established is

Theorem 1 (Main Theorem). *Let θ_c be the critical angle of the given data such that the corresponding wedge boundary Γ^2_{wedge} passes the intersection point of the two sonic circles of the given Riemann data. Then there exists an $\alpha = \alpha(\rho_0, \rho_1, u_0, \gamma) \in (0, \frac{1}{2})$ such that, when $\theta_w \in (\theta_c, \pi)$, there exists a pair of global self-similar solutions:*

$$\rho(\mathbf{x}, t) = \Big(- (\gamma - 1)(\partial_t \Phi(\mathbf{x}, t) + \frac{1}{2}|\nabla_{\mathbf{x}} \Phi(\mathbf{x}, t)|^2) \Big)^{\frac{1}{\gamma - 1}},$$

$$(u, v)(\mathbf{x}, t) = \nabla_{\mathbf{x}} \Phi(\mathbf{x}, t) \qquad \text{for } \frac{\mathbf{x}}{t} \in \Lambda, \ t > 0$$

for the shock diffraction by the wedge, with $\psi(\mathbf{x}, t)$ defined by (5) which satisfies

$$\Phi(\mathbf{x}, t) = t\psi\left(\frac{\mathbf{x}}{t}\right) + \frac{|\mathbf{x}|^2}{t} \qquad \text{for } \frac{\mathbf{x}}{t} \in \Lambda, \ t > 0.$$

Equivalently, (U, V) *with the pseudo-potential velocity* ψ *solving Problem 1 satisfies that, for* $(\xi, \eta) = \frac{\mathbf{x}}{t}$,

$$(U, V) \in C^\infty(\Omega) \cap C^\alpha(\overline{\Omega}),$$

$$(\rho, U, V) = \begin{cases} (\rho_0, u_0 - \xi, -\eta) & \text{for } \xi < \xi_1 \text{ and above the sonic circle } \widehat{P_1P_2}, \\ (\rho_1, -\xi, -\eta) & \text{on the right of or below the diffracted shock.} \end{cases} \tag{20}$$

In addition, the corresponding pseudo-potential velocity ψ *is* $C^{1,1}$ *across the part* $\widehat{P_0P_1}$ *including the endpoints* P_0 *and* P_1*, the* $C^{1,1}$*-regularity is optimal, and the limit of* $D^2\psi$ *at* P_1 *does not exist. The transonic shock* $\widehat{P_1P_2}$ *is* C^2 *at* P_1 *and* C^∞ *except at* P_1*. Furthermore, the solution pair* (U, V) *is stable with respect to the wedge-angle* θ_w*, i.e.* ψ*, as well as* (U, V)*, converges to the unique incident plane shock solution at* $\xi = \xi_1$ *as* $\theta_w \to \pi$*.*

We remark here that, when the wedge-angle $\theta_w \le \frac{\pi}{2}$, it needs a transformation and other technical arguments in order to prove the existence of the solutions. To illustrate the key ideas more directly, we will restrict our sketch of the proof to the case when $\theta_w > \frac{\pi}{2}$, for which such a transformation is not needed.

3 Admissible Solutions and *A Priori* Estimates

In this section, we introduce some notions of admissible solutions and weighted Hölder norms, and present some *a priori* estimates of admissible solutions in the Hölder norms.

3.1 Weighted Hölder Spaces and Norms

Let $\mathscr{P}$ denote the corner points of $\partial\Omega$, and let $B_\delta(\mathscr{P})$ be the union of the balls of radius δ centered at the corner points in $\mathscr{P}$. We then define the following weighted Hölder norms and Hölder spaces:

$$\begin{aligned}
[u]_{k,\Omega}^{(-\sigma),\mathscr{P}} &= [u]_{k,0,\Omega}^{(-\sigma),\mathscr{P}} = \sup_{\delta>0} \sup_{\substack{x \in \Omega\setminus B_\delta(\mathscr{P}) \\ |\beta| = k}} \left(\delta^{k-\sigma}|D^\beta u(x)|\right); \\
[u]_{k+\alpha,\Omega}^{(-\sigma),\mathscr{P}} &= \sup_{\delta>0} \sup_{\substack{x,y \in \Omega\setminus B_\delta(\mathscr{P}) \\ |\beta| = k}} \left(\delta^{k+\alpha-\sigma}\frac{|D^\beta u(x) - D^\beta u(y)|}{|x-y|^\alpha}\right); \\
\|u\|_{k,\Omega}^{(-\sigma),\mathscr{P}} &= \textstyle\sum_{j=0}^k [u]_{j,\Omega}^{(-\sigma),\mathscr{P}}; \\
\|u\|_{k+\alpha,\Omega}^{(-\sigma),\mathscr{P}} &= \|u\|_{k,\Omega}^{(-\sigma),\mathscr{P}} + [u]_{k+\alpha,\Omega}^{(-\sigma),\mathscr{P}}; \\
C_{k+\alpha,\Omega}^{(-\sigma),\mathscr{P}} &:= \{u : u \in C^{k,\alpha}(\Omega) \text{ and } \|u\|_{k+\alpha,\Omega}^{(-\sigma),\mathscr{P}} < \infty\},
\end{aligned} \tag{21}$$

where k is an integer and $0 < \alpha < 1$. We remark that the weight near the wedge corner O will be dealt with separately from the others since the angle is bigger than π here. It is easy to verify that

$$\|fg\|_{0+\alpha,\Omega}^{(\tau_1+\tau_2),\mathcal{P}} \leq \|f\|_{0+\alpha,\Omega}^{(\tau_1),\mathcal{P}} \|g\|_{0+\alpha,\Omega}^{(\tau_2),\mathcal{P}} \qquad \text{for } \tau_1+\tau_2 \geq 0. \tag{22}$$

As in [16], there are two important properties of these norms:

(A) $\|u\|_{\alpha,\Omega}^{(-\sigma),\mathcal{P}} \leq C\|u\|_{\sigma,\Omega}^{(-\sigma),\mathcal{P}} = C\|u\|_{\sigma,\Omega}$ for $0 < \alpha \leq \sigma$, where $\|u\|_{\sigma,\Omega}$ is the non-weighted Hölder norm for u.

(B) If $a \geq b > 0$ and if $\{u_m\}$ is a bounded sequence in $C_a^{(-b),\mathcal{P}}$, then there is a subsequence $\{u_{m_j}\}$ which converges in any $C_{a'}^{(-b'),\mathcal{P}}$, with $0 < b' < b$, $0 < a' < a$, and $a' \geq b'$.

Before introducing the parabolic norm near the sonic circle, first we define Ω' and Ω'' for any domain Ω as

$$\begin{aligned} \Omega' &:= \Omega \cap \{(\xi,\eta) \,:\, \text{dist}\{(\xi,\eta),\Gamma_{\text{sonic}}\} < 2\epsilon_0\}, \\ \Omega'' &:= \Omega \cap \{(\xi,\eta) \,:\, \text{dist}\{(\xi,\eta),\Gamma_{\text{sonic}}\} > \epsilon_0\} \end{aligned} \tag{23}$$

with a small constant $\epsilon_0 > 0$. Obviously, $\Omega = \Omega' \cup \Omega''$, and it will be seen later that the equation studied is uniformly elliptic in Ω'' and elliptic in Ω', in fact degenerate on $\Gamma_{sonic} := \Omega \cap \{(\xi,\eta) \,:\, \sqrt{(\xi-u_0)^2+\eta^2} = c_0\}$.

In Ω', the equation has elliptic degeneracy, for which the Hölder norms with parabolic scaling are natural. We define the norm $||\psi||_{2,\alpha,\Omega'}^{\text{par}}$ as follows: First introduce new coordinates (x,y) in Ω' as

$$x = c_0 - \sqrt{(\xi-u_0)^2+\eta^2}, \qquad y = \arctan\Big(\frac{\eta}{\xi-u_0}\Big).$$

With $z = (x,y)$ and $\bar{z} = (\bar{x},\bar{y})$ where $x,\bar{x} \in (0,2\epsilon_0)$ and

$$\delta_\alpha^{\text{par}}(z,\bar{z}) := \big(|x-\bar{x}|^2 + \min\{x,\bar{x}\}|y-\bar{y}|^2\big)^{\frac{\alpha}{2}},$$

and for $\psi \in C^2(\Omega')$ in the (x,y)-coordinates, we define

$$\begin{aligned} \|\psi\|_{2,0,\Omega'}^{\text{par}} &:= \textstyle\sum_{0\leq m+l\leq 2} \sup_{z\in\Omega'} \big(x^{m+\frac{l}{2}-2}|\partial_x^m\partial_y^l u(z)|\big), \\ [\psi]_{2,\alpha,\Omega'}^{\text{par}} &:= \textstyle\sum_{m+l=2} \sup_{z,\bar{z}\in\Omega', z\neq\bar{z}} \Big(\big(\min\{x,\bar{x}\}\big)^{\alpha-\frac{l}{2}} \frac{|\partial_x^m\partial_y^l u(z)-\partial_x^m\partial_y^l u(\bar{z})|}{\delta_\alpha^{\text{par}}(z,\bar{z})}\Big), \\ \|\psi\|_{2,\alpha,\Omega'}^{\text{par}} &:= \|u\|_{2,0,\Omega'}^{\text{par}} + [u]_{2,\alpha,\Omega'}^{\text{par}}. \end{aligned} \tag{24}$$

We refer to [6] for more details and for the motivation of this definition.

3.2 Notion of Admissible Solutions

The proof of Theorem 1 is based on the local existence and the uniform *a priori* estimates of admissible solutions. More precisely, we define the set

$$I \subset [0, \pi] \tag{25}$$

so that, for any $\theta_w \in I$, there exists an admissible solution $(\rho^{(\theta_w)}, U^{(\theta_w)}, V^{(\theta_w)})$ for the shock diffraction problem. Here, the admissible solutions are defined as follows:

Definition 2. Let $\gamma > 1$, $\rho_0 > \rho_1 > 0$, and $u_0 < c_0$, and let (ρ_0, ρ_1, u_0) satisfy (17) and (18). For any wedge-angle $\theta_w \in (\theta_c, \pi)$ and function $W = (U, V) \in (C^\alpha(\Lambda))^2$, $\theta_w \in I$ if and only if:

(i) The function W is a weak solution to the shock diffraction problem, *i.e.* W satisfies Definition 1 and the Rankine-Hugoniot conditions (12) and (13).
(ii) The free boundary Γ_{shock}, with endpoints $P_1 = (\xi_1, \eta_1)$ and $P_2 = (\xi_2, \eta_2)$, lies between the two sonic circles of state (0) and state (1), *i.e.*, $(\rho_0, u_0 - \xi, -\eta)$ and $(\rho_1, -\xi, -\eta)$ respectively, and meets the wedge at P_2 perpendicularly. In addition, Γ_{shock} is C^∞ everywhere except at the point P_1.
(iii) (U, V) satisfies (20) outside of Ω, and

$$(U, V) \in \left(C^\alpha(\overline{\Omega}) \cap C^1(\overline{\Omega}\backslash\overline{OP_0P_1}) \cap C^\infty(\overline{\Omega}\backslash\overline{\Gamma_{\text{sonic}} \cup \{O\}})\right)^2,$$

where $\alpha \in (0, 1)$ depends only on θ_w and the given data.
(iv) Equation (8) is strictly elliptic in $\overline{\Omega\backslash\Gamma_{\text{sonic}}}$, that is,

$$|\nabla\psi| < c_\star(\psi, \gamma) := \sqrt{-\frac{2(\gamma - 1)}{\gamma + 1}\psi}.$$

(v) $u = U + \xi > 0$ and $v = V + \eta < 0$ in Ω.

In fact, admissible solutions have the following additional properties. Some of these properties require technical proofs, which can be found in Chen-Xiang [8].

Remark 1 (Extension of the background solutions to a smaller wedge-angle). The property that Γ_{shock} meets the wedge at P_2 perpendicularly in (ii) of Definition 2 and the slip boundary condition yield that, for any $\theta_w \in I$ and any $\Theta_w < \theta_w$, there are functions $\tilde{W} = (\tilde{U}, \tilde{V})$ such that they satisfy Eqs. (11) in $\Omega^{(\Theta_w)}$ and $\tilde{W} = W$ in $\Omega^{(\theta_w)}$, where $\Omega^{(\theta_w)}$ is the domain corresponding to the wedge-angle θ_w. We call $\tilde{W}$ the extension of the admissible solution W, which will be used as a background solution in our proof of Theorem 1.

Remark 2 (Existence of the shock up to the wedge). The property that $v < 0$ in (v) of Definition 2 and the fact that $v = 0$ on the right-hand side mean that Γ_{shock} exists up to the wedge boundary due to the jump of the velocity v.

Remark 3 (Positivity of the horizontal speed u along Γ_{shock}). Property (v) of Definition 2 imply that, along Γ_{shock}, the horizontal velocity u is positive.

Remark 4 (Uniform estimates of the size of the domain Ω). The property that the shock lies between two sonic circles in (ii) and the fact that Γ_{shock} exists up to the wedge boundary mean that the size of the domain Ω is bounded.

Remark 5 (The entropy condition). Properties (i), (iv), and (v) of Definition 2 imply that

$$\partial_{\boldsymbol{\nu}}\varphi_1 > \partial_{\boldsymbol{\nu}}\varphi > 0 \qquad \text{on } \Gamma_{\text{shock}},$$

where $\boldsymbol{\nu}$ is the unit normal to Γ_{shock} interior to Ω.

Remark 6 (The shape of Γ_{shock}). Properties (i) and (v) imply that, if $\Gamma_{\text{shock}} = \{(\xi,\eta) : \xi = \xi(\eta)\} = \{(r,\theta) : r = r(\theta)\}$, then

$$\xi'(\eta) \geq 0, \quad r'(\theta) \geq 0.$$

Remark 7 (I is non-empty). Based on the proof of the existence of the solutions to the wedge-angle near π, we have $\theta_w \in I$ when $\pi - \theta_w$ is small. Thus, $I \neq \emptyset$. Then Theorem 1 is established if we can prove that the subset I is both open and closed.

3.3 A Second-Order Equation for v and the Boundary Conditions

It is important to deduce first an equation for v from the potential flow equation for our study. To do so, we first introduce an elliptic cut-off function which will be given in detail later, take the derivative on the equation of the conservation of mass with respect to η, and then use the irrotationality to obtain a second-order equation for v in Ω as

$$\begin{aligned} Q(v;u) &= \bar{a}_{11}v_{\xi\xi} + 2\bar{a}_{12}v_{\xi\eta} + \bar{a}_{22}v_{\eta\eta} + b_{11}v_\xi^2 + b_{12}v_\xi v_\eta + b_{22}v_\eta^2 + c_1 v_\xi + c_2 v_\eta \\ &= 0, \end{aligned} \tag{26}$$

where

$$|b_{11}| + |b_{12}| + |b_{22}| \leq \frac{C}{a_{11}}$$

with C depending on the C^1-bounds of $\hat{\psi}$ and the cut-off functions ζ_i and ζ_M, while

$$d_O^\alpha(|c_1| + |c_2|) \leq \frac{C}{a_{11}}$$

with C depending on $||\hat{\psi}||_{2,\alpha,\Omega''}^{(-1-\alpha),\{O,P_0\}}$, where $d_O(X) = \text{dist}\{X, O\}$.

Modify the Rankine-Hugoniot condition $F(u, v, \varphi, \eta) = 0$ to be

$$G := \zeta_s F + (1 - \zeta_s)(L_1(u - \hat{u}) + L_2(v - \hat{v})),$$

where ζ_s is a special cut-off function such that $(\zeta_s)_u(u - \hat{u}) + (\zeta_s)_v(v - \hat{v})$ is a small term, and L_2 is chosen to be close to $F_v(\hat{u}, \hat{v}, \hat{\varphi}, \hat{\eta})$ and L_1 is appropriately determined by $F_v(\hat{u}, \hat{v}, \hat{\varphi}, \hat{\eta})$ and $\frac{F_u(\hat{u},\hat{v},\hat{\varphi},\hat{\eta})}{F_v(\hat{u},\hat{v},\hat{\varphi},\hat{\eta})}$. Then, differentiating it along the shock, we have the following boundary condition on Γ_{shock}:

$$M^{(2)}v := \beta_1^s v_\xi + \beta_2^s v_\eta = \bar{a}_{11} A_{s,1} v + g_s(u, v, \varphi) \qquad \text{on } \Gamma_{\text{shock}}. \tag{27}$$

Note that ζ_s, L_1, and L_2 are defined in this way to ensure that $\bar{a}_{11} A_{s,1} \geq 0$, and $\|g_s\|_\infty \leq C$, independent of s.

On the other hand, taking the derivative on the slip boundary condition along the boundary, we have the following boundary condition on Γ^2_{wedge}:

$$M^{(1)}v = \beta_1^{(1)} v_\xi + \beta_2^{(1)} v_\eta = 0 \qquad \text{on } \Gamma^2_{\text{wedge}}. \tag{28}$$

Moreover, v satisfies the Dirichlet boundary condition:

$$v = 0 \qquad \text{on } \Gamma_{\text{sonic}} \cup \Gamma^1_{\text{wedge}}, \tag{29}$$

and the one point boundary condition:

$$v = -g(\xi_w, \theta_w) \tan(\pi - \theta_w) \tag{30}$$

to guarantee the equivalence of the deduced equations and the original equations. The one point boundary condition is obtained from the slip boundary condition and the Rankine-Hugoniot condition.

3.4 Uniform Estimates of the Obliqueness Along Γ_{shock}

The crucial result guaranteeing the obliqueness of the operator $M^{(2)}$ is that, if $(\hat{u}, \hat{v}, \hat{\varphi})$ is the solution in the sense of Definition 2, then

$$F_u(\hat{u}, \hat{v}, \hat{\varphi}, \eta) > 0 \qquad \text{along } \Gamma_{\text{shock}}.$$

With this result, after careful calculation, we can prove that the operators $M^{(i)}$ are oblique along Γ^2_{wedge} and Γ_{shock} respectively. Here the fact that $\hat{u} > 0$ and $\hat{v} < 0$ along Γ_{shock} plays a fundamental role. At the same time, $-\bar{a}_{11} A_{s,1} \leq 0$ is important for the maximum principle.

3.5 *Uniform Estimates of the Approximate Solutions Near the Origin*

Consider the approximate solutions v^ϵ governed by

$$Q(v^\epsilon; u^\epsilon) + \epsilon \Delta v^\epsilon = 0,$$

and the boundary conditions (27) and (30), where Q is defined in (26). We prove that there exist $\sigma^* > 0$ and $\alpha_0 > 0$ such that, for each $\sigma \le \sigma^*$ and $\alpha \le \alpha_0$, and for any approximate solution v^ϵ, near the wedge corner O, we have

$$\|v^\epsilon\|^{(-\sigma)}_{2+\alpha,\Omega} \le C(\lambda, \theta_w, \Lambda, \Omega)\big(\|g(\xi_w, \theta_w)\| + \|g_s\|_\infty\big). \tag{31}$$

Furthermore, if the solution $(u^\epsilon, v^\epsilon, \varphi^\epsilon)$ is sufficiently close to the background solution $(\hat{u}, \hat{v}, \hat{\varphi})$, then the boundary condition on Γ_{shock} will not be involved in the inhomogeneous term:

$$M^{(2)} v^\epsilon = \beta_1^{(2)} v^\epsilon_\xi + \beta_2^{(2)} v^\epsilon_\eta = 0;$$

thus the solution v^ϵ has a better estimate:

$$-\, g(\xi_w, \theta_w) \tan(\pi - \theta_w) \le v^\epsilon \le 0. \tag{32}$$

3.6 *Impossibility of Γ_{shock} Meeting the Sonic Circle of State* (1) *and the Sonic Circle of State* (0) *Except at P_1*

We prove that $r'(\theta) \ge 0$ along Γ_{shock}, which means that Γ_{shock} will not meet the sonic circle of state (0) again away from P_1. Next, we prove that there exists a constant $C > 0$ such that

$$\operatorname{dist}\{\Gamma_{\text{shock}}, B_{c_1}(O)\} > \frac{1}{C},$$

for any solution in the sense of Definition 2, where c_1 is the sonic speed of state (1), *i.e.* the right-state. These estimates are crucial to guarantee the ellipticity in the domain Ω.

3.7 Uniform Hölder Estimates of (u^ϵ, v^ϵ) Near Γ_{sonic}, and Uniform Upper and Lower Estimates of Density ρ^ϵ

In order to pass to the limit as $\epsilon \to 0$, we need uniform estimates of the approximate solutions near $\Gamma_{\rm sonic}$, where the ellipticity may degenerate. In fact, we prove the uniform estimates near $\Gamma_{\rm sonic}$ by scaling,

$$|v^\epsilon| \leq A(c_0 - r)^{1/4}, \tag{33}$$

$$|u^\epsilon - u_0| + |\rho^\epsilon - \rho_0| \leq A(c_0 - r)^{\frac{1}{6}} \qquad \text{for } 0 \leq c_0 - r \leq m. \tag{34}$$

As in Sect. 3.5, if the solution $(u^\epsilon, v^\epsilon, \varphi^\epsilon)$ is sufficiently close to the background solution $(\hat{u}, \hat{v}, \hat{\varphi})$, we have

$$-A(c_0 - r)^{\frac{1}{4}} \leq v^\epsilon \leq 0 \qquad \text{for } 0 \leq c_0 - r \leq m. \tag{35}$$

From the uniform estimates away from $\Gamma_{\rm sonic}$, $\|u^\epsilon\|_0$ and $\|v^\epsilon\|_0$ are uniformly bounded, hence $\|\varphi^\epsilon\|_{C^{0,1}}$ is also uniformly bounded, and

$$\left(\frac{2}{\gamma+1}\right)^{\frac{1}{\gamma-1}} \rho_1 \leq \rho^\epsilon \leq C \qquad \text{in } \Omega.$$

3.8 Monotonicity of the Solution v Along Γ_{shock}

From now on, we consider the solutions without the viscosity term $\epsilon\Delta v$, *i.e.* after passing to the limit as $\epsilon \to 0$. What we can actually prove for the monotonicity of v is that, if the solution (u, v, φ) is sufficiently close to the background solution $(\hat{u}, \hat{v}, \hat{\varphi})$, then the solution v is monotonically increasing along $\Gamma_{\rm shock}$.

3.9 Uniform Estimates of the Ellipticity in Ω Up to Γ_{shock}

Note that the Mach number

$$M^2 = \frac{(u-\xi)^2 + (v-\eta)^2}{c^2} \in C^\alpha(\overline{\Omega}) \cap C^\infty(\overline{\Omega}\backslash(\overline{\Gamma_{\rm sonic} \cup \{P_1\}})).$$

Then we can show that there exists a constant $\mu > 0$ such that, for any $\theta_w \in (\theta_c, \pi)$, we have

$$M^2(\xi, \eta) \leq 1 - \mu d \qquad \text{for all } (\xi, \eta) \in \Omega,$$

where $d = \text{dist}\{(\xi, \eta), \Gamma_{\text{sonic}}\}$. This means that, for all $\boldsymbol{\xi} = (\xi_1, \xi_2) \in \mathbb{R}^2$, we have

$$C^{-1} d |\boldsymbol{\xi}|^2 \le \sum_{i,j=1}^{2} a_{ij}\xi_i\xi_j \le C|\boldsymbol{\xi}|^2.$$

3.10 Regularity Away from and Near Γ_{sonic}

For the regularity away from Γ_{sonic}, we employ the weighted Hölder norms and a transformation to control the behaviour of the quadratic nonlinear terms to estimate the solution v near the corners and other points. Next we use the irrotationality to obtain the regularity of u and then the regularity of ρ.

For the regularity near Γ_{sonic}, we use the parabolic norms and a scaling to make the equation non-degenerate. We introduce new coordinates

$$(x, y) = (c_0 - r, \theta - \theta_1)$$

to flatten Γ_{sonic}, where (r, θ) are the polar coordinates, c_0 is the sonic speed of state (0), and (r_1, θ_1) is P_1. Then, following the procedures in [1] exactly, we can derive the following property:

Theorem 2 (Optimal regularity). *Let ψ be a solution obtained as before. Then we have*

(i) *ψ cannot be C^2 across the pseudo-sonic circle Γ_{sonic};*
(ii) *$\varphi = \psi - \psi_0$ is $C^{2+\alpha}$ in Ω up to Γ_{sonic} away from the point P_1 for any $\alpha \in (0, 1)$;*
(iii) *For any $(\xi_0, \eta_0) \in \Gamma_{sonic} \backslash \{P_1\}$,*

$$\lim_{\substack{(\xi,\eta)\to(\xi_0,\eta_0)\\(\xi,\eta)\in\Omega}} D_{rr}\varphi = \frac{1}{\gamma+1}, \qquad \lim_{\substack{(\xi,\eta)\to(\xi_0,\eta_0)\\(\xi,\eta)\in\Omega}} D_{\theta\theta}\varphi = 0, \qquad \lim_{\substack{(\xi,\eta)\to(\xi_0,\eta_0)\\(\xi,\eta)\in\Omega}} D_{r\theta}\varphi = 0;$$

(iv) *$D^2\varphi$ has a jump across Γ_{sonic}: for any $(\xi_0, \eta_0) \in \Gamma_{sonic} \backslash \{P_1\}$,*

$$\lim_{\substack{(\xi,\eta)\to(\xi_0,\eta_0)\\(\xi,\eta)\in\Omega}} D_{rr}\varphi - \lim_{\substack{(\xi,\eta)\to(\xi_0,\eta_0)\\(\xi,\eta)\in\Lambda\backslash\Omega}} D_{rr}\varphi = \frac{1}{\gamma+1};$$

(v) *The limit $\lim_{\substack{(\xi,\eta)\to P_1\\(\xi,\eta)\in\Omega}} D^2\varphi$ does not exist.*

4 Existence of the Shock Diffraction Configuration

Once the *a priori* estimates are proved, the existence of the shock diffraction configuration can be established by a topological argument. Thanks to the uniform estimates in Sect. 3, the set I is obviously closed. Then the remaining task is to prove that the set I is open.

The main idea of the existence proof is that, instead of studying the potential flow equation of φ, we study a system for (ρ, u, v) directly. In order to do that, we first introduce the degenerate elliptic cut-off, the higher order cut-off near the pseudo-sonic circle, and the uniform elliptic cut-off away from the pseudo-sonic circle, and introduce the modified Rankine-Hugoniot condition along Γ_{shock}. Then differentiate them to obtain a second-order equation for v with the oblique boundary conditions on Γ_{shock} and Γ^2_{wedge}. Once the existence of v is obtained, we use the irrotational equation to recover u by v. Next, we pass to the limit to obtain a solution (u, v, φ) which is actually equivalent to the original potential flow equation of φ. Using this scalar equation, we can obtain a better regularity to remove the introduced cut-off functions and prove that the solution we have obtained is actually sufficiently close to the background solution, if the wedge-angle is sufficiently close to the background wedge-angle. For the main part, the existence of the modified free boundary problem for v, we in fact have the following theorem.

Theorem 3 (Modified free boundary problem). *Assume that $\Theta_w \in I$. Then there exist $\delta_0 = \delta(\rho_0, \rho_1, u_0, \gamma, \Theta_w) > 0$ small enough, $\sigma^* > 0$, $\alpha_0 > 0$, and $\epsilon^* > 0$ such that, for each $\theta_w \in [\Theta_w - \delta_0, \Theta_w)$, $\sigma < \sigma^*$, $\alpha < \alpha_0$, and $\epsilon \in (0, \epsilon^*)$, there exists a solution $(u^\epsilon, v^\epsilon, \xi^\epsilon(\eta)) \in \left(C^{2+\alpha}_{(-\sigma)}(\Omega^\epsilon)\right)^2 \times C^{2+\alpha}$ to the regularized free boundary problem:*

$$\begin{cases} Q^\epsilon(v; u) := Q(v; u) + \epsilon \Delta v = 0, \\ u_\eta = v_\xi, \end{cases} \tag{36}$$

with the free boundary position:

$$\xi' = -\zeta_s \frac{v}{u} - (1 - \zeta_s) \frac{\hat{v}}{\hat{u}} \tag{37}$$

and the following boundary conditions:

$$(u, v, \psi) = (u_0, 0, \psi_0) \qquad \text{on } \Gamma_{sonic}, \tag{38}$$

$$v = 0 \qquad \text{on } \Gamma^1_{wedge}, \tag{39}$$

$$M^{(1)} v = 0 \qquad \text{on } \Gamma^2_{wedge}, \tag{40}$$

$$M^{(2)} v - \bar{a}_{11} A_{s,1} v = g_s(u, v, \varphi) \qquad \text{on } \Gamma_{shock}, \tag{41}$$

$$v = -g(\xi_w, \theta_w) \tan(\pi - \theta_w) \qquad \text{at } P_2. \tag{42}$$

In addition, the solution satisfies the following estimates:

$$|\xi(\eta)-\hat{\xi}(\eta)|\le\delta_1,\qquad 0\le\xi'(\eta)\le K_2, \tag{43}$$

$$a_{11}^{\epsilon}u_{\xi}^{\epsilon}+2a_{12}^{\epsilon}v_{\xi}^{\epsilon}+a_{22}^{\epsilon}v_{\eta}^{\epsilon}=C_1(\epsilon)\to 0\qquad \text{when } \epsilon\to 0, \tag{44}$$

$$|v^{\epsilon}|\le A(c_0-r)^{\frac{1}{4}}\qquad \text{for } 0\le c_0-r\le m, \tag{45}$$

$$|u^{\epsilon}-u_0|+|\rho^{\epsilon}-\rho_0|\le A(c_0-r)^{\frac{1}{6}}\qquad \text{for } 0\le c_0-r\le m, \tag{46}$$

$$\|(u^{\epsilon},v^{\epsilon})\|_{2+\alpha,\Omega}^{(-\sigma)}+\|(u^{\epsilon},v^{\epsilon})\|_{2+\alpha,\Omega\setminus B_{d_0}(O)}^{(-\sigma-1)}\le C_2(\epsilon), \tag{47}$$

$$\|v^{\epsilon}\|_{2+\alpha,\Omega\cap\{c_0-r\ge s\}}^{(-\sigma)}+\|v^{\epsilon}\|_{2+\alpha,\Omega\cap\{c_0-r\ge s\}\setminus B_{d_0}(O)}^{(-\sigma-1)}\le C(s), \tag{48}$$

and

$$\|u^{\epsilon}\|_{1+\alpha,\Omega\cap\{c_0-r\ge s\}}^{(-\sigma)}+\|u^{\epsilon}\|_{1+\alpha,\Omega\cap\{c_0-r\ge s\}\setminus B_{d_0}(O)}^{(-\sigma-1)}\le C(s) \tag{49}$$

for some small positive constants δ_1 and K_2, while $C_1(\epsilon)$, $C_2(\epsilon)$ and $C(s)$ depend only on the data, the background solution, as well as ϵ and s, respectively. Meanwhile, A and m are independent of θ, and ϵ_0 is chosen such that $\epsilon_0<m$.

The proof of this theorem is long and technical. Thus, instead of proving it here, we would like to illustrate the ideas by proving a simpler case when the wedge-angle is near π. In this case, the background solution is constant, namely, $(\hat{u},\hat{v})=(u_0,0)$. Then the inhomogeneous terms vanish, and the uniform estimate of the smallness between the solution and the background solution can easily be obtained. In fact, the constants on the right-hand side of inequalities (47)–(49) are all multiplied by a small term $\pi-\theta_w$. We now illustrate the proof below.

4.1 The Degenerate Elliptic Cut-off Near the Pseudo-sonic Circle

First define the regions Ω' and Ω'' for any domain Ω as

$$\begin{aligned}\Omega'&:=\Omega\cap\{(\xi,\eta)\ :\ \operatorname{dist}\{(\xi,\eta),\Gamma_{\text{sonic}}\}<2\epsilon_0\},\\ \Omega''&:=\Omega\cap\{(\xi,\eta)\ :\ \operatorname{dist}\{(\xi,\eta),\Gamma_{\text{sonic}}\}>\epsilon_0\}\end{aligned} \tag{50}$$

with a small constant $\epsilon_0>0$. Obviously, $\Omega=\Omega'\cup\Omega''$. In this subsection, we will introduce a degenerate elliptic cut-off function ζ_1 and also a cut-off function ζ_M of higher order smallness in Ω'. Since the equation we study requires more precise estimates near Γ_{sonic}, the elliptic cut-off function introduced in this subsection is more accurate in comparison with that given in [6]. In addition, the elliptic cut-off function does not take its values simply on φ_x, but on $\varphi_x - a$ in order to remove the elliptic cut-off function, where a is some constant which is defined in the following statement.

The leading term of the second-order elliptic equation for v is of the following form:

$$\big(c^2 - (u-\xi)^2\big)v_{\xi\xi} - 2(u-\xi)(v-\eta)v_{\xi\eta} + \big(c^2 - (v-\eta)^2\big)v_{\eta\eta}. \tag{51}$$

Thus, in polar coordinates, we introduce the cut-off function ζ_M for small quantities of higher order as

$$\zeta_M = \begin{cases} s & \text{for } |s| \le M, \\ M+1 & \text{for } |s| \ge M+2, \end{cases}$$

so that

$$\zeta_M(-s) = -\zeta_M(s), \quad 0 \le \zeta_M'(s) \le 1 \qquad \text{on } \mathbb{R},$$

for some constant M to be determined later. We then rewrite the above form by plugging the cut-off function into the terms involving higher order small quantities as

$$\begin{aligned} &\Big((c_0 - r)c_0 + (\gamma+1)\big((u-u_0)\cos\theta + v\sin\theta + \tfrac{c_0 - r}{\gamma+1}\big)r + O_1\Big)v_{rr} \\ &\quad + \tfrac{2}{r}O_3 v_{r\theta} + \tfrac{1}{r^2}(c_0^2 + O_2)v_{\theta\theta} + \tfrac{1}{r}(c_0^2 + O_2)v_r - \tfrac{2}{r^2}O_3 v_\theta, \end{aligned}$$

with

$$\begin{aligned} O_1 &= (c_0 - r)^2\zeta_M\big(\tfrac{O_1^\varphi}{(c_0-r)^2}\big), \\ O_2 &= (c_0 - r)\zeta_M\big(\tfrac{\hat{c}^2 - c_0^2 - (O_2^\varphi)^2}{(c_0 - r)}\big), \\ O_3 &= (c_0 - r)^{\frac{3}{2}}\zeta_M\big(\tfrac{-O_2^\varphi r + O_3^\varphi}{(c_0-r)^{3/2}}\big). \end{aligned}$$

Therefore, the ellipticity of this form is equivalent to

$$(c_0 - r)c_0 + (\gamma+1)\Big((u-u_0)\cos\theta + v\sin\theta + \frac{c_0 - r}{\gamma+1}\Big)r > 0 \quad \text{and} \quad c^2 > 0.$$

Next, for the degenerate elliptic cut-off, let $\zeta_1 \in C^\infty(\mathbb{R})$ satisfy

$$\zeta_1(s) = \begin{cases} s & \text{if } -\frac{1}{3(\gamma+1)} < s < \frac{7}{6(\gamma+1)}, \\ -\frac{2}{3(\gamma+1)} & \text{if } s < -\frac{1}{(\gamma+1)}, \\ \frac{5}{4(\gamma+1)} & \text{if } s > \frac{4}{3(\gamma+1)}, \end{cases} \tag{52}$$

so that

$$\zeta_1'(s) \geq 0 \qquad \text{on } \mathbb{R}, \tag{53}$$

$$\pm\zeta_1''(s) \geq 0 \qquad \text{on } \{\pm s \leq 0\}. \tag{54}$$

The value s that the cut-off function ζ_1 takes on is

$$\frac{(u-u_0)\cos\theta + v\sin\theta}{c_0 - r} + \frac{1}{\gamma+1}.$$

Then (51) becomes the following modified form:

$$A_{11}v_{\xi\xi} + 2A_{12}v_{\xi\eta} + A_{22}v_{\eta\eta},$$

where

$$\begin{aligned}
A_{11} = {} & c_0^2 - (\xi - u_0)^2 + (\gamma - 1)(c_0 - r)r\Big(\zeta_1\big(\tfrac{(u-u_0)\cos\theta+v\sin\theta}{c_0-r} + \tfrac{1}{\gamma+1}\big) - \tfrac{1}{\gamma+1}\Big) \\
& + \tfrac{2(c_0-r)(\xi-u_0)^2}{r}\Big(\zeta_1\big(\tfrac{(u-u_0)\cos\theta+v\sin\theta}{c_0-r} + \tfrac{1}{\gamma+1}\big) - \tfrac{1}{\gamma+1}\Big) \\
& + \tfrac{1}{r^2}\big(O_1(\xi - u_0)^2 - 2O_3(\xi - u_0)\eta + O_2\eta^2\big), \\
A_{12} = {} & -(\xi - u_0)\eta + \tfrac{2(c_0-r)(\xi-u_0)\eta}{r}\Big(\zeta_1\big(\tfrac{(u-u_0)\cos\theta+v\sin\theta}{c_0-r} + \tfrac{1}{\gamma+1}\big) - \tfrac{1}{\gamma+1}\Big) \\
& + \tfrac{1}{r^2}\big((O_1 - O_2)(\xi - u_0)\eta + O_3(\xi - u_0)^2 - O_3\eta^2\big), \\
A_{22} = {} & c_0^2 - \eta^2 + (\gamma - 1)(c_0 - r)r\Big(\zeta_1\big(\tfrac{(u-u_0)\cos\theta+v\sin\theta}{c_0-r} + \tfrac{1}{\gamma+1}\big) - \tfrac{1}{\gamma+1}\Big) \\
& + \tfrac{2(c_0-r)\eta^2}{r}\Big(\zeta_1\big(\tfrac{(u-u_0)\cos\theta+v\sin\theta}{c_0-r} + \tfrac{1}{\gamma+1}\big) - \tfrac{1}{\gamma+1}\Big) \\
& + \tfrac{1}{r^2}\big(O_1\eta^2 + 2O_3(\xi - u_0)\eta + O_2(\xi - u_0)^2\big).
\end{aligned}$$

4.2 *The Uniform Elliptic Cut-off Away from the Pseudo-sonic Circle*

Let $\zeta_2 \in C^\infty$ be a smooth increasing function such that

$$\zeta_2(s) = \begin{cases} s & \text{if } s \geq \epsilon_1, \\ \frac{1}{2}\epsilon_1 & \text{if } s < 0, \end{cases} \tag{55}$$

and $|\zeta_2'(s)| \leq 1$. Let ζ_2 be evaluated at $c^2 - U^2 - V^2$. In Ω', consider the following modified system:

$$\begin{cases} \frac{U^2\zeta_2+V^2c^2}{U^2+V^2}u_\xi + \frac{2UV}{U^2+V^2}(\zeta_2 - c^2)u_\eta + \frac{V^2\zeta_2+U^2c^2}{U^2+V^2}v_\eta = 0, \\ v_\xi = u_\eta, \\ c^2 = -\frac{\gamma-1}{2}(U^2+V^2) - (\gamma-1)\psi. \end{cases} \tag{56}$$

Finally, we combine the coefficients introduced above in $\mathcal{D}$ as follows. Let $\zeta_3 \in C^\infty(\mathbb{R})$ satisfy

$$\zeta_3(s) = \begin{cases} 0 & \text{if } s \leq 2\epsilon_0, \\ 1 & \text{if } s \geq 4\epsilon_0, \end{cases} \qquad 0 \leq \zeta_3'(s) \leq \frac{10}{\epsilon_0} \quad \text{on } \mathbb{R}.$$

Then we define, for $(\rho, u, v) \in \mathbb{R}^3$ and $(\xi, \eta) \in \mathcal{D}$,

$$\begin{aligned} \bar{a}_{11} &= \zeta_3(c_0 - r)\tfrac{U^2\zeta_2+V^2c^2}{U^2+V^2} + \big(1 - \zeta_3(c_0 - r)\big)A_{11}, \\ \bar{a}_{12} &= \zeta_3(c_0 - r)\tfrac{UV}{U^2+V^2}(\zeta_2 - c^2) + \big(1 - \zeta_3(c_0 - r)\big)A_{12}, \\ \bar{a}_{22} &= \zeta_3(c_0 - r)\tfrac{V^2\zeta_2+U^2c^2}{U^2+V^2} + \big(1 - \zeta_3(c_0 - r)\big)A_{22}. \end{aligned} \tag{57}$$

This transforms system (11) into the following modified system:

$$\begin{cases} \bar{a}_{11}u_\xi + 2\bar{a}_{12}u_\eta + \bar{a}_{22}v_\eta = 0, \\ v_\xi = u_\eta, \\ D(\psi - \psi_0) = (u - u_0,\ v), \\ \rho = \big(-\frac{\gamma-1}{2}(U^2+V^2) - (\gamma-1)\psi\big)^{\frac{1}{\gamma-1}}. \end{cases} \tag{58}$$

4.3 A Second-Order Equation for v

In order to study the existence of solutions to system (58), we introduce a second-order equation from this system for v, $Q(v; u)$, by taking the derivative of the first equation with respect to η and then using the other equations to replace the unknown terms. We have

$$\begin{aligned} Q(v; u) &:= \bar{a}_{11}v_{\xi\xi} + 2\bar{a}_{12}v_{\xi\eta} + \bar{a}_{22}v_{\eta\eta} + b_{11}v_\xi^2 + b_{12}v_\xi v_\eta + b_{22}v_\eta^2 + c_1 v_\xi + c_2 v_\eta \\ &= 0, \end{aligned} \tag{59}$$

where

$$|b_{11}| + |b_{12}| + |b_{22}| < \frac{C}{a_{11}}$$

with C depending on the C^1-bounds of $\hat{\psi}$ and the cut-off functions ζ_i and ζ_M, while

$$d_O^{\alpha}(|c_1| + |c_2|) < \frac{C}{a_{11}}$$

with C depending on $\|\hat{\psi}\|_{2,\alpha,\Omega''}^{(-1-\alpha),\{O,P_0\}}$ and $d_O(X) = \text{dist}\{X, O\}$.

Near Γ_{sonic}, in the (r,θ)-coordinates, this equation reads

$$\begin{aligned}
&(c_0 - r)\big(1 + (\gamma+1)\zeta_1\big)v_{rr} + \tfrac{1}{c_0}v_{\theta\theta} + b_r v_r \\
&+\tfrac{(\gamma+1)c_0^2 \sin\theta \zeta_1'}{a_{11}}\Big(v_r^2 + \tfrac{\cos\theta}{c_0}(-(u-u_0)\sin\theta + v\cos\theta)v_r + \tfrac{(u-u_0)\cos^2\theta + v\sin\theta\cos\theta}{c_0 - r}v_r\Big) \\
&-\tfrac{(2c_0^2+(\gamma-1)r^2)\cos\theta\zeta_1'}{a_{11}r}v_r v_\theta + O_1 v_r + O_2 v_r v_\theta + O_3 v_\theta = 0,
\end{aligned} \tag{60}$$

where

$$\begin{aligned}
b_r := \frac{1}{a_{11}}\Big(&(\sin^2\theta + 1)\big(a_{11}\cos^2\theta + 2a_{12}\sin\theta\cos\theta + a_{22}\sin^2\theta\big) \\
&-(\gamma+1)\big(c_0 + O(1)(c_0 - r)\big)r\sin^2\theta\zeta_1\Big).
\end{aligned} \tag{61}$$

Lemma 1. *If*

$$\zeta_1 \geq -\frac{2}{3(\gamma+1)},$$

then there exists an $\epsilon_0 > 0$ such that, for any $0 \leq c_0 - r \leq \epsilon_0$, we have

$$-\frac{9}{8}(\gamma+1)\max\{\zeta_1, 0\} \leq b_r \leq C, \tag{62}$$

where C is a uniform constant independent of θ, u, and v.

This lemma is crucial for the proof of the uniform Hölder estimate of v near Γ_{sonic}. Finally, Eq. (60) can be rewritten in the divergent form by scaling as follows:

$$\begin{aligned}
&\Big((c_0 - r)\big(1 + (\gamma+1)\zeta_1\big)v_r\Big)_r + \Big(\frac{1}{c_0}v_\theta\Big)_\theta + O_1 v_r + O_2(c_0 - r)(v_r)^2 \\
&+O_3(c_0 - r)v_r v_\theta + O_4 v_\theta = 0,
\end{aligned} \tag{63}$$

with $|O_i| \leq C$, provided that $\sin\theta > 0$.

On the other hand, away from Γ_{sonic}, we notice that the equation is strictly and uniformly elliptic with the bounded coefficients depending only on δ_0 and C.

4.4 The Different Boundary Conditions from Those Stated in Theorem 3

The difference comes out at the free boundary. First, the condition for the free boundary position can simply be proposed as

$$\xi' = -\frac{v}{u}. \tag{64}$$

Then take the derivative on the Rankine-Hugoniot condition along Γ_{shock} and use (64) to yield the oblique boundary condition on Γ_{shock}:

$$M^{(2)}v = \beta_1^{(2)} v_\xi + \beta_2^{(2)} v_\eta = 0 \qquad \text{on } \Gamma_{\text{shock}}, \tag{65}$$

with

$$\begin{aligned}
\beta_1^{(2)} &= (-\bar{a}_{11} + 2\bar{a}_{12}\xi')F_u - \bar{a}_{11}F_v\xi', \\
\beta_2^{(2)} &= -\bar{a}_{11}F_v + \bar{a}_{22}F_u\xi',
\end{aligned}$$

where, along Γ_{shock}, $F(u, v, \varphi, \eta) = 0$.

4.5 Existence of Solutions for the Linearized Viscous Fixed Boundary Problem for v

We now linearize the modified problem for v, and first show the local existence of solutions near the wedge corner O (where Γ^1_{wedge} and Γ^2_{wedge} meet) by the method of continuity. Next we show the local existence near P_2, where Γ_{shock} and Γ^2_{wedge} meet. With this local solvability, we focus on the proof of the global existence of solutions by the Perron method, as used in [14, 16, 17].

Before proving the existence of solutions, we introduce some notations which are important in the Perron method. The linearized problem is called locally solvable if, for each $y \in \bar{\Omega}$, there is a neighborhood $N = O(y) \cap \Omega$ such that, for any $h \in C(\bar{N})$, there is a solution $v \in C^2(N) \cap C(\bar{N})$ of the problem:

$$\begin{cases} Lv = 0 \quad \text{in } N, \\ N^{(1)}v\big|_{\bar{N}\cap\Gamma^2_{\text{wedge}}} = 0, \\ N^{(2)}v\big|_{\bar{N}\cap\Gamma_{\text{shock}}} = 0, \\ v\big|_{\partial N'} = h, \\ v\big|_{P_2} = -g(\xi_w, \theta_w)\tan(\pi - \theta_w), \end{cases}$$

where $\partial N' = \partial N \cap \Omega$. For brevity, as in [16], denote this function v by $(h)_y$ to emphasize its dependence on h and y. Denote by $S^-(S^+)$ the set of all subsolutions (supersolutions) of the problem. A subsolution or supersolution $w \in S^\pm$ of the linearized problem is a function $w \in C(\bar{\Omega})$ satisfying

$$\pm\big(g(\xi_w, \theta_w)\tan(\pi - \theta_w) + w\big) \le 0 \qquad \text{at } P_2$$

and

$$\pm w \le 0 \qquad \text{on } \bar{N} \cap (\Gamma_{\text{sonic}} \cup \Gamma^1_{\text{wedge}}),$$

such that, for any $y \in \bar{\Omega}$, if $\pm(h - w) \ge 0$ on $\partial N'$, then

$$\pm\big((h)_y - w\big) \ge 0 \qquad \text{in } N(y).$$

Then we show properties (i)–(vii) listed below to prove the global existence for the linearized problem:

(i) If $u_1, u_2 \in S^-$, then $\max\{u_1, u_2\} \in S^-$.
(ii) If $u_1 \in S^-$ and $y \in \bar{\Omega}$, and if $\bar{u}_1$ is given by $\bar{u}_1 = u_1$ in $\bar{\Omega}\backslash N(y)$ and $\bar{u}_1 = (u_1)_y$ in $N(y)$, then $\bar{u}_1 \in S^-$.
(iii) If $w^\pm \in S^\pm$, then $w^+ \ge w^-$ in Ω.
(iv) If $w^\pm \in C^2(N) \cap C(\bar{N})$ satisfy $Lw^+ = Lw^-$ in $N \cap \Omega$, $\tilde{M}w^+ = \tilde{M}w^-$ on $N \cap \Gamma_{\text{wedge}}$, and $w^+ \ge w^-$ in $N \cap \Omega$, then either $w^+ = w^-$ in N or else $w^+ > w^-$ in N.
(v) $S^\pm$ are non-empty.
(vi) Let $\{u_m\}$ be a bounded sequence of $C^2(N) \cap C(\bar{N})$-solutions of $Lu_m = 0$ in $N \cap \Omega$ and $\tilde{M}u_m = 0$ on $N \cap \Gamma_{\text{wedge}}$. Then there is a convergent subsequence $\{u_m\}$ such that $u = \lim u_{m_i}$ is a $C^2(N)$-solution of $Lu = 0$ in $N \cap \Omega$ and $\tilde{M}u = 0$ on $N \cap \Gamma_{\text{wedge}}$.
(vii) For each $x_0 \in \Gamma^1_{\text{wedge}} \cup \Gamma_{\text{sonic}} \cup \{P_2\}$, there are sequences $\{w^\pm_m\}$ of subsolutions and supersolutions such that $\lim w^\pm_m(x_0) = u(x_0)$.

4.6 Existence of Solutions for the Modified Nonlinear Fixed Boundary Problem for v

Once the linearized problem is solved, the existence for the modified nonlinear problem can be proved by the Leray-Schauder fixed point theorem (cf. Theorem 11.3 in [14]).

To achieve this, we first introduce the sets $\mathscr{H}^\epsilon$ defined in a bounded domain Ω and $\mathscr{K}^\epsilon$ in a bounded domain $(-\xi_1 \tan(\pi - \theta_w), \eta_1]$, depending on the given values θ_w, ρ_0, ρ_1 and u_0, as follows:

Definition 3. The elements of $\mathscr{H}^\epsilon \in C^{2+\alpha}_{(-\nu)}$ satisfy:

(H1) $u = u_0$ on Γ_{sonic};
(H2) $|u - u_0| \le A_0(c_0 - r)^{1/6}$ when $|c_0 - r|$ is small, independent of θ;
(H3) $\|u\|^{(-\nu)}_{2+\alpha} \le A_1(\epsilon)$;
(H4) $\|u\|^{(-\nu-1)}_{2+\alpha} \le A_2(\epsilon)$ away from the wedge-angle O.

Definition 4. The elements of $\mathscr{K}^\epsilon \in C^{2+\alpha}$ satisfy:

(K1) $\xi(\eta_1) = \xi_1$;
(K2) $\xi'(\eta_1) = 0$;
(K3) $|\xi(\eta) - \hat{\xi}(\eta)| \le \delta_*$;
(K4) $0 \le \xi'(\eta) \le K_2$.

The weighted Hölder space is defined in (21). The values of α, $\nu \in (0, 1)$, as well as K_i, δ_1, and A_i, will be specified later. Obviously, $\mathscr{H}^\epsilon$ and $\mathscr{K}^\epsilon$ are closed, bounded, and convex.

The crucial step to apply the fixed point theorem is then to prove the following uniform estimates which are also stated in Sect. 3:

Lemma 2. *For given K_i, δ_1, and A_i for $\mathscr{K}^\epsilon$ and $\mathscr{H}^\epsilon$, there exist σ^*, $\alpha_0 \in (0, 1)$, and $d_0 > 0$ such that any solution $v \in C^{2+\alpha}_{(-\sigma)}(\Omega) \cap C^{2+\alpha}_{(-\sigma-1)}(\Omega \backslash B_{d_0}(O))$ to the nonlinear problem $v = \sigma \mathbf{T} v$ with $\alpha \le \alpha_0$, $\sigma \le \sigma^*$, and $\sigma \in [0, 1]$ satisfies*

$$\|v\|^{(-\sigma)}_{2+\alpha,\Omega} \le C \tan(\pi - \theta_w), \tag{66}$$

$$\|v\|^{(-1-\sigma)}_{2+\alpha,\{\Omega \backslash B_{d_0}(P_0)\}} \le C \tan(\pi - \theta_w), \tag{67}$$

and

$$- g(\xi_w, \theta_w) \tan(\pi - \theta_w) \le v \le 0, \tag{68}$$

where C is independent of v.

Finally, we can show that the solution obtained in this subsection is unique by the maximum principle, which will be used to demonstrate that the mapping introduced in Sect. 4.7 is well-defined.

4.7 Existence of Solutions for the Modified Nonlinear Fixed Boundary Problem for (ρ, u, v)

Thanks to the uniform estimates of v and then u near Γ_{sonic} stated in Sect. 3, we can prove the existence for the modified nonlinear fixed boundary problem (36) and (38)–(42) by the Leray-Schauder fixed point theorem.

From Sect. 4.6, for every $u \in \mathscr{H}^{\epsilon}$, there exists a unique $v \in C^{(-\sigma)}_{2+\alpha}$ satisfying $\|v\|^{(-\sigma)}_{2+\alpha} < C \tan(\pi - \theta_w)$. Thus, we can define a mapping for u as

$$\mathbf{S} : u \to \bar{u},$$

in the following:

$$\bar{u}(\xi, \eta) = \mathbf{S}u := u_0 + \int_{\eta(\xi)}^{\eta} v_{\xi}(\xi, s)\mathrm{d}s, \tag{69}$$

where $(\xi, \eta(\xi))$ denotes the point on the sonic circle Γ_{sonic}. For the other quantities ρ and ψ, we can obtain them once the nonlinear problem for u and v is established as follows:

$$\begin{aligned} &\psi_{\xi} = U = u - \xi, \quad \psi_{\eta} = V = v - \eta, \\ &\rho = \big(- (\gamma - 1)\psi - \tfrac{\gamma-1}{2}(U^2 + V^2)\big)^{\frac{1}{\gamma-1}}. \end{aligned} \tag{70}$$

Furthermore, for the solutions to the nonlinear equations, we can prove v and then φ is monotone along Γ_{shock} by a contradiction argument.

4.8 The Free Boundary Problem

We can now prove the existence of solutions to the free boundary value problem. As indicated above, for any given boundary $\xi = \xi(\eta) \in \mathscr{K}^{\epsilon}$, which is a small perturbation of the background solution $\xi = \xi_1$, we solve the fixed boundary problem and then give an updated boundary by

$$\frac{J(\xi)(\eta)}{d\eta} = -\frac{v^{\epsilon}}{u^{\epsilon}} \quad \text{with } J(\xi)(\eta_1) = \xi_1. \tag{71}$$

The fixed point theorem which will be used here is the standard Schauder theorem (cf. Corollary 11.2, [14]). Then Theorem 3 is proved.

4.9 The Limiting Solution and the Equivalence to the Original System

We now study the limiting solution, as the elliptic regularization parameter ϵ tends to zero, to obtain a solution to system (58) and then to the original system, *i.e.* the potential flow equation, which we will study next to remove the elliptic cut-off. In fact, we can establish the following existence result.

Proposition 1. *There exist constants $\sigma^* > 0$, $\alpha_0 > 0$, and $\delta_0 > 0$ small enough such that, for any $\sigma < \sigma^*$, $\alpha < \alpha_0$, and $\pi - \delta_0 \leq \theta_w < \pi$, there exists a solution*

$$(u, v) \in (C^\alpha(\overline{\Omega}) \cap \ldots)^2$$

with $(u - \xi, v - \eta) = (\psi_\xi, \psi_\eta)$ to problem (58), (38)–(40), (42), (64), *and* (65), *i.e.*

$$\begin{cases} \bar{a}_{11} u_\xi + 2\bar{a}_{12} u_\eta + \bar{a}_{22} v_\eta = 0, \\ v_\xi = u_\eta, \\ D(\psi - \psi_0) = (u - u_0, v), \end{cases} \tag{72}$$

so that the velocity potential ψ satisfies (8) *in Ω, i.e.,*

$$\operatorname{div}\big(\rho(|\nabla\psi|^2, \psi) D\psi\big) + 2\rho(|\nabla\psi|^2, \psi) = 0, \tag{73}$$

the slip boundary condition on Γ_{wedge} with $\varphi = \psi - \psi_0$:

$$\varphi_\nu = 0 \tag{74}$$

with ν the normal direction and the following boundary conditions on Γ_{shock}:

$$\varphi = \varphi_1, \tag{75}$$

$$F(\varphi_\xi, \varphi_\eta, \varphi, \eta) = 0, \tag{76}$$

where $F(\mathbf{p}, z, \eta) = 0$ comes from the Rankine-Hugoniot condition satisfying $F(\mathbf{0}, 0, \eta) = 0$, $D_{\mathbf{p}} F \cdot \boldsymbol{\nu} \neq 0$, and $D_z F \neq 0$. Moreover, on Γ_{sonic}, the velocity potential ψ satisfies the Dirichlet boundary condition:

$$\psi = \psi_0. \tag{77}$$

4.10 Removal of the Cut-off Function ζ_M for the Higher Order Smallness

It is convenient to study this problem in the new coordinates introduced by $(x, y) = (c_0 - r, \theta - \theta_1)$ near Γ_{sonic}. Then the equation reads

$$\begin{aligned}&\Big(c_0 x + (\gamma+1)c_0 x\zeta\big(\tfrac{1}{\gamma+1} - \tfrac{\varphi_x}{x}\big) + O_1\Big)\varphi_{xx} + O_2\varphi_{xy} + (1+O_3)\varphi_{yy}\\&\quad -(c_0+O_3)\varphi_x - O_2\varphi_y = 0,\end{aligned} \tag{78}$$

with

$$O_1 \le (M+1)|x|^2, \quad O_2 \le (M+1)|x|^{\frac{3}{2}}, \quad |O_3| \le (M+1)|x|,$$

due to the cut-off function ζ_M. By a scaling argument, we have the following estimates to remove the cut-off function ζ_M for the higher order smallness:

$$0 \le \varphi \le \frac{3}{5(\gamma+1)}x^2 \qquad \text{in } \Omega \cap \{c_0 - r \le 2\epsilon_0\} \tag{79}$$

and

$$\|\varphi\|^{(-1-\alpha)}_{2+\alpha,\Omega\cap\{c_0-r\ge s\}} \le C(s)(\pi - \theta_w) \tag{80}$$

for all $s \in (0, 8\epsilon_0)$ with $C(s)$ depending only on the data and s.

4.11 Removal of the Degenerate Elliptic Cut-off

Now we remove the degenerate elliptic cut-off ζ_1 in the (x, y)-coordinates with

$$(x, y) = (c_0 - r, \theta - \theta_1) \qquad \text{in } \Omega \cap \{c_0 - r < 4\epsilon_0\}.$$

In this subsection, we let $|\pi - \theta_w|$ be sufficiently small, depending only on the data, so that φ is a solution of the shock diffraction problem. Since the elliptic cut-off introduced here is more precise than that given in [6] and since $\sin\theta$ may be 0 at P_0, the argument cannot be applied directly; we need a more careful argument to re-control it.

First we bound φ_x near P_1 by the following lemma:

Lemma 3. *For $|\pi - \theta_w|$ sufficiently small, we have*

$$-\frac{x}{6(\gamma+1)} \le \varphi_x \le \frac{4x}{3(\gamma+1)} \qquad \textit{in } \Omega \cap \{x \le 4\epsilon_0\} \cap \{y \le 4\epsilon_2\}. \tag{81}$$

Next, away from P_1, we bound φ_x with an additional assumption, by the following lemma:

Lemma 4. *Assume that*

$$|\varphi - \frac{x^2}{2(\gamma+1)}| \le C_1 x^{2+\alpha} \qquad \text{in } \Omega \cap \{x \le 2\epsilon_0\} \cap \{y \ge 2\epsilon_2\}. \tag{82}$$

Then, for $|\pi - \theta_w|$ sufficiently small, we have

$$-\frac{x}{3(\gamma+1)} \le \varphi_x - \frac{x}{\gamma+1} \le \frac{x}{3(\gamma+1)} \qquad \text{in } \Omega \cap \{x \le 4\epsilon_0\} \cap \{x \ge 2\epsilon_2\}. \tag{83}$$

For this lemma, we first prove that the cut-off function can be removed when x is small enough, which may depend on y. Then, in this domain, rewrite this equation in a more convenient form and scale it to obtain a uniform estimate to guarantee that the removal can be extended to $x = 2\epsilon_0$ independent of y.

With this proposition in hand, the remaining task is to show that (82) holds for some $\alpha < \frac{1}{2}$, which is proved in the following lemma.

Lemma 5. *For $|\pi - \theta_w|$ sufficiently small, we have*

$$|\varphi - \frac{x^2}{2(\gamma+1)}| \le C_1 x^{2+\alpha} \qquad \text{in } \Omega \cap \{x \le \epsilon'\} \cap \{y \ge 2\epsilon_2\}, \tag{84}$$

where C_1 and ϵ' depend only on the data.

This completes the proof of the existence theory of the shock diffraction configuration with the required properties stated in Definition 2 when $\pi - \theta_w$ is small. If it is large, using the same idea but with more technicalities, we can obtain that, for any $\Theta_w \in I$, there exists a constant $\delta_0 > 0$ such that, for any $\Theta_w - \delta_0 < \theta_w \le \Theta_w$, there is a solution $W^{(\theta_w)} = (U^{(\theta_w)}, V^{(\theta_w)})$ close to $W^{(\Theta_w)}$. Then, from the estimates stated above, we obtain that $(\theta_w, W^{(\theta_w)})$ belongs to the solution set defined in Definition 2. This means that the set I is open. Thus, from the fact that I is closed and nonempty, we then finally have $(\theta_c, \pi) \subset I$. For further details, see Chen-Xiang [8].

Acknowledgements The research of Gui-Qiang G. Chen was supported in part by the National Science Foundation under Grant DMS-0807551, the UK EPSRC Science and Innovation Award to the Oxford Centre for Nonlinear PDE (EP/E035027/1), the NSFC under a joint project Grant 10728101, and the Royal Society–Wolfson Research Merit Award (UK). Wei Xiang was supported in part by the China Scholarship Council No. 2008631071 while visiting the University of Oxford and the Doctoral Program Foundation of the Ministry Education of China.

References

1. M. Bae, G.-Q. Chen, M. Feldman, Regularity of solutions to regular shock reflection for potential flow. Invent. Math. **175**, 505–543 (2009)
2. M. Bae, G.-Q. Chen, M. Feldman, Global solutions to the Prandtl-Meyer reflection for supersonic flow impinging onto a solid wedge. Quart. Appl. Math. **71**, 583–600 (2013)
3. V. Bargman, On nearly glancing reflection of shocks. Off. Sci. Res. Dev. Rep. No. **5117** (1945)
4. S. Čanić, B.L. Keyfitz, E.H. Kim, Free boundary problems for the unsteady transonic small disturbance equation: transonic regular reflection. Methods Appl. Anal. **7**, 313–336 (2000)
5. S. Čanić, B.L. Keyfitz, E.H. Kim, Free foundary problems for nonlinear wave systems: Mach stems for interacting shocks. SIAM J. Math. Anal. **37**, 1947–1977 (2006)
6. G.-Q. Chen, M. Feldman, Global solutions of shock reflection by large-angle wedges for potential flow. Ann. Math. **171**(2), 1067–1182 (2010)
7. G.-Q. Chen, M. Feldman, Mathematics of shock reflection-diffraction and von Neumanns conjectures. Research Monograph (2013 preprint)
8. G.-Q. Chen, W. Xiang, Global solutions of the shock diffraction problem by wedges for potential flow (2013 preprint)
9. G.-Q. Chen, X. Deng, W. Xiang, The global existence and optimal regularity of solutions for shock diffraction problem to the nonlinear wave systems. Arch. Ration. Mech. Anal. (2013 to appear)
10. R. Courant, K.O. Friedrichs, *Supersonic Flow and Shock Waves* (Reprinting of the 1948 original). Applied Mathematical Sciences, vol 21 (Springer, New York/Heidelberg, 1976)
11. V. Elling, T.-P. Liu, Supersonic flow onto a solid wedge. Comm. Pure Appl. Math. **61**, 1347–1448 (2008)
12. C.H. Fletcher, D.K. Weimer, W. Bleakney, Pressure behind a shock wave diffracted through a small angle. Phys. Rev. **78**(5), 634–635 (1950)
13. C.H. Fletcher, A.H. Taub, W. Bleakney, The Mach reflection of shock waves at nearly glancing incidence. Rev. Mod. Phys. **23**(3), 271–286 (1951)
14. D. Gilbarg, N. Trudinger, *Elliptic Partial Differential Equations of Second Order*, 3rd edn. (Springer, Berlin, 1998)
15. E.H. Kim, Global sub-sonic solution to an interacting transonic shock of the self-similar nonlinear wave equation. J. Differ. Equ. **248**, 2906–2930 (2010)
16. G. Lieberman, The Perron process applied to oblique derivative problems. Adv. Math. **55**, 161–172 (1985)
17. G.M. Lieberman, Mixed boundary value problems for elliptic and parabolic differential equations of second order. J. Math. Anal. Appl. **113**, 422–440 (1986)
18. M.J. Lighthill, The diffraction of blast I. Proc. R. Soc. **198A**, 454–470 (1949)
19. M.J. Lighthill, The diffraction of blast II. Proc. R. Soc. **200A**, 554–565 (1950)
20. G.B. Whitham, *Linear and Nonlinear Waves* (Reprint of the 1974 original). Pure and Applied Mathematics (Wiley, New York, 1999)
21. Y. Zheng, Two-dimensional regular shock reflection for the pressure gradient system of conservation laws. Acta Math. Appl. Sinica (English Ser) **22**, 177–210 (2006)

Some Wellposedness Results for the Ostrovsky–Hunter Equation

G.M. Coclite, L. di Ruvo, and K.H. Karlsen

Abstract The Ostrovsky-Hunter equation provides a model for small-amplitude long waves in a rotating fluid of finite depth. It is a nonlinear evolution equation. In this paper the welposedness of the Cauchy problem and of an initial boundary value problem associated to this equation is studied.

Keywords Existence • Uniqueness • Stability • Entropy solutions • Conservation laws • Ostrovsky-Hunter equation • Boundary value problems • Cauchy problems

2000 Mathematics Subject Classification 35G25, 35G15, 35L65, 35L05, 35A05

1 Introduction

The non-linear evolution equation

$$\partial_x(\partial_t u + \partial_x f(u) - \beta \partial^3_{xxx} u) = \gamma u, \tag{1}$$

with $\beta, \gamma \in \mathbb{R}$ and $f(u) = \frac{u^2}{2}$ was derived by Ostrovsky [20] to model small-amplitude long waves in a rotating fluid of finite depth. This equation generalizes the

G.M. Coclite (✉) · L. di Ruvo
Department of Mathematics, University of Bari, via E. Orabona 4, I–70125 Bari, Italy
e-mail: giuseppemaria.coclite@uniba.it; coclitegm@dm.uniba.it; diruvo@dm.uniba.it

K.H. Karlsen
Centre of Mathematics for Applications (CMA), University of Oslo, P.O. Box 1053, Blindern, N–0316 Oslo, Norway
e-mail: kennethk@math.uio.no

G.-Q.G. Chen et al. (eds.), *Hyperbolic Conservation Laws and Related Analysis with Applications*, Springer Proceedings in Mathematics & Statistics 49,
DOI 10.1007/978-3-642-39007-4_7,

Korteweg-deVries equation (corresponding to $\gamma = 0$) by an additional term induced by the Coriolis force. It is deduced by considering two asymptotic expansions of the shallow water equations, first with respect to the rotation frequency and then with respect to the amplitude of the waves [8].

Mathematical properties of the Ostrovsky equation (1) have been studied recently in great depth, including the local and global well-posedness in energy space [7, 12, 14, 25], stability of solitary waves [10, 13, 15], and convergence of solutions in the limit of the Korteweg-deVries equation [11, 15]. We shall consider the limit of the no high-frequency dispersion $\beta = 0$, therefore (1) reads

$$\partial_x(\partial_t u + \partial_x f(u)) = \gamma u. \tag{2}$$

In this form, Eq. (2) is known under various different names such as the reduced Ostrovsky equation [21, 23], the Ostrovsky-Hunter equation [3], the short-wave equation [8], and the Vakhnenko equation [18, 22].

Integrating (2) with respect to x we obtain the integro-differential formulation of (2) (see [16])

$$\partial_t u + \partial_x f(u) = \gamma \int^x u(t, y)dy,$$

which is equivalent to

$$\partial_t u + \partial_x f(u) = \gamma P, \qquad \partial_x P = u.$$

Due to the regularizing effect of the P equation we have that

$$u \in L^\infty_{loc} \Longrightarrow P \in L^\infty((0, T); W^{1,\infty}_{loc}),\ T > 0.$$

The flux f is assumed to be smooth, Lipschitz continuous, and *genuinely nonlinear*, i.e.:

$$f \in C^2(\mathbb{R}), \qquad |\{u \in \mathbb{R}; f''(u) = 0\}| = 0, \qquad f'(0) = 0, \qquad |f'(\cdot)| \le L, \tag{3}$$

and the constant γ is assumed to be real

Since the solutions are merely locally bounded, the Lipschitz continuity of the flux f assumed in (3) guarantees the finite speed of propagation of the solutions of (2).

This paper is devoted to the wellposedness of the initial-boundary value problem (see Sect. 2) and the Cauchy problem (see Sect. 3) for (2). Our existence argument is based on a passage to the limit using a compensated compactness argument [24] in a vanishing viscosity approximation of (8):

$$\partial_t u_\varepsilon + \partial_x f(u_\varepsilon) = \gamma P_\varepsilon + \varepsilon \partial^2_{xx} u_\varepsilon, \qquad \partial_x P_\varepsilon = u_\varepsilon.$$

On the other hand we use the method of [9] for the uniqueness and stability of the solutions of (2).

2 The Initial Boundary Value Problem

In this section, we augment (2) with the boundary condition

$$u(t,0) = 0, \qquad t > 0, \tag{4}$$

and the initial datum

$$u(0,x) = u_0(x), \qquad x > 0. \tag{5}$$

We assume that

$$u_0 \in L^2(0,\infty) \cap L^\infty_{loc}(0,\infty), \quad \int_0^\infty u_0(x)dx = 0. \tag{6}$$

The zero mean assumption on the initial condition is motivated by (2). Indeed, integrating both sides of (2) we have that $u(t,\cdot)$ has zero mean for every $t > 0$, therefore it is natural to assume the same on the initial condition.

Integrating (2) on $(0,x)$ we obtain the integro-differential formulation of the initial-boundary value problem (2), (4), (5) (see [16])

$$\begin{cases} \partial_t u + \partial_x f(u) = \gamma \int_0^x u(t,y)dy, & t > 0,\ x > 0, \\ u(t,0) = 0, & t > 0, \\ u(0,x) = u_0(x), & x > 0. \end{cases} \tag{7}$$

This is equivalent to

$$\begin{cases} \partial_t u + \partial_x f(u) = \gamma P, & t > 0,\ x > 0, \\ \partial_x P = u, & t > 0,\ x > 0, \\ u(t,0) = P(t,0) = 0, & t > 0, \\ u(0,x) = u_0(x), & x > 0. \end{cases} \tag{8}$$

Due to the regularizing effect of the P equation in (8) we have that

$$u \in L^\infty_{loc}((0,\infty)^2) \Longrightarrow P \in L^\infty_{loc}((0,\infty); W^{1,\infty}_{loc}(0,\infty)). \tag{9}$$

Therefore, if a map $u \in L^\infty_{loc}((0,\infty)^2)$ satisfies, for every convex map $\eta \in C^2(\mathbb{R})$,

$$\partial_t \eta(u) + \partial_x q(u) - \gamma \eta'(u) P \leq 0, \qquad q(u) = \int^u f'(\xi)\eta'(\xi)\, d\xi, \tag{10}$$

in the sense of distributions, then [5, Theorem 1.1] provides the existence of a strong trace u_0^τ on the boundary $x = 0$.

Definition 1. We say that $u \in L^\infty_{loc}((0,\infty)^2)$ is an entropy solution of the initial-boundary value problem (2), (4), and (5) if:

(i) u is a distributional solution of (7) or equivalently of (8);
(ii) for every convex function $\eta \in C^2(\mathbb{R})$ the entropy inequality (10) holds in the sense of distributions in $(0,\infty) \times (0,\infty)$;
(iii) for every convex function $\eta \in C^2(\mathbb{R})$ with corresponding q defined by $q' = f'\eta'$, the boundary entropy condition

$$q(u_0^\tau(t)) - q(0) - \eta'(0)\frac{(u_0^\tau(t))^2}{2} \leq 0 \tag{11}$$

holds for a.e. $t \in (0,\infty)$, where $u_0^\tau(t)$ is the trace of u on the boundary $x = 0$.

We observe that the previous definition is equivalent to the following inequality (see [2]):

$$\begin{aligned}\int_0^\infty \int_0^\infty (|u-c|\partial_t\phi + \operatorname{sign}(u-c)\,(f(u)-f(c))\partial_x\phi)dt\,dx \\ + \gamma \int_0^\infty \int_0^\infty \operatorname{sign}(u-c)\,P\,dt\,dx \\ - \int_0^\infty \operatorname{sign}(c)\,(f(u_0^\tau(t)) - f(c))dt \\ + \int_0^\infty |u_0(x)-c|\phi(0,x)dx \geq 0,\end{aligned}$$

for every non-negative $\phi \in C^\infty(\mathbb{R}^2)$ with compact support, and for every $c \in \mathbb{R}$.

The main result of this section is the following theorem.

Theorem 1. *Assume* (3), (5), *and* (6). *The initial-boundary value problem* (2), (4), *and* (5) *possesses a unique entropy solution u in the sense of Definition 1. Moreover, if u and v are two entropy solutions* (2), (4), (5) *in the sense of Definition 1 the following inequality holds*

$$\|u(t,\cdot) - v(t,\cdot)\|_{L^1(0,R)} \leq e^{Ct}\, \|u(0,\cdot) - v(0,\cdot)\|_{L^1(0,R+Lt)}, \tag{12}$$

for almost every $t > 0$, $R, T > 0$, and a suitable constant $C > 0$.

Our existence argument is based on a passage to the limit in a vanishing viscosity approximation of (8). Fix a small number $\varepsilon > 0$, and let $u_\varepsilon = u_\varepsilon(t,x)$ be the unique classical solution of the following mixed problem

$$\begin{cases} \partial_t u_\varepsilon + \partial_x f(u_\varepsilon) = \gamma P_\varepsilon + \varepsilon \partial_{xx}^2 u_\varepsilon, & t > 0,\ x > 0, \\ \partial_x P_\varepsilon = u_\varepsilon, & t > 0,\ x > 0, \\ u_\varepsilon(t,0) = P_\varepsilon(t,0) = 0, & t > 0, \\ u_\varepsilon(0,x) = u_{\varepsilon,0}(x), & x > 0, \end{cases} \tag{13}$$

where $u_{\varepsilon,0}$ is a $C^\infty(0,\infty)$ approximation of u_0 such that

$$\|u_{\varepsilon,0}\|_{L^2(0,\infty)} \leq \|u_0\|_{L^2(0,\infty)}, \quad \int_0^\infty u_{\varepsilon,0}(x)dx = 0. \tag{14}$$

Clearly, (13) is equivalent to the integro-differential problem

$$\begin{cases} \partial_t u_\varepsilon + \partial_x f(u_\varepsilon) = \gamma \int_0^x u_\varepsilon(t,y)dy + \varepsilon \partial_{xx}^2 u_\varepsilon, & t > 0,\ x > 0, \\ u_\varepsilon(t,0) = 0, & t > 0, \\ u_\varepsilon(0,x) = u_{\varepsilon,0}(x), & x > 0. \end{cases} \tag{15}$$

The existence of such solutions can be obtained by fixing a small number $\delta > 0$ and considering the further approximation of (13) (see [4])

$$\begin{cases} \partial_t u_{\varepsilon,\delta} + \partial_x f(u_{\varepsilon,\delta}) = \gamma P_{\varepsilon,\delta} + \varepsilon \partial_{xx}^2 u_{\varepsilon,\delta}, & t > 0,\ x > 0, \\ -\delta \partial_{xx}^2 P_{\varepsilon,\delta} + \partial_x P_{\varepsilon,\delta} = u_{\varepsilon,\delta}, & t > 0,\ x > 0, \\ u_{\varepsilon,\delta}(t,0) = P_{\varepsilon,\delta}(t,0) = \partial_x P_{\varepsilon,\delta}(t,0) = 0, & t > 0, \\ u_{\varepsilon,\delta}(0,x) = u_{\varepsilon,0}(x), & x > 0, \end{cases}$$

and then sending $\delta \to 0$.

Let us prove some a priori estimates on u_ε.

Lemma 1. *The following statements are equivalent*

$$\int_0^\infty u_\varepsilon(t,x)dx = 0, \quad t \geq 0, \tag{16}$$

$$\frac{d}{dt}\int_0^\infty u_\varepsilon^2 dx + 2\varepsilon \int_0^\infty (\partial_x u_\varepsilon)^2 dx = 0, \quad t > 0. \tag{17}$$

Proof. Let $t > 0$. We begin by proving that (16) implies (17). Multiplying (15) by u_ε and integrating over $(0,\infty)$ gives

$$\begin{aligned}\frac{1}{2}\frac{d}{dt}\int_0^\infty u_\varepsilon^2 dx &= \int_0^\infty u_\varepsilon \partial_t u_\varepsilon dx\\ &= \varepsilon\int_0^\infty u_\varepsilon \partial_{xx}^2 u_\varepsilon dx - \int_0^\infty u_\varepsilon f'(u_\varepsilon)\partial_x u_\varepsilon dx + \gamma\int_0^\infty u_\varepsilon\Big(\int_0^x u_\varepsilon dy\Big)dx\\ &= -\varepsilon\int_0^\infty (\partial_x u_\varepsilon)^2 dx + \gamma\int_0^\infty u_\varepsilon\Big(\int_0^x u_\varepsilon dy\Big)dx.\end{aligned}$$

For (13),

$$\int_0^\infty u_\varepsilon\Big(\int_0^x u_\varepsilon dy\Big)dx = \int_0^\infty P_\varepsilon(t,x)\partial_x P_\varepsilon(t,x)dx = \frac{1}{2}P_\varepsilon^2(t,\infty).$$

Then,

$$\frac{d}{dt}\int_0^\infty u_\varepsilon^2 dx + 2\varepsilon\int_0^\infty (\partial_x u_\varepsilon)^2 dx = \gamma P_\varepsilon^2(t,\infty). \tag{18}$$

Thanks to (16),

$$\lim_{x\to\infty} P_\varepsilon^2(t,x) = \Big(\int_0^\infty u_\varepsilon(t,x)dx\Big)^2 = 0. \tag{19}$$

Now (18) and (19) give (17).

Let us show that (17) implies (16). We assume by contraddiction that (16) does not hold, namely:

$$\int_0^\infty u_\varepsilon(t,x)dx \neq 0.$$

For (13),

$$P_\varepsilon^2(t,\infty) = \Big(\int_0^\infty u_\varepsilon(t,x)dx\Big)^2 \neq 0.$$

Therefore, (18) gives

$$\frac{d}{dt}\int_0^\infty u_\varepsilon^2 dx + 2\varepsilon\int_0^\infty (\partial_x u_\varepsilon)^2 dx \neq 0,$$

which contradicts (17). □

Lemma 2. *For each $t \geq 0$, (16) holds true. In particular, we have that*

$$\|u_\varepsilon(t,\cdot)\|_{L^2(0,\infty)}^2 + 2\varepsilon\int_0^t \|\partial_x u_\varepsilon(s,\cdot)\|_{L^2(0,\infty)}^2 ds \leq \|u_0\|_{L^2(0,\infty)}^2 . \tag{20}$$

Proof. We begin by observing that $u_\varepsilon(t,0) = 0$ implies $\partial_t u_\varepsilon(t,0) = 0$. Thus, thanks to (3),

$$\varepsilon\partial_{xx}^2 u_\varepsilon(t,0) = \partial_t u_\varepsilon(t,0) + f'(u_\varepsilon(t,0))\partial_x u_\varepsilon(t,0) - \gamma \int_0^0 u_\varepsilon(t,x)dx = 0. \quad (21)$$

Differentiating (15) with respect to x, we have

$$\partial_x(\partial_t u_\varepsilon + \partial_x f(u_\varepsilon) - \varepsilon\partial_{xx}^2 u_\varepsilon) = \gamma u_\varepsilon.$$

For (21) and the smoothness of u_ε, an integration over $(0,\infty)$ gives (16). Lemma 1 says that (17) also holds true. Therefore, integrating (17) on $(0,t)$, for (14), we have (20). □

Lemma 3. *We have that*

$$\{u_\varepsilon\}_{\varepsilon>0} \quad \text{is bounded in} \quad L^\infty_{loc}((0,\infty)^2). \quad (22)$$

Consequently,

$$\{P_\varepsilon\}_{\varepsilon>0} \quad \text{is bounded in} \quad L^\infty_{loc}((0,\infty)^2). \quad (23)$$

Proof. Thanks to (15), (20), and the Hölder inequality,

$$\begin{aligned}\partial_t u_\varepsilon + \partial_x f(u_\varepsilon) - \varepsilon\partial_{xx}^2 u_\varepsilon &= \gamma \int_0^x u_\varepsilon(t,y)dy \le \gamma\left|\int_0^x u_\varepsilon(t,y)dy\right| \\ &\le \gamma \int_0^x |u_\varepsilon(t,y)|dy \le \gamma\sqrt{x}\,\|u_\varepsilon(t,\cdot)\|_{L^2(0,\infty)} \\ &\le \gamma\sqrt{x}\,\|u_0\|_{L^2(0,\infty)}\,.\end{aligned}$$

Let v, w, v_ε, and w_ε be the solutions of the following equations:

$$\begin{cases}\partial_t v + \partial_x f(v) = \gamma\,\|u_0\|_{L^2(0,\infty)}\sqrt{x}, & t>0, x>0,\\ v(t,0) = 0, & t>0,\\ v(0,x) = u_0(x), & x>0,\end{cases}$$

$$\begin{cases}\partial_t w + \partial_x f(w) = -\gamma\,\|u_0\|_{L^2(0,\infty)}\sqrt{x}, & t>0, x>0,\\ w(t,0) = 0, & t>0,\\ w(0,x) = u_0(x), & x>0,\end{cases}$$

$$\begin{cases} \partial_t v_\varepsilon + \partial_x f(v_\varepsilon) = \gamma \left\|u_0\right\|_{L^2(0,\infty)} \sqrt{x} + \varepsilon \partial_{xx}^2 v_\varepsilon, & t > 0, x > 0, \\ v_\varepsilon(t,0) = 0, & t > 0, \\ v_\varepsilon(0,x) = u_{\varepsilon,0}(x), & x > 0, \end{cases}$$

$$\begin{cases} \partial_t w_\varepsilon + \partial_x f(w_\varepsilon) = -\gamma \left\|u_0\right\|_{L^2(0,\infty)} \sqrt{x} + \varepsilon \partial_{xx}^2 w_\varepsilon, & t > 0, x > 0, \\ w_\varepsilon(t,0) = 0, & t > 0, \\ w_\varepsilon(0,x) = u_{\varepsilon,0}(x), & x > 0, \end{cases}$$

respectively. Then u_ε, v_ε, and w_ε are respectively a solution, a supersolution, and a subsolution of the parabolic problem

$$\begin{cases} \partial_t q + \partial_x f(q) = \gamma \int_0^x u_\varepsilon(t,y)dy + \varepsilon \partial_{xx}^2 q, & t > 0, \ x > 0, \\ q(t,0) = 0, & t > 0, \\ q(0,x) = u_{\varepsilon,0}(x), & x > 0. \end{cases}$$

Thus, see [6, Chap. 2, Theorem 9],

$$w_\varepsilon \le u_\varepsilon \le v_\varepsilon.$$

Moreover, $\{w_\varepsilon\}_{\varepsilon>0}$ and $\{v_\varepsilon\}_{\varepsilon>0}$ are uniformly bounded in $L^\infty_{loc}((0,\infty)^2)$ and converge to w and v respectively, see [1, 17]. Therefore the two functions

$$W = \inf_{\varepsilon>0} w_\varepsilon, \qquad V = \sup_{\varepsilon>0} v_\varepsilon$$

belong to $L^\infty_{loc}((0,\infty)^2)$ and satisfy

$$W \le w_\varepsilon \le u_\varepsilon \le v_\varepsilon \le V. \tag{24}$$

This gives (22). Since

$$|P_\varepsilon(t,x)| = \left| \int_0^x u_\varepsilon(t,y)dy \right| \le \int_0^x |u_\varepsilon(t,y)|dy,$$

(23) follows from (22). □

Let us continue by proving the existence of a distributional solution to (2), (4), and (5) satisfying (10).

Lemma 4. *There exists a function $u \in L^\infty_{loc}((0,\infty)^2)$ that is a distributional solution of* (8) *and satisfies* (10) *for every convex entropy $\eta \in C^2(\mathbb{R})$.*

We construct a solution by passing to the limit in a sequence $\{u_\varepsilon\}_{\varepsilon>0}$ of viscosity approximations (13). We use the compensated compactness method [24].

Lemma 5. *There exist a subsequence* $\{u_{\varepsilon_k}\}_{k\in\mathbb{N}}$ *of* $\{u_\varepsilon\}_{\varepsilon>0}$ *and a limit function* $u \in L^\infty_{loc}((0,\infty)^2)$ *such that*

$$u_{\varepsilon_k} \to u \text{ a.e. and in } L^p_{loc}((0,\infty)^2),\ 1 \le p < \infty. \tag{25}$$

Moreover, we have

$$P_{\varepsilon_k} \to P \text{ a.e. and in } L^p_{loc}(0,\infty; W^{1,p}_{loc}(0,\infty)),\ 1 \le p < \infty, \tag{26}$$

where

$$P(t,x) = \int_0^x u(t,y)dy, \qquad t \ge 0, \quad x \ge 0.$$

Proof. Let $\eta : \mathbb{R} \to \mathbb{R}$ be any convex C^2 entropy function, and $q : \mathbb{R} \to \mathbb{R}$ be the corresponding entropy flux defined by $q' = f'\eta'$. By multiplying the first equation in (13) by $\eta'(u_\varepsilon)$ and using the chain rule, we get

$$\partial_t \eta(u_\varepsilon) + \partial_x q(u_\varepsilon) = \underbrace{\varepsilon \partial^2_{xx}\eta(u_\varepsilon)}_{=:\mathcal{L}_{1,\varepsilon}} \underbrace{-\varepsilon\eta''(u_\varepsilon)\,(\partial_x u_\varepsilon)^2}_{=:\mathcal{L}_{2,\varepsilon}} \underbrace{+\gamma\eta'(u_\varepsilon)P_\varepsilon}_{=:\mathcal{L}_{3,\varepsilon}},$$

where $\mathcal{L}_{1,\varepsilon}$, $\mathcal{L}_{2,\varepsilon}$, $\mathcal{L}_{3,\varepsilon}$ are distributions.

Thanks to Lemma 2

$$\mathcal{L}_{1,\varepsilon} \to 0 \text{ in } H^{-1}_{loc}((0,\infty)^2),$$
$$\{\mathcal{L}_{2,\varepsilon}\}_{\varepsilon>0} \text{ is uniformly bounded in } L^1_{loc}((0,\infty)^2).$$

We prove that

$$\{\mathcal{L}_{3,\varepsilon}\}_{\varepsilon>0} \text{ is uniformly bounded in } L^1_{loc}((0,\infty)^2).$$

Let K be a compact subset of $(0,\infty)^2$. For Lemma 3,

$$\begin{aligned}\left\|\gamma\eta'(u_\varepsilon)P_\varepsilon\right\|_{L^1(K)} &= \gamma\iint_K |\eta'(u_\varepsilon)||P_\varepsilon|dt\,dx\\ &\le \gamma\left\|\eta'(u_\varepsilon)\right\|_{L^\infty(K)}\|P_\varepsilon\|_{L^\infty(K)}|K|.\end{aligned}$$

Therefore, Murat's lemma [19] implies that

$$\{\partial_t\eta(u_\varepsilon) + \partial_x q(u_\varepsilon)\}_{\varepsilon>0} \text{ lies in a compact subset of } H^{-1}_{\mathrm{loc}}((0,\infty)^2). \tag{27}$$

The L^∞_{loc} bound stated in Lemma 3, (27), and Tartar's compensated compactness method [24] give the existence of a subsequence $\{u_{\varepsilon_k}\}_{k\in\mathbb{N}}$ and a limit function $u \in L^\infty_{loc}((0,\infty)^2)$ such that (25) holds.

Finally, (26) follows from (25), the Hölder inequality, and the identities

$$P_{\varepsilon_k}(t,x) = \int_0^x u_{\varepsilon_k}(t,y)dy, \qquad \partial_x P_{\varepsilon_k} = u_{\varepsilon_k}.$$

Moreover, [5, Theorem 1.1] tells us that the limit u admits a strong boundary trace u_0^τ at $(0,\infty) \times \{x = 0\}$. Since, arguing as in [5, Sect. 3.1] (indeed our solution is obtained as the vanishing viscosity limit of (8)), [5, Lemma 3.2] and the boundedness of the source term P (cf. (9)) imply (11). □

We are now ready for the proof of Theorem 1.

Proof (Proof of Theorem 1). Lemma (5) gives the existence of an entropy solution $u(t,x)$ of (7), or equivalently (8).

Let us show that $u(t,x)$ is unique, and that (12) holds true. Since our solutions is only locally bounded we use the doubling of variables method and get local estimates based on the finite speed of propagation of the waves generated by (2). Let u, v be two entropy solutions of (7), or equivalently of (8), and $0 < t < T$. By arguing as in [2,9], using the fact that the two solutions satisfy the same boundary conditions, we can prove that

$$\partial_t(|u-v|) + \partial_x((f(u)-f(v))\text{sign}\,(u-v)) - \gamma\text{sign}\,(u-v)\,(P_u - P_v) \le 0$$

holds in the sense of distributions in $(0,\infty) \times (0,\infty)$, and

$$\begin{aligned}\|u(t,\cdot)-v(t,\cdot)\|_{I(t)} &\le \|u_0 - v_0\|_{I(0)} \\ &\quad + \gamma \int_0^t \int_{I(s)} \text{sign}\,(u-v)\,(P_u - P_v) ds\,dx, \qquad 0 < t < T, \end{aligned} \tag{28}$$

where

$$P_u(t,x) = \int_0^x u(t,y)dy, \quad P_v = \int_0^x v(t,y)dy, \quad I(s) = (0, R + L(t-s)),$$

and L is the Lipschitz constant of the flux f.

Since

$$\begin{aligned}\gamma \int_0^t \int_{I(s)} \text{sign}\,(u-v)\,(P_u - P_v) ds\,dx &\le \gamma \int_0^t \int_{I(s)} |P_u - P_v| ds\,dx \\ &\le \gamma \int_0^t \int_{I(s)} \left(\left| \int_0^x |u-v| dy \right| \right) ds\,dx\end{aligned}$$

$$\leq \gamma \int_0^t \int_{I(s)} \left(\left| \int_{I(s)} |u - v| dy \right| \right) ds\, dx$$
$$= \gamma \int_0^t |I(s)| \, \|u(s,\cdot) - v(s,\cdot)\|_{L^1(I(s))} \, ds, \tag{29}$$

and

$$|I(s)| = R + L(t - s) \leq R + Lt \leq R + LT, \tag{30}$$

we can consider the following continuous function:

$$G(t) = \|u(t,\cdot) - v(t,\cdot)\|_{L^1(I(t))}, \quad t \geq 0. \tag{31}$$

Using this notation, it follows from (28)–(30) that

$$G(t) \leq G(0) + C \int_0^t G(s) ds,$$

where $C = \gamma(R + LT)$. Gronwall's inequality and (31) give

$$\|u(t,\cdot) - v(t,\cdot)\|_{L^1(0,R)} \leq e^{Ct} \, \|u_0 - v_0\|_{L^1(0,R+Lt)},$$

that is (12). □

3 The Cauchy Problem

Let us consider now the Cauchy problem associated to (2). Since the arguments are similar to those of the previous section we simply sketch them, highlighting only the differences between the two problems.

In this section we augment (2) with the initial datum

$$u(0,x) = u_0(x), \qquad x \in \mathbb{R}. \tag{32}$$

We assume that

$$u_0 \in L^2(\mathbb{R}) \cap L^\infty_{loc}(\mathbb{R}), \qquad \int_{\mathbb{R}} u_0(x) dx = 0. \tag{33}$$

Indeed, integrating both sides of (2) we have that $u(t,\cdot)$ has zero mean for every $t > 0$, therefore it is natural to assume the same on the initial condition. We rewrite the Cauchy problem (2), (32) in the following way

$$\begin{cases} \partial_t u + \partial_x f(u) = \gamma \int_0^x u(t,y)dy, & t > 0,\ x \in \mathbb{R}, \\ u(0,x) = u_0(x), & x \in \mathbb{R}, \end{cases} \tag{34}$$

or equivalently

$$\begin{cases} \partial_t u + \partial_x f(u) = \gamma P, & t > 0,\ x \in \mathbb{R}, \\ \partial_x P = u, & t > 0,\ x \in \mathbb{R}, \\ P(t,0) = 0, & t > 0, \\ u(0,x) = u_0(x), & x \in \mathbb{R}. \end{cases} \tag{35}$$

Due to the regularizing effect of the P equation in (35) we have that

$$u \in L^\infty_{loc}((0,\infty)\times\mathbb{R}) \Longrightarrow P \in L^\infty_{loc}((0,\infty); W^{1,\infty}_{loc}(\mathbb{R})).$$

Definition 2. We say that $u \in L^\infty_{loc}((0,\infty)\times\mathbb{R})$ is an entropy solution of the initial value problem (2), and (32) if:

(i) u is a distributional solution of (34) or equivalently of (35);
(ii) For every convex function $\eta \in C^2(\mathbb{R})$ the entropy inequality

$$\partial_t \eta(u) + \partial_x q(u) - \gamma\eta'(u)P \le 0, \qquad q(u) = \int^u f'(\xi)\eta'(\xi)\,d\xi, \tag{36}$$

holds in the sense of distributions in $(0,\infty)\times\mathbb{R}$.

The main result of this section is the following theorem.

Theorem 2. *Assume* (32) *and* (33)*. The initial value problem* (2) *and* (32) *possesses a unique entropy solution u in the sense of Definition 2. Moreover, if u and v are two entropy solutions* (2) *and* (32)*, in the sense of Definition 2 the following inequality holds*

$$\|u(t,\cdot)-v(t,\cdot)\|_{L^1(-R,R)} \le e^{Ct}\,\|u(0,\cdot)-v(0,\cdot)\|_{L^1(-R-Lt,R+Lt)}, \tag{37}$$

for almost every $t > 0$, $R, T > 0$, *and a suitable constant* $C > 0$.

Our existence argument is based on a passage to the limit in a vanishing viscosity approximation of (35).

Fix a small number $\varepsilon > 0$, and let $u_\varepsilon = u_\varepsilon(t,x)$ be the unique classical solution of the following mixed problem

$$\begin{cases} \partial_t u_\varepsilon + \partial_x f(u_\varepsilon) = \gamma P_\varepsilon + \varepsilon\partial^2_{xx} u_\varepsilon, & t > 0,\ x \in \mathbb{R}, \\ \partial_x P_\varepsilon = u_\varepsilon, & t > 0,\ x \in \mathbb{R}, \\ P_\varepsilon(t,0) = 0, & t > 0, \\ u_\varepsilon(0,x) = u_{\varepsilon,0}(x), & x \in \mathbb{R}, \end{cases} \tag{38}$$

where $u_{\varepsilon,0}$ is a $C^\infty(\mathbb{R})$ approximation of u_0 such that

$$\|u_{\varepsilon,0}\|_{L^2(\mathbb{R})} \leq \|u_0\|_{L^2(\mathbb{R})}, \qquad \int_{\mathbb{R}} u_{\varepsilon,0}(x)dx = 0. \tag{39}$$

Clearly, (38) is equivalent to the integro-differential problem

$$\begin{cases} \partial_t u_\varepsilon + \partial_x f(u_\varepsilon) = \gamma \int_0^x u_\varepsilon(t,y)dy + \varepsilon \partial_{xx}^2 u_\varepsilon, & t > 0,\ x \in \mathbb{R}, \\ u_\varepsilon(0,x) = u_{\varepsilon,0}(x), & x \in \mathbb{R}. \end{cases} \tag{40}$$

The existence of such solutions can be obtained by fixing a small number $\delta > 0$ and considering the further approximation of (38) (see [4])

$$\begin{cases} \partial_t u_{\varepsilon,\delta} + \partial_x f(u_{\varepsilon,\delta}) = \gamma P_{\varepsilon,\delta} + \varepsilon \partial_{xx}^2 u_{\varepsilon,\delta}, & t > 0,\ x \in \mathbb{R}, \\ -\delta \partial_{xx}^2 P_{\varepsilon,\delta} + \partial_x P_{\varepsilon,\delta} = u_{\varepsilon,\delta}, & t > 0,\ x \in \mathbb{R}, \\ P_{\varepsilon,\delta}(t,0) = 0, & t > 0, \\ u_{\varepsilon,\delta}(0,x) = u_{\varepsilon,0}(x), & x \in \mathbb{R}, \end{cases}$$

and then sending $\delta \to 0$.

Let us prove some a priori estimates on u_ε. Arguing as in Lemma 1 we have the following.

Lemma 6. *Let us suppose that*

$$P_\varepsilon(t,-\infty) = 0, \quad t \geq 0, \quad (or \quad P_\varepsilon(t,\infty) = 0), \tag{41}$$

where $P_\varepsilon(t,x)$ is defined in (38)*. Then the following statements are equivalent*

$$\int_{\mathbb{R}} u_\varepsilon(t,x)dx = 0, \quad t \geq 0, \tag{42}$$

$$\frac{d}{dt}\int_{\mathbb{R}} u_\varepsilon^2 dx + 2\varepsilon \int_{\mathbb{R}} (\partial_x u_\varepsilon)^2 dx = 0, \quad t > 0. \tag{43}$$

Lemma 7. *For each $t \geq 0$,* (42) *holds true, and*

$$P_\varepsilon(t,\infty) = P_\varepsilon(t,-\infty) = 0. \tag{44}$$

In particular, we have that

$$\|u_\varepsilon(t,\cdot)\|_{L^2(\mathbb{R})}^2 + 2\varepsilon \int_0^t \|\partial_x u_\varepsilon(s,\cdot)\|_{L^2(\mathbb{R})}^2 ds \leq \|u_0\|_{L^2(\mathbb{R})}^2. \tag{45}$$

Proof. Differentiating (40) with respect to x, we have

$$\partial_x(\partial_t u_\varepsilon + \partial_x f(u_\varepsilon) - \varepsilon\partial_{xx}^2 u_\varepsilon) = u_\varepsilon.$$

Since u_ε is a smooth solution of (40), an integration over $\mathbb{R}$ gives (42).
Again for the regularity of u_ε, from (38), we get

$$\lim_{x\to-\infty}(\partial_t u_\varepsilon + \partial_x f(u_\varepsilon) - \varepsilon\partial_{xx}^2 u_\varepsilon) = \gamma\int_0^{-\infty} u_\varepsilon(t,x)dx = \gamma P_\varepsilon(t,-\infty) = 0,$$

$$\lim_{x\to\infty}(\partial_t s u_\varepsilon + \partial_x f(u_\varepsilon) - \varepsilon\partial_{xx}^2 u_\varepsilon) = \gamma\int_0^{\infty} u_\varepsilon(t,x)dx = \gamma P_\varepsilon(t,\infty) = 0,$$

that is (44).

Lemma 6 says that (43) also holds true. Therefore, integrating (43) on $(0,t)$, for (39), we have (45). □

Arguing as in Lemma 3 we obtain the following lemma:

Lemma 8. *We have that*

$$\{u_\varepsilon\}_{\varepsilon>0} \quad \text{is bounded in} \quad L^\infty_{loc}((0,\infty)\times\mathbb{R}). \tag{46}$$

Consequently,

$$\{P_\varepsilon\}_{\varepsilon>0} \quad \text{is bounded in} \quad L^\infty_{loc}((0,\infty)\times\mathbb{R}). \tag{47}$$

Let us continue by proving the existence of a distributional solution to (2) and (5) satisfying (36).

Lemma 9. *There exists a function* $u \in L^\infty_{loc}((0,\infty)\times\mathbb{R})$ *that is a distributional solution of* (35) *and satisfies* (36) *for every convex entropy* $\eta \in C^2(\mathbb{R})$.

We construct a solution by passing to the limit in a sequence $\{u_\varepsilon\}_{\varepsilon>0}$ of viscosity approximations (38). We use the compensated compactness method [24].

Lemma 10. *There exists a subsequence* $\{u_{\varepsilon_k}\}_{k\in\mathbb{N}}$ *of* $\{u_\varepsilon\}_{\varepsilon>0}$ *and a limit function* $u \in L^\infty_{loc}((0,\infty)\times\mathbb{R})$ *such that*

$$u_{\varepsilon_k} \to u \text{ a.e. and in } L^p_{loc}((0,\infty)\times\mathbb{R}),\ 1 \le p < \infty. \tag{48}$$

Moreover, we have

$$P_{\varepsilon_k} \to P \text{ a.e. and in } L^p_{loc}((0,\infty); W^{1,p}_{loc}(\mathbb{R})),\ 1 \le p < \infty, \tag{49}$$

where

$$P(t,x) = \int_0^x u(t,y)dy, \qquad t \ge 0, \quad x \in \mathbb{R}.$$

Proof. Let $\eta : \mathbb{R} \to \mathbb{R}$ be any convex C^2 entropy function, and $q : \mathbb{R} \to \mathbb{R}$ be the corresponding entropy flux defined by $q' = f'\eta'$. By multiplying the first equation in (38) by $\eta'(u_\varepsilon)$ and using the chain rule, we get

$$\partial_t \eta(u_\varepsilon) + \partial_x q(u_\varepsilon) = \underbrace{\varepsilon \partial_{xx}^2 \eta(u_\varepsilon)}_{=:\mathcal{L}_{1,\varepsilon}} \underbrace{-\varepsilon \eta''(u_\varepsilon)\,(\partial_x u_\varepsilon)^2}_{=:\mathcal{L}_{2,\varepsilon}} \underbrace{+\gamma \eta'(u_\varepsilon) P_\varepsilon}_{=:\mathcal{L}_{3,\varepsilon}},$$

where $\mathcal{L}_{1,\varepsilon}$, $\mathcal{L}_{2,\varepsilon}$, $\mathcal{L}_{3,\varepsilon}$ are distributions.

Arguing as in Lemma 5, we have that

$$\begin{aligned}&\mathcal{L}_{1,\varepsilon} \to 0 \text{ in } H_{loc}^{-1}((0,\infty)\times\mathbb{R}),\\ &\{\mathcal{L}_{2,\varepsilon}\}_{\varepsilon>0} \text{ and } \{\mathcal{L}_{3,\varepsilon}\}_{\varepsilon>0} \text{ are uniformly bounded in } L_{loc}^1((0,\infty)\times\mathbb{R}).\end{aligned}$$

Therefore, Murat's lemma [19] implies that

$$\{\partial_t \eta(u_\varepsilon) + \partial_x q(u_\varepsilon)\}_{\varepsilon>0} \text{ lies in a compact subset of } H_{\mathrm{loc}}^{-1}((0,\infty)\times\mathbb{R}). \tag{50}$$

The L_{loc}^∞ bound stated in Lemma 8, (50), and Tartar's compensated compactness method [24] imply the existence of a subsequence $\{u_{\varepsilon_k}\}_{k\in\mathbb{N}}$ and a limit function $u \in L_{loc}^\infty((0,\infty)\times\mathbb{R})$ such that (48) holds.

Finally, (49) follows from (48), the Hölder inequality, and the identities

$$P_{\varepsilon_k}(t,x) = \int_0^x u_{\varepsilon_k}(t,y)dy, \qquad \partial_x P_{\varepsilon_k} = u_{\varepsilon_k}.$$

□

We are now ready for the proof of Theorem 2.

Proof (Proof of Theorem 2). Lemma (10) gives the existence of an entropy solution u of (7), or equivalently (35).

Let us show that u is unique, and that (37) holds true. Let u, v be two entropy solutions of (7) or equivalently of (35) and $0 < t < T$. Arguing as in [9] we can prove that

$$\begin{aligned}\|u(t,\cdot) - v(t,\cdot)\|_{I(t)} &\le \|u_0 - v_0\|_{I(0)}\\ &\quad + \gamma \int_0^t \int_{I(s)} \mathrm{sign}\,(u-v)\,(P_u - P_v) ds\, dx \qquad 0 < t < T,\end{aligned} \tag{51}$$

where

$$P_u(t,x) = \int_0^x u(t,y)dy, \quad P_v = \int_0^x v(t,y)dy, \quad I(s) = (-R - L(t-s), R + L(t-s)),$$

and L is the Lipschitz constant of the flux f.

Since

$$\begin{aligned}\gamma\int_0^t\int_{I(s)}\operatorname{sign}(u-v)\,(P_u-P_v)ds\,dx &\le \gamma\int_0^t\int_{I(s)}|P_u-P_v|ds\,dx\\ &\le \gamma\int_0^t\int_{I(s)}\left(\left|\int_0^x|u-v|dy\right|\right)ds\,dx\\ &\le \gamma\int_0^t\int_{I(s)}\left(\left|\int_{I(s)}|u-v|dy\right|\right)ds\,dx\\ &= \gamma\int_0^t|I(s)|\,\|u(s,\cdot)-v(s,\cdot)\|_{L^1(I(s))}\,ds,\end{aligned} \tag{52}$$

and

$$|I(s)| = 2R + 2L(t-s) \le 2R + 2Lt \le 2R + 2LT, \tag{53}$$

we can consider the following continuous function:

$$G(t) = \|u(t,\cdot)-v(t,\cdot)\|_{L^1(I(t))}, \quad t \ge 0. \tag{54}$$

It follows from (51) to (53) that

$$G(t) \le G(0) + C\int_0^t G(s)ds,$$

where $C = \gamma(2R + 2LT)$.

Gronwall's inequality and (54) give

$$\|u(t,\cdot)-v(t,\cdot)\|_{L^1(-R,R)} \le e^{Ct}\,\|u_0-v_0\|_{L^1(-R-Lt,R+Lt)},$$

that is (37). □

References

1. D. Amadori, L. Gosse, G. Guerra, Godunov-type approximation for a general resonant balance law with large data. J. Differ. Equ. **198**, 233–274 (2004)
2. C. Bardos, A.Y. le Roux, J.C. Nédélec, First order quasilinear equations with boundary conditions. Commun. Partial Differ. Equ. 4 **9**, 1017–1034 (1979)
3. J. Boyd, Ostrovsky and Hunters generic wave equation for weakly dispersive waves: matched asymptotic and pseudospectral study of the paraboloidal travelling waves (corner and near-corner waves). Eur. J. Appl. Math. **16**(1), 65–81 (2005)

4. G.M. Coclite, H. Holden, K.H. Karlsen, Wellposedness for a parabolic-elliptic system. Discret. Contin. Dyn. Syst. **13**(3), 659–682 (2005)
5. G.M. Coclite, K.H. Karlsen, Y.-S. Kwon, Initial-boundary value problems for conservation laws with source terms and the Degasperis-Procesi equation. J. Funct. Anal. **257**(12), 3823–3857 (2009)
6. A. Friedman, *Partial Differential Equations of Parabolic Type* (Dover Books on Mathematics, New York, 2008)
7. G. Gui, Y. Liu, On the Cauchy problem for the Ostrovsky equation with positive dispersion. Commun. Partial Differ. Equ. **32**(10–12), 1895–1916 (2007)
8. J. Hunter, Numerical solutions of some nonlinear fispersive wave equations. (Computational solution of nonlinear systems of equations (Fort Collins, CO, 1988)). Lect. Appl. Math. (American Mathematical Society, Providence) **26**, 301–316 (1990)
9. S.N. Kružkov, First order quasilinear equations with several independent variables. Mat. Sb. (N.S.), **81**(123), 228–255 (1970)
10. S. Levandosky, Y. Liu, Stability of solitary waves of a generalized Ostrovsky equation. SIAM J. Math. Anal. **38**(3), 985–1011 (2006)
11. S. Levandosky, Y. Liu, Stability and weak rotation limit of solitary waves of the Ostrovsky equation. Discret. Contin. Dyn. Syst. B **7**(7), 793–806 (2007)
12. F. Linares, A. Milanes, Local and global well-posedness for the Ostrovsky equation. J. Differ. Equ. **222**(2), 325–340 (2006)
13. Y. Liu, On the stability of solitary waves for the Ostrovsky equation. Quart. Appl. Math. **65**(3), 571–589 (2007)
14. Y. Liu, V. Varlamov, Cauchy problem for the Ostrovsky equation. Discret. Contin. Dyn. Syst. **10**(3), 731–753 (2004)
15. Y. Liu, V. Varlamov, Stability of solitary waves and weak rotation limit for the Ostrovsky equation. J. Differ. Equ. **203**(1), 159–183 (2004)
16. Y. Liu, D. Pelinovsky, A. Sakovich, Wave breaking in the Ostrovsky–Hunter equation. SIAM J. Math. Anal. **42**(5), 1967–1985 (2010)
17. J. Málek, J. Nevcas, M. Rokyta, M. Rocircuvzivcka, *Weak and Measure-Valued Solutions to Evolutionary PDEs*. Applied Mathematics and Mathematical Computation, vol. 13 (Chapman-Hall, London, 1996)
18. A.J. Morrison, E.J. Parkes, V.O. Vakhnenko, The N loop soliton solutions of the Vakhnenko equation. Nonlinearity **12**(5), 1427–1437 (1999)
19. F. Murat, L'injection du cône positif de H^{-1} dans $W^{-1,q}$ est compacte pour tout $q < 2$. J. Math. Pures Appl. (9), **60**(3), 309–322 (1981)
20. L.A. Ostrovsky, Nonlinear internal waves in a rotating ocean. Okeanologia **18**, 181–191 (1978)
21. E.J. Parkes, Explicit solutions of the reduced Ostrovsky equation. Chaos Solitons and Fractals **31**(3), 602–610 (2007)
22. E.J. Parkes, V.O. Vakhnenko, The calculation of multi-soliton solutions of the Vakhnenko equation by the inverse scattering method. Chaos, Solitons and Fractals **13**(9), 1819–1826 (2002)
23. Y.A. Stepanyants, On stationary solutions of the reduced Ostrovsky equation: periodic waves, compactons and compound solitons. Chaos, Solitons and Fractals **28**(1), 193–204 (2006)
24. L. Tartar, Compensated compactness and applications to partial differential equations. In: *Nonlinear Analysis and Mechanics: Heriot-Watt Symposium*, vol. IV (Pitman, Boston, 1979), pp. 136–212
25. K. Tsugawa, Well-posedness and weak rotation limit for the Ostrovsky equation. J. Differ. Equ. **247**(12), 3163–3180 (2009)

An Overview of Piston Problems in Fluid Dynamics

Min Ding and Yachun Li

Abstract The piston problem is analyzed from the mathematical point of view. Some features and phenomena caused by the motion of the piston are revealed. Some developments for piston problems are reviewed. We discuss some piston problems for both classical Euler equations and relativistic Euler equations of compressible fluids. In particular, we focus on shock front solutions.

Keywords Piston problem • Euler equations • Relativistic Euler equations • Shock front solution • Shock waves • Rarefaction waves • Glimm scheme • Interaction of waves • Newton iteration scheme

2010 Mathematics Subject Classification Primary: 35A01, 35B20, 35D30, 76N15, 76Y05; Secondary: 35L04, 35L65, 35L67, 83A05.

1 Mathematical Models and Physical Phenomena

The piston problem is a special initial-boundary value problem in fluid dynamics (see [13, 42, 46]), which can be described as follows. In a thin long tube closed at one end by a piston and open at the other end, any motion of the piston causes the corresponding motion of the gas in the tube. More precisely, if the piston is pushed

M. Ding
Department of Mathematics, Shanghai Jiao Tong University, Shanghai 200240, China

Mathematical Institute, University of Oxford, Oxford OX1 3LB, UK
e-mail: minding@sjtu.edu.cn; dingm@maths.ox.ac.uk

Y. Li (✉)
Department of Mathematics and Key Lab of Scientific and Engineering Computing (MOE), Shanghai Jiao Tong University, Shanghai 200240, China
e-mail: ycli@sjtu.edu.cn

G.-Q.G. Chen et al. (eds.), *Hyperbolic Conservation Laws and Related Analysis with Applications*, Springer Proceedings in Mathematics & Statistics 49,
DOI 10.1007/978-3-642-39007-4_8,

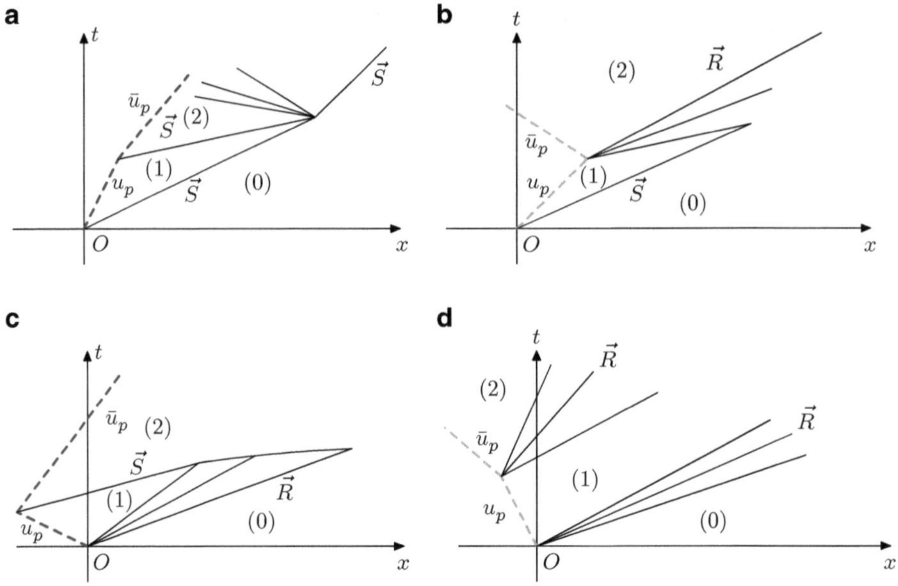

Fig. 1 (**a**) One shock overtakes another. (**b**) Rarefaction wave overtakes shock wave. (**c**) Shock overtakes rarefaction wave. (**d**) Forward rarefaction waves never meet. The *dash line*: the piston boundary. $\vec{S}$: a forward shock wave. $\vec{R}$: a forward rarefaction wave

forward relative to the gas, a shock appears and moves forward faster than the piston. Otherwise, a rarefaction wave appears.

We describe these phenomena in detail for the classical compressible flow. If the piston is initially located at $x = 0$ and moves forward with the velocity u_p into the initial static gas in the tube, then after a time t_0, the velocity of the piston is changed into $\bar{u}_p$ in an infinitely small time interval, i.e., instantaneously. Now we consider the motion of the gas caused by the piston in four cases. (All the velocities are observed relative to the gas in the region (1) between the two waves.)

Case 1. $u_p > 0$, $\bar{u}_p > u_p$. Since $u_p > 0$, the piston is pushed forward, a forward shock appears. $\bar{u}_p > u_p$ leads to the occurrence of another forward shock. The former forward shock travels with subsonic speed and the latter one travels with supersonic speed. Therefore, the latter forward shock will overtake the former one (see Fig. 1a).

Case 2. $u_p > 0$, $\bar{u}_p < u_p$. Since $\bar{u}_p < u_p$, a rarefaction wave appears. The head of the rarefaction wave travels with sonic speed, and the shock front travels with subsonic speed. Therefore, the rarefaction wave overtakes the forward shock (see Fig. 1b).

Case 3. $u_p < 0$, $\bar{u}_p > u_p$. Since $u_p < 0$, the piston is pulled back, a forward rarefaction wave appears. $\bar{u}_p > u_p$ leads to the occurrence of a forward shock. The tail of the rarefaction wave travels with sonic speed while the shock front travels with supersonic speed. So the shock front overtakes the rarefaction wave (see Fig. 1c).

Case 4. $u_p < 0$, $\bar{u}_p < u_p$. Since $u_p < 0$, a rarefaction wave appears. $\bar{u}_p < u_p$ leads to the occurrence of another forward rarefaction wave. Since the tail of one travels at the same speed as the head of the other, these two forward rarefaction waves never meet (see Fig. 1d).

1.1 Mathematical Models

Some partial differential equations in fluid dynamics come from conservation laws of some physical quantities, such as, mass, momentum and energy, etc. The general conservation laws can be written in divergence form

$$\partial_t u + \nabla \cdot f(u) = 0, \quad u \in \mathbb{R}^m, \tag{1}$$

where $(x,t) \in \mathbb{R}^{d+1} = \mathbb{R}^d \times \mathbb{R}^+$, $\nabla = (\partial_{x_1}, \dots, \partial_{x_d})$, and $f = (f_1, \dots, f_d) : \mathbb{R}^m \to (\mathbb{R}^m)^d$ is a nonlinear mapping with $f_i : \mathbb{R}^m \to \mathbb{R}^m$ for $i = 1, \dots, d$.

The hyperbolicity for (1) means that, for any $\omega \in \mathcal{S}^{d-1}$ and any u belonging to the state domain $\mathcal{D}$,

$$(\nabla_u f(u) \cdot \omega)_{m \times m} \text{ has } m \text{ real eigenvalues } \lambda_i(u, \omega), \quad 1 \le i \le m. \tag{2}$$

The jth-characteristic family of (1) in $\mathcal{D}$ is called genuinely nonlinear if, for any $\omega \in \mathcal{S}^{d-1}$, the jth-eigenvalue $\lambda_j(u;\omega)$ and the corresponding eigenvector $r_j(u,\omega)$ satisfy

$$\nabla_u \lambda_j(u,\omega) \cdot r_j(u,\omega) \neq 0 \qquad \text{for any } u \in \mathcal{D},\ \omega \in \mathcal{S}^{d-1}.$$

The jth-characteristic family of (1) in $\mathcal{D}$ is called linearly degenerate if

$$\nabla_u \lambda_j(u,\omega) \cdot r_j(u,\omega) = 0 \qquad \text{for any } u \in \mathcal{D},\ \omega \in \mathcal{S}^{d-1}.$$

An entropy-entropy flux pair (η, q) is a pair of C^1 functions satisfying

$$\nabla \eta(u) \nabla F(u) = \nabla q(u),$$

where $\eta : \mathcal{D} \to \mathbb{R}$ is called an entropy of (1), and $q = (q_1, \dots, q_d) : \mathcal{D} \to \mathbb{R}^d$ is called the corresponding entropy flux.

As we know, no matter how smooth the initial data are, the nonlinearity of (1) leads to the appearance of the singularities in a finite time. In consideration of weak solutions, the uniqueness is lost. In order to single out the physical unique discontinuous solution, we need the following Lax entropy inequality

$$\partial_t \eta(u) + \partial_x q(u) \le 0. \tag{3}$$

The weak solution satisfying (3) in the sense of distributions is called the entropy solution.

A basic prototype of nonlinear conservation laws is the system of Euler equations for compressible fluids, which is described as

$$\begin{cases} \partial_t \rho + \nabla \cdot (\rho \vec{u}) = 0, \\ \partial_t (\rho \vec{u}) + \nabla \cdot (\rho \vec{u} \otimes \vec{u}) + \nabla p = 0, \\ \partial_t (\rho E) + \nabla \cdot \left(\rho \vec{u} \left(E + \frac{p}{\rho} \right) \right) = 0, \end{cases} \tag{4}$$

where ρ is the density, $\vec{u} = (u_1, \ldots, u_d)$ is the velocity, p is the pressure and $E = \frac{1}{2}|\vec{u}|^2 + e$ is the total energy with e the internal energy. System (4) is closed by the state equation $p = p(\rho, e)$.

When the flow is isentropic, the Euler system takes a simpler form

$$\begin{cases} \partial_t \rho + \nabla \cdot (\rho \vec{u}) = 0, \\ \partial_t (\rho \vec{u}) + \nabla \cdot (\rho \vec{u} \otimes \vec{u}) + \nabla p = 0, \end{cases} \tag{5}$$

where the state equation is $p = p(\rho)$ satisfying

$$p'(\rho) > 0 \qquad \text{(hyperbolicity)}, \tag{6}$$

$$\rho p''(\rho) + 2p'(\rho) > 0 \qquad \text{(genuine nonlinearity)}. \tag{7}$$

The mechanical entropy-entropy flux pair of system (5) is

$$\eta_* = \rho \left(\frac{1}{2}|\vec{u}|^2 + \int^\rho \frac{p(s)}{s^2} \mathrm{d}s \right), \quad q_* = \left(\frac{1}{2}|\vec{u}|^2 + \int^\rho \frac{p(s)}{s^2} \mathrm{d}s + \frac{p(\rho)}{\rho} \right) \rho \vec{u}. \tag{8}$$

For our use, we represent the eigenvalues of isentropic compressible Euler equations (5) in the one-dimensional case as

$$\lambda_1 = u - c_s, \quad \lambda_2 = u + c_s, \tag{9}$$

where $c_s = \sqrt{p'(\rho)}$ is the local sonic speed, and the corresponding Riemann invariants as

$$W = u + l(\rho), \quad Z = u - l(\rho), \text{ where } l(\rho) := \int^\rho \frac{c_s}{\rho} \mathrm{d}\rho. \tag{10}$$

Comparing with the classical Euler equations, we consider the relativistic Euler equations. When the macroscopic velocity of the fluid or the velocity of the microscopic particles of the fluid is very close to the speed of light, the relativistic effect has to be taken into account. In this regime, the classical Euler equations are no longer valid and have to be replaced by the relativistic Euler equations. Many efforts have been made to understand the following two sub-systems of relativistic Euler equations (see [1, 5, 20, 21, 24, 30, 33–35, 38, 39] etc.).

Model I. Conservation system of momentum and energy:

$$\begin{cases} \partial_t \left(\dfrac{(\rho + p/c^2)\vec{u}}{1 - |\vec{u}|^2/c^2} \right) + \nabla \cdot \left(\dfrac{(\rho + p/c^2)\vec{u} \otimes \vec{u}}{1 - |\vec{u}|^2/c^2} \right) + \nabla p = 0, \\ \partial_t \left(\dfrac{(\rho + p/c^2)|\vec{u}|^2}{c^2 - |\vec{u}|^2} + \rho \right) + \nabla \cdot \left(\dfrac{(\rho + p/c^2)\vec{u}}{1 - |\vec{u}|^2/c^2} \right) = 0, \end{cases} \tag{11}$$

where ρ and $\vec{u} = (u_1, u_2, \ldots, u_d)$ are the mass-energy density and the velocity, respectively. p is the pressure and c is the speed of light.

Model II. Isentropic conservation system of baryon numbers and momentum, which is more physically significant:

$$\begin{cases} \partial_t \left(\dfrac{n}{\sqrt{1 - |\vec{u}|^2/c^2}} \right) + \nabla \cdot \left(\dfrac{n\vec{u}}{\sqrt{1 - |\vec{u}|^2/c^2}} \right) = 0, \\ \partial_t \left(\dfrac{(\rho + p/c^2)\vec{u}}{1 - |\vec{u}|^2/c^2} \right) + \nabla \cdot \left(\dfrac{(\rho + p/c^2)\vec{u} \otimes \vec{u}}{1 - |\vec{u}|^2/c^2} \right) + \nabla p = 0, \end{cases} \tag{12}$$

where n is the proper number density of baryons. $p = p(\rho)$ satisfies

$$p(\rho) > 0, \quad p'(\rho) > 0 \qquad \text{(hyperbolicity)}, \tag{13}$$

$$\rho p''(\rho) + 2p'(\rho) + (pp''(\rho) - 2p'^2(\rho))/c^2 > 0 \qquad \text{(genuine nonlinearity)}. \tag{14}$$

System (12) is strictly hyperbolic and genuinely nonlinear on the physical region $\sum = \{(\rho, u) : 0 < \rho < \rho_{\max}, |u| < c\}$, where $\rho_{\max} = \sup\{\rho : 0 < p'(\rho) \le c^2\}$.

The proper number density of baryons is determined by the first law of thermodynamics:

$$\theta dS = \frac{d\rho}{n} - \frac{\rho + p/c^2}{n^2} dn,$$

where θ is the temperature and S is the entropy per baryon. In particular, for isentropic fluids ($S = const.$), we have

$$\frac{dn}{n} = \frac{d\rho}{\rho + p/c^2},$$

that is,

$$n(\rho) = n_* e^{\int_1^\rho \frac{ds}{s + p(s)/c^2}}.$$

One of the motivations to study system (12) is that the Newtonian limit of system (12) is the classical system of isentropic Euler equations (5).

The physical entropy-entropy flux pair of system (12) is given by

$$\eta_* = \frac{\rho + p|\vec{u}|^2/c^4}{1 - |\vec{u}|^2/c^2} - \frac{n}{n_*\sqrt{1 - |\vec{u}|^2/c^2}}, \quad q_* = \left(\frac{(\rho + p/c^2)}{1 - |\vec{u}|^2/c^2} - \frac{n}{n_*\sqrt{1 - |\vec{u}|^2/c^2}} \right) \vec{u}.$$

It is easy to check that, as $c \to +\infty$, $(c^2\eta_*, c^2 q_*)$ converges to the physical entropy-entropy flux pair (8) of the corresponding classical system (5).

For the one-dimensional case, the two eigenvalues of (12) are

$$\lambda_1 = \frac{u - \sqrt{p'(\rho)}}{1 - u\sqrt{p'(\rho)}/c^2}, \quad \lambda_2 = \frac{u + \sqrt{p'(\rho)}}{1 + u\sqrt{p'(\rho)}/c^2}, \tag{15}$$

which converge to the eigenvalues (9) of the classical non-relativistic system (5), respectively, as $c \to +\infty$.

The Riemann invariants of (12) in the one-dimensional case are

$$W = \frac{c}{2} \ln \frac{c + u}{c - u} + \int^{\rho} \frac{\sqrt{p'(s)}}{s + p(s)/c^2} \mathrm{d}s, \quad Z = \frac{c}{2} \ln \frac{c + u}{c - u} - \int^{\rho} \frac{\sqrt{p'(s)}}{s + p(s)/c^2} \mathrm{d}s,$$

which converge to the Riemann invariants (10) of system (5) as $c \to +\infty$.

1.2 Some Physical Phenomena Caused by the Motion of the Piston

No matter how slow the piston recedes from or advances into the gas, a wave proceeds from the piston into the gas and only the particles which have been reached by the wave front are disturbed. We take the one-dimensional compressible Euler equations (5) as an example to demonstrate the physical phenomena mathematically.

1.2.1 Rarefaction Waves

First, we consider the continuous wave motion when the piston recedes from the piston. Assume that the gas in the tube is initially static with constant density ρ_0. Suppose that the piston which is located at the origin, and initially static, is pulled back with decreasing speed until the velocity reaches a constant u_*. Let the path of the piston be denoted by $x = b(t)$. Along the piston the velocity of the gas is the same as the piston velocity $b'(t)$.

Since the C_+-characteristic velocity $u + c_s$ is greater than the particle velocity u, the gas enters each characteristic from the right, i.e., comes from the side with greater values of x, a forward-facing simple wave appears, and in the wave region,

Fig. 2 $u_* > -l_0$

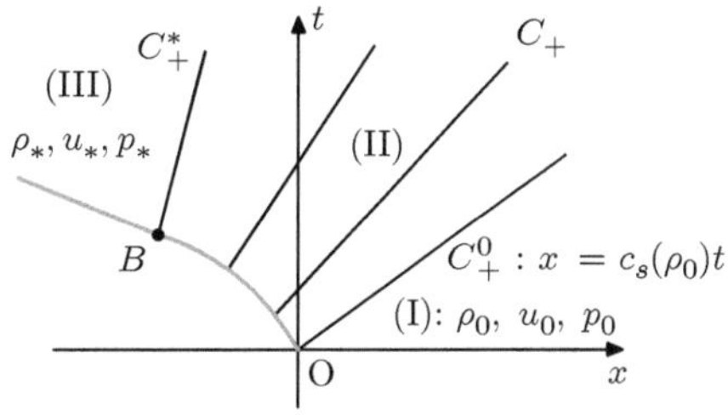

the Riemann invariant Z remains unchanged, i.e., $u - l = u_0 - l(\rho_0)$, then $l(\rho) = l(\rho_0) + b'(t)$. Since $\frac{\mathrm{d}l}{\mathrm{d}\rho} > 0$ and $p'(\rho) > 0$, ρ, p and c_s are determined by l, and thus along each characteristic line $\frac{\mathrm{d}x}{\mathrm{d}t} = u + c_s$ issuing from the piston, u and ρ are constants. So this simple wave is determined as a whole. Now we verify that this simple wave is a rarefaction wave. Due to (6) and (7), we have

$$\frac{\mathrm{d}\lambda_2}{\mathrm{d}u} = \frac{\mathrm{d}(u + c_s)}{\mathrm{d}u} = \frac{\mathrm{d}(l + c_s)}{\mathrm{d}l} = \frac{\rho p''(\rho) + 2p'(\rho)}{2c_s^2} > 0, \tag{16}$$

which implies that $u + c_s$ is increasing with respect to u, and the characteristics diverge. In addition, ρ and p decrease across this wave, since the piston is pulled back with the velocity $b'(t)$ decreasing. Therefore, this wave is a forward rarefaction wave.

From $u = l + u_0 - l(\rho_0)$ and $l > 0$, we have $|u| < l_0$. $l(\rho_0)$ is called the escape speed of the originally static gas.

When u_* does not reach the escape speed, i.e., $u_* > -l(\rho_0)$, the (x, t)-space is divided into three regions (see Fig. 2). Region (I): $\rho = \rho_0$, $u = 0$, $p = p_0$. Region (II) is covered by the simple wave consisting of the straight C_+-characteristics through every point on the piston from O and B. The characteristic lines in (II) fan out. The rarefaction wave ends at the C_+^*-characteristic through the point B of the piston with $u = u_*$. It is followed by the region (III) of constant state, $\rho = \rho_*$, $u = u_*$ and $p = p_*$.

When $-u_*$ reaches the escape speed, i.e., $u_* \leq -l_0$, the rarefaction wave rarefies the gas to density zero; then the pressure and the sonic speed are decreased to zero. If the rarefaction wave extends to this stage, it is called a complete rarefaction wave as it then ends in a vacuum.

If $u_* = -l_0$, the C_+^*-characteristic through the point B is tangent to the curve of the piston, since at the point B, the slope of the piston is u_*, while that of the C_+^*-characteristic is $\frac{\mathrm{d}x}{\mathrm{d}t} = u_* + c_* = u_*$. The rarefaction wave is just completed at the piston (see Fig. 3).

If $u_* < -l_0$, there exists a point B_* on the piston curve between O and B on which the $C_+^{B_*}$-characteristic is tangent to the piston curve. Beyond it, region (III) of cavitation forms a vacuum between the tail of the wave and the piston (see Fig. 4).

Fig. 3 $u_* = -l_0$

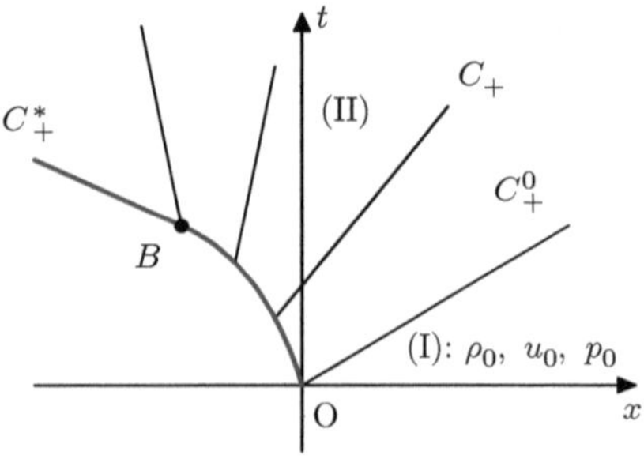

Fig. 4 $u_* < -l_0$

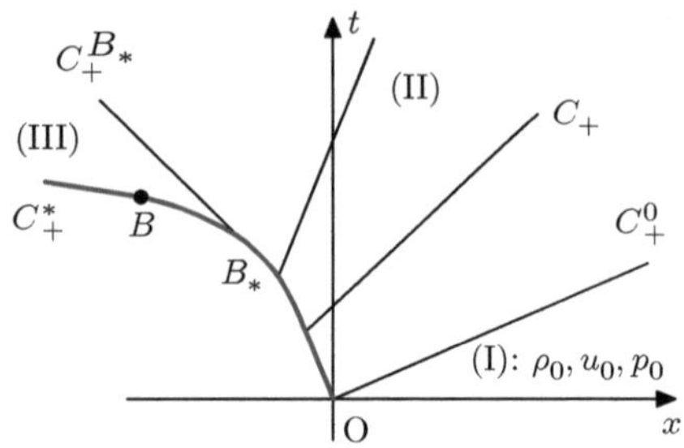

In particular, if the acceleration of the piston from rest to a constant terminal velocity u_* happens instantaneously, the family of the simple wave degenerates into a centered rarefaction wave from the origin O: $x = 0, t = 0$. The justification is similar.

1.2.2 Compression Waves

Now we consider a more complicated case when the piston is not pulled back, but moves into the gas, or when a receding piston is slowed down or stopped, and a compression wave is produced. The formulas pertaining to rarefaction waves still apply to the compression waves, except that the density and pressure at the piston increase and that the C_+-characteristics would converge and form an envelope on which the values of u would conflict, thus the simple wave does not exist for all time. At the earliest time $t = t_c$, an envelope occurs and it forms a cusp at some point $t = t_c$. Continuity of the flow through the simple wave beyond time t_c is impossible.

The development of discontinuities is formed by a piston moving into the static gas with a speed ultimately exceeding the sonic speed relative to the static gas on the right. If the flow remains continuous, the gas would be static in the zone $x \geq c_s(\rho_0)t$, which cannot be reached from the original position of the piston with a speed less than that of sound. Since the piston moves with a speed greater than that of sound, it enters this zone. Consequently, the motion cannot remain continuous.

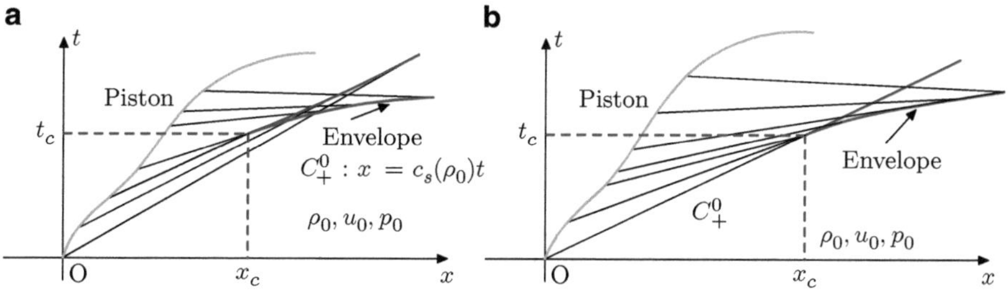

Fig. 5 (**a**) $b''(0) = 0$. (**b**) $b''(0) > 0$

The forward-facing simple wave is written in the form

$$x = \xi(\beta) + \omega(\beta)t, \qquad \omega(\beta) = u(\beta) + c_s(\beta).$$

Differentiating with respect to the parameter β, we have

$$t = -\frac{\mathrm{d}\xi}{\mathrm{d}\omega}, \qquad x = \xi - \omega\frac{\mathrm{d}\xi}{\mathrm{d}\omega}, \tag{17}$$

which is the parametric representation of the envelope, which is admissible if $\omega'(\beta) \neq 0$ in the region.

Now we determine the earliest time t_c when the envelope forms. Assume that $\xi(\beta) \in C^2$. If $\frac{\mathrm{d}\xi}{\mathrm{d}\omega}$ has an extremum for some β_0, then $\frac{\mathrm{d}^2\xi}{\mathrm{d}\omega^2} = -\frac{\mathrm{d}t}{\mathrm{d}\omega}$ changes sign at $\beta = \beta_0$ while $\frac{\mathrm{d}x}{\mathrm{d}\omega} = -\omega\frac{\mathrm{d}^2\xi}{\mathrm{d}\omega^2}$ also changes sign at such point if $\omega'(\beta) \neq 0$. Such an extremum of t occurs at a cusp of the envelope and at the cusp we have

$$\frac{\mathrm{d}^2\xi}{\mathrm{d}\omega^2} = 0.$$

The formation of the envelope in the compression wave is caused by the motion of the piston described by $x = b(t)$.

In the sequel, we show that an envelope will form if the piston moves into the gas with positive acceleration and that this envelope always forms a cusp inside the wave region if the piston begins with acceleration zero (see Fig. 5a).

Assume that $b(0) = 0$, $b'(0) = 0$, and the state of the static gas is denoted by $u = 0$, $\rho = \rho_0$. The simple wave through the point of the piston is described as

$$x = b(\beta) + (u(\beta) + c_s(\beta))(t - \beta),$$

where $u(\beta) + c_s(\beta) = l(\beta) - l(\rho_0) + c_s(\beta) = \omega(\beta)$. Due to (6) and (7), we have

$$\frac{\mathrm{d}\omega}{\mathrm{d}u} = \frac{\mathrm{d}(l + c_s)}{\mathrm{d}l} = \frac{\mathrm{d}(\rho c_s)}{c_s \mathrm{d}\rho} = \frac{\rho p''(\rho) + 2p'(\rho)}{2c_s^2} := h(\beta) > 0.$$

Thus,

$$\omega'(\beta) = h(\beta)b''(\beta).$$

Then the envelope can be represented by

$$t^E(\beta) = \beta + \frac{c_s(\beta)}{h(\beta)b''(\beta)}, \qquad x^E(\beta) = b(\beta) + \omega(\beta)(t^E(\beta) - \beta). \tag{18}$$

Since $c_s(\beta) > 0$, for $\beta > 0$ with $b''(\beta) > 0$, we have

$$t^E(\beta) > \beta, \quad x^E(\beta) > b(\beta).$$

Therefore, an envelope is formed in the flow region.

When the initial acceleration of the piston is zero, i.e., $b''(0) = 0$, then $t^E(0) = \infty$. Meanwhile, either of the following two cases holds:

(i) If $b''(\beta) > 0$ for any $\beta > 0$, then $t^E(\beta) \to +\infty$ as $\beta \to +\infty$.
(ii) If $b''(t_1) = 0$ for some time $t_1 > 0$, then $t^E(\beta) \to +\infty$ as $\beta \to t_1$.

Thus, we obtain that $t^E(\beta)$ first decreases and then increases. Consequently, $t^E(\beta)$ has a minimum t_c for some β, and the envelope forms a cusp at the time t_c.

When the initial acceleration of the piston is positive, i.e., $b''(0) > 0$, the cusp of the envelope is on the straight C_+-characteristic: $x = c_s(\rho_0)t$ (see Fig. 5b), satisfying

$$t_c = \frac{c_s(0)}{h_0 b''(0)},$$

where $h_0 := h(0) = \frac{\mathrm{d}(l+c_s)}{\mathrm{d}l} > 0$.

In other words, the C_+-characteristics form an envelope if the piston is accelerated in forward motion or decelerated in backward motion. For a decelerated piston, $b''(\beta) < 0$, there is no point of the envelope in the domain $x > b(\beta)$.

2 One-Dimensional Piston Problems

For one-dimensional fluid dynamics, some well-posed or inverse piston problems have been solved. In this section, we mainly focus on the piston problems in compressible Euler equations and relativistic Euler equations.

2.1 Euler Equations

Consider the full Euler equations (4) in Lagrangian coordinates in the one-dimensional case:

$$\begin{cases} \partial_t \tau - \partial_x u = 0, \\ \partial_t u + \partial_x p = 0, \\ \partial_t E + \partial_x (pu) = 0, \end{cases} \tag{19}$$

where u, τ, p and e are the velocity, specific volume, pressure and internal energy of the gas, respectively, and $E = e + \frac{1}{2}u^2$ is the total energy.

In Liu [27], the initial data are given as

$$(u(x,0), \tau(x,0), E(x,0)) = (u_0(x), \tau_0(x), E_0(x)), \quad u_0(0+) = u_0(0-). \tag{20}$$

Although the position of the piston is unknown, in Lagrangian coordinates the piston is fixed at $x = 0$. The piston is accelerated by the difference between the pressure on each side of the piston, the boundary conditions are described by

$$\begin{cases} \dfrac{\mathrm{d}u(0,t)}{\mathrm{d}t} = M_p(p(0-,t) - p(0+,t)), \quad t > 0, \\ u(0+,t) = u(0-,t) = u(0,t), \end{cases} \tag{21}$$

where constant M_p is the mass of the piston.

Theorem 1 ([27]). *Suppose that the total variation Σ_T of the initial data* (20) *is sufficiently small. Then the free piston problem* (19)–(21) *has a global solution $U(x,t) = (u(x,t), \tau(x,t), E(x,t))$ satisfying*

$$\begin{aligned} &\mathrm{TV}\{U(x,t) : -\infty < x < \infty\} \le C\,\Sigma_T, \\ &\mathrm{TV}\{u(0,t) : t \ge 0\} \le C\,\Sigma_T\,, \\ &\mathrm{TV}\{p(0\pm,t) : t \ge 0\} \le C\,\Sigma_T, \end{aligned}$$

for constant C depending only on the system, where

$$\begin{aligned} \Sigma_T =& \mathrm{TV}\{(u_0(x), \tau_0(x), E_0(x)) : x > 0\} \\ &+ \mathrm{TV}\{(u_0(x), \tau_0(x), E_0(x)) : x < 0\} + |p_0(0+) - p_0(0-)|. \end{aligned}$$

Furthermore, Liu studied the asymptotic behavior of the solution, which converges pointwise to the linear superposition of the traveling waves, shock waves and rarefaction waves as time tends to infinity. In particular, the asymptotic behavior of the velocity and pressure of the gas depends only on the value of the initial data at far field. Consequently, the difference between the pressure on each side of the piston and the acceleration of the piston tend to zero. To deal with the boundary condition (21), Liu modified the Glimm scheme [18] at $x = 0$ to establish the global existence of the solution. The asymptotic behavior was studied based on Liu [26], in which Liu studied the asymptotic behavior of solutions of the Cauchy problem for general conservation laws.

Li-Wang [22] considered system (19) with different conditions on the piston which is originally located at the origin and moves with speed $\phi(t)$. They solved an inverse piston problem for one-dimensional gas dynamics: under the assumption that the original state of the gas on the right of the piston and the position of the forward shock are known, they obtained the global existence and uniqueness of the C^1 solution on the corresponding maximum determinate domain, i.e., the piston speed can be globally and uniquely determined. Li-Wang [23] also solved an inverse problem for one-dimensional isentropic flow (5).

Takeno [36] studied the following piston problems for one-dimensional isentropic gas dynamics (5) with the polytropic gas $p(\rho) = \rho^\gamma/\gamma$, $1 < \gamma \le 5/3$:

Problem 1. Piston problem in $D_1 = \{(x,t) : x > b_1(t), t > 0\}$ with conditions:

$$\begin{cases} (\rho, u)(x,0) = (\rho_0(x), u_0(x)), & x > 0, \\ \rho(b_1(t),t)\,(u(b_1(t) - u_1(t)) = 0, & t > 0. \end{cases} \tag{22}$$

Problem 2. Piston problem in $D_2 = \{(x,t) : b_1(t) < x < b_2(t), t > 0\}$ with conditions:

$$\begin{cases} (\rho, u)(x,0) = (\rho_0(x), u_0(x)), & 0 < x < L, \\ \rho(b_i(t),t)\,(u(b_i(t) - u_i(t)) = 0, & t > 0,\ \ i = 1,2, \end{cases} \tag{23}$$

where $b_1(t) = \int_0^t u_1(s)\mathrm{d}s$ and $b_2(t) = L + \int_0^t u_2(s)\mathrm{d}s$ are moving boundaries. The boundary speeds $u_1(t)$ and $u_2(t)$ are bounded and measurable. $\rho_0(x)$ and $u_0(x)$ are bounded and measurable. For Problem 2, assume that

$$b_2(t) - b_1(t) \ge \delta(t), \qquad t > 0, \tag{24}$$

where $\delta(t)$ is some positive continuous function.

Theorem 2 ([36]). *Each of the initial boundary value Problems 1 and 2 has a global generalized solution $(\rho, m)(x,t)$, which is a locally bounded measurable function pair and satisfies*

$$\begin{cases} 0 \le \rho(x,t) \le B_1 \\ |m(x,t)| \le B_2\rho(x,t) \end{cases} \qquad a.e. \text{ on } \ \{(x,t) : 0 < t < T\},$$

where $B_i > 0$ are constants. For Problem 1, B_1 and B_2 do not depend on T. For Problem 2, if $A_1 \le A_2$, B_1 and B_2 do not depend on T; otherwise, B_1 and B_2 depend on T and satisfy

$$B_1^\theta, B_2 \le B_3 \exp\left(2(A_1 - A_2)\int_0^T \frac{dt}{\delta(t)}\right),$$

where B_3 is a constant, $\theta = \frac{\gamma-1}{2}$, $A_1 = \operatorname{ess\,sup}_{t>0} u_1(t)$, and $A_2 = \operatorname{ess\,inf}_{t>0} u_2(t)$.

The proof of this theorem mainly follows Diperna [17] and Ding-Chen-Luo [2, 4], and was carried out using a Godunov scheme and compensated compactness.

Following the method in [36], Takeno [37] also considered a free piston problem for one-dimensional isentropic gas dynamics with the state equation

$$p(\rho) = \begin{cases} p_1(\rho), & x < b(t),\ t > 0, \\ p_2(\rho), & x > b(t),\ t > 0, \end{cases} \qquad p_i(\rho) = a_i \rho^{\gamma_i},\quad a_i > 0,$$

where γ_i is the adiabatic exponent, $1 \le \gamma_i \le 5/3$, $i = 1, 2$. $b(t)$ is the position of the free piston boundary. The corresponding boundary conditions are

$$\begin{cases} b''(t) = -k\big(p_2(\rho(b(t)+0,t)) - p_1(\rho(b(t)-0,t))\big), \\ \rho(b(t)+0,t)\,(u(b(t)+0,t) - b'(t)) = 0, \\ b(0) = 0, \qquad b'(0) = \bar{u}_0, \end{cases} \tag{25}$$

where $\bar{u}_0$ is a constant and $k = A_s/M_p$ is a positive constant, here A_s is the sectional area of the piston and M_p is the mass of the piston.

Theorem 3 ([37]). *For the one-dimensional free piston problems* (5) *and* (25) *with bounded measurable functions $(\rho_0(x), u_0(x))$ as the initial state, there exists a generalized solution $(\rho, u)(x,t)$, which is bounded and measurable, satisfying*

$$\begin{cases} 0 \le \rho(x,t) \le C_1, \\ |u(x,t)| \le C_2, \end{cases} \quad a.e. \text{ on } \{(x,t) : t > 0\},$$

where C_1 and C_2 are constants. The piston path $x(t)$ is continuously differentiable with respect to t and the derivative of $x(t)$ is Lipschitz continuous and bounded. $(\rho, u)(x,t)$ satisfies an entropy inequality in each side of the piston in the sense of distributions.

In [41], Wang established the existence of strong shock front solutions to the one-dimensional piston problem of the compressible Euler equations (5) for a polytropic gas $p(\rho) = \rho^\gamma$. Suppose that the path of the piston is $x = b(t)$, the velocity of the gas next to the piston is the same as that of the piston, i.e.,

$$u = b'(t) \qquad \text{on } x = b(t), \tag{26}$$

and the corresponding curve of the shock is $x = s(t)$. A shock appears when the piston moves into the initial static gas $(\rho, u)(x, 0) = (\rho_0, 0)$. This physical phenomenon is demonstrated in the following theorem:

Theorem 4 ([41]). *Assume that $b(t) \in C^2$, $b(0) = 0$, $b'(0) = b_0$. Then when $\int_0^{\infty} |b''(t)|dt$ is sufficiently small, problems (5) and (26) admits a global weak solution, which includes a strong shock, which is a small perturbation of $x = s_0 t$. The gas between the piston and the shock is also a perturbation of the solution when the piston moves with constant velocity b_0.*

A modified Glimm scheme is employed to solve such a piston problem (26).

2.2 Relativistic Euler Equations

For system (11) of Model I, [43] studied the one-dimensional piston problem with static constant initial state and the speed of piston a small perturbation of a constant, and established global existence of weak solutions.

In [15], we established the global existence of the shock front solutions of the 1-D piston problem for system (12) of Model II. Differing from [41, 43], we considered not only a small perturbation of the piston speed, but also a small perturbation of a constant initial state. On this occasion, the position of the strong shock is no longer fixed on the rightmost but varies, so the interactions between the strong 2-shocks and weak waves from the left and from the right become more complicated, and the Glimm functional in [41, 43] is no longer applicable. So we had to redefine the approaching waves, including not only the approach between weak waves, but also the approach between weak waves and strong 2-shocks as well as the approach between weak waves and the piston boundary. Then we constructed a new Glimm functional, containing some additional terms regarding weak waves approaching both the strong shocks and the piston boundary. In this functional, we employed a weighed strength for weak waves in order to remove the restrictions on reflection coefficients of weak waves on the strong 2-shock from the right.

Suppose that the initial gas satisfies $\rho(x, 0) = \rho_0(x)$, $u(x, 0) = u_0(x)$, and the piston moves with a speed depending only on time t. Let the movement curve of the piston be $x = b(t)$ and the shock be $x = s(t)$ with speed $b'(t)$ and $s'(t)$, respectively. We study the state of the gas in domain $\Omega = \{(x, t) : x > b(t), t > 0\}$ with $\Gamma = \{(x, t) : x = b(t), t > 0\}$ (as shown in Fig. 6). Then the initial-boundary conditions for the piston problem can be described as

$$\begin{cases} (\rho, u)(x, 0) = (\rho_0(x), u_0(x)), \\ u = b'(t) \qquad \text{on } x = b(t). \end{cases} \tag{27}$$

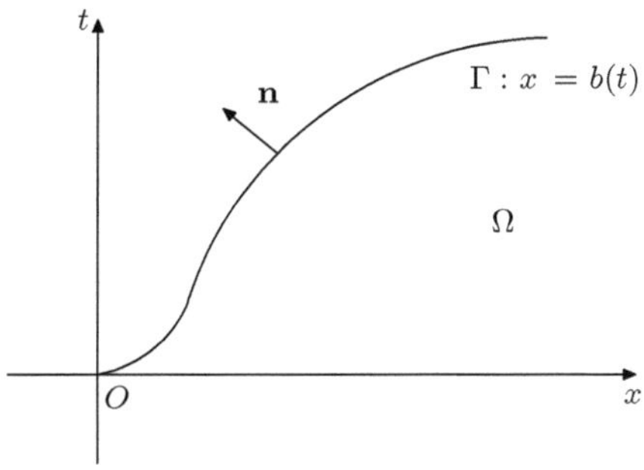

Fig. 6 Definition domain

Theorem 5 ([15]). *Assume that* $b(0) = 0$ *and* $b'(0+) = b_0$. *If the total variations* $\mathrm{TV}\{b'(\cdot)\}$ *and* $\mathrm{TV}\{\rho_0, u_0\}$ *are sufficiently small, there exists an* L^∞ *entropy solution* (ρ, u) *of problems* (12) *and* (27)*, satisfying*

$$\mathrm{TV}\,(\rho, u)\,(\cdot,\ t) \le N$$

for all $t \ge 0$*, containing a strong shock, which is a small perturbation of* $x = s_0 t$*, where* N *is a constant depending on the initial data, the background solution and* $\mathrm{TV}\{b'(\cdot)\}$.

The proof of this main conclusion is based on Sects. 2.2.1–2.2.4.

Remark 1. In addition to the global existence of shock fronts, we also consider the non-relativistic global limits of entropy solutions as the light speed $c \to +\infty$. So we make every effort to establish uniform (independent of large c) estimates on the interactions of perturbation waves and their reflections on the strong shock and the piston. Based on these estimates, we prove the convergence of entropy solutions to the corresponding entropy solutions of the classical non-relativistic Euler equations (5) as $c \to +\infty$.

2.2.1 The Background Solution and Piston Riemann Problem

When the piston moves with a constant speed b_0 and the initial data are given by constant values $(\bar{\rho}_0, \bar{u}_0)$, the piston problem can be described as

$$(\mathrm{I}) \begin{cases} (12), \\ (\rho, u)(x, 0) = (\bar{\rho}_0, \bar{u}_0), \\ u = b_0 \quad \text{on} \ \ x = b_0 t, \ \text{where} \ b_0 > \bar{u}_0. \end{cases}$$

We call the solution of (I) the background solution. The solvability of (I) is given by

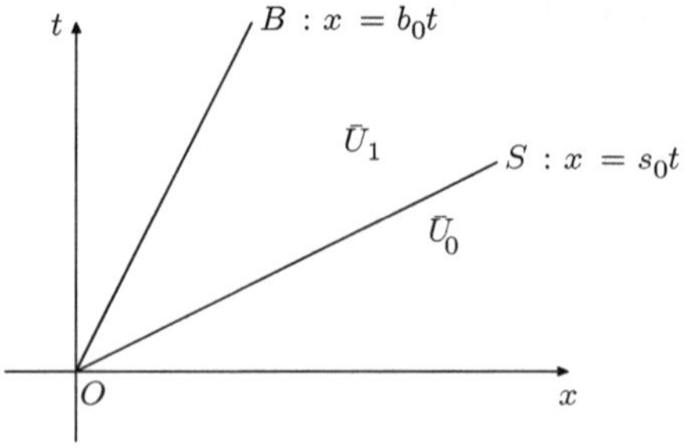

Fig. 7 Background solution

Lemma 1 ([15]). *There exists a unique solution of problem* (I), *which consists of two constant states* $\bar{U}_0 = (\bar{\rho}_0, \bar{u}_0)$ *and* $\bar{U}_1 = (\bar{\rho}_1, \bar{u}_1)$ *connected by a strong 2-shock wave with velocity* s_0, *and* $\bar{u}_1 = b_0$ *(see Fig. 7).*

For given initial data

$$U|_{t=t_0} := (\rho, u)|_{t=t_0} = \begin{cases} U_L, & x < x_0, \\ U_R, & x > x_0, \end{cases} \tag{28}$$

where $U_L = (\rho_L, u_L)$ and $U_R = (\rho_R, u_R)$ represent the left and right constant states, respectively. The solvability of the piston Riemann problems (12) and (28) is given in the following lemma:

Lemma 2 ([15]). *For any right state* $U_R \in O_\epsilon(\bar{U}_0)$, *there exists a constant* $\delta > 0$ *such that, for* $U_L \in O_\delta(S_2^{-1}(U_R)) \cap O_\epsilon(\bar{U}_1)$, *problems* (12) *and* (28) *admits a unique solution containing a weak 1-wave* α_1 *and a strong 2-shock* s, *where* $S_2^{-1}(U_R)$ *denotes the inverse shock curves based on the given state* U_R, *and* O_ϵ *stands for a small neighborhood.*

2.2.2 Construction of the Approximate Solutions

We constructed approximate solutions of problems (12) and (27) by a modified Glimm scheme, where the mesh grids are chosen to follow the slope of the piston so that the piston problem contains only a 2-wave issuing from the mesh points on the boundary. Let Δx denote a mesh length in x and Δt a mesh length in t. We introduce the notation $t_k = k\Delta t$, $x_h(k) = x_0(k) + h\Delta x$, $p_k = \dfrac{x_0(k+1) - x_0(k)}{\Delta t}$, $k, h = 0, 1, 2, \cdots$, and define $(x_h(k), t_k)$ as the grids of the scheme. Δt and Δx satisfy the C-F-L condition:

$$\frac{\Delta x}{\Delta t} \geq \sup_{k \geq 0} p_k + |s_0| + \max_{i=1,2} \left(\sup_U |\lambda_i(U)| \right).$$

The movement curve of the piston is approximated by a piecewise linear function denoted as:

$$x = b_\Delta(t) = x_0(k) + p_k(t - t_k), \quad t_k \le t < t_{k+1}, \ k = 0, 1, \cdots.$$

In addition, denote by ω_k the angle between the straight line $x = x_0(k) + p_k(t - t_k)$ and t-axis and by $\theta_k = \omega_k - \omega_{k-1}$ the angle between the straight line $x = x_0(k) + p_k(t - t_k)$ and $x = x_0(k-1) + p_{k-1}(t - t_{k-1})$. Let

$$\Omega_\Delta = \{(x,t) : x > b_\Delta(t), t > 0\}, \quad \Gamma_\Delta = \{(x,t) : x = b_\Delta(t), t > 0\}.$$

By induction, we construct the approximate solutions $U_\Delta(x,t)$ in the region Ω_Δ. For $k = 1$, $U_\Delta(x,t)$ on $\{0 \le t < \Delta t\} \cap \Omega_\Delta$ can be constructed by solving a series of (piston) Riemann problems, which can be carried out in the same way as the construction of U_Δ in $\{t_k \le t < t_{k+1}\}$ by taking the given data $U_0 = (\rho_0(x), u_0(x))$ instead of the approximate solution $U_\Delta(x, t_k)$ as the initial data (see below). Suppose that the approximate solutions U_Δ have already been defined in $\{0 < t < t_k\} \cap \Omega_\Delta (k \ge 1)$ and define

$$U_\Delta(x, t_k) = U_\Delta(a_{k,h}, t_k-), \qquad x_{2h}(k) < x < x_{2h+2}(k), h = 0, 1, \cdots, \tag{29}$$

where $a_{k,h} = x_0(k) + (2h + 1 + a_k)\Delta x$ is a random point in $(x_{2h}(k), x_{2h+2}(k))$, and $\{a_k\}_{k\ge 0}$ is an equi-distributed sequence in $(-1, 1)$. Then we define the approximate solutions in $\{t_k \le t < t_{k+1}\} \cap \Omega_\Delta$ in two cases.

Case 1. Near the corner point $(x_0(k), t_k)$ of the approximate curve of the piston, we solve (12) with the following initial-boundary conditions:

$$\begin{cases} U_\Delta(x,t) = U_\Delta(x_0(k)+, t_k) & \text{for } t = t_k, \ x_0(k) < x < x_1(k), \\ u(x,t) = p_k & \text{on } x = x_0(k) + p_k(t - t_k), \end{cases} \tag{30}$$

where the value of $U_\Delta(x_0(k)+, t_k)$ is given by (29). By the choice of the random points, the 2-wave of the solution of problem (12) and (30) may be a strong 2-shock, or a weak 2-rarefaction or shock wave, determined by the relation between p_k and p_{k-1}, and the reflection of the 1-wave from the grid $(x_2(k-1), t_{k-1})$ on the piston or on the strong 2-shock from $(x_0(k-1), t_{k-1})$. Hereinafter, the state on the left of the piston is defined to be the same as the state next to the piston on the right whenever needed.

Case 2. At each grid point $(x_{2h}(k), t_k), h = 1, 2, \cdots$, we solve a series of Riemann problems of (12) in the region $\{t_k \le t < t_{k+1}\}$ with the initial data:

$$U = \begin{cases} U_\Delta(x_{2h}(k)-, t_k), & x_{2h-1}(k) < x < x_{2h}(k), \\ U_\Delta(x_{2h}(k)+, t_k), & x_{2h}(k) < x < x_{2h+1}(k). \end{cases} \tag{31}$$

The solution can be given in two cases:

(i) $U_\Delta(x_{2h}(k)\pm, t_k) \in O_\epsilon(\bar{U}_1)(O_\epsilon(\bar{U}_0))$, and $U_\Delta(x_{2h}(k)-, t_k) \neq U_\Delta(x_{2h}(k)+, t_k)$, the solution contains two weak waves (shock or rarefaction waves);

(ii) $U_\Delta(x_{2h}(k)-, t_k) \in O_\epsilon(\bar{U}_1)$ and $U_\Delta(x_{2h}(k)+, t_k) \in O_\epsilon(\bar{U}_0)$, the solution contains a weak 1-wave (shock or rarefaction wave) and a strong 2-shock denoted by s_{k+1} which also denotes the speed of the 2-shock provided that no confusion occurs. We approximate the strong 2-shock curves in $t_k \leq t < t_{k+1}, k = 0, 1, \cdots$, as

$$x = s_\Delta(t) = x_h(k) + s_{k+1}(t - t_k), \quad t_k \leq t < t_{k+1}.$$

Thus, we obtain the approximate solutions in the region $\{t_k \leq t < t_{k+1}\} \cap \Omega_\Delta$. Then we extend to the whole region Ω_Δ by induction.

2.2.3 Estimates on Local Interactions

We use the velocity s to parameterize the strong 2-shock. We denote by $\alpha_i, \beta_i, \gamma_i$, and δ_i the parameters of the corresponding i-waves ($i = 1, 2$), and by their absolute values the corresponding strengths of the waves. Sometimes we also use the parameters to represent the corresponding waves provided that no confusion occurs.

Let (U_l, U_r) denote the nonlinear waves solving the Riemann problem with the left state U_l and the right state U_r.

Based on the construction of approximate solutions, we can obtain some space-like curves, composed of the segments connecting the mesh points $a_{k,h}$ and $a_{k+1,h+1}$(or $a_{k-1,h+1}$). These space-like curves divide the region Ω_Δ into two parts: I^- and I^+, where I^- denotes the part containing the x-axis. Suppose that I and J are two space-like curves, and $J > I$ if every mesh point of J is either on I or contained in I^+. In particular, J is called an immediate successor to I if $J > I$ and every mesh point of J except one is on I.

Suppose that I and J are two space-like mesh curves and $J > I$. In order to obtain the estimates of the local interactions, it suffices to discuss the case that J is an immediate successor to I. Suppose that there exists a diamond denoted by Δ between I and J. According to the different positions of the diamond, we divide the discussion into five cases.

Case 1: Δ covers neither $x = b_\Delta(t)$ nor $x = s_\Delta(t)$;

Let α and β be the waves entering Δ from $(x_{h-1}(k-1), t_{k-1})$ and $(x_{h+1}(k-1), t_{k-1})$, respectively, and γ the wave generated from $(x_{h-1}(k), t_k)$. Let $\alpha_1 = (U_6, U_5)$, $\alpha_2 = (U_5, U_4)$, $\beta_1 = (U_4, U_3)$, $\beta_2 = (U_3, U_2)$, $\gamma_1 = (U_6, U_7)$, and $\gamma_2 = (U_7, U_2)$ (see Fig. 8).

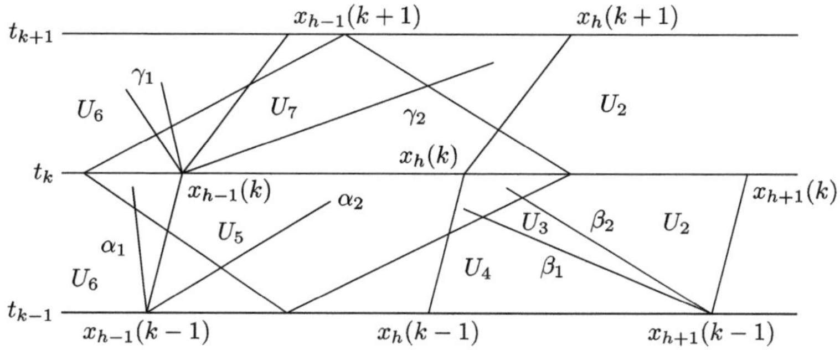

Fig. 8 Δ covers neither $x = b_\Delta(t)$ nor $x = s_\Delta(t)$

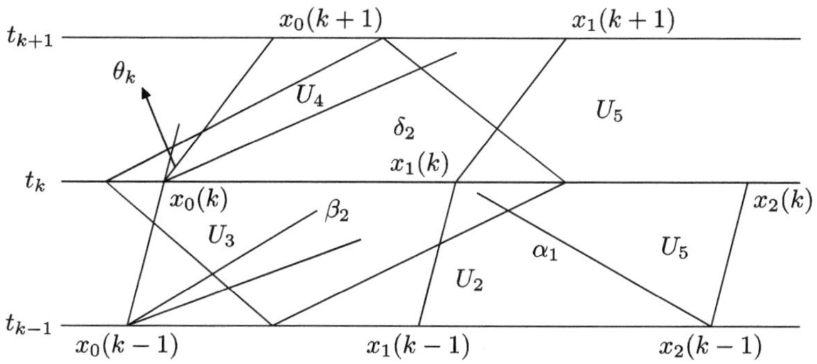

Fig. 9 Δ covers part of $x = b_\Delta(t)$ but none of $x = s_\Delta(t)$

Case 2: Δ covers part of $x = b_\Delta(t)$ but none of $x = s_\Delta(t)$;

Let α_1 and β_2 be the waves entering Δ from $(x_2(k-1), t_{k-1})$ and $(x_0(k-1), t_{k-1})$, respectively. Let δ_2 be the 2-wave generated from $(x_0(k), t_k)$ (see Fig. 9). Let U_3 and U_4 denote, respectively, the states next to the segments $x = x_0(k-1) + p_{k-1}(t - t_{k-1})$ and $x = x_0(k) + p_k(t - t_k)$, $t_{k-1} < t < t_{k+1}$. Let $\alpha_1 = (U_2, U_5)$, $\beta_2 = (U_3, U_2)$, $\delta_2 = (U_4, U_5)$, and $U_4, U_3, U_2, U_5 \in O_\epsilon(\bar{U}_1)$.

Case 3: Δ covers part of both $x = b_\Delta(t)$ and $x = s_\Delta(t)$;

Let s_k be the strong 2-shock issuing from the point $(x_0(k-1), t_{k-1})$ and entering Δ, and s_{k+1} be the strong 2-shock generated from $(x_0(k), t_k)$. Let α_1 and α_2 be the 1-wave and 2-wave issuing from the point $(x_2(k-1), t_{k-1})$ and entering Δ, and $\alpha_1 = (U_2, U_3)$ and $\alpha_2 = (U_3, U_4)$ (see Fig. 10). Assume that $U(s_k), U(s_{k+1}) \in O_\epsilon(\bar{U}_1)$, and $U_2, U_3, U_4 \in O_\epsilon(\bar{U}_0)$.

Case 4: Δ covers part of $x = s_\Delta(t)$ but none of $x = b_\Delta(t)$ (I): weak waves interact with the strong shock from the left;

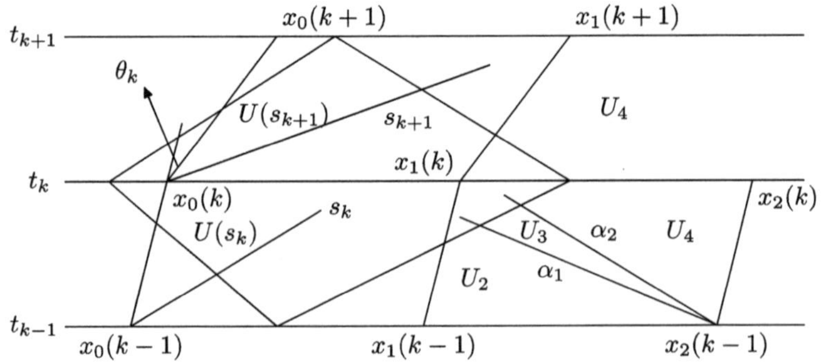

Fig. 10 Δ covers part of both $x = b_\Delta(t)$ and $x = s_\Delta(t)$

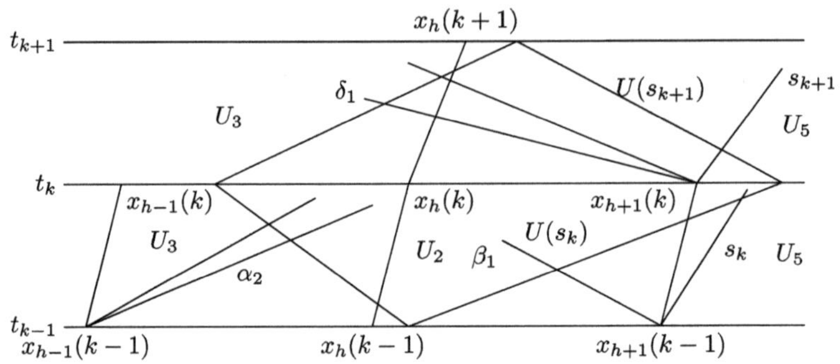

Fig. 11 Δ covers part of $x = s_\Delta(t)$ but none of $x = b_\Delta(t)$: (I)

Let s_k and s_{k+1} be the strong 2-shocks issuing, respectively, from $(x_{h+1}(k-1), t_{k-1})$ and $(x_{h+1}(k), t_k)$, β_1 and α_2 the waves entering Δ from $(x_{h+1}(k-1), t_{k-1})$ and $(x_{h-1}(k-1), t_{k-1})$, respectively, and δ_1 the 1-wave generated from $(x_{h+1}(k), t_k)$. Let $\alpha_2 = (U_3, U_2)$, $\beta_1 = (U_2, U(s_k))$ and $\delta_1 = (U_3, U(s_{k+1}))$ (see Fig. 11). Suppose that $U_3, U_2, U(s_k), U(s_{k+1}) \in O_\epsilon(\bar{U}_1)$, and $U_5 \in O_\epsilon(\bar{U}_0)$.

Case 5: Δ covers part of $x = s_\Delta(t)$ but none of $x = b_\Delta(t)$ (II): weak waves interact with the strong shock from the right.

Let s_k and s_{k+1} be the strong 2-shocks issuing, respectively, from $(x_{h-1}(k-1), t_{k-1})$ and $(x_{h-1}(k), t_k)$, β_1 be the 1-wave from $(x_{h-1}(k-1), t_{k-1})$, α_1 and α_2 the 1- and 2-waves from $(x_{h+1}(k-1), t_{k-1})$ entering Δ, and δ_1 the 1-wave generated from $(x_{h-1}(k), t_k)$. Let $\beta_1 = (U_3, U(s_k))$, $\alpha_1 = (U_4, U_2)$, $\alpha_2 = (U_2, U_5)$ and $\delta_1 = (U_3, U(s_{k+1}))$ (see Fig. 12). Suppose that $U_3, U(s_k)$ and $U(s_{k+1}) \in O_\epsilon(\bar{U}_1)$, and $U_4, U_2, U_5 \in O_\epsilon(\bar{U}_0)$.

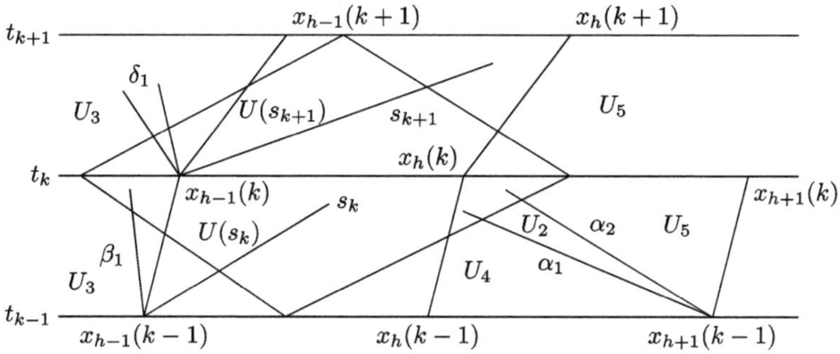

Fig. 12 Δ covers part of $x = s_\Delta(t)$ but none of $x = b_\Delta(t)$: (II)

Lemma 3 ([15]). *For cases 1–5, the following estimates hold respectively,*

(i) $\gamma_i = \alpha_i + \beta_i + O(1)Q_0(\alpha, \beta), \qquad i = 1, 2,$

(ii) $\delta_2 = K_{b_1}\alpha_1 + K_{b_0}\theta_k + \beta_2,$

(iii) $s_{k+1} = s_k + K_{b_{s_1}}\alpha_1 + K_{b_{s_2}}\alpha_2 + K_{b_{s_0}}\theta_k,$

(iv) $\delta_1 = \beta_1 + C_*\alpha_2, \quad s_{k+1} = s_k + K_{s_2}\alpha_2, \quad k = 1, 2, \cdots,$

(v) $\delta_1 = \beta_1 + \tilde{K}_{s_3}\alpha_1 + \tilde{K}_{s_4}\alpha_2, \quad s_{k+1} = s_k + \tilde{K}_{s_1}\alpha_1 + \tilde{K}_{s_2}\alpha_2, \quad k = 1, 2, \cdots,$

where $Q_0(\alpha, \beta) = \{\Sigma|\alpha_i||\beta_j| : \alpha_i$ *and* β_j *are approaching,* $i, j = 1, 2\}$*, and* $O(1)$ *is bounded.* K_{b_1} *and* K_{b_0} *are* C^2*-function of* $(\alpha_1, \beta_2, \theta_k)$*,* K_{b_1}*,* K_{b_0}*,* $K_{b_{s_0}}$*,* $K_{b_{s_1}}$*,* $K_{b_{s_2}}$*,* K_{s_2}*,* C_* *and* $\tilde{K}_{s_i}$*,* $i = 1, 2, 3, 4$ *are bounded. Furthermore,* $K_{b_1}C_* \in (-1, 1)$*.*

Remark 2. In addition, we have also proved that the coefficients $O(1)$, K_{b_1}, K_{b_0}, $K_{b_{s_0}}$, $K_{b_{s_1}}$, $K_{b_{s_2}}$, K_{s_2}, C_* and $\tilde{K}_{s_i}$, $i = 1, 2, 3, 4$ are uniformly bounded and independent of large $c \geq c_0$ in [15, 17].

Remark 3. In (ii), the coefficient K_{b_1} of α_1 can be regarded as the reflection coefficient of α_1 on the piston. (ii) indicates that the 1-wave turns into a 2-wave after being reflected by the piston. Moreover, the existence of the term $K_{b_0}\theta_k$ is caused by the change of the speed of the piston. Controlling this term is one of the main differences in the Glimm scheme between the boundary value problem and the Cauchy problem.

Remark 4. The coefficient C_* in (iv) can be regarded as the reflection coefficient of α_2 on the strong shock. The first equality of (iv) indicates that, after being reflected by the strong shock, the 2-wave turns into a 1-wave.

Remark 5. Condition $K_{b_1}C_* \in (-1, 1)$ is important for the monotonicity of the Glimm functional.

2.2.4 Monotonicity of the Glimm Functional and Convergence of Approximate Solutions

Since the initial state is not constant, the interactions between the strong 2-shocks and weak waves from the left and from the right should be taken into consideration in the Glimm functional, in particular, in the approaching waves, which are redefined as follows.

Definition 1. (Approaching waves)

- $(\alpha_i, \beta_j) \in \mathcal{A}_1$: two weak waves α_i and β_j $(i, j \in \{1, 2\})$ located at points x_α and x_β respectively, with $x_\alpha < x_\beta$, satisfy the following condition:
  Either $i > j$ or $i = j$ and at least one of them is a shock.
- $\alpha \in \mathcal{A}_s$: a weak i-wave α is approaching a strong 2-shock if $\alpha \in \Omega_-$, $i = 2$ or $\alpha \in \Omega_+$, $i = \{1, 2\}$, where

$$\Omega_- = \{(x,t) : b(t) < x < s(t), t > 0\}, \quad \Omega_+ = \{(x,t) : x > s(t), t > 0\}.$$

- $\alpha \in \mathcal{A}_b$: a weak i-wave α is approaching the boundary if $\alpha \in \Omega_-$ and $i = 1$.

Remark 6. The approaching wave in $\mathcal{A}_1$ is in fact the original approaching wave between weak waves.

Denote the set of the corner points A_k lying in J^+ by

$$\Gamma_J = \{A_k : A_k \in J^+,\ A_k = (x_0(k), t_k)\}.$$

We define the Glimm functional $F(J)$ on the mesh curve J as

$$F(J) = L(J) + KQ(J),$$

where

$$\begin{aligned}
L(J) &= L_0(J) + L_1(J) + L_2(J),\\
L_0(J) &= \{|s^J - s_0| : s^J \text{ denotes the speed of the strong shock passing } J\},\\
L_1(J) &= \Sigma\{|b_\alpha| : \alpha \text{ is the 1-wave passing } J\},\\
L_2(J) &= \Sigma\{|b_\alpha| : \alpha \text{ is the 2-wave (not including the strong shock) passing } J\},\\
Q(J) &= \sum_{(\alpha_i,\beta_j)\in\mathcal{A}_1} |b_{\alpha_i}||b_{\beta_j}| + K_1^* \sum_{\alpha\in\mathcal{A}_s} |b_\alpha| + \sum_{\beta\in\mathcal{A}_b} |b_\beta| + K_2^* \sum_{A_k\in\Gamma_J} |\theta_k|,
\end{aligned}$$

and $\theta_k = \omega_k - \omega_{k-1}$, K is sufficiently large, K_1^* and K_2^* are constants satisfying

$$K_1^*|K_{b_1}| - 1 < 0, \qquad K_1^*|K_{b_0}| - K_2^* < 0, \tag{32}$$

$$|C_*| < K_1^*, \tag{33}$$

$$K_0 K_1^* > |\tilde{K}_{s_3}|, \qquad K_0 K_1^* > |\tilde{K}_{s_4}|, \tag{34}$$

$$b_\alpha = \begin{cases} \alpha & \text{if } x \in \Omega_-, \\ K_0\alpha & \text{if } x \in \Omega_+, \end{cases} \qquad K_0 = 2\max\{|K_{b_1}||\tilde{K}_{s_3}|, |K_{b_1}||\tilde{K}_{s_4}|\}.$$

Remark 7. In fact, K_1^* and K_2^* can be determined from (32) to (34) at the background solution. Since K_{b_1}, K_{b_0}, C_*, $\tilde{K}_{s_3}$ and $\tilde{K}_{s_4}$ are continuous with respect to U, uniformly for large c, i.e., under some small perturbations (independent of large c) of the background solution, (32)–(34) are still valid.

Remark 8. A weighted strength b_α for a weak wave α on the right of the strong 2-shock is introduced to remove the restriction on the reflection coefficient of α on the strong 2-shock from the right.

Let O denote the space-like mesh curve in $0 \leq t \leq \Delta t$. Under the assumption of the smallness of $L(O)$ and $\mathrm{TV}\{b'(\cdot)\}$, we obtain the monotonicity of the Glimm functional and the equivalence between the Glimm functional and the total variation of U.

Lemma 4 ([15]). *Assume that* $\mathrm{TV}\{U_0\}$ *and* $\mathrm{TV}\{b'(\cdot)\}$ *are small, then for any mesh curves* $J > I$, *we have*

$$\frac{1}{C}\mathrm{TV}_J\{U\} \leq F(J) \leq F(I) \leq F(O) \leq C\left(\mathrm{TV}\{U_0\} + \mathrm{TV}\{b'(\cdot)\}\right), \tag{35}$$

where C *is a positive constant depending on the initial data and background solution.*

From Lemma 4, the convergence of the approximate solutions can be proved by a standard procedure (see [11, 17, 18, 32, 44, 45]).

As for other applications of the Glimm scheme to the Cauchy problem as well as to the initial-boundary value problems, we refer to [6, 11, 17–19, 31, 41, 43–45] and the references cited therein.

3 Multidimensional Piston Problems

For multidimensional piston problems, the piston is replaced by a cylinder body. Assume that there is uniform static gas filling up the whole space outside a given body with moving boundary. Starting from $t = 0$, the piston expands and its boundary moves into the air as in the one-dimensional case. Then there is a shock front moving into the air. Not all parts of the gas are affected. Ahead of the shock front the state of the air is kept unchanged, the location of the shock and the flow between the shock and the piston is to be determined. In addition, we have many multidimensional piston models, for example, the surface layer of an inflatable balloon behaves as a spherically symmetric piston. The gas outside is compressed by the expansion of the balloon, then a shock appears. When the location of the piston initially degenerates into a single point, this phenomenon is related to explosive waves in Physics.

The multidimensional piston problem is challenging and remains open. However, some special multidimensional cases have been studied. For instance, the spherically symmetric piston problem. Let $r = |\vec{x}|$, then a spherically symmetric solution is denoted by

$$\rho(\vec{x},t) = \rho(r,t), \quad \vec{u}(\vec{x},t) = u(r,t)\frac{\vec{x}}{r}.$$

Assume that the initial state of the gas is denoted by $\rho = \rho_*, u = 0$. Starting from $t = 0$, the piston is located at the origin, and expands into the static gas. Suppose that the velocity of the piston depends only on t and the path of the piston is described as $r = b(t)$, and the velocity of the gas next to the piston is the same as that of the piston. Then the initial-boundary conditions for the spherically symmetric piston problem are formulated as

$$\begin{cases} u = b'(t) & \text{on } \ r = b(t), \\ (\rho, u) = (\rho_*, 0) & \text{on } \ t = 0. \end{cases} \tag{36}$$

3.1 Non-relativistic Fluids

The spherically symmetric isentropic Euler equations read as

$$\begin{cases} \partial_t(\rho u) + \partial_r(\rho u^2 + p) + \frac{(d-1)\rho u^2}{r} = 0, \\ \partial_t \rho + \partial_r(\rho u) + \frac{(d-1)\rho u}{r} = 0. \end{cases} \tag{37}$$

For $d = 2$, under the assumption that the velocity of the piston and the density of the gas outside at initial time, and TV$b'(\cdot)$ are small, Chen-Wang-Zhang [11] established the global existence of a BV solution to the axially symmetric piston problem (36) and (37) by a modified Glimm scheme with the state equation $p(\rho) = \rho^\gamma, 1 < \gamma < 3$.

Theorem 6 ([11]). *Suppose that $b(t) \in C^\infty$, $b(0) = 0$, $b'(0) = b_0 > 0$, $b^{(k)}(0) = 0$ for $2 \le k \le M$ where M is a large integer. If $\rho_* + b_0 + ||b(t) - b_0 t||_{C^K([0,1])} + \int_T^\infty |b''(t)|dt$ is sufficiently small for some positive constants T and K, then there exists a global* BV *solution to problem* (36) *and* (37).

Remark 9. Moreover, this shock front is next to the uniform flow and is a small perturbation of $r = s_0 t$, and the solution between the shock front and the piston is a small perturbation of the self similar solution of problem (36) and (37) with constant piston velocity b_0.

Remark 10. Based on [11], Chen-Wang-Zhang [12] removed the restriction on the strength of the leading shock, but required that the velocity of the piston is rather fast or the density at the initial time is small. Some more precise estimates near the leading shock were presented, which is essential for establishing the global existence of shock front solution.

In [3], the global L^∞ entropy solution for the spherically symmetric piston problem to the Euler equations (5) was established. A local shock front solution was approximated by using a finite expansion. The convergence of the approximate solution was shown by a Newton iteration scheme for a polytropic gas $p(\rho) = \rho^\gamma/\gamma$, $\gamma \geq 1$. Based on the local existence, it was extended to a global entropy solution by a shock capturing approach and the method of compensated compactness for isothermal gas $p = \rho$.

For the full Euler equations (4), Wang [40] obtained the local existence of a shock front solution to the axi-symmetrical piston problem for a polytropic gas $p = (\gamma - 1)\rho e$ by energy estimates and the Newton iteration scheme.

Chen [10] studied a special piston problem for the unsteady potential flow equations in two space dimensions under the assumption that the piston expands with a velocity depending on $\theta = \arctan y/x$ and independent of time t. In order to fix the free boundary, by the partial hodograph transformation $\xi = x/t$, $\eta = y/t$ and coordinate transformation $r = (\xi^2 + \eta^2)^{\frac{1}{2}}$, $\theta = \arctan \eta/\xi$, the annular domain was decomposed into a set of overlapping domains, on which a set of auxiliary boundary value problems were solved by employing the nonlinear alternating iteration. The existence and stability of shock front solutions for the original problem were established under small perturbation.

Some related wedge problems are referred to in [7–9, 28, 29].

3.2 *Relativistic Fluids*

A multidimensional piston problem for (11) with the state equation $p(\rho) = a^2\rho$, where a is a positive constant, has been studied in Ding-Li [14], where the local existence and non-relativistic limits of shock front solutions were established for the spherically symmetric piston problem. For convenience, let $\epsilon := 1/c^2$. The spherically symmetric solution $(\rho(r,t), u(r,t))$ satisfies

$$\begin{cases} \partial_t \left(\dfrac{(\rho + p\epsilon)u}{1 - u^2\epsilon} \right) + \partial_r \left(\dfrac{(\rho + p\epsilon)u^2}{1 - u^2\epsilon} + p \right) + \dfrac{(d-1)(\rho + p\epsilon)u^2}{(1 - u^2\epsilon)r} = 0, \\ \partial_t \left(\dfrac{(\rho + p\epsilon)u^2\epsilon}{1 - u^2\epsilon} + \rho \right) + \partial_r \left(\dfrac{(\rho + p\epsilon)u}{1 - u^2\epsilon} \right) + \dfrac{(d-1)(\rho + p\epsilon)u}{(1 - u^2\epsilon)r} = 0. \end{cases} \tag{38}$$

Theorem 7 ([14]). *Assume that $b(t) \in C^\infty$, $b(0)=0$, $b'(0)=b_0$, and $b^{(k)}(0)=0$ for $2 \le k \le K$, where K is a given large integer depending only on the background solution. There exists a constant $T_0 > 0$ such that there exists a shock front solution (ρ^c, u^c) to problem* (36) *and* (38) *for $t \le T_0$. Moreover, there exists a subsequence $\{c_n\}$, $c_n \to +\infty$, as $n \to +\infty$, such that*

$$(\rho^{c_n}, u^{c_n}) \to (\rho, u), \qquad a.e.,$$

and the limit (ρ, u) is the corresponding solution for the classical non-relativistic system (37).

The proof of this main conclusion is based on Sects. 3.2.1–3.2.3.

3.2.1 Background Solution

When the piston velocity is a constant denoted by b_0, the velocity of the shock front is also a constant denoted by s_0. Since problem (36) and (38) keeps invariant under the scaling $r \to \lambda r$, $t \to \lambda t$, it admits a self-similar solution. By the self-similar transformation $\nu = r/t$, problem (36) and (38) reduces to

$$\text{(I)}\begin{cases} \frac{a^2+u^2-u\nu(1+a^2\epsilon)}{1-u^2\epsilon}\rho_\nu + \frac{(\rho+p\epsilon)(2u-\nu-\nu u^2\epsilon)}{(1-u^2\epsilon)^2}u_\nu + \frac{(d-1)(\rho+p\epsilon)u^2}{(1-u^2\epsilon)\nu} = 0, \\ \frac{(1+a^2\epsilon)u-\nu(1+a^2u^2\epsilon^2)}{1-u^2\epsilon}\rho_\nu + \frac{(\rho+p\epsilon)(1+u^2\epsilon-2\nu u\epsilon)}{(1-u^2\epsilon)^2}u_\nu + \frac{(d-1)(\rho+p\epsilon)u}{(1-u^2\epsilon)\nu} = 0, \\ u = b_0, \qquad \text{on} \qquad r = b_0 t, \\ (\rho, u) = (\rho_*, 0), \qquad \text{on} \qquad t = 0. \end{cases}$$

The solution of problem (I) is called the background solution. From the first two equations, we have

$$u_\nu = \frac{a^2u(1-u\nu\epsilon)(1-u^2\epsilon)}{\nu\left((1-au\epsilon)\nu - u + a\right)\left((1+au\epsilon)\nu - u - a\right)}, \tag{39}$$

$$\rho_\nu = \frac{(\rho+p\epsilon)u(\nu-u)}{\nu\left((1-au\epsilon)\nu - u + a\right)\left((1+au\epsilon)\nu - u - a\right)}. \tag{40}$$

The curve satisfying (39) and (40) which describes the state of the gas between the piston and the shock is referred to as the solution curve in the (ρ, u)-plane.

From the Rankine-Hugoniot conditions of (38), we obtain that

$$u^2 = \frac{a^2(\rho-\rho_a)^2}{(1+a^2\epsilon)^2\rho\rho_a + a^2(\rho-\rho_a)^2\epsilon}, \quad s_0^2 = \frac{a^2(\rho + a^2\rho_a\epsilon)}{\rho_a + a^2\rho\epsilon}. \tag{41}$$

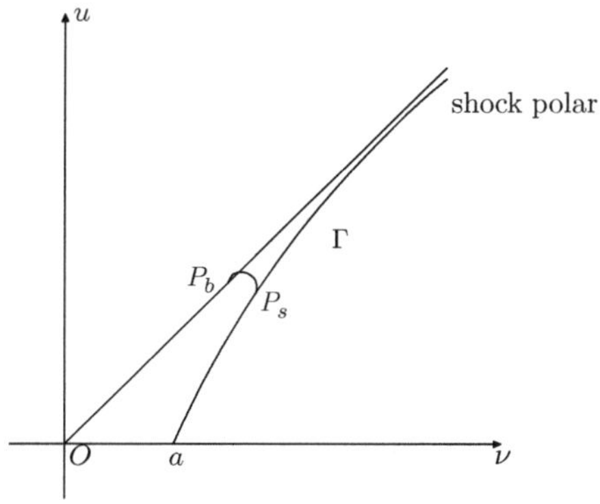

Fig. 13 Shock polar

Hence, $s_0 \geq a$. When s varies in $[a, +\infty)$, we obtain a curve $\Gamma : u = u(s)$,

$$u = \frac{s^2 - a^2}{s(1 - a^2\epsilon)}, \tag{42}$$

which is called the shock polar. We construct the solution of problem (I) as follows.

Step 1. Solve ODE (39) with the initial data $u|_{\nu=b_0} = b_0$;
Step 2. Consider the solution of (39) intersecting with the shock polar, which is described by (42);
Step 3. From (41) and (42), we obtain the speed of the shock wave and the left state $U_1 = (\rho_1, u_1)$ which is connected to the given right state U_0;
Step 4. Solve ODE (40) with the initial data $\rho|_{\nu=s_0} = \rho_1$ where u is given by Step 1.

Thus, we obtain the existence of background solution of problem (I):

Lemma 5 ([14]). *There exists a solution curve of problem (I) starting from point P_s: $\nu = s_0$ on the shock polar Γ, monotonically decreasing with respect to ν and intersecting with the diagonal at the point P_b: $u = \nu$ (see Fig. 13).*

3.2.2 Approximate Solutions and Energy Estimates

By introducing a new coordinate transformation $x = t$, $\alpha = r/t$, we construct an N-th order approximate solution (ρ, u, s) to problem (36) and (38) in (x, α)-coordinates by the following finite expansions with error $O(x^{N+1})$:

$$u(\alpha, x) = \sum_{n=0}^{N} u_n(\alpha)x^n, \quad \rho(\alpha, x) = \sum_{n=0}^{N} \rho_n(\alpha)x^n, \tag{43}$$

$$s(x) = \sum_{n=0}^{N} s_n x^{n+1}, \qquad b(x) = \sum_{n=0}^{N} b_n x^{n+1}. \tag{44}$$

We establish the existence and uniqueness of the approximate solution using some properties of symmetric hyperbolic systems.

For convenience, we rewrite $s(x) = x\sigma(x)$ and $b(x) = x\beta(x)$. We introduce a transformation

$$y = x, \qquad \theta = \frac{\alpha - \beta(x)}{\sigma(x) - \beta(x)}$$

to fix the boundaries of the piston and the shock as $\theta = 0$ and $\theta = 1$. Meanwhile, we take another transformation $y = e^{\tau}$ to remove the singularity of the system at $y = 0$. Then problem (36) and (38) reduces to

$$\phi_\tau + A\phi_\theta + W(\phi) = 0, \tag{45}$$

$$u = \beta + \beta' \qquad \text{on } \theta = 0, \tag{46}$$

where

$$A = \begin{pmatrix} a_{11} & a_{12} \\ a_{21} & a_{22} \end{pmatrix}, \quad W = \left(\frac{\rho u(1 + a^2\epsilon)}{\alpha(1 - a^2u^2\epsilon^2)}, -\frac{a^2u^2(1 - u^2\epsilon)\epsilon}{\alpha(1 - a^2u^2\epsilon^2)} \right)^{\mathrm{T}},$$

$$a_{11} = a_{22} = e^{\tau} D + \frac{u - \alpha - a^2u\epsilon + \alpha a^2u^2\epsilon^2}{(\sigma - \beta)(1 - a^2u^2\epsilon^2)},$$

$$a_{12} = \frac{\rho(1 + a^2\epsilon)}{(\sigma - \beta)(1 - a^2u^2\epsilon^2)}, \quad a_{21} = \frac{a^2(1 - u^2\epsilon)^2}{\rho(1 + a^2\epsilon)(\sigma - \beta)(1 - a^2u^2\epsilon^2)},$$

$$D := \frac{-\beta'(\sigma - \beta) - (\alpha - \beta)(\sigma' - \beta')}{(\sigma - \beta)^2}.$$

Linearizing problem (45) and (46) at the approximate solution $(\tilde{\phi}, \tilde{\sigma})$, we define an η-weighted norm in the domain $[0, 1] \times (-\infty, T]$ as follows:

$$||f||^2_{k,\eta,T} \equiv \sum_{j+i_1+i_2=k} \int_{-\infty}^{T} \int_0^1 e^{-2\eta\tau} \eta^{2j} \left| \frac{\partial^{i_1+i_2} f}{\partial\tau^{i_1} \partial\theta^{i_2}} \right|^2 \mathrm{d}\theta \mathrm{d}\tau. \tag{47}$$

We define the norms on the boundaries $\theta = 0,\ 1$:

$$< f >^2_{k,\eta,T,\theta=0,1} \equiv \sum_{j+i=k} \int_{-\infty}^{T} e^{-2\eta\tau} \eta^{2j} \left| \frac{\partial^i f}{\partial\tau^i} \right|^2_{\theta=0,1} \mathrm{d}\tau, \tag{48}$$

in which only the normal derivatives are involved. We also define

$$\ll f \gg^2_{k,\eta,T} \equiv \sum_{j=0}^{k} < \partial^j_\theta f >^2_{k-j,\eta,T,\theta=0} + \sum_{j=0}^{k} < \partial^j_\theta f >^2_{k-j,\eta,T,\theta=1} . \tag{49}$$

Energy estimates for the error terms were established for the linearized problem in the following theorem.

Theorem 8 ([14]). *For any fixed* $T < +\infty$ *and integer* $k > 0$, *there exists an* $\eta_0 > 0$ *such that for any* $\eta > \eta_0$, *the solution* $(\dot{\phi}, \dot{\sigma})$ *to the linearized problem at* $(\tilde{\phi}, \tilde{\sigma})$ *satisfies*

$$|||\dot{\phi}|||^2_{k,\eta,T} \equiv \eta \parallel \dot{\phi} \parallel^2_{k,\eta,T} + \ll \dot{\phi} \gg^2_{k,\eta,T} + < \dot{\sigma} >^2_{k+1,\eta,T} \tag{50}$$

$$\leq C_k \left(\frac{1}{\eta} ||\dot{f}||^2_{k,\eta,T} + < \dot{g} >^2_{k,\eta,T,\theta=0} + < \dot{h}_1 >^2_{k,\eta,T,\theta=1} + < \dot{h}_2 >^2_{k,\eta,T,\theta=1} \right).$$

In addition, for $k \geq 2$, *the constant* C_k *depends only on* $\| \cdot \|_{k,\eta,T}$ *of the coefficients, which in turn depends only on* $\| \tilde{\phi} \|_{k,\eta,T}$, $\ll \tilde{\phi} \gg_{k,\eta,T}$, *and* $< \tilde{\sigma} >_{k+1,\eta,T}$, *where*

$$|||\tilde{\phi} - \phi_0|||_{k,\eta,T} < \delta,$$

for some small positive constant δ, *and* $\dot{f}$, $\dot{g}$, $\dot{h}_1$ *and* $\dot{h}_2$ *are the higher order terms resulting from the linearization of the equation and the boundaries at* $(\tilde{\phi}, \tilde{\sigma})$.

3.2.3 Local Existence by Iteration and Non-relativistic Limits

The proof of Theorem 7 was carried out by the Newton iteration scheme, involving the boundness of the higher order norm and the contraction of the lower order norm. In addition, we established the convergence of shock front solutions as $c \to +\infty$.

By the Newton iteration scheme, we constructed a shock front solution depending on the light speed c, denoted by (ρ^c, u^c). From the expressions of $\dot{f}$, $\dot{g}$, $\dot{h}_1$ and $\dot{h}_2$ represented in [14], which are uniformly bounded and independent of large c, and from the energy estimates (50), there exists a subsequence $\{c_n\}$, $c_n \to +\infty$, as $n \to +\infty$, such that

$$(\rho^{c_n}, u^{c_n}) \to (\rho, u), \qquad a.e.,$$

and the limit (ρ, u) is the corresponding solution for the classical non-relativistic system (37).

Remark 11. In the sequel, we will extend local existence of the spherically symmetric piston problem to global existence by a modified Glimm scheme, which originates from Lien-Liu [25] where the nonlinear stability of a self-similar three-dimensional gas flow past an infinite cone with a small vertex angle for the full Euler equations (4) has been established. See also [11, 12].

Meanwhile, we will study mathematically the physical phenomena that a big rarefaction wave exists when the piston recedes away from the gas. However, we may need to employ wave front tracking methods instead of the Glimm scheme and give the estimates on the interactions between the big rarefaction wave and other waves.

Acknowledgements Min Ding would like to thank Professor Gui-Qiang G. Chen for his precious guidance and suggestions, and also expresses her sincere gratitude to OxPDE Center, Mathematical Institute for its hospitality whenever she visits the University of Oxford. Min Ding's research was supported in part by China Scholarship Council (No: 2009623053), and the EPSRC Science and Innovation Award to the Oxford Center for nonlinear PDE (No: EP/E035027/1). Yachun Li's research was supported in part by the National Natural Science Foundation of China under grants 10971135 and 11231006.

References

1. A.M. Anile, *Relativistic Fluids and Magneto-Fluids*. Cambridge Monographs on Mathematical Physics (Cambridge University Press, Cambridge, 1989)
2. G. Chen, Convergence of the Lax-Friedichs scheme for isentropic gas dynamics (III). Acta Math. Sci. **6**, 75–120 (1986) (Chinese edition **8**, 101–134 (1988))
3. G. Chen, S. Chen, D. Wang, Z. Wang, A multidimensional piston problem for the Euler equations for compressible flow. Discret. Contin. Dyn. Syst. **12**, 361–383 (2005)
4. G. Chen, X. Ding, P. Luo, Convergence of the Lax-Friedichs scheme for isentropic gas dynamics (I)–(II). Acta Math. Sci. **5**, 415–432, 433–472 (1985) (Chinese edition: **7**, 467–481 (1987); **8**, 61–94 (1988))
5. G. Chen, Y. Li, Relativistic Euler equations for isentropic fluids: stability of Riemann solutions with large oscillation. Z. Angew. Math. Phys. **55**, 903–926 (2004)
6. G. Chen, Y. Zhang, D. Zhu, Existence and stability of supersonic Euler flows past Lipschitz wedges. Arch. Ration. Mech. Anal. **181**, 261–310 (2006)
7. S. Chen, Existence of local solution to supersonic flow past a three-dimensional wing. Adv. Appl. Math. **13**, 273–304 (1992)
8. S. Chen, Existence of stationary supersonic flow past a pointed body. Arch. Ration. Mech. Anal. **156**, 141–181 (2001)
9. S. Chen, A free boundary value problem of Euler system arising in supersonic flow past a curved cone. Tohoku Math. J. **54**, 105–120 (2002)
10. S. Chen, A singular multidimensional piston problem in compressible flow. J. Differ. Equ. **189**, 292–317 (2003)
11. S. Chen, Z. Wang, Y. Zhang, Global existence of shock front solutions for the axially symmetric piston problem for compressible fluids. J. Hyper. Differ. Equ. **1**, 51–84 (2004)
12. S. Chen, Z. Wang, Y. Zhang, Global existence of shock front solutions for the axially symmetric piston problem for compressible flow. Z. Angew. Math. Phys. **59**, 434–456 (2008)
13. R. Courant, K.O. Friedrichs, *Supersonic Flow and Shock Waves*. Applied Mathematical Sciences, vol 12 (Wiley-Interscience, New York, 1948)
14. M. Ding, Y. Li, Local existence and non-relativistic limits of shock solutions to a multidimensional piston problem for the relativistic Euler equations. Z. Angew. Math. Phys. **64**, 101–121 (2013)
15. M. Ding, Y. Li, Global existence and non-relativistic global limits of entropy solutions to 1-D piston problem for the relativistic Euler equations. J. Math. Phys. **54**, 031506 (2013)
16. M. Ding, Global existence and non-relativistic global limits of entropy solutions to some piston problem for the relativistic Euler equations. Ph.D. Dissertation, Shanghai Jiao Tong University, 2012 (preprint)
17. R. DiPerna, Convergence of the viscosity method for isentropic gas dynamics. Commun. Math. Phys. **91**, 1–30 (1983)

18. J. Glimm, Solutions in the large for nonlinear hyperbolic systems of equations. Commun. Pure Appl. Math. **18**, 697–715 (1965)
19. J. Goodman, Initial boundary value problems for hyperboilic systems of conservation laws. Ph.D. Dissertation, Stanford University, p. 60, 1983
20. L. Landau, E. Lifchitz, *Fluid Mechnics*, 2nd edn. (Pergamon Press, Oxford, 1987)
21. T.-T. Li, T. Qin, *Physics and Partial Differential Equations*, 2nd edn. (Higher Education Press, Beijing, 2005)
22. T.-T. Li, L. Wang, Existence and uniqueness of global solution to an inverse problem. Inverse Probl. **23**, 683–694 (2007)
23. T.-T. Li, L. Wang, Inverse piston problem for the system of one-dimensional isentropic flow. Chin. Ann. Math. B **28**, 265–282 (2007)
24. E.P.T. Liang, Relativistic simple waves: shock damping and entropy production. Astrophys. J. **211**, 361–376 (1977)
25. W. Lien, T. Liu, Nonlinear stability of a self-similar 3-dimensional gas flow. Commun. Math. Phys. **204**, 525–549 (1999)
26. T. Liu, Linear and nonlinear large-time behavior of nonlinear hyperbolic conservation laws. Commun. Pure Appl. Math. **55**, 163–177 (1977)
27. T. Liu, The free piston problem for gas dynamics. J. Differ. Equ. **30**, 175–191 (1978)
28. A. Majda, The stability of multidimensional shock fronts. Mem. Am. Math. Soc. **273**, 1–95 (1983)
29. A. Majda, The existence of multidimensional shock fronts. Mem. Am. Math. Soc. **281**, 1–93 (1983)
30. V. Pant, Global entropy solutions for isentropic relativistic fluid dynamics. Commun. Partial Differ. Equ. **21**, 1609–1641 (1996)
31. M. Sablé-Tougeron, Méthode Glimm et problème mixte (French. English, French summary) [Glimm method and mixed problem]. Ann. Inst. H. Poincaré Anal. Non Linéaire **10**(4), 423–443 (1993)
32. J. Smoller, *Shock Waves and Reaction-Diffusion Equations* (Springer, New York, 1983)
33. J. Smoller, B. Temple, Global solutions of the relativistic Euler equations. Commun. Math. Phys. **156**, 67–99 (1993)
34. A. Taub, Relativistic Rankine-Hügoniot equations. Phys. Rev. **74**, 328–334 (1948)
35. A.H. Taub, Approximate solutions of the Einstein equations for isentropic motions of plane symmetric distributions of perfect fluids. Phys. Rev. **107**, 884–900 (1957)
36. S. Takeno, Initial boundary value problems for isentropic gas dynamics. Proc. R. Soc. Edinb. **120A**, 1–23 (1992)
37. S. Takeno, Free boundary problem for isentropic gas dynamics. Jpn. J. Indust. Appl. Math. **12**, 163–194 (1995)
38. K. Thompson, The special relativistic shock tube. J. Fluid Mech. **171**, 365–375 (1986)
39. K. Thorne, Relativistic shocks: the Taub adiabt. Astrophys. J. **179**, 897–907 (1973)
40. Z. Wang, Local existence of shock solutions to 2-dimensional piston problem in isentropic compressible flow. Acta Math. Sini. **20**, 589–604 (2004)
41. Z. Wang, Global existence of shock front solution to 1-dimensional piston problem (Chinese). Chin. Ann. Math. Ser. A **26**, 549–560 (2005)
42. B. Whitham, *Linear and Nonlinear Waves* (Wiley, New York/London/Sydney/Toronto, 1974)
43. Y. Xu, Y. Dou, Global existence of shock front solutions in 1-dimensional piston problem in the relativistic equations. Z. Angew. Math. Phys. **59**, 244–263 (2008)
44. Y. Zhang, Global existence of steady supersonic potential flow past a curved wedge with piecewise smooth boundary. SIAM J. Math. Anal. **31**, 166–183 (1999)
45. Y. Zhang, Steady supersonic flow past an almost straight wedge with large vertex angle. J. Differ. Equ. **192**, 1–46 (2003)
46. Y. Zhou, *One-Dimensional Unsteady Fluid Dynamics*, 2nd edn., Chinese (Science Press, Beijing, 1998)

The Quasineutral Limit for the Navier-Stokes-Fourier-Poisson System

Donatella Donatelli and Pierangelo Marcati

Abstract This paper is a first attempt to describe the quasineutral limit for a Navier-Stokes-Poisson system where the thermal effects are taken into consideration. In the framework of weak solutions and ill-prepared data, we show that as $\lambda \to 0$ the velocity field u^λ strongly converges towards an incompressible velocity vector field u, the density fluctuation $n^\lambda - 1$ weakly converges to zero and the temperature equation converges towards the so called Fourier equation. We shall provide a detailed mathematical description of the convergence process by analyzing the acoustic equations, by using microlocal defect measures and by developing an explicit correctors analysis.

2010 Mathematics Subject Classification Primary: 35Q35; Secondary: 35Q30, 82D10

1 Introduction

This paper deals with the analysis of the quasineutral limit for a hydrodynamical model for plasma dynamics. In numerical simulations the hydrodynamical models represent an acceptable compromise between accuracy and computational

D. Donatelli
Dipartimento di Ingegneria e Scienze dell'Informazione e Matematica, Università degli Studi dell'Aquila, Via Vetoio, 67010 L'Aquila, Italy
e-mail: donatell@univaq.it

P. Marcati (✉)
Dipartimento di Ingegneria e Scienze dell'Informazione e Matematica, Università degli Studi dell'Aquila, Via Vetoio, 67010 L'Aquila, Italy

Gran Sasso Science Institute, 67010 L'Aquila, Italy
e-mail: marcati@univaq.it

G.-Q.G. Chen et al. (eds.), *Hyperbolic Conservation Laws and Related Analysis with Applications*, Springer Proceedings in Mathematics & Statistics 49,
DOI 10.1007/978-3-642-39007-4_9,

efficiency. Their common feature is the fact that the number of independent variables is reduced, then the hydrodynamical models are obtained from the infinite hierarchy of moment equations of the Boltzmann transport equation by suitable truncation procedures. One of the most general models reads as follows (see for example [2] and [16]):

$$\partial_t n + \operatorname{div}(nu) = 0, \tag{1}$$

$$\partial_t (nu) + \operatorname{div}(nu \otimes u) + \nabla p = \mu \Delta u + (\nu + \mu)\nabla \operatorname{div} u + n \nabla V, \tag{2}$$

$$\partial_t (n\overline{e}) - \operatorname{div}(nu(\overline{e} + k_B \theta^\lambda)) - nu\nabla V = \kappa \Delta \theta + p \operatorname{div} u + n|u|^2, \tag{3}$$

$$\lambda^2 \Delta V = n - C(x), \tag{4}$$

where

$$\overline{e} = \frac{|nu|^2}{(n)^2} + \frac{3}{2} k_B \theta^\lambda \qquad p = p(n,\theta)$$

and $x \in \mathbb{R}^3$, $t \geq 0$, denote the space and time variables, $n(x,t)$ the *negative charge density*, $m(x,t) = n(x,t)u(x,t)$ the *current density*, $u(x,t)$ the *velocity field*, $V(x,t)$ the *electrostatic potential*, θ^λ the *temperature*, μ and η the *shear viscosity* and *bulk viscosity* respectively, k_B the *Boltzmann constant* and $C(x)$ is a *charged ion background density*. The parameter λ is the so called *Debye length* (up to a constant factor), which in terms of physical variables can be expressed as

$$\lambda = \lambda_D / L \qquad \lambda_D = \sqrt{\frac{\varepsilon_0 k_B T}{e^2 n_0}}, \tag{5}$$

where L is the macroscopic length scale, ε_0 is the vacuum permittivity, T the average plasma temperature, e the absolute electron charge and n_0 the average plasma density. In many cases the Debye length is very small compared to the macroscopic length $\lambda_D \ll L$ and so it makes sense to consider the quasineutral limit $\lambda \to 0$ of the system (6)–(8). In this setting the particle density is constrained to be close to the background density (we will consider $C(x) = 1$) of the oppositely charged particle. The limit $\lambda \to 0$ is called the quasineutral limit since the charge density almost vanishes identically. The velocity of the fluid then evolves according to an incompressible flow dynamics. This type of limit has attracted the attention of many people. See in the case of the Euler-Poisson system the results of Cordier and Grenier [3], Grenier [12], Cordier, Degond, Markowich and Schmeiser [4], Loeper [18], Peng, Wang and Yong [19]. Concerning the viscous case the system was analyzed without the energy balance equation (3), namely the first two Eqs. (1) and (2) coupled with the Poisson equation (4) and with the following structural hypothesis on the pressure law: $p(n,\theta) = n^\gamma$, $\gamma \geq 3/2$. See for instance Wang [23] and Jiang and Wang [13]. In fact Wang [23] studied the quasineutral limit for the smooth solution with well-prepared initial data. Wang and Jiang [13] studied the combined quasineutral and inviscid limit of the compressible Navier-Stokes-Poisson system for the weak solution and obtained the convergence of the Navier-Stokes-Poisson system to the incompressible Euler equations with general initial data. Moreover in [13] the vanishing of the viscosity coefficient was required in order to take the quasineutral limit and no convergence rate was derived therein. The authors in [7] investigated the quasineutral limit of the isentropic Navier-Stokes-Poisson system in the whole space and obtained the convergence of the weak solution of the Navier-Stokes-Poisson system to the weak solution of the incompressible Navier-Stokes equations by means of dispersive estimates of Strichartz's type under the assumption that the Mach number is related to the Debye length. Ju, Li and Wang [14] studied the quasineutral limit of the isentropic Navier-Stokes-Poisson system both in the whole space and in the torus without the restriction on the viscous coefficient with well prepared initial data. In the framework of weak solutions and general ill-prepared data there is the authors' paper [8]. A common feature of these kinds of limits in the ill-prepared data framework is the high plasma oscillations, namely the presence of high frequency time oscillations along the acoustic waves (see [6]). In these phenomena the various vector fields in the model exhibit different behaviors, and it is particularly important to understand the relationship between high frequency interacting waves, dispersive behavior and the different roles of time and space oscillations. In [8] the authors provide a detailed mathematical description of the convergence process. Since the velocity fields both disperse and oscillate and the dispersion behavior dominates on the high frequency time oscillations, Strichartz estimates are sufficient to pass to the limit of the convective term, however the presence of quadratic terms on the electric field (e.g. $n\nabla V$) cannot be analyzed in the same way since the dispersive behavior no longer dominates on high frequency time wave packets, so by using microlocal defect measures, and by developing an explicit correctors analysis, an explicit pseudo parabolic pde satisfied by the leading correctors terms is identified. Concerning the quasineutral limit for the full Navier-Stokes system (1)–(4), there are very few results. For example in [15] the limit is analyzed in the framework of smooth solutions and by providing an asymptotic expansion. At the moment there are no known results for weak solutions and ill-prepared data for the full system (1)–(4). This paper is a first attempt in that direction. In particular we study a simplified version of the Eqs. (1)–(4), namely the so called Navier-Stokes-Fourier-Poisson system

$$\partial_t n^\lambda + \operatorname{div}(n^\lambda u^\lambda) = 0, \tag{6}$$

$$\partial_t (n^\lambda u^\lambda) + \operatorname{div}(n^\lambda u^\lambda \otimes u^\lambda + (n^\lambda)^\gamma \mathbb{I}) = \mu \Delta u^\lambda + (\nu + \mu)\nabla \operatorname{div} u^\lambda + n^\lambda \nabla V^\lambda, \tag{7}$$

$$\partial_t (n^\lambda \theta^\lambda) + \operatorname{div}(n^\lambda u^\lambda \theta^\lambda) = \kappa \Delta \theta^\lambda, \tag{8}$$

$$\lambda^2 \Delta V^\lambda = n^\lambda - 1. \tag{9}$$

with the following initial data

$$n^\lambda_{t=0} = n^\lambda_0 \geq 0, \quad \theta^\lambda|_{t=0} = \theta^\lambda_0, \quad V^\lambda|_{t=0} = V^\lambda_0, \tag{ID}$$

$$n^\lambda u^\lambda|_{t=0} = m^\lambda_0, \quad m^\lambda_0 = 0 \text{ on } \{x \in \mathbb{R}^3 \mid n^\lambda_0(x) = 0\},$$

$$\int_{\mathbb{R}^3} \left(\pi^\lambda|_{t=0} + \frac{|m^\lambda_0|^2}{2n^\lambda_0} + n^\lambda_0 \theta^\lambda_0 + n^\lambda_0 (\theta^\lambda_0)^2 + \lambda^2 |V^\lambda_0| \right) dx \leq C_0.$$

To simplify our notation from now on we will set

$$\pi^\lambda = \frac{(n^\lambda)^\gamma - 1 - \gamma(n^\lambda - 1)}{(\gamma - 1)} \qquad \sigma^\lambda = n^\lambda - 1 \qquad \mu = \nu = \kappa = 1.$$

Also in this case the main difficulties in approaching this problem will be the fast oscillation of the acoustic waves and the presence of quadratic terms in the electric field, $\lambda \nabla V^\lambda \otimes \lambda \nabla V^\lambda$. The latter problem will be analyzed by observing that the density fluctuation $\sigma^\lambda = n^\lambda - 1$ satisfies a Klein-Gordon equation, so the acoustic waves analysis follows by treating the system as a dispersive equation and we will obtain uniform estimates, while the latter will be overcome by noticing that $\lambda \nabla V^\lambda$ is bounded in $L^\infty_t L^2_x$ and so we can define the microlocal defect measure ν^E introduced by P. Gérard in [11] and by L. Tartar (H-measure) in [22] with correctors E^+ and E^- to handle time oscillations at frequency $1/\lambda$.

The existence of weak solutions for the system (6)–(9) may be proved as in [5, 9, 10]. By following the same line of argument as in [8] we are able to prove the following theorem.

Theorem 1. *Let $(n^\lambda, u^\lambda, \theta^\lambda, V^\lambda)$ be a sequence of weak solutions in $\mathbb{R}^3$ of the system (6)–(9) and assume that the initial data satisfy (ID). Then*

(*i*) $n^\lambda \longrightarrow 1$ *weakly in $L^\infty([0,T]; L^k_2(\mathbb{R}^3))$.*

(*ii*) *There exists a $u \in L^\infty([0,T]; L^2(\mathbb{R}^3)) \cap L^2([0,T]; \dot{H}^1(\mathbb{R}^3))$ such that*

$$u^\lambda \rightharpoonup u \quad \textit{weakly in } L^2([0,T]; \dot{H}^1(\mathbb{R}^3)).$$

(*iii*) *The gradient component Qu^λ of the vector field u^λ satisfies*

$$Qu^\lambda \longrightarrow 0 \quad \textit{strongly in } L^2([0,T]; L^p(\mathbb{R}^3)), \textit{for any } p \in [4,6).$$

(*iv*) *The divergence free component Pu^λ of the vector field u^λ satisfies*

$$Pu^\lambda \longrightarrow Pu = u \quad \textit{strongly in } L^2([0,T]; L^2_{loc}(\mathbb{R}^3)).$$

(*v*) *There exists a $\theta \in L^\infty([0,T]; L^2(\mathbb{R}^3)) \cap L^2([0,T]; \dot{H}(\mathbb{R}^3))$ such that*

$$\theta^\lambda \longrightarrow \theta \quad \textit{strongly in } L^2([0,T]; L^2_{loc}(\mathbb{R}^3)),$$

$$\nabla\theta^\lambda \rightharpoonup \nabla\theta \quad \textit{weakly in } L^2([0,T] \times \mathbb{R}^3).$$

(*vi*) *There exist correctors E^+, E^- in $L^\infty((0,T),L^2(\mathbb{R}^3))$ and a positive microlocal defect measure ν^E on $\mathbb{R}^3 \times S^2$ depending measurably on t, associated to the electric field $E^\lambda = \nabla V^\lambda$, such that for all pseudodifferential operators $A \in \psi^0_{comp}(\mathbb{R}^3, \mathcal{K}(\mathbb{R}^3))$ of symbol $a(x,\xi)$ and for all $\phi \in \mathcal{D}(0,t)$ one has*

$$\lim_{\lambda\to 0}\int dt\phi(t)\lambda^2(AE^\lambda, E^\lambda) = \int dt\phi(t)(AE^+,E^+) + \int dt\phi(t)(AE^-,E^-) + \int dt\phi(t)\int_{\mathbb{R}^3\times S^2} tr\left(a(x,\xi)\frac{\xi\otimes\xi}{|\xi|^2}\right)d\nu^E. \tag{10}$$

(*vii*) *$u = Pu$ and θ satisfy the following equations respectively*

$$P\Big(\partial_t u - \Delta u + (u\cdot\nabla)u - \operatorname{div}(E^+\otimes E^+ + E^-\otimes E^-) - \operatorname{div}\Big\langle \nu^E, \frac{\xi\otimes\xi}{|\xi|^2}\Big\rangle\Big) = 0, \tag{11}$$

$$\partial_t\theta + u\cdot\nabla\theta - \Delta\theta = 0 \tag{12}$$

in $\mathcal{D}'([0,T]\times\mathbb{R}^3)$.

The paper is organized as follows. In Sect. 2 we present all the definitions and technical tools which will be used later in the paper, then the rest of the paper is devoted to the proof of Theorem 1. In Sects. 3 and 4 we recover all the a priori bounds, in Sects. 5 and 6 we provide the convergence analysis of the velocity field and of the temperature. Finally in Sect. 7 we give a formal proof of the equation satisfied by the correctors $E^\pm$.

2 Notations

For convenience we record here the basic notations that we shall be using later. We will denote by $\mathcal{D}(\Omega\times\mathbb{R}_+)$ the space of test function $C_0^\infty(\Omega\times\mathbb{R}_+)$, by $\mathcal{D}'(\Omega\times\mathbb{R}_+)$ the space of Schwartz distributions and $\langle\cdot,\cdot\rangle$ the duality bracket between $\mathcal{D}'$ and $\mathcal{D}$. Moreover $W^{k,p}(\Omega) = (I-\Delta)^{-\frac{k}{2}}L^p(\Omega)$ and $H^k(\Omega) = W^{k,2}(\Omega)$ denote the nonhomogeneous Sobolev spaces for any $1\le p\le\infty$ and $k\in\mathbb{R}$. The notations $L^p_tL^q_x$ and $L^p_tW^{k,q}_x$ will abbreviate respectively the spaces $L^p([0,T];L^q(\Omega))$ and $L^p([0,T];W^{k,q}(\Omega))$. We denote by $L^p_2(\mathbb{R}^d)$ the Orlicz space defined as follows

$$L^p_2(\mathbb{R}^d) = \{f\in L^1_{loc}(\mathbb{R}^d) \mid |f|\chi_{|f|\le\frac12}\in L^2(\mathbb{R}^d),\ |f|\chi_{|f|>\frac12}\in L^p(\mathbb{R}^d)\}. \tag{13}$$

See [1, 17] for more details. We denote by $\mathcal{L}(\mathbb{R}^3)$ the space of bounded operators, $\mathcal{K}(\mathbb{R}^3)$ the space of compact operators and if X, Y are Banach spaces, $\mathcal{L}(X,Y)$

is the space of bounded operators. If Ω is an open set in $\mathbb{R}^3$, we denote by $\psi^m_{comp}(\Omega, \mathcal{L}(H))$, respectively, $\psi_{comp}(\Omega, \mathcal{K}(H))$ the space of polyhomogeneous pseudo-differential operators of order m on Ω, with values in $\mathcal{L}(H)$, respectively $\mathcal{K}(H)$, whose kernel is compactly supported in $\Omega \times \Omega$, moreover we recall that if $A \in \psi^m_{comp}(\Omega, \mathcal{L}(H))$, then its symbol $a(x,\xi)$ ($A = OP(a(x,\xi))$) is a linear application from $\psi^m_{comp}(\Omega, \mathcal{L}(H))$ to $C_0^\infty(S^*\Omega, \mathcal{L}(H))$, where $S^*\Omega = S^{d-1} \times \Omega$.

Following P. Gérard we say that μ is the *microlocal defect measure* (or following L. Tartar the *H-measure*) for a bounded sequence w_k in L^2 if for any $A \in \psi^0_{comp}(\omega, \mathcal{K}(H))$ one has (up to subsequences)

$$\lim_{k\to\infty} (A(w_k - w), (w_k - w)) = \int_{S^*\Omega} tr(a(x,\xi)\mu(dxd\xi)),$$

where $A = OP(a(x,\xi))$.

We define the Leray's projector P on the space of divergence-free vector fields and Q on the space of gradient vector fields by

$$Q = \nabla\Delta^{-1}\operatorname{div} \qquad P = I - Q. \tag{14}$$

Let us describe the dispersive estimate we are going to use later on. We recall that if w is a solution of the following Klein-Gordon equation in the space $[0,T] \times \mathbb{R}^d$

$$\left(-\frac{\partial^2}{\partial t} + \Delta - m^2\right) w(t,x) = F(t,x)$$

with Cauchy data

$$w(0,\cdot) = f, \quad \partial_t w(0,\cdot) = g,$$

where $m > 0$ is the mass and $0 < T < \infty$, then for any $s > 0$, w satisfies the following Strichartz estimates, (see [21]),

$$\|w\|_{L^4_t W^{-s,4}_x} + \|\partial_t w\|_{L^4_t W^{-1-s,4}_x} \lesssim \|f\|_{H^{1/2-s}_x} + \|g\|_{H^{-1/2-s}_x} + \|F\|_{L^1_t H^{-s}_x}. \tag{15}$$

Here we state the following elementary lemma which will be used later on.

Lemma 1. *Let us consider a smoothing kernel $j \in C_0^\infty(\mathbb{R}^d)$, such that $j \geq 0$, $\int_{\mathbb{R}^d} j dx = 1$, and let us define*

$$j_\alpha(x) = \alpha^{-d} j\left(\frac{x}{\alpha}\right).$$

Then for any $f \in \dot{H}^1(\mathbb{R}^d)$, one has

$$\|f - f * j_\alpha\|_{L^p(\mathbb{R}^d)} \leq C_p \alpha^{1-d\left(\frac{1}{2}-\frac{1}{p}\right)} \|\nabla f\|_{L^2(\mathbb{R}^d)}, \tag{16}$$

where

$$p \in [2,\infty) \quad \text{if } d = 2, \qquad p \in [2,6] \quad \text{if } d = 3.$$

Moreover the following Young type inequality holds

$$\|f * j_\alpha\|_{L^p(\mathbb{R}^d)} \le C\alpha^{s-d\left(\frac{1}{q}-\frac{1}{p}\right)}\|f\|_{W^{-s,q}(\mathbb{R}^d)}, \tag{17}$$

for any $p, q \in [1,\infty]$, $q \le p$, $s \ge 0$, $\alpha \in (0,1)$.

3 A Priori Estimate

By a standard method we can prove the following energy inequalities hold for almost every $t \ge 0$:

$$\int_{\mathbb{R}^3}\left(n^\lambda\frac{|u^\lambda|^2}{2} + \pi^\lambda + n^\lambda\theta^\lambda + \lambda^2|\nabla V^\lambda|^2\right)dx + \int_0^t\int_{\mathbb{R}^3}\left(\mu|\nabla u^\lambda|^2 + (\nu+\mu)|\operatorname{div} u^\lambda|^2\right)dx\,ds \le C_0. \tag{18}$$

$$\frac{1}{2}\int_{\mathbb{R}^3} n^\lambda\frac{|\theta^\lambda|^2}{2}dx + \int_0^t\int_{\mathbb{R}^3}(|\nabla\theta^\lambda(x,s)|^2)dx\,ds = \frac{1}{2}\int_{\mathbb{R}^3} n_0^\lambda(\theta_0^\lambda)^2dx. \tag{19}$$

As a consequence we get the following bounds:

$$\sigma^\lambda \quad \text{is bounded in } L^\infty([0,T];L_2^k(\mathbb{R}^3)),\ k=\min(\gamma,2), \tag{20}$$

$$\nabla u^\lambda \quad \text{is bounded in } L^2_{t,x}, \quad u^\lambda \quad \text{is bounded in } L^2_{t,x}\cap L^2_tL^6_x, \tag{21}$$

$$\nabla\theta^\lambda \quad \text{is bounded in } L^2_{t,x}, \quad \theta^\lambda \quad \text{is bounded in } L^2_{t,x}\cap L^2_tL^6_x, \tag{22}$$

$$\sigma^\lambda u^\lambda, \sigma^\lambda\theta^\lambda \quad \text{are bounded in } L^2_tH^{-1}_x. \tag{23}$$

4 Density Fluctuation Acoustic Equation

From the estimates of Sect. 3 we get only the weak convergence of the velocity field and unfortunately this will not be sufficient to pass to the limit in the nonlinear terms (such as the convective term $\operatorname{div}(n^\lambda u^\lambda \otimes u^\lambda)$). In particular this weak convergence is induced by the rapid time oscillation of the acoustic waves or by the so called plasma oscillations. In order to overcome this problem we will estimate the density fluctuation σ^λ uniformly with respect to λ. So we derive the so called acoustic equation which governs the time evolution of σ^λ. First of all we rewrite the continuity equation (6) and the momentum equation (7) in the following way

$$\partial_t \sigma^\lambda + \operatorname{div}(n^\lambda u^\lambda) = 0, \tag{24}$$

$$\begin{aligned} \partial_t (n^\lambda u^\lambda) + \nabla \sigma^\lambda &= \mu \Delta u^\lambda + (\nu + \mu) \nabla \operatorname{div} u^\lambda - \operatorname{div}(n^\lambda u^\lambda \otimes u^\lambda) \\ &\quad - (\gamma - 1) \nabla \pi^\lambda + \sigma^\lambda \nabla V^\lambda + \nabla V^\lambda, \end{aligned} \tag{25}$$

$$\lambda^2 \Delta V^\lambda = \sigma^\lambda. \tag{26}$$

Then, by differentiating Eq. (24) with respect to time, taking the divergence of (25) and by using (26) we get that σ^λ satisfies the following equation

$$\begin{aligned} \partial_{tt} \sigma^\lambda - \Delta \sigma^\lambda + \frac{\sigma^\lambda}{\lambda^2} &= - \operatorname{div}(\mu \Delta u^\lambda + (\nu + \mu) \nabla \operatorname{div} u^\lambda) \\ &\quad + \operatorname{div}\left(\operatorname{div}(n^\lambda u^\lambda \otimes u^\lambda) + (\gamma - 1) \nabla \pi^\lambda + \sigma^\lambda \nabla V^\lambda\right). \end{aligned} \tag{27}$$

It turns out that (27) is a nonhomogeneous Klein-Gordon equation with mass $1/\lambda$. In order to get some more uniform estimates on σ^λ we apply to (27) the Strichartz estimates (15) and we are able to prove the following estimate for any $s_0 \geq 3/2$:

$$\begin{aligned} &\lambda^{-\frac{1}{2}} \|\sigma^\lambda\|_{L_t^4 W_x^{-s_0-2,4}} + \lambda^{-\frac{1}{2}} \|\partial_t \sigma^\lambda\|_{L_t^4 W_x^{-s_0-3,4}} \\ &\lesssim \lambda^{s_0 - \frac{1}{2}} \|\sigma_0^\lambda\|_{H_x^{-3/2}} + \lambda^{s_0 - \frac{1}{2}} \|m_0^\lambda\|_{H_x^{-5/2}} \\ &+ T \| \operatorname{div}(\operatorname{div}(n^\lambda u^\lambda \otimes u^\lambda) - (\gamma - 1) \nabla \pi^\lambda) \|_{L_t^\infty H_x^{-s_0-2}} \\ &+ \lambda^{s_0} \| \operatorname{div} \Delta u^\lambda + \nabla \operatorname{div} u^\lambda \|_{L_t^2 H_x^{-2}} + T \| \operatorname{div}(\sigma^\lambda V^\lambda) \|_{L_t^\infty H_x^{-s_0-2}}. \end{aligned} \tag{28}$$

For the details of the proof see [8].

5 Strong Convergence of Qu^λ and Pu^λ

To prove the strong convergence of Qu^λ and Pu^λ one follows the same line of argument as Sect. 5 of [8]. The main tool is that Qu^λ can be written in terms of the density fluctuation, namely $Q(n^\lambda u^\lambda) = \nabla \Delta^{-1} \partial_t \sigma^\lambda$, and by using the estimate (28) we are able to prove that

$$Qu^\lambda \longrightarrow 0 \quad \text{strongly in } L_t^2 L_x^p \text{ for any } p \in [4, 6).$$

By proving equicontinuity in time properties for Pu^λ we get the following convergence result

$$Pu^\lambda \longrightarrow Pu \quad \text{strongly in } L_t^2 L_{loc,x}^2.$$

6 Strong Convergence of θ^λ

It remains to prove the strong compactness of the temperature θ^λ. To achieve this goal, as for Pu^λ we need to look for some time regularity properties of θ^λ. This will be done in the next lemma.

Lemma 2. *Let us consider the solution* $(n^\lambda, u^\lambda, \theta^\lambda, V^\lambda)$ *of the Cauchy problem for the system* (6)–(9). *Assume that the hypotheses (ID) hold. Then for all* $h \in (0,1)$, *we have*

$$\|\theta^\lambda(t+h) - \theta^\lambda(t)\|_{L^2([0,T]\times\mathbb{R}^3)} \leq C_T h^{2/5}. \tag{29}$$

Proof. Let us set $\Theta^\lambda = \theta^\lambda(t+h) - \theta^\lambda(t)$. We have

$$\begin{aligned}\|\theta^\lambda(t+h) - \theta^\lambda(t)\|^2_{L^2_{t,x}} &= \int_0^T\int_{\mathbb{R}^3} dt\,dx(\Theta^\lambda)\cdot(\Theta^\lambda - \Theta^\lambda * j_\alpha)\\ &+ \int_0^T\int_{\mathbb{R}^3} dt\,dx(\Theta^\lambda)\cdot(\Theta^\lambda * j_\alpha) = I_1 + I_2.\end{aligned} \tag{30}$$

By using (16) together with (22) we can estimate I_1 in the following way

$$I_1 \leq \|\Theta^\lambda\|_{L^2_{t,x}}\|\Theta^\lambda(t) - (\Theta^\lambda * j_\alpha)(t)\|_{L^2} \lesssim \alpha\|u^\lambda\|_{L^2_{t,x}}\|\nabla u^\lambda\|_{L^2_{t,x}}. \tag{31}$$

In order to estimate I_2 we split it as follows

$$\begin{aligned}I_2 &= \int_0^T\int_{\mathbb{R}^3} dt\,dx(n^\lambda\Theta^\lambda)\cdot(\Theta^\lambda * j_\alpha) + \int_0^T\int_{\mathbb{R}^3} dt\,dx(\sigma^\lambda\Theta^\lambda)\cdot(\Theta^\lambda * j_\alpha)\\ &= I_{2,1} + I_{2,2}.\end{aligned} \tag{32}$$

$I_{2,2}$ can be estimated by taking into account (18), (22) and (17) so we have

$$\begin{aligned}I_{2,2} =&\lambda^2\int_0^T\int_{\mathbb{R}^3} dt\,dx(\Delta V^\lambda\Theta^\lambda(\Theta^\lambda * j_\alpha)\\ &= \lambda^2\int_0^T\int_{\mathbb{R}^3} dt\,dx\left[(\nabla V^\lambda\Theta^\lambda)(\nabla\Theta^\lambda * j_\alpha) + \nabla V^\lambda\nabla\Theta^\lambda(\Theta^\lambda * j_\alpha)\right]\\ &\leq \lambda\alpha^{-3/2}\|\nabla u^\lambda\|_{L^2_{t,x}}\|\lambda\nabla V^\lambda\theta^\lambda + \lambda\nabla V^\lambda\nabla\theta^\lambda\|_{L^2_tL^1_x}.\end{aligned} \tag{33}$$

Now we estimate $I_{2,1}$. Let us reformulate $n^\lambda\Theta^\lambda$ in integral form by using Eq. (9),

$$\begin{aligned}
I_{2,1} &\leq \left| \int_0^T dt \int_{\mathbb{R}^3} dx \int_t^{t+h} ds (\operatorname{div}(n^\lambda u^\lambda \theta^\lambda) + \Delta\theta^\lambda)(s,x) \cdot (\Theta^\lambda * j_\alpha)(t,x) \right| \\
&= \left| \int_0^T dt \int_{\mathbb{R}^3} dx \int_t^{t+h} ds (\operatorname{div}(n^\lambda u^\lambda \otimes u^\lambda) + \Delta u^\lambda) \cdot (Pz^\lambda * j_\alpha)(t,x) \right| \\
&\leq h\|\nabla\theta^\lambda\|^2_{L^2_{t,x}} + C\alpha^{-3/2}T^{1/2}h\|\nabla\theta^\lambda\|_{L^2_{t,x}} \|n^\lambda|u^\lambda|^2\|_{L^\infty_t L^1_x}. \qquad (34)
\end{aligned}$$

Summing up I_1, $I_{2,1}$, $I_{2,2}$ and by taking into account the bounds in Sect. 3 we have

$$\|\theta^\lambda(t+h) - \theta^\lambda(t)\|^2_{L^2([0,T]\times\mathbb{R}^3)} \leq C(\alpha + h + h\alpha^{-3/2}T^{1/2}),$$

and by choosing $\alpha = h^{2/5}$, we end up with (29). □

By using Lemma 2 and standard compactness arguments (see [20]) we get (6).

$$\theta^\lambda \longrightarrow \theta \qquad \text{strongly in } L^2(0,T; L^2_{loc}(\mathbb{R}^3)). \qquad (35)$$

7 Convergence of the Electric Field

This section is addressed to the study of the convergence of the electric field $E^\lambda = \nabla V^\lambda$. By the a priori estimate (18) we only know that λE^λ is bounded in $L^\infty_t L^2_x$, which does not give enough information to pass to the limit in the quadratic term $n^\lambda \nabla V^\lambda = \operatorname{div}(\lambda E^\lambda \otimes \lambda E^\lambda) - 1/2\nabla|\lambda E^\lambda|^2$ appearing in the right-hand side of (7). Hence the problem is how to recover the weak continuity of quadratic forms in L^2. Since λE^λ is bounded in $L^\infty_t L^2_x$ we can define the so called microlocal defect measure introduced by P. Gérard in [11] and by L. Tartar in [22] (H-measures), but in order to handle time oscillations we need to introduce correctors. In this section we will be able to prove the following theorem.

Theorem 2. *Let $(n^\lambda, u^\lambda, \theta^\lambda, E^\lambda)$ be a sequence of solutions of the Navier-Stokes-Fourier-Poisson system (6)–(9), then*

(i) There exists E^+, E^- in $L^\infty((0,T), L^2(\mathbb{R}^3))$,
(ii) There exists a positive measure ν^E on $\mathbb{R}^3 \times S^2$ depending measurably on t

such that for all pseudodifferential operators $A \in \psi^0_{comp}(\mathbb{R}^3, \mathcal{K}(\mathbb{R}^3))$ of symbol $a(x,\xi)$ and for all $\phi \in \mathcal{D}(0,t)$ one has

$$\lim_{\lambda\to 0}\int dt\phi(t)\lambda^2(AE^\lambda, E^\lambda) = \int dt\phi(t)(AE^+, E^+) + \int dt\phi(t)(AE^-, E^-)$$
$$+\int dt\phi(t)\int_{\mathbb{R}^3\times S^2} tr\left(a(x,\xi)\frac{\xi\otimes\xi}{|\xi|^2}\right) d\nu^E. \tag{36}$$

Proof. First we rewrite (7) in terms of E^λ, namely

$$\lambda^2\partial_{tt}E^\lambda + E^\lambda = \operatorname{div}\Delta^{-1}\nabla\operatorname{div}\left(n^\lambda u^\lambda\otimes u^\lambda + (n^\lambda)^\gamma\mathbb{I} - \lambda^2 E^\lambda\otimes E^\lambda\right)$$
$$+\frac{\lambda^2}{2}\operatorname{div}\left(|E^\lambda|^2\mathbb{I}\right) - 2\nabla\operatorname{div} u^\lambda = F^\lambda, \tag{37}$$

then we observe that by using (28) and the uniqueness of the weak limit we have

$$\lambda\nabla V^\lambda \rightharpoonup 0 \quad \text{weakly in } L^2(0,T;L^2(\mathbb{R}^3)). \tag{38}$$

By (38) we see that we are in precisely the framework described by P. Gérard, but we have to pay attention to one fact. In our case, in the quadratic form $\lambda^2\langle AE^\lambda, E^\lambda\rangle$, A is a pseudodifferential operator homogeneous only with respect to the x variable and in the general case we cannot extend it to a pseudodifferential operator homogeneous in (x,t). Hence we have to work on λE^λ in order to isolate the components that oscillate fast in time. For this reason we introduce what we call the correctors of the electric field. In order to understand how to isolate the oscillating terms let us consider Eq. (37) in the case when F^λ does not depend on x. Then, if we take the Fourier transform with respect to time we have ($\hat{E}$ denotes the Fourier transform with respect to time)

$$\lambda\hat{E^\lambda} = \frac{\lambda}{1-\lambda^2|\tau|^2}\hat{F^\lambda},$$

and we can see that all the L^2-mass of λE^λ is concentrated in $\tau = \pm 1/\lambda$ as $\lambda\to 0$. This simple fact leads us to introduce correctors in time of order $1/\lambda$. So we define

$$E_+^\lambda = \lambda e^{-it/\lambda}E^\lambda \qquad E_-^\lambda = \lambda e^{it/\lambda}E^\lambda. \tag{39}$$

In particular they take into account the L^2-mass of λE^λ around $1/\lambda$. By construction it easily follows that E_+^λ and E_-^λ are bounded in $L^2_{t,x}$ and converge weakly to E^+ and E^- respectively. So, if we look at the limit of $\lambda E^\lambda - e^{it/\lambda}E^+ - e^{-it/\lambda}E^-$ as $\lambda\to 0$, we expect to take away the L^2-mass of λE^λ which concentrates around $1/\lambda$. Now we can define

$$\widetilde{E^\lambda} = E^\lambda - e^{it/\lambda}\frac{E^+}{\lambda} - e^{-it/\lambda}\frac{E^-}{\lambda}, \tag{40}$$

and we can prove the following lemma.

Lemma 3. *Let $(n^\lambda, u^\lambda, \theta^\lambda, E^\lambda)$ be a sequence of solutions of the Navier-Stokes-Poisson system* (6) *and* (7) *which satisfy (ID), then*

$$\lambda \widetilde{E^\lambda} \rightharpoonup 0 \qquad \textit{weakly in } L^2(0, T, L^2(\mathbb{R}^3)).$$

Proof. The proof follows by taking into account (38) and the fact that $\lambda \widetilde{E^\lambda}$ is bounded in $L^2_{t,x}$. □

At this point we can hope that the weak convergence of $\lambda \widetilde{E^\lambda}$ is caused only by spatial oscillations, which allows us to introduce the microlocal defect measure in space. In order to do this, since the solutions are defined only in $(0, T)$, we need to extend E^λ and F^λ to 0 and to cut-off the frequencies greater than a certain quantity. Now, by using the same strategy as in [8] we can finish the proof of the Theorem.

□

8 The Equation for the Correctors

We conclude this paper with a short remark concerning the equation satisfied by the correctors. If we assume that the solutions of the system (6)–(9) are smooth enough then we are able to prove the following result.

Theorem 3. *Let $(n^\lambda, u^\lambda, \theta^\lambda, V^\lambda)$ be a sequence of solutions of the Navier-Stokes-Fourier-Poisson system* (6)–(9) *satisfying for $s \geq 4$*

$$\|n^\lambda - 1\|_{L^\infty(0,T;H^s(\mathbb{R}^3))} \leq C \qquad \|\lambda E^\lambda\|_{L^\infty(0,T;H^s(\mathbb{R}^3))} \leq C \tag{41}$$

then, for all $s' < s - 2$

$$u^\lambda - \frac{1}{i} e^{-it/\lambda} E^+ - \frac{1}{i} e^{it/\lambda} E^- \longrightarrow v \quad \textit{strongly in } C^0(0, T, H^{s'-1}_{loc}(\mathbb{R}^3)), \tag{42}$$

$$\lambda(E^\lambda - e^{-it/\lambda} E^+ - e^{it/\lambda} E^-) \longrightarrow 0 \quad \textit{strongly in } C^0(0, T, H^{s'-1}_{loc}(\mathbb{R}^3)), \tag{43}$$

and $E^\pm$ satisfy

$$\partial_t E^\pm - \Delta E^\pm + Q \operatorname{div}(v \otimes E^\pm) = 0, \qquad PE^\pm = 0, \tag{44}$$

where v is defined by (42).

Here we will only sketch a formal proof. For rigorous details we refer to [8].

Proof. By the previous section we know that we can decompose the electric field and the velocity in the following way:

$$E^\lambda \sim \frac{E^+}{\lambda} e^{it/\lambda} + \frac{E^-}{\lambda} e^{-it/\lambda} \qquad u^\lambda \sim v - ie^{it/\lambda} E^+ - ie^{-it/\lambda} E^-,$$

where v is a divergence-free vector field. Now if we substitute this decomposition into Eq. (37) we get

$$2i\partial_t E^+ e^{it/\lambda} - 2i\partial_t E^- e^{-it/\lambda} + \lambda\partial_{tt}E^+ e^{it/\lambda} + \lambda\partial_{tt}E^- e^{-it/\lambda} =$$

$$\operatorname{div}\Big[\Delta^{-1}\nabla\operatorname{div}\Big(n^\lambda(v - ie^{it/\lambda}E^+ - ie^{-it/\lambda}E^-)\otimes(v - ie^{it/\lambda}E^+ - ie^{-it/\lambda}E^-)$$

$$+(n^\lambda)^\gamma\mathbb{I} - \lambda^2 E^\lambda\otimes E^\lambda\Big) + \frac{\lambda^2}{2}|E^\lambda|^2\mathbb{I}\Big] - 2\nabla\operatorname{div}(v - ie^{it/\lambda}E^+ - ie^{-it/\lambda}E^-).$$

If we consider only the oscillatory part of the electric field we get

$$\partial_t E^\pm + \operatorname{div}(v\otimes E^\pm) - \nabla\operatorname{div}E^\pm = 0, \qquad PE^\pm = 0,$$

or equivalently

$$\partial_t E^\pm - \Delta E^\pm + Q\operatorname{div}(v\otimes E^\pm) = 0.$$

□

9 Proof of Theorem 1

The proof of Theorem 1 is obtained by combining the results of Sects. 5, 6 and Theorem 2.

References

1. R.A. Adams, *Sobolev Spaces* (Academic, New York, 1975)
2. A. Anile, S. Pennisi, Thermodynamic derivation of the hydrodynamical model for charge transport in semiconductors. Phys. Rev. B, **46**(20), 13186–13193 (1992)
3. S. Cordier, E. Grenier, Quasineutral limit of an Euler-Poisson system arising from plasma physics. Commun. Partial Differ. Equ. **25**(5–6), 1099–1113 (2000)
4. S. Cordier, P. Degond, P. Markowich, C. Schmeiser, Travelling wave analysis of an isothermal Euler-Poisson model. Ann. Fac. Sci. Toulouse Math. (6) **5**(4), 599–643 (1996)
5. D. Donatelli, Local and global existence for the coupled Navier-Stokes-Poisson problem. Q. Appl. Math. **61**(2), 345–361 (2003)
6. D. Donatelli, P. Marcati, A dispersive approach to the artificial compressibility approximations of the Navier Stokes equations in 3d. J. Hyperbolic Differ. Equ. **3**(3), 575–588 (2006)
7. D. Donatelli, P. Marcati, A quasineutral type limit for the Navier-Stokes-Poisson system with large data. Nonlinearity **21**(1), 135–148 (2008)
8. D. Donatelli, P. Marcati, Analysis of oscillations and defect measures for the quasineutral limit in plasma Physics. Arch. Ration. Mech. Anal. **206**(1), 159–188 (2012)
9. B. Ducomet, E. Feireisl, H. Petzeltová, I. Straškraba, Existence globale pour un fluide barotrope autogravitant. C. R. Acad. Sci. Paris Sér. I Math. **332**(7), 627–632 (2001)
10. B. Ducomet, E. Feireisl, H. Petzeltová, I. Straškraba, Global in time weak solutions for compressible barotropic self-gravitating fluids. Discret. Contin. Dyn. Syst. **11**(1), 113–130 (2004)

11. P. Gérard, Microlocal defect measures. Commun. Partial Differ. Equ. **16**(11), 1761–1794 (1991)
12. E. Grenier, Oscillations in quasineutral plasmas. Commun. Partial Differ. Equ. **21**(3–4), 363–394 (1996)
13. S. Jiang, S. Wang, The convergence of the Navier-Stokes-Poisson system to the incompressible Euler equations. Commun. Partial Differ. Equ. **31**(4–6), 571–591 (2006)
14. Q. Ju, F. Li, S. Wang, Convergence of the Navier-Stokes-Poisson system to the incompressible Navier-Stokes equations. J. Math. Phys. **49**(7), 073515, 8 (2008)
15. Q. Ju, F. Li, H. Li, The quasineutral limit of compressible Navier-Stokes-Poisson system with heat conductivity and general initial data. J. Differ. Equ. **247**(1), 203–224 (2009)
16. A. Jüngel, *Transport Equations for Semiconductors*. Lecture Notes in Physics, vol. 773 (Springer, Berlin, 2009)
17. P.-L. Lions, *Mathematical Topics in Fluid Dynamics, Incompressible Models* (Claredon/Oxford Science Publications, New York, 1996)
18. G. Loeper, Quasi-neutral limit of the Euler-Poisson and Euler-Monge-Ampère systems. Commun. Partial Differ. Equ. **30**(7–9), 1141–1167 (2005)
19. Y.-J. Peng, Y.-G. Wang, W.-A. Yong, Quasi-neutral limit of the non-isentropic Euler-Poisson system. Proc. R. Soc. Edinb. Sect. A **136**(5), 1013–1026 (2006)
20. J. Simon, Compact sets in the space $L^p(0,T;B)$. Ann. Math. Pura Appl. (4) **146**, 65–96 (1987)
21. R.S. Strichartz, Restrictions of Fourier transforms to quadratic surfaces and decay of solutions of wave equations. Duke Math. J. **44**(3), 705–714 (1977)
22. L. Tartar, H-measures, a new approach for studying homogenisation, oscillations and concentration effects in partial differential equations. Proc. R. Soc. Edinb. Sect. A **115**,(3–4), 193–230 (1990)
23. S. Wang, Quasineutral limit of Euler-Poisson system with and without viscosity. Commun. Partial Differ. Equ. **29**(3–4), 419–456 (2004)

Divergence-Measure Fields on Domains with Lipschitz Boundary

Hermano Frid

Abstract In this work we are particularly interested in analyzing some consequences of the additional assumption that the domain has a Lipschitz boundary, in the investigation of the properties of the divergence-measure fields, especially those which are vector-valued (Radon) measures whose divergence is a signed (Radon) measure.

Keywords Divergence-measure fields • Normal traces • Gauss-Green theorem • Product rule

1991 Mathematics Subject Classification Primary: 26B20,28C05, 35L65, 35B35; Secondary: 26B35, 26B12, 35L67

1 Introduction

The purpose of this paper is to establish further properties of the (extended) divergence-measure fields introduced by Chen and Frid [2–4], whose theory was further developed by Silhavý [10, 11], under the additional assumption that the underlying domain has a Lipschitz boundary. We begin by briefly reviewing the basic theory, and then we make the assumption that the domain possesses a Lipchitz deformable boundary, analyzing some consequences of this assumption. We refer to [9] for a more detailed review of the theory of the divergence-measure fields up to this date. We also refer to [6] and the papers already mentioned for a more

H. Frid (✉)
Instituto de Matemática Pura e Aplicada – IMPA, Estrada Dona Castorina, 110, Rio de Janeiro, RJ 22460-320, Brazil
e-mail: hermano@impa.br

G.-Q.G. Chen et al. (eds.), *Hyperbolic Conservation Laws and Related Analysis with Applications*, Springer Proceedings in Mathematics & Statistics 49,
DOI 10.1007/978-3-642-39007-4_10,

complete bibliography on the theory of divergence-measure fields, as well as many of its possible applications.

2 Divergence-Measure Fields

We begin by recalling the definition of the divergence-measure fields.

Definition 1. Let $U \subset \mathbb{R}^N$ be open. For $F \in L^p(U;\mathbb{R}^N)$, $1 \le p \le \infty$, or $F \in \mathcal{M}(U;\mathbb{R}^N)$, set

$$|\mathrm{div}\, F|(U) := \sup\{ \int_U \nabla\varphi \cdot F \ :\ \varphi \in C_0^1(U),\ |\varphi(x)| \le 1,\ x \in U \}. \tag{1}$$

For $1 \le p \le \infty$, we say that F is an L^p-divergence-measure field over U, i.e., $F \in \mathcal{DM}^p(U)$, if $F \in L^p(U;\mathbb{R}^N)$ and

$$\|F\|_{\mathcal{DM}^p(U)} := \|F\|_{L^p(U;\mathbb{R}^N)} + |\mathrm{div}\, F|(U) < \infty. \tag{2}$$

We say that F is an extended divergence-measure field over D, i.e., $F \in \mathcal{DM}^{ext}(U)$, if $F \in \mathcal{M}(U;\mathbb{R}^N)$ and

$$\|F\|_{\mathcal{DM}^{ext}(U)} := |F|(U) + |\mathrm{div}\, F|(U) < \infty. \tag{3}$$

If $F \in \mathcal{DM}^*(U)$ for any open set $U \Subset \mathbb{R}^N$, then we say $F \in \mathcal{DM}^*_{loc}(\mathbb{R}^N)$.

In order to introduce notation and go directly to the heart of the matter, we recall the following product rule proved in [2], whose proof is almost entirely transposed to prove the main product rule that we will state subsequently, which is the key to establishing the Gauss-Green formula (see Theorem 3 below).

Theorem 1 (Chen and Frid [2]). *Given $F \in \mathcal{DM}^\infty(U)$ and $g \in BV(U) \cap L^\infty(U)$, then $gF \in \mathcal{DM}^\infty(U)$ and*

$$\mathrm{div}\,(gF) = \bar{g}\mathrm{div}\, F + \overline{F \cdot \nabla g}, \tag{4}$$

in the sense of Radon measures in U, where $\bar{g}$ (equal to g a.e.) is the limit of a mollified sequence for g through a symmetric mollifier, and $\overline{F \cdot \nabla g}$ is a Radon measure absolutely continuous with respect to $|\nabla g|$, whose absolutely continuous part with respect to the Lebesgue measure in U satisfies

$$(\overline{F \cdot \nabla g})_{ac} = F \cdot (\nabla g)_{ac}, \qquad \text{a.e. in } U. \tag{5}$$

Moreover, $|\overline{F \cdot \nabla g}|(U) \le \|F\|_\infty |\nabla g|(U)$.

Proof. Let $g_\delta = \omega_\delta * g$, where $\omega_\delta(x) = \delta^{-N}\eta(\frac{x}{\delta})$ with a positive symmetric mollifier ω. One easily deduces that

$$\mathrm{div}\,(g_\delta F) = g_\delta \mathrm{div}\, F + F \cdot \nabla g_\delta. \tag{6}$$

Now, it is well known that g_δ converges to a Borel function $\bar{g}$, $\mathcal{H}^{N-1}$-a.e. in U (this function equals g a.e. in U).

We claim that, for a Borel set $A \subset U$, $\mathcal{H}^{N-1}(A) = 0$ implies $|\,\mathrm{div}\, F|(A) = 0$. Indeed, since $|\mathrm{div}\, F|$ is a Radon measure, we may assume that A is compact. Also, we may assume that $\mathrm{div}\, F(A) = |\mathrm{div}\, F|(A)$. Hence, given $\varepsilon > 0$, we may cover A with a finite number of balls $B_i = B(x_i; r_i)$, $i = 1, \ldots, J$,

$$A \subset A_\varepsilon := \cup_{i=1}^{J} B_i, \text{ such that } \sum_{i=1}^{J} r_i^{N-1} \leq \varepsilon. \tag{7}$$

We may also assume that $|\mathrm{div}\, F|(\partial B_i) = 0$, $i = 1, \ldots, J$, since otherwise we can modify r_i slightly to satisfy this property and (7). By using an approximation of the identity sequence, we obtain a sequence $F_\delta \in C^\infty(U; \mathbb{R}^N)$ such that $F_\delta \to F$ a.e. in U, and $|\mathrm{div}\, F_\delta| \rightharpoonup |\mathrm{div}\, F|$ in $\mathcal{M}(U)$. Again, we may assume that $F_\delta \to F$ a.e. in ∂B_i, $i = 1, \ldots, J$. Now, by the usual Gauss-Green formula for smooth vector fields and domains with Lipschitz boundaries, we have

$$\int_{A_\varepsilon} \mathrm{div}\, F_\delta \, dx = \int_{\partial A_\varepsilon} F_\delta \cdot \nu \, d\mathcal{H}^{N-1},$$

so that, passing to the limit when $\delta \to 0$, we obtain

$$\int_{A_\varepsilon} \mathrm{div}\, F = \int_{\partial A_\varepsilon} F \cdot \nu \, d\mathcal{H}^{N-1} \leq c\|F\|_\infty \sum_{i=1}^{J} r_i^{N-1} \leq c\|F\|_\infty \varepsilon.$$

Since A is compact, $\chi_{A_\varepsilon} \to \chi_A$ everywhere in U, and by dominated convergence applied to the measure $|\mathrm{div}\, F|$, we get $|\mathrm{div}\, F|(A) = \mathrm{div}\, F(A) = 0$, which proves the claim.

Then, using the claim we just proved, we get

$$g_\delta \mathrm{div}\, F \rightharpoonup \bar{g} \mathrm{div}\, F, \qquad \text{in } \mathcal{M}(U),$$

as a consequence of dominated convergence applied to the measure $\mathrm{div}\, F$.

On the other hand, we claim that $\{\mathrm{div}\,(g_\delta F)\}$ is uniformly bounded in $\mathcal{M}(U)$. Indeed, this follows from

$$\begin{aligned}\langle \mathrm{div}\,(g_\delta F), \phi \rangle &= -\int_U g_\delta F \cdot \nabla \phi \, dx = -\int_U F \cdot \nabla(g_\delta \phi) \, dx + \int_U \phi F \cdot \nabla g_\delta \, dx \\ &\leq \|g\|_\infty |\mathrm{div}\, F|(U) + \|F\|_\infty |\nabla g|(U),\end{aligned}$$

for all $\phi \in C_c^\infty(U)$, with $\|\phi\|_\infty = 1$.

Now, $\operatorname{div}(g_\delta F)$ converges to $\operatorname{div}(gF)$, in the sense of distributions over U. Then, $\operatorname{div}(g_\delta F) \rightharpoonup \operatorname{div}(gF)$ in $\mathcal{M}(U)$. Hence,

$$F \cdot \nabla g_\delta \rightharpoonup \overline{F \cdot \nabla g} := \operatorname{div}(gF) - \bar{g}\operatorname{div} F.$$

Now we prove that $\overline{F \cdot \nabla g}$ is absolutely continuous w.r.t. $|\nabla g|$. Let $A \subset D$ be such that $|\nabla g|(A) = 0$. We are going to prove that $|\overline{F \cdot \nabla g}|(A) = 0$. It suffices to consider any compact set A with $|\nabla g|(A) = 0$. Given $\varepsilon > 0$, we can cover A by a finite number, J, of balls so that

$$A \subset \cup_{i=1}^J B(x_i; r_i), \quad r_i < \varepsilon; \qquad |\nabla g|\big(\cup_{i=1}^J B(x_i; r_i)\big) < \varepsilon.$$

We may assume that $|\nabla g|(\partial B(x_i; r_i)) = 0$, $i = 1, \cdots, J$. Let $\phi \in C_0(\cup_{i=1}^J B(x_i; r_i))$. Thus

$$\begin{aligned}\langle \overline{F \cdot \nabla g}, \phi \rangle &= \lim_{\delta \to 0} \int \phi(x)\, F(x) \cdot \nabla g_\delta(x)\, dx \\ &= \|\phi\|_\infty \|F\|_\infty |\nabla g|\big(\cup_{i=1}^J B(x_i; r_i)\big) \le \varepsilon \|\phi\|_\infty \|F\|_\infty,\end{aligned}$$

from the fact that $|\nabla g_\delta|(B) \to |\nabla g|(B)$, for all open sets $B \subset D$ with $|\nabla g|(\partial B) = 0$. Hence, we obtain

$$|\overline{F \cdot \nabla g}|(A) \le |\overline{F \cdot \nabla g}|\left(\cup_{i=1}^J B(x_i; r_i)\right) \le \varepsilon \|F\|_\infty.$$

The proof of (5) is a little more technical and, for that, we simply refer to [2] since it escapes our purposes here. □

We now recall a result of Silhavý in [11] that is in some sense a dual formulation for the previous result, in the sense that it compensates a relaxation on the regularity of the vector field F, which now may be just a vector measure, by imposing more regularity on the function g, which now is assumed to be in $W^{1,\infty}(U)$. As we will see, its proof follows exactly the same lines as that of Theorem 1 just recalled.

Theorem 2 (Silhavý [11]). *Given $F \in \mathcal{DM}^{ext}(U)$ and $g \in W^{1,\infty}(U)$, then $gF \in \mathcal{DM}^{ext}(U)$ and*

$$\operatorname{div}(gF) = g\operatorname{div} F + \overline{\nabla g \cdot F}, \tag{8}$$

in the sense of Radon measures in U, where $\overline{\nabla g \cdot F}$ is a Radon measure absolutely continuous with respect to $|F|$. Moreover,

(i) $|\overline{\nabla g \cdot F}|(U) \le \|\nabla g\|_\infty |F|(U)$.
(ii) *If $h \in W^{1,\infty}(U)$, $\overline{\nabla(gh) \cdot F} = h\overline{\nabla g \cdot F} + g\overline{\nabla h \cdot F} = \overline{\nabla g \cdot hF} + \overline{\nabla h \cdot gF}$.*
(iii) *If $V \subset U$ is an open set, then $(\overline{\nabla g|V \cdot F\lfloor V})_V = \overline{\nabla g \cdot F}_U \lfloor V$.*
(iv) $(\overline{\nabla g \cdot F})_{ac} = \nabla g \cdot (F)_{ac}$.

Proof. We again define g_δ as above and obtain (6). We have that g_δ converges locally uniformly to g so that the first term on the right-hand side of (6) converges to $g\,\mathrm{div}\,F$, in the sense of Radon measures. It is also easy to see that $\nabla g_\delta \cdot F$ is uniformly bounded in $\mathcal{M}(U)$. Therefore, the left-hand side of (6) is also compact in $\mathcal{M}(U)$, in the weak star topology, and since it converges to $\mathrm{div}\,(gF)$ in the sense of distributions, it follows that $\mathrm{div}\,(gF)$ is indeed a Radon measure and the whole sequence $\mathrm{div}\,(g_\delta F)$ converges to $\mathrm{div}\,(gF)$. Hence, the whole sequence $\nabla g_\delta \cdot F$ converges to the Radon measure

$$\overline{\nabla g \cdot F} := \mathrm{div}\,(gF) - g\,\mathrm{div}\;F.$$

The assertions (i)–(ii) are proved in the standard way. Assertion (iii) is called the localization property in [11]; it follows trivially from the definitions. Finally, the proof of (iv) is entirely similar to that of the analogous assertion in Theorem 1. □

We recall the Gauss-Green formula for general divergence-measure fields, first proved in [3, 4] and extended by Silhavý in [11].

Theorem 3 (Chen and Frid [3, 4], Silhavý [11]). *If $F \in \mathcal{DM}^{ext}(U)$ then there exists a linear functional $F \cdot \nu : \mathrm{Lip}(\partial U) \to \mathbb{R}$ such that*

$$F \cdot \nu(g|\partial U) = \int_U \overline{\nabla g \cdot F} + \int_U g\,\mathrm{div}\,F, \tag{9}$$

for every $g \in \mathrm{Lip}(\mathbb{R}^N) \cap L^\infty(\mathbb{R}^N)$. Moreover,

$$|F \cdot \nu(h)| \le |F|_{\mathcal{DM}(U)} |h|_{Lip(\partial U)}, \tag{10}$$

for all $h \in Lip(\partial U)$, where we use the notation

$$|g|_{Lip(C)} := \sup_{x \in C} |g(x)| + Lip_C(g).$$

Proof. A major step in the proof of this result is to prove that the right-hand side of (9) depends only on the values of g restricted to ∂U, that is, that if $g \in \mathrm{Lip}(\mathbb{R}^N)$, with $g(x) = 0$, for $x \in \partial U$, then

$$\int_U \overline{\nabla g \cdot F} + \int_U g\,\mathrm{div}\,F = 0. \tag{11}$$

Clearly, we may as well assume $g(x) = 0$, for $x \in \mathbb{R}^N \setminus U$ (cf. Lemma 3.2 in [11]). We first prove (11) in the case where $\mathrm{supp}\,g$ is a compact subset of U. In this case, for $\delta > 0$ sufficiently small we have $\mathrm{supp}\,g_\delta \subset U$, where, as above, $g_\delta = g * \omega_\delta$. Then, by the definition of the divergence of the (vector-valued) distribution F, we have

$$\int_U \nabla g_\delta \cdot F + \int_U g_\delta \operatorname{div} F = 0. \tag{12}$$

Hence, taking the limit when $\delta \to 0$ in (12), using the definition of $\overline{\nabla g \cdot F}$, we obtain (11) in this case. We now consider the case where $g \in \mathrm{Lip}(\mathbb{R}^N)$ and $g(x) = 0$, for $x \in \mathbb{R}^N \setminus U$. Let $\zeta : \mathbb{R} \to \mathbb{R}$ be given by

$$\zeta(t) := \begin{cases} 0, & \text{if } t < 1/2, \\ 2(t - 1/2), & \text{if } 1/2 \le t \le 1, \\ 1, & \text{if } t > 1, \end{cases}$$

and for each $\varepsilon > 0$ let $h_\varepsilon : \mathbb{R}^N \to \mathbb{R}$ be defined by

$$h_\varepsilon(x) := \begin{cases} \zeta(\varepsilon^{-1} \operatorname{dist}(x, \partial U)), & x \in U, \\ 0, & x \in \mathbb{R}^N \setminus U. \end{cases}$$

Observe that h_ε is a Lipschitz function satisfying $h_\varepsilon(x) = 1$, if $x \in U_\varepsilon := \{x \in U : \operatorname{dist}(x, \partial U) \ge \varepsilon\}$. Then the function $h_\varepsilon g$ is a Lipschitz function which coincides with g on U_ε and

$$\overline{\nabla(h_\varepsilon g) \cdot F} = h_\varepsilon \overline{\nabla g \cdot F} + g \overline{\nabla h_\varepsilon \cdot F}.$$

By what we have already proved, we have

$$\int_U h_\varepsilon \overline{\nabla g \cdot F} + \int_U g \overline{\nabla h_\varepsilon \cdot F} + \int_U h_\varepsilon g \operatorname{div} F = 0. \tag{13}$$

Now, we have

$$\int_U g \overline{\nabla h_\varepsilon \cdot F} = \int_{U \setminus U_{2\varepsilon}} g \overline{\nabla h_\varepsilon \cdot F},$$

since $\nabla h_\varepsilon \equiv 0$ in U_ε. Also, $|\nabla h_\varepsilon| \le 2\varepsilon^{-1}$, and $|g(x)| \le 2\mathrm{Lip}(g)\varepsilon$, for $x \in U \setminus U_{2\varepsilon}$. Therefore,

$$\lim_{\varepsilon \to 0} \int_U g \overline{\nabla h_\varepsilon \cdot F} = \lim_{\varepsilon \to 0} \int_{U \setminus U_{2\varepsilon}} g \overline{\nabla h_\varepsilon \cdot F} = 0,$$

by dominated convergence. Hence, letting $\varepsilon \to 0$ in (13), since $h_\varepsilon \to 1$, as $\varepsilon \to 0$, everywhere in U, we finally get (11).

The assertion just proved shows that the right-hand side of (9) depends only on $g|\partial U$. Also, the inequality (10) is clear from (9), in the case where $h = H|\partial U$, where $H \in \mathrm{Lip}(\mathbb{R}^N)$, and

$$|H|_{\mathrm{Lip}(\mathbb{R}^N)} = |h|_{\mathrm{Lip}(\partial U)}.$$

Now, Kirszbraun's Theorem (see, e.g., [7, 8]) guarantees, for any $h \in \mathrm{Lip}(\partial U)$, the existence of $H \in \mathrm{Lip}(\mathbb{R}^N)$ such that $H|\partial U = h$ and $\mathrm{Lip}_{\mathbb{R}^N}(H) = \mathrm{Lip}_{\partial U}(h)$. Moreover, a trivial cut-off procedure ensures that $\|H\|_{L^\infty(\mathbb{R}^N)} = \|h\|_{L^\infty(\partial U)}$; this completes the proof. □

We now discuss a direct way of defining the normal trace functional $F \cdot \nu : \mathrm{Lip}(\partial U) \to \mathbb{R}$. The formula was first obtained in [3, 4], under regularity restrictions on the boundary, and in [11], for general boundaries. Before stating the corresponding result, we recall the following lemma, which is a slight modification of Lemma 3.3 of [11].

Lemma 1 (Silhavý [11]). *If $F \in \mathcal{DM}^{ext}(U)$, $m \in Lip(U)$, $t \in \mathbb{R}$ and if $T \subset m^{-1}(t)$ is a compact subset of U, then the restriction $\overline{\nabla m \cdot F}\lfloor T$ of $\overline{\nabla m \cdot F}$ to T satisfies*

$$\overline{\nabla m \cdot F}\lfloor T = 0. \tag{14}$$

Proof. Clearly, we can take $t = 0$. Also, multiplying m by a suitable function in $C_0^\infty(U)$, if necessary, we can assume that m has compact support in U. Therefore, we can assume that m is a Lipshitz function vanishing on $\mathbb{R}^N \setminus W$, with $W = U \setminus T$, and, in particular, also on $\mathbb{R}^N \setminus U$. Therefore, for any $\eta \in C_0^\infty(U)$, we have

$$\int_W \overline{\nabla(\eta m) \cdot F} + \int_W \eta m \,\mathrm{div}\, F = 0, \tag{15}$$

$$\int_U \overline{\nabla(\eta m) \cdot F} + \int_U \eta m \,\mathrm{div}\, F = 0. \tag{16}$$

Subtracting (15) from (16), since ηm vanishes on T, we get

$$0 = \int_T \overline{\nabla(\eta m) \cdot F} = \int_T \eta \overline{\nabla m \cdot F} + \int_T m \overline{\nabla \eta \cdot F} = \int_T \eta \overline{\nabla m \cdot F},$$

and so, since η is arbitrary, we arrive at (14). □

The following result gives a simple formula to compute the normal trace of $\mathcal{DM}$-fields. This formula, displayed in (i) of the statement below, was first obtained in [3,4] under some regularity restrictions on the boundary, and later was extended to general domains in [11]. Item (ii) gives a useful necessary condition for the normal trace to be a measure over ∂U established by Silhavý [11].

Theorem 4. *Let $F \in \mathcal{DM}^{ext}(U)$ and $m : \mathbb{R}^N \to \mathbb{R}$ be a nonnegative Lipschitz function with $\mathrm{supp}\, m \subset \bar{U}$ which is strictly positive on U, and for each $\varepsilon > 0$ let $L_\varepsilon = \{x \in U : 0 < m(x) < \varepsilon\}$. Then:*

(i) (*cf.* [3,4] and [11]) *If $g \in Lip(\mathbb{R}^N) \cap L^\infty(\mathbb{R}^N)$, we have*

$$F \cdot \nu(g|\partial U) = -\lim_{\varepsilon \to 0} \varepsilon^{-1} \int_{L_\varepsilon} g\, d\,(\overline{\nabla m \cdot F}). \tag{17}$$

(ii) (cf. [11]*) If*

$$\liminf_{\varepsilon\to 0} \varepsilon^{-1}|\overline{\nabla m\cdot F}|(L_\varepsilon)<\infty, \tag{18}$$

then $F\cdot\nu$ is a measure over ∂U.

Proof. We repeat the proof given in [11].

(i) For each $\varepsilon>0$ we define $m_\varepsilon(x)=\varepsilon^{-1}\min\{m(x),\varepsilon\}$. We see that m_ε is a Lipschitz function vanishing on ∂U. We have that $gm_\varepsilon\in \mathrm{Lip}(\mathbb{R}^N)$ and

$$\overline{\nabla(gm_\varepsilon)\cdot F}=m_\varepsilon\overline{\nabla g\cdot F}+g\overline{\nabla m_\varepsilon\cdot F},$$

by the properties of the pairing $\overline{\nabla g\cdot F}$. Since gm_ε vanishes on ∂U, we have

$$\int_U m_\varepsilon\, d(\overline{\nabla g\cdot F})+\int_U g\, d(\overline{\nabla m_\varepsilon\cdot F})+\int_U gm_\varepsilon \operatorname{div} F=0. \tag{19}$$

Now, $m_\varepsilon(x)\to 1$ everywhere in U, so that dominated convergence implies

$$\int_U m_\varepsilon\, d(\overline{\nabla g\cdot F})\to\int_U d(\overline{\nabla g\cdot F})$$

and $\int_U gm_\varepsilon \operatorname{div} F\to\int_U g\operatorname{div} F$. On the other hand, we have $m_\varepsilon=\varepsilon^{-1}m$ in L_ε, so $\nabla m_\varepsilon=\varepsilon^{-1}\nabla m$, a.e. in L_ε, which gives $\overline{\nabla m_\varepsilon\cdot F}=\varepsilon^{-1}\overline{\nabla m\cdot F}$, over L_ε. Moreover, since $U\setminus L_\varepsilon=m_\varepsilon^{-1}(1)$, by Lemma 1, we have $\overline{\nabla m_\varepsilon\cdot F}=0$ on $U\setminus L_\varepsilon$. Hence, we obtain (17) from (19) when $\varepsilon\to 0$, by the definition of $F\cdot\nu(g|\partial U)$ in (9).

(ii) By (18), we have $|\overline{\nabla m\cdot F}|(L_\varepsilon)\le C\varepsilon$, for some $C>0$ independent of ε, at least for a subsequence of $\varepsilon\to 0$, so that

$$\left|\varepsilon^{-1}\int_{L_\varepsilon} g\, d(\overline{\nabla m\cdot F})\right|\le C\|g\|_{L^\infty(\mathbb{R}^N)},$$

for each $g\in \mathrm{Lip}(\mathbb{R}^N)\cap L^\infty(\mathbb{R}^N)$. Therefore, given $g\in\mathrm{Lip}(\partial U)$, we may extend g to a Lipschitz function on $\mathbb{R}^N$ so that $\|g|\partial U\|_{L^\infty(\partial U)}=\|g\|_{L^\infty(\mathbb{R}^N)}$, and so, by (17), we deduce that $|F\cdot\nu(g)|\le C\|g\|_{L^\infty(\partial U)}$, which implies, by the Riesz representation theorem, that $F\cdot\nu$ is a measure on ∂U, as asserted. □

Remark 1. A typical example of m in the statement of Theorem 4 is provided by $m(x)=\operatorname{dist}(x,\partial U)$, for $x\in U$, and $m(x)=0$, for $x\in\mathbb{R}^N\setminus U$.

Remark 2. The following interesting example from [11] shows cases where $F\cdot\nu$ is a measure over ∂U and cases where $F\cdot\nu$ fails to be a measure. Namely, for $1\le\alpha<3$, let $F:\mathbb{R}^2\setminus\{0\}\to\mathbb{R}^2$ be defined by

$$F(x) = \frac{1}{|x|^\alpha}(x_2, -x_1),$$

and let $U = \{x = (x_1, x_2) \in \mathbb{R}^2 : |x| < 1,\ x_2 < 0\}$. Clearly, $\operatorname{div} F = 0$, in $\mathbb{R}^2 \setminus \{0\}$, and we easily verify that

$$F \in \mathcal{DM}^p(U; \mathbb{R}^2) \quad \text{with } 1 \le p < 2/(\alpha - 1), \text{ for } 1 < \alpha < 3, \text{ and } p = \infty, \text{ for } \alpha = 1.$$

Now, if $g \in \operatorname{Lip}(\partial U)$ and $\operatorname{supp} g \subset \{(x_1, x_2) \in \mathbb{R}^2 : x_2 = 0,\ |x_1| < 1\}$, we may use (18) with m satisfying $m(x) = -x_2$, for $-\varepsilon_0 < x_2 \le 0$, and $|x_1| < 1 - \varepsilon_0$, with $\varepsilon_0 > 0$ small enough so that $g(x) = 0$, if $|x_1| \ge 1 - \varepsilon_0$. Also, we may consider an extension of g to $\mathbb{R}^2$ such that $g(x_1, x_2) = g(x_1, 0)$, for $|x_2| < \varepsilon_0$. Applying (18) with m and the extension of g so defined, we get

$$F \cdot \nu(g) = \lim_{\varepsilon \to 0} \frac{1}{\varepsilon} \int_0^\varepsilon \int_{-1}^1 g(t, s) \frac{t}{(t^2 + s^2)^{\frac{\alpha}{2}}}\, dt\, ds,$$

which gives

$$F \cdot \nu(g) = \int_{-1}^1 g(t, 0) \operatorname{sgn}(t) |t|^{1-\alpha}\, dt, \qquad \text{for } 1 \le \alpha < 2,$$

and

$$F \cdot \nu(g) = \lim_{\varepsilon \to 0} \int_{|t| > \varepsilon} g(t, 0) \operatorname{sgn}(t) |t|^{1-\alpha}\, dt, \qquad \text{for } 2 \le \alpha < 3.$$

This shows that, for $1 \le \alpha < 2$, $F \cdot \nu$ is a measure, while, for $2 \le \alpha < 3$, $F \cdot \nu$ is not a measure on ∂U.

Remark 3. For $\varepsilon_0 > 0$ sufficiently small and $0 < s < \varepsilon_0$, we may consider the open set $U_s := \{x \in U : m(x) > s\}$, for m as in Theorem 4. By Theorem 4, for the normal trace $F \cdot \nu | \partial U_s$, we have the following formula similar to (17),

$$F \cdot \nu(g | \partial U_s) = -\lim_{\varepsilon \to 0} \varepsilon^{-1} \int_{\{s < m(x) < s + \varepsilon\}} g\, d\,(\overline{\nabla m \cdot F}), \tag{20}$$

and, again, we have that the condition

$$\liminf_{\varepsilon \to 0} \varepsilon^{-1} |\overline{\nabla m \cdot F}|(\{s < m(x) < s + \varepsilon\}) < \infty, \tag{21}$$

implies that $F \cdot \nu | \partial U_s$ is a measure on ∂U_s. If we consider the monotone function $V(s) = |\overline{\nabla m \cdot F}|(\{0 < m(x) < s\})$, for $s \in (0, \varepsilon_0)$, we see that the left-hand side of (21) is the right-derivative of V at s, except possibly for a countable subset of $(0, \varepsilon_0)$, and we know that it exists for a.e. $s \in (0, \varepsilon_0)$. Therefore, $F \cdot \nu | \partial U_s$ is a measure for a.e. $s \in (0, \varepsilon_0)$, and in this sense we may assert that "for almost all boundaries" ∂U the normal trace $F \cdot \nu|\partial U$ is a measure.

3 Domains with a Lipschitz Deformable Boundary

We now enter into the main subject of the present paper, beginning with the definition of a deformable Lipschitz boundary.

Definition 2. Let $\Omega \subset \mathbb{R}^N$ be an open set. We say that $\partial\Omega$ is a *deformable Lipschitz boundary* if the following hold:

(i) For each $x \in \partial\Omega$, there exist an $r > 0$ and a Lipschitz mapping $\gamma : \mathbb{R}^{N-1} \to \mathbb{R}$ such that, upon rotating and relabeling the coordinate axis if necessary,

$$\Omega \cap Q(x,r) = \{\, y \in \mathbb{R}^N \,:\, \gamma(y_1, \cdots, y_{N-1}) < y_N \,\} \cap Q(x,r),$$

where $Q(x,r) = \{\, y \in \mathbb{R}^N \,:\, |y_i - x_i| \le r,\ i = 1, \cdots, N \,\}$. We denote by $\tilde{\gamma}$ the map $\tilde{y} \mapsto (\tilde{y}, \gamma(\tilde{y}))$, $\tilde{y} = (y_1, \cdots, y_{N-1})$.

(ii) There exists a map $\Psi : \partial\Omega \times [0,1] \to \bar{\Omega}$ such that Ψ is a bi-Lipschitz homeomorphism over its image and $\Psi(x,0) = x$, for all $x \in \partial\Omega$. For $s \in [0,1]$, we denote by Ψ_s the mapping from $\partial\Omega$ to $\bar{\Omega}$ given by $\Psi_s(x) = \Psi(x,s)$, and set $\partial\Omega_s := \Psi_s(\partial\Omega)$.

We say that the Lipschitz deformation $\Psi : \partial\Omega \times [0,1] \to \bar{\Omega}$ is *regular*, and that Ω has a *regular Lipschitz deformable boundary*, if, besides (i) and (ii), we have

(iii) $J[\nabla\Psi_s \circ \tilde{\gamma}] \rightharpoonup J[\nabla\tilde{\gamma}]$, as $s \to 0$, in the weak star topology of $L^\infty(B)$ for any bounded open set $B \subset \mathbb{R}^{N-1}$ such that $\tilde{\gamma}(B) \subset \partial\Omega$, with $\tilde{\gamma}$ as in (i), where $J[\nabla g]$ denotes the Jacobian of ∇g (see, e.g., [7]).

Remark 4. In [2] the additional condition (iii) for defining a regular Lipschitz deformation was stated in a slightly stronger way, asking that $\nabla\Psi_s \circ \tilde{\gamma} \to \nabla\tilde{\gamma}$, as $s \to 0$, in $L^1(B)$. Nevertheless, the weak convergence of the Jacobian is already enough to guarantee the validity of the formula

$$F \cdot \nu\big|_{\partial\Omega} = \text{ess.}\lim (F \cdot \nu_s) \circ \Psi_s, \quad \text{in the weak star topology of } L^\infty(\partial\Omega, \mathcal{H}^{N-1}), \tag{22}$$

which holds for $\mathcal{DM}^\infty$-fields, as established in [2].

Actually, condition (iii) is equivalent to $J[d\Psi_s] := \det(d\Psi_s^*\, d\Psi_s)^{1/2} \rightharpoonup 1$ in the weak star topology of $L^\infty(\partial\Omega)$, where, for each $\omega \in \partial\Omega$, $d\Psi_s(\omega) : T_\omega(\partial\Omega) \to \mathbb{R}^N$ is the differential mapping of Ψ_s at $\omega \in \partial\Omega$ and $d\Psi_s^*(\omega) : \mathbb{R}^N \to T_\omega(\partial\Omega)$ denotes the adjoint mapping. This follows from the Cauchy-Binet formula for the Jacobian (see, e.g., [7]).

We start our discussion by introducing the *level set function* $h : \mathbb{R}^N \to \mathbb{R}$, defined by

$$h(x) = \begin{cases} 0, & \text{for } x \in \mathbb{R}^N \setminus \bar{\Omega}, \\ s, & \text{for } x \in \partial\Omega_s, \\ 1, & \text{for } x \in \Omega \setminus \Psi(\partial\Omega \times [0,1]). \end{cases}$$

By formula (17) we have

$$F \cdot \nu(g|\partial U) = -\lim_{\varepsilon \to 0} \varepsilon^{-1} \int_{L_\varepsilon} g \, d\,(\overline{\nabla h \cdot F}), \tag{23}$$

for any $F \in \mathcal{DM}^{ext}(\Omega)$, and any $g \in \mathrm{Lip}(\mathbb{R}^N) \cap L^\infty(\mathbb{R}^N)$, with $L_\varepsilon = \{x \in \Omega : 0 < h(x) < \varepsilon\}$.

Remark 5. The following standard example of a domain with a regular deformable Lipschitz boundary shows that, for the sake of studying local properties of the normal trace operator, any domain with a Lipschitz boundary may be viewed as a domain with a regular deformable Lipschitz boundary. So, let

$$U := \{ x \in \mathbb{R}^N \,:\, \gamma(x_1, \cdots, x_{N-1}) < x_N \}, \tag{24}$$

where $\gamma : \mathbb{R}^{N-1} \to \mathbb{R}$ is a Lipschitz function. U is then an unbounded open set, ∂U is the graph of γ, $\partial U = \{(\tilde{x}, x_N) \in \mathbb{R}^N : \tilde{x} \in \mathbb{R}^{N-1},\ x_N = \gamma(\tilde{x})\}$, and it is very easy to define a regular Lipschitz deformation for ∂U by simply setting $\Psi((\tilde{x}, \gamma(\tilde{x})), s) = (\tilde{x}, \gamma(\tilde{x}) + s\delta)$, $\tilde{x} \in \mathbb{R}^{N-1}$, $s \in [0,1]$, where $\delta > 0$ is arbitrary. It turns out that, by property (i) in Definition 2, for test functions g, as in (9), with support contained in a sufficiently small neighborhood, say, a neighborhood like those appearing in Definition 2(i), the normal trace operator given by (9) may be defined using (23) where h is the level set function associated to this trivial standard deformation. More specifically, in this case, the level set function is simply defined by

$$h(x) = \begin{cases} 0, & \text{if } x_N < \gamma(\tilde{x}), \\ s, & \text{if } x_N = \gamma(\tilde{x}) + s\delta, \text{ for } s \in [0,1], \\ 1, & \text{if } x_N \geq \gamma(\tilde{x}) + \delta. \end{cases}$$

Therefore, considered as distributions in $\mathbb{R}^N$ with support contained in $\partial\Omega$, the normal trace operators associated to $\mathcal{DM}^{ext}$-fields can always be split in a countable sum of distributions, whose supports possess the finite intersection property, each of which may be defined like the normal trace operator for a standard domain as just described. Indeed, it suffices to employ a partition of unity subordinate to a suitable covering of $\partial\Omega$.

The above remark is important in connection, for instance, with the theory of hyperbolic systems of conservation laws (see, e.g., [6]). Namely, if $\mathbb{R}^N$ is the space-time space $\mathbb{R}^{n+1}$, so $N = n+1$, with points denoted (x,t), suppose $F(x,t) = (\eta(u(x,t)), q(u(x,t)))$ where $\eta : \mathbb{R}^m \to \mathbb{R}$ and $q : \mathbb{R}^m \to \mathbb{R}^n$ form an entropy-entropy flux pair for a hyperbolic system of conservation laws, and $u : \mathbb{R}^n \times (0,\infty) \to \mathbb{R}^m$ is a weak entropy solution for this system, and let $\Omega \subset \mathbb{R}^{n+1}$ be an open set. It is an important question to determine whether there is a measurable function $u_\tau : \partial\Omega \to \mathbb{R}^m$ such that the normal trace operator may be represented by $(\eta(u_\tau(\omega)), q(u_\tau(\omega))) \cdot \nu(\omega)$, where $\nu(\omega)$ is the outer unit normal vector at $\omega \in \partial\Omega$. Through the splitting of the normal trace operator mentioned in the above remark, this question, for a general domain with Lipschitz boundary, may be reduced to the corresponding one for a hyper-graph domain as U in (24).

For simplicity, in what follows, we will always assume that Ω is a bounded open set with a regular deformable Lipschitz boundary. We emphasize that, for the purpose of getting local information about the normal trace operator, as has been already mentioned, this assumption does not represent any additional restriction beyond that of possessing a Lipschitz boundary. The fact that Ω is bounded allows us to restrict our discussion to just two cases, namely, that for fields in $\mathcal{DM}^1(\Omega)$ and that for fields in $\mathcal{DM}^{ext}(\Omega)$, since the boundedess of Ω implies $\mathcal{DM}^p(\Omega) \subset \mathcal{DM}^1(\Omega)$, for all $1 < p \le \infty$. Let us then focus our attention in these two cases.

Theorem 5. *Let Ω be a bounded open set with a deformable Lipschitz boundary and $F \in \mathcal{DM}^1(\Omega)$. Let $\Psi : \partial\Omega \times [0,1] \to \bar{\Omega}$ be a Lipschitz deformation of $\partial\Omega$. Then, for almost all $s \in [0,1]$, and all $\phi \in C_0^\infty(\mathbb{R}^N)$,*

$$\int_{\Omega_s} \phi \,\mathrm{div}\, F = \int_{\partial\Omega_s} \phi(\omega) F(\omega) \cdot \nu_s(\omega)\, d\mathcal{H}^{N-1}(\omega) - \int_{\Omega_s} F(x) \cdot \nabla\phi(x)\, dx, \tag{25}$$

where ν_s is the unit outward normal field defined $\mathcal{H}^{N-1}$-almost everywhere in $\partial\Omega_s$, and Ω_s is the open subset of Ω bounded by $\partial\Omega_s$.

Proof. For $\phi \in C_0^\infty(\mathbb{R}^N)$, let

$$\zeta_\phi(s) = \int_{\partial\Omega_s} \phi(\omega) F(\omega) \cdot \nu_s(\omega)\, d\mathcal{H}^{N-1}(\omega), \quad s \in [0,1],$$

where ν_s is as in the statement. Let $s_0 \in [0,1]$ be a Lebesgue point for ζ_ϕ, for ϕ in a countable dense set in $C_0^\infty(\mathbb{R}^N)$. For $\delta > 0$ sufficiently small, let $g_\delta : \mathbb{R} \to \mathbb{R}$ be defined as

$$g_\delta(s) = \begin{cases} 0, & s < s_0 - \delta, \\ \frac{s - s_0 + \delta}{2\delta}, & s_0 - \delta \le s \le s_0 + \delta, \\ 1, & s > s_0 + \delta. \end{cases}$$

Set $\psi_\delta = g_\delta \circ h\,\phi$, where h is the level set function associated to the Lipschitz deformation Ψ. By the Gauss-Green formula, we have

$$
\begin{aligned}
0 &= \int_\Omega F \cdot \nabla \psi_\delta \, dx + \int_\Omega \psi_\delta \operatorname{div} F \\
&= \int_\Omega \phi g_\delta'(h(x)) F \cdot \nabla h \, dx + \int_\Omega g_\delta(h(x)) F \cdot \nabla \phi \, dx + \int_\Omega \psi_\delta \operatorname{div} F,
\end{aligned}
$$

which gives, by the coarea formula,

$$
0 = -\frac{1}{2\delta} \int_{s_0-\delta}^{s_0+\delta} \int_{\partial\Omega_s} \phi F \cdot \nu_s \, d\mathcal{H}^{N-1}(\omega) \, ds + \int_\Omega g_\delta(h(x)) F \cdot \nabla \phi \, dx + \int_\Omega \psi_\delta \operatorname{div} F.
$$

Letting $\delta \to 0$, we obtain (25) for $s = s_0$, where s_0 is an arbitrary Lebesgue point of ζ_ϕ, for ϕ in a countable dense subset of $C_0^\infty(\mathbb{R}^N)$, and, so, (25) holds for almost all $s \in [0, 1]$ as was to be proved. □

From Theorem 5 and the Gauss-Green formula (9), when $F \in \mathcal{DM}^1(\Omega)$, it follows that, for any $g \in \operatorname{Lip}(\mathbb{R}^N) \cap L^\infty(\mathbb{R}^N)$, we have the following formula for the normal trace functional $F \cdot \nu : \operatorname{Lip}(\partial\Omega) \to \mathbb{R}$,

$$
\langle F \cdot \nu, g | \partial\Omega \rangle = \text{ess.} \lim_{s \to 0} \int_{\partial\Omega_s} g F(\omega) \cdot \nu(\omega) \, d\mathcal{H}^{N-1}(\omega), \tag{26}
$$

where the limit on the right-hand side exists by applying dominated convergence to the other two terms in (25). Therefore, for any $\phi \in \operatorname{Lip}(\partial\Omega)$, we have

$$
\langle F \cdot \nu, \phi \rangle = \text{ess.} \lim_{s \to 0} \int_{\partial\Omega_s} \phi \circ \Psi_s^{-1}(\omega) F(\omega) \cdot \nu_s(\omega) \, d\mathcal{H}^{N-1}(\omega),
$$

or, by using the area formula,

$$
\begin{aligned}
\langle F \cdot \nu, \phi \rangle &= \text{ess.} \lim_{s \to 0} \int_{\partial\Omega} \phi(\omega) F \circ \Psi_s(\omega) \cdot \nu_s(\Psi_s(\omega)) J[\Psi_s] \, d\mathcal{H}^{N-1}(\omega) \\
&= \text{ess.} \lim_{s \to 0} \int_{\partial\Omega} \phi(\omega) F \circ \Psi_s(\omega) \cdot \nu(\Psi_s(\omega)) \, d\mathcal{H}^{N-1}(\omega),
\end{aligned}
$$

where we have used the fact that Ψ is a regular Lipschitz deformation. Therefore we have proved the following formula for the normal trace for a $\mathcal{DM}^1$-field.

Theorem 6. *Let $F \in \mathcal{DM}^1(\Omega)$, where Ω is a bounded open set with a Lipschitz boundary admitting a regular deformation $\Psi : \partial\Omega \times [0, 1] \to \bar{\Omega}$. Denoting by $F \cdot \nu|_{\partial\Omega}$ the continuous linear functional Lip$(\partial\Omega) \to \mathbb{R}$ given by the normal trace of F at $\partial\Omega$, we have the formula*

$$
F \cdot \nu|_{\partial\Omega} = \text{ess.} \lim_{s \to 0} F \circ \Psi_s(\cdot) \cdot \nu_s(\Psi_s(\cdot)), \tag{27}
$$

with equality in the sense of $(Lip(\partial\Omega))^*$, *where on the right-hand side the functionals are given by ordinary functions in* $L^1(\partial\Omega)$.

We now turn to the case where $F \in \mathcal{DM}^{ext}(\Omega)$. Let us again consider the level set function h associated to the regular Lipschitz deformation $\Psi : \partial\Omega \times [0,1] \to \bar{\Omega}$. Let us consider the measure μ over $\Psi(\partial\Omega \times [0,1])$ given by

$$\mu := |\overline{\nabla h \cdot F}| \lfloor \Psi(\partial\Omega \times [0,1]).$$

We consider the pull back of μ by Ψ, $\Psi^\sharp\mu$, which is the measure on $\partial\Omega \times [0,1]$ defined by

$$\langle \Psi^\sharp\mu, \varphi\rangle = \langle \mu, \varphi \circ \Psi^{-1}\rangle, \qquad \forall \varphi \in C(\partial\Omega \times [0,1]).$$

We may apply the disintegration process to $\Psi^\sharp\mu$ (see, e.g., Theorem 2.28, p. 57 in [1]) to write $\Psi^\sharp\mu = \sigma \otimes \tilde{\mu}_s$, for the Radon measure σ on $[0,1]$ given by the projection of $\Psi^\sharp\mu$ onto $[0,1]$, and so $\sigma(E) = \Psi^\sharp\mu(\partial\Omega \times E)$ for any Borel set $E \subset [0,1]$, and Radon measures $\tilde{\mu}_s$ such that $s \mapsto \tilde{\mu}_s$ is σ-measurable, $\tilde{\mu}_s(\partial\Omega) = 1$, σ-a.e. in $[0,1]$, so that we have

$$\int_{\partial\Omega\times[0,1]} \varphi(\omega,s)\, d\Psi^\sharp\mu = \int_{[0,1]} \left(\int_{\partial\Omega} \varphi(\omega,s)\, d\tilde{\mu}_s(\omega)\right) d\sigma(s),\ \forall\varphi \in C(\partial\Omega\times[0,1]). \tag{28}$$

Therefore, by pushing forward the equation $\Psi^\sharp\mu = \sigma \otimes \tilde{\mu}_s$ by Ψ, we obtain

$$\mu = \sigma \otimes \mu_s, \quad \mu_s := (\Psi_s)_\sharp\tilde{\mu}_s,$$

where, for any $\zeta \in C(\partial\Omega_s)$,

$$\langle (\Psi_s)_\sharp\tilde{\mu}_s, \zeta\rangle = \langle \tilde{\mu}_s, \zeta \circ \Psi_s\rangle.$$

In particular, for any $\phi \in C_0^\infty(\mathbb{R}^N)$, with $\operatorname{supp}\phi \cap \Omega \subset \Psi(\partial\Omega \times [0,1])$, we have

$$\int_\Omega \phi(x)\, d(\overline{\nabla h \cdot F}) = \int_{[0,1]} \left(\int_{\partial\Omega_s} \phi(x)\theta(x)\, d\mu_s\right) d\sigma(s), \tag{29}$$

where θ is the μ-measurable function, with $|\theta| = 1$, μ-a.e., such that

$$\theta\mu = \overline{\nabla h \cdot F} \lfloor \Psi(\partial\Omega \times [0,1]).$$

Now, we have the decomposition $\sigma = H(s)\, ds + \sigma_{\text{sing}}$, for some non-negative $H \in L^1([0,1])$, and $\sigma_{\text{sing}} = \sigma\lfloor\mathcal{N}$, for some Borel set $\mathcal{N} \subset [0,1]$ of one-dimensional Lebesgue measure zero, by the Lebesgue decomposition theorem (see, e.g., [7], p. 42). We then define

$$(\overline{\nabla h \cdot F})_s := \theta\, H(s)\, \mu_s. \tag{30}$$

We have the following analogue of Theorem 5 when $F \in \mathcal{DM}^{ext}(\Omega)$.

Theorem 7. *Let Ω be a bounded open set with a deformable Lipschitz boundary and $F \in \mathcal{DM}^{ext}(\Omega)$. Let $\Psi : \partial\Omega \times [0,1] \to \bar{\Omega}$ be a Lipschitz deformation of $\partial\Omega$. Then, for almost all $s \in [0,1]$, and all $\phi \in C_0^\infty(\mathbb{R}^N)$,*

$$\int_{\Omega_s} \phi \,\mathrm{div}\, F = \int_{\partial\Omega_s} \phi(\omega)\, d(\overline{\nabla h \cdot F})_s - \int_{\Omega_s} \nabla\phi(x) \cdot F. \tag{31}$$

Proof. The proof is nearly identical to that of Theorem 5, the only difference being that now we must choose $s_0 \in [0,1]\backslash\mathcal{N}$, with $\mathcal{N}$ as above, such that s_0 is a Lebesgue point of

$$\zeta_\phi(s) := \int_{\partial\Omega_s} \phi\, d(\overline{\nabla h \cdot F})_s$$

for ϕ in a countable dense set in $C_0^\infty(\mathbb{R}^N)$, and we take g_δ only for small $\delta > 0$ such that $|(\overline{\nabla h \cdot F})|(\partial\Omega_{s_0 \pm \delta}) = 0$. □

Similarly to what was done for $\mathcal{DM}^1$-fields, from Theorem 7 we get the following result.

Theorem 8. *Let $F \in \mathcal{DM}^{ext}(\Omega)$, where Ω is a bounded open set with a Lipschitz boundary admitting a regular deformation $\Psi : \partial\Omega \times [0,1] \to \bar{\Omega}$. Denoting by $F \cdot \nu|_{\partial\Omega}$ the continuous linear functional $Lip(\partial\Omega) \to \mathbb{R}$ given by the normal trace of F at $\partial\Omega$, we have the formula*

$$F \cdot \nu|_{\partial\Omega} = \text{ess.}\lim_{s\to 0} \Psi_s^\sharp\, d(\overline{\nabla h \cdot F})_s, \tag{32}$$

with equality in the sense of $(Lip(\partial\Omega))^*$, where on the right-hand side the functionals are given by the pull back by Ψ_s of the measures $d(\overline{\nabla h \cdot F})_s$, resulting from the disintegration of $d(\overline{\nabla h \cdot F})$.

4 Application to Time-Regularity of Entropy Solutions to Hyperbolic Conservation Laws

Let $n, d \in \mathbb{N}$, $\mathbb{R}_+^{d+1} = \mathbb{R}^d \times (0,\infty)$, and $\mathcal{U} \subset \mathbb{R}^n$ be an open and convex set. We consider the N-dimensional system of n conservation laws

$$\partial_t U + \sum_{\alpha=1}^{d} \partial_\alpha F^\alpha(U) = 0, \qquad \text{in } \mathbb{R}_+^{N+1}, \tag{33}$$

with $U(x,t) \in \mathcal{U}$ and $F^\alpha : \mathcal{U} \to \mathbb{R}^n$, where ∂_α denotes the partial derivative with respect to x_α.

Together with (33), we consider the initial data

$$U(x,0) = U_0(x). \tag{34}$$

The following result provides time-regularity information about entropy solutions of the problem (33) and (34). It extends a result established in [6] (Theorem 4.5.1), which follows from the theory for L^∞ divergence-measure fields.

Theorem 9. *Let $U_0 \in L^1_{loc}(\mathbb{R}^d)$, and let $U \in L^1_{loc}(\mathbb{R}^d \times [0,\infty))$ be a weak solution of (33) and (34), in the sense that, for any $\phi \in C_c^\infty(\mathbb{R}^{d+1})$, we have*

$$\int_{\mathbb{R}^{d+1}_+} U(x,t)\partial_t\phi + \sum_{\alpha=1}^{d} F^\alpha(U)\partial_\alpha\phi \, dx\,dt + \int_{\mathbb{R}^d} U_0(x)\phi(x,0)\,dx = 0. \tag{35}$$

Let $\eta : \mathcal{U} \to [0,\infty)$ be a strictly convex function, with $\eta(U) \ge c_1|U| + c_2$, for some $c_1 > 0$, $c_2 \in \mathbb{R}$, such that $\eta(U(x,t)) \in L^p(K \cap \mathbb{R}^{d+1}_+)$, for any compact set $K \subset \mathbb{R}^{d+1}$, for some $p > 1$. Suppose that there exists a vector measure $Q \in \mathcal{M}(K \cap \mathbb{R}^{d+1}_+; \mathbb{R}^d)$, for any compact set $K \subset \mathbb{R}^{d+1}$, such that $\eta(U(x,t))$ satisfies

$$\partial_t\eta(U) + div_x Q \le 0, \qquad in\ \mathbb{R}^{d+1}_+, \tag{36}$$

in the sense of distributions, where $\mathcal{M}(\Omega;\mathbb{R}^d)$ denotes the $\mathbb{R}^d$-valued Radon measures with finite total variation on Ω. Then,

$$U \in C((0,\infty) \setminus S; L^1_{loc}(\mathbb{R}^d)), \tag{37}$$

for some at most countable set $S \subset (0,\infty)$. Moreover, if we have, for all nonnegative $\psi \in C_c^\infty(\mathbb{R}^{d+1})$,

$$\int_{\mathbb{R}^{d+1}_+} \{\eta(U(x,t))\,\partial_t\psi\,dx\,dt + \nabla_x\psi \cdot dQ\} + \int_{\mathbb{R}^d} \eta(U_0(x))\psi(x,0)\,dx \ge 0, \tag{38}$$

then the above strong continuity holds on the right for $t = 0$.

Proof. The result follows by applying Theorem 8 to the domains $\Omega[t_0+] := \{(x,t) : t > t_0\}$, $t_0 > 0$, with regular Lipschitz deformation $\Omega[t_0+]_s = \Omega[(t_0+s)+]$, $s \in [0,1]$, and $\Omega[t_0-] := \{(x,t) : -\infty < t < t_0\}$, $t_0 > 0$, with regular Lipschitz deformation $\Omega[t_0-]_s = \Omega[(t_0-s)-]$, $s \in [0,1]$. Here, for simplicity, we may view $U(x,t)$ as extended to $t < 0$ as 0, as well as $\eta(U(x,t))$, as $\eta(0)$, and Q as the null measure, for $t < 0$. We then obtain that, for a.e. $t_0 > 0$,

$$\begin{cases} \int_{\mathbb{R}^d} U(x,t_0)\,\phi(x)\,dx = \text{ess.}\lim_{t\to t_0} \int_{\mathbb{R}^d} U(x,t)\phi(x)\,dx, \\ \int_{\mathbb{R}^d} \eta(U(x,t_0))\phi(x)\,dx = \text{ess.}\lim_{t\to t_0} \int_{\mathbb{R}^d} \eta(U(x,t))\phi(x)\,dx, \end{cases} \tag{39}$$

for all $\phi \in C_0^\infty(\mathbb{R}^d)$. Now, using (37), for almost all $0 < \delta < s < t < T$, and $R > 0$, there exists an $A(R, \delta, T) > 0$ such that

$$\int_{|x|<R} \eta(U(x,t))\,dx \le \int_{|x|<R} \eta(U(x,s))\,dx + A(R,\delta,T). \tag{40}$$

This gives that, for any $\delta > 0$, $\int_{|x|<R} \eta(U(x,t))\,dx$ is uniformly bounded for $\delta < t < T$, for almost all $R > 0$. Using the assumptions on η, we conclude that $\int_{|x|<R} |U(x,t)|\,dx$ is also uniformly bounded for $t > \delta$, for almost all $R > 0$. Hence, we may take $\phi \in L^{p'}(\mathbb{R}^d)$, with compact support, in (41), with $p' = p/(p-1)$. Now, since η is strictly convex, we conclude the proof in a standard way. □

As an example, in [5], Chen and Perepelitsa prove the convergence of the solutions $(\rho^\varepsilon, \rho^\varepsilon u^\varepsilon)$ to the Cauchy problem for the Navier-Stokes equations

$$\begin{cases} \rho_t + (\rho u)_x = 0, \\ (\rho u)_t + (\rho u^2 + \kappa \rho^\gamma)_x = \varepsilon u_{xx}, \end{cases} \tag{41}$$

with initial data

$$\rho(x,0) = \rho_0(x), \qquad u(x,0) = u_0(x), \tag{42}$$

where $\gamma > 1$ and $\kappa > 0$, by a scaling defined by $\kappa = (\gamma-1)^2/4\gamma$. Using energy estimates and compensated compactness with Young measures with unbounded support, they prove the convergence in $L^1(K \cap \mathbb{R}^2_+)$ of $(\rho^\varepsilon, m^\varepsilon)$, with $m^\varepsilon = \rho^\varepsilon u^\varepsilon$, to some $(\rho(x,t), m(x,t)) \in L^1_{\mathrm{loc}}(\mathbb{R}^2_+)$, and also the convergence in $L^1(K \cap \mathbb{R}^2_+)$ of $\eta^*(\rho^\varepsilon(x,t), m^\varepsilon(x,t))$ to $\eta^*(\rho(x,t), m(x,t))$, where

$$\eta^*(\rho, m) = \frac{1}{2}\frac{m^2}{\rho} + \rho e(\rho), \qquad e(\rho) = \frac{\kappa}{\gamma - 1}\rho^{\gamma-1},$$

for any compact K. Nevertheless, passing to the limit in the inequality

$$\partial_t \eta^*(\rho^\varepsilon, m^\varepsilon) + \partial_x q^*(\rho^\varepsilon, m^\varepsilon) \le \frac{\varepsilon}{2}\partial_x^2 (u^\varepsilon)^2, \tag{43}$$

which holds for each $\varepsilon > 0$, with

$$q^*(\rho, m) = \frac{1}{2}\frac{m^3}{\rho^2} + m e(\rho) + \rho m e'(\rho),$$

we obtain an inequality of the form

$$\partial_t \eta^*(\rho(x,t), m(x,t)) + \partial_x Q(x,t) \le 0, \qquad \text{in } \mathbb{R}^2_+, \tag{44}$$

in the sense of distributions, for some $Q \in \mathcal{M}(K \cap \mathbb{R}^2_+)$, for any compact $K \subset \mathbb{R}^2$. Because of the presence of the term $\rho^\varepsilon (u^\varepsilon)^3$ in $q^*(\rho^\varepsilon, m^\varepsilon)$, whose estimates obtained in [5] only guarantee the uniform boundedness in $L^1(K \cap \mathbb{R}^2_+)$, for any compact $K \subset \mathbb{R}^2$, we can only deduce that

$$\langle q^*(\rho^\varepsilon, m^\varepsilon), \partial_x \psi \rangle \to \langle Q, \partial_x \psi \rangle,$$

for any $\psi \in C_c^\infty(\mathbb{R}^2_+)$, for some (signed) Radon measure Q in $\mathbb{R}^2_+$, with finite total variation in $K \cap \mathbb{R}^2_+$, for any compact $K \subset \mathbb{R}^2$.

Actually, from the results in [5] we obtain

$$\int_{\mathbb{R}^2_+} \{\psi_t \eta^*(\rho, m)\, dx\, dt + \psi_x\, dQ\} + \int_{\mathbb{R}} \psi(x,0)\eta^*(\rho_0(x), m_0(x))\, dx \geq 0, \quad (45)$$

for all nonnegative $\psi \in C_c^\infty(\mathbb{R}^2)$, under suitable conditions on the initial data.

We can then apply Theorem 9 to conclude that the weak solution of the compressible isentropic Euler equations, $(\rho(x,t), m(x,t))$, obtained in [5] as the limit of the vanishing viscosity solutions of the corresponding Navier-Stokes equations, satisfies

$$(\rho, m) \in C((0,\infty) \setminus S; L^1_{\text{loc}}(\mathbb{R})),$$

for some at most countable subset $S \subset (0,\infty)$, and, moreover, $(\rho(\cdot,t), m(\cdot,t)) \to (\rho_0(\cdot), m(\cdot))$, as $t \downarrow 0$, in $L^1_{\text{loc}}(\mathbb{R})$.

Acknowledgements The author gratefully acknowledges the support from CNPq, through grant proc. 303950/2009-9, and FAPERJ, through grant E-26/103.019/2011.

References

1. L. Ambrosio, N. Fusco, D. Pallara, *Functions of Bounded Variation and Free Discontinuity Problems*. Oxford Mathematical Monographs (Oxford University Press, Oxford, 2000)
2. G.-Q. Chen, H. Frid, Divergence-measure fields and hyperbolic conservation laws. Arch. Ration. Mech. Anal. **147**(2), 89–118 (1999)
3. G.-Q. Chen, H. Frid, On the theory of divergence-measure fields and its applications. Bol. Soc. Brasil. Mat. (N.S.) **32**(3), 401–433 (2001)
4. G.-Q. Chen, H. Frid, Extended divergence-measure fields and the Euler equations for gas dynamics. Commun. Math. Phys. **236**(2), 251–280 (2003)
5. G.-Q. Chen, M. Perepelitsa, Vanishing viscosity limit of the Navier-Stokes equations to the Euler equations for compressible fluid flow. Commun. Pure Appl. Math. **63**(11), 1469–1504 (2010)
6. C.M. Dafermos, *Hyperbolic Conservation Laws in Continuum Physics*, 3rd edn. (Springer, Berlin/Heidelberg, 1999/2005/2010)
7. L.C. Evans, R.F. Gariepy, *Lecture Notes on Measure Theory and Fine Properties of Functions* (CRC, Boca Raton, 1992)

8. H. Federer, *Geometric Measure Theory* (Springer, Berlin/Heidelberg/New York, 1969)
9. H. Frid, Remarks on the theory of the (extended) divergence-measure fields. Q. Appl. Math. **70**(3), 579–596 (2012)
10. M. Silhavý, Normal currents: struture, duality pairings and div-curl lemmas. Milan J. Math. **76**, 275–306 (2008)
11. M. Silhavý, The divergence theorem for divergence measure vectorfields on sets with fractal boundaries. Math. Mech. Solids **14**(5), 445–455 (2009)

On Strong Local Alignment in the Kinetic Cucker-Smale Model

Trygve K. Karper, Antoine Mellet, and Konstantina Trivisa

Abstract In this paper, we rigorously derive a kinetic Cucker-Smale model with strong local alignment. The local alignment term is obtained by considering the limit of a non-local alignment term recently proposed by Motsch and Tadmor. The main difficulty in the analysis is presented by the non-symmetry of the Motsch-Tadmor term as well as the behavior of the velocity when the density vanishes (vacuum). Tools involved are the averaging lemma and several L^p estimates.

Keywords Flocking • Kinetic equations • Existence • Velocity averaging • Cucker-Smale • Self-organized dynamics

1991 Mathematics Subject Classification Primary: 35Q84; Secondary: 35D30

1 Introduction

In [7] Motsch and Tadmor identify an undesirable feature of the widely studied Cucker-Smale flocking model (cf. [2–4]): In the Cucker-Smale model, the alignment of each individual is scaled with the total mass so that the effect of alignment is almost negligible in sparsely populated regions. To avoid this effect, they propose a new model in which the alignment term is normalized with a local average

---

T.K. Karper (✉)
Center for Scientific Computation and Mathematical Modeling, University of Maryland, College Park, MD 20742, USA
e-mail: karper@gmail.com; http://folk.uio.no/~trygvekk

A. Mellet · K. Trivisa
Department of Mathematics, University of Maryland, College Park, MD 20742, USA
e-mail: mellet@math.umd.edu; trivisa@math.umd.edu; http://www.math.umd.edu/~mellet

G.-Q.G. Chen et al. (eds.), *Hyperbolic Conservation Laws and Related Analysis with Applications*, Springer Proceedings in Mathematics & Statistics 49,
DOI 10.1007/978-3-642-39007-4_11, © Springer-Verlag Berlin Heidelberg 2014

density instead of the total mass. Motivated by this work, the authors of the present paper proposed in [5] to combine the Cucker-Smale and Motsch-Tadmor models, letting the usual Cucker-Smale alignment term dominate the large scale dynamics and the Motsch-Tadmor term dominate the small scale dynamics. This remedies the aforementioned deficiency while maintaining the large scale dynamics of the Cucker-Smale model. At the mesoscopic level, the proposed model takes the following form

$$f_t + \mathrm{div}_x(vf) + \mathrm{div}_v(fF[f]) + \mathrm{div}_v(fL^r[f]) = 0. \tag{1}$$

Here, the unknown is the distribution function $f := f(t,x,v)$.

The first alignment term $F[\cdot]$ is the standard Cucker-Smale alignment term given by

$$F[f(x,v)] = \int_{\mathbb{R}^{2d}} \Phi(x-y) f(y,w)(w-v)\, dw\, dy, \tag{2}$$

where $\Phi(x)$ is an influence function satisfying

$$0 \le \Phi \in L^\infty(\mathbb{R}), \qquad \Phi(x) = \Phi(-x), \qquad \int_{R^d} \Phi(x)\, dx = 1.$$

A typical example is $\Phi(x) \sim 1/(1+|x|^2)^\gamma$ for some $\gamma > 0$ (cf. [1]). The second alignment term $L^r[\cdot]$ in (1) is the Motsch-Tadmor alignment term given by (see [7]):

$$L^r[f(x,v)] = \frac{\int_{\mathbb{R}^{2d}} K^r(x-y) f(y,w)(w-v)\, dw\, dy}{\int_{\mathbb{R}^{2d}} K^r(x-y) f(y,w)\, dw\, dy}, \tag{3}$$

where the index r denotes the radius of influence of K^r (see (4) below for the definition of K^r).

The only fundamental difference between (2) and (3) is the renormalization by the local average density $\int_{\mathbb{R}^{2d}} K^r(x-y) f(y,w)\, dw\, dy$. We can also write L^r as follows:

$$L^r(f) = \tilde{u}^r - v$$

where

$$\tilde{u}^r(x) = \frac{\int_{\mathbb{R}^{2d}} K^r(x-y) w f(y,w)\, dw\, dy}{\int_{\mathbb{R}^{2d}} K^r(x-y) f(y,w)\, dw\, dy}.$$

In this form, it is obvious that the strength of the alignment force is now independent of the total mass, which was the original intention of [7]. Another effect of this renormalization is to break the symmetry of the alignment. As a consequence, (1) does not conserve momentum nor energy, and the derivation of an energy bound will be one of the main difficulties in the analysis of (1).

The purpose of this paper is to study the limit as $r \to 0$ in Eq. (1) when the function K^r converges to the Dirac distribution δ_0. In other words, we study the limit of (1) when the Motsch-Tadmor term $\mathrm{div}_v(fL^r[f])$ becomes a local (in space) alignment term. For the sake of simplicity, we assume that K^r is derived from a given function K through the scaling

$$K^r(x) = r^{-d} K\left(\frac{x}{r}\right), \tag{4}$$

where K is required to satisfy

$$0 \leq K \in C_c(\mathbb{R}^d), \quad K(0) > 0, \quad \int_{\mathbb{R}^d} K(x)\, dx = 1. \tag{5}$$

When $r \to 0$, we then formally expect to have

$$\tilde{u}^r(t,x) \overset{r\to 0}{\longrightarrow} u(t,x) := \frac{\int_{\mathbb{R}^d} w f(t,x,w)\, dw}{\int_{\mathbb{R}^d} f(t,x,w)\, dw}$$

and so

$$L^r[f](t,x,v) \overset{r\to 0}{\longrightarrow} \frac{\int_{\mathbb{R}^d} f(t,x,w)(w-v)\, dw}{\int_{\mathbb{R}^d} f(t,x,w)\, dw} := u(t,x) - v. \tag{6}$$

Passing to the limit in (1), we thus obtain the equation

$$f_t + \mathrm{div}_x(vf) + \mathrm{div}_v(fF[f]) + \mathrm{div}_v(f(u-v)) = 0. \tag{7}$$

Note that u is not well defined when $\int_{\mathbb{R}^d} f(t,x,v)\, dv = 0$. We thus set $u(t,x) = \frac{\int_{\mathbb{R}^d} v f(t,x,v)\, dv}{\int_{\mathbb{R}^d} f(t,x,v)\, dv}$ if $\int_{\mathbb{R}^d} f(t,x,v)\, dv \neq 0$ and $u(t,x) = 0$ otherwise. Equation (7) is studied in [5, 6]: In [5] we prove the existence of weak solutions and establish various entropy inequalities. In [6], we investigate a singular limit corresponding to strong local alignment and (rigorously) derive an Euler-Flocking type model. The purpose of the present paper is to rigorously justify the convergence of (1)–(7) when $r \to 0$.

More precisely, we will prove the following theorem:

Theorem 1. *Let $0 \leq f_0 \in L^1(\mathbb{R}^{2d}) \cap L^\infty(\mathbb{R}^{2d})$ be given and T be a finite final time. For each $r > 0$, let $f^r(t,x,v)$ be a weak solution of* (1) *in the sense that*

$$\int_{\mathbb{R}^{2d+1}} -f^r\phi_t - vf^r\nabla_x\phi - f^r F[f^r]\nabla_v\phi\, dv\, dx\, dt \tag{8}$$
$$- \int_{\mathbb{R}^{2d+1}} f^r L^r[f^r]\nabla_v\phi\, dv\, dx\, dt = \int_{\mathbb{R}^{2d}} f_0\phi(0,\cdot)\, dv\, dx, \quad \forall \phi \in C_c^\infty([0,T)\times\mathbb{R}^{2d}),$$

where L^r *is given by* (3) *and* K^r *is given by* (4)*. Then, as* $r \to 0$,

$$f^r \overset{\star}{\rightharpoonup} f \quad \text{in } L^\infty(0,T;L^p(\mathbb{R}^{2d})) \quad \text{for any } p \in (1,\infty),$$
$$f^r L^r[f^r] \rightharpoonup f(u-v) \quad \text{in } L^q((0,T)\times\mathbb{R}^{2d}),\ q < \frac{d+2}{d+1},$$

with $u(t,x)$ *defined by*

$$u(t,x) = \begin{cases} \dfrac{\int_{\mathbb{R}^d} v\, f(t,x,v)\,dv}{\int_{\mathbb{R}^d} f(t,x,v)\,dv} & \text{if } \displaystyle\int_{\mathbb{R}^d} f(t,x,v)\,dv \neq 0 \\ 0 & \text{otherwise.} \end{cases}$$

Furthermore, the limit $f(t,x,v)$ *is a weak solution of* (7) *in the sense that*

$$\int_{\mathbb{R}^{2d+1}} -f\phi_t - vf\nabla_x\phi - fF[f]\nabla_v\phi \,dv\,dx\,dt \tag{9}$$
$$-\int_{\mathbb{R}^{2d+1}} f(u-v)\nabla_v\phi\,dv\,dx\,dt = \int_{\mathbb{R}^{2d}} f^0\phi(0,\cdot)\,dv\,dx, \quad \forall\phi\in C_c^\infty([0,T)\times\mathbb{R}^{2d}).$$

2 Preliminary Material

In this section we have gathered some results that will be needed to prove Theorem 1. We begin by introducing some convenient notations. We denote the moments of f, and their K^r weighted counterparts, as follows:

$$\rho(t,x) = \int_{\mathbb{R}^d} f(t,x,v)\,dv, \qquad \tilde{\rho}^r(t,x) = \int_{\mathbb{R}^{2d}} K^r(x-y)f(t,y,v)\,dv\,dy,$$
$$j(t,x) = \int_{\mathbb{R}^d} v\,f(t,x,v)\,dv, \qquad \tilde{j}^r(t,x) = \int_{\mathbb{R}^{2d}} K^r(x-y)\,v\,f(t,y,v)\,dv\,dy$$

(note that $\tilde{\rho}^r = K^r \star \rho$ and $\tilde{j}^r = K^r \star j$). With these notations, we can rewrite the definition of the velocity as

$$u(t,x) = \begin{cases} \dfrac{j(t,x)}{\rho(t,x)} & \text{if } \rho(t,x) \neq 0 \\ 0 & \text{if } \rho(t,x) = 0 \end{cases} \tag{10}$$

(and similarly for $\tilde{u}^r(t,x)$). Since we have

$$j(t,x) \leq \left(\int |v|^2 f(t,x,v)\,dv\right)^{1/2} \rho(t,x)^{1/2},$$

a bound on the kinetic energy of f (see (13) below) will imply that $j = 0$ whenever $\rho = 0$ and so (10) implies in particular $j = \rho u$.

With the above notation, we have $L^r[f] = \tilde{u}^r - v$, and (1) can be written as

$$f_t + \text{div}_x(fv) + \text{div}_v(fF[f]) + \text{div}_v(f(\tilde{u}^r - v)) = 0. \tag{11}$$

The following proposition states that (11) is well-posed in the sense of weak solutions (see [5] for the proof).

Proposition 1. *Assume that* $0 \le f_0 \in L^\infty \cap L^1(\mathbb{R}^{2d})$ *with* $(x^2+v^2)f_0 \in L^1(\mathbb{R}^{2d})$. *For a given* $T < +\infty$ *and for any* $r > 0$, (11) *admits a weak solution* $0 \le f \in C(0,T; L^1(\mathbb{R}^{2d}))$. *Moreover, for all* $p \in [1,\infty]$, f *satisfies*

$$\|f\|_{L^\infty(0,T;L^p(\mathbb{R}^{2d}))} \le \|f_0\|_{L^p(\mathbb{R}^{2d})} e^{\frac{p-1}{p}CT}, \tag{12}$$

$$\mathcal{E}(t) := \int_{\mathbb{R}^{2d}} \left(|v|^2 + |x|^2\right) f(t,x,v)\,dv\,dx \le Ce^{CT}\mathcal{E}(0), \tag{13}$$

where the constant C *might depend on* r.

To conclude this section, we recall the following classical lemma, which will be used to derive the L^p integrability of ρ and j (see [5] for the proof):

Lemma 1. *Assume that* $f(t,x,v)$, $t \in (0,T)$, $(x,v) \in \mathbb{R}^{2d}$ *satisfies*

$$\|f\|_{L^\infty([0,T]\times\mathbb{R}^{2d})} \le M, \quad \text{and} \quad \sup_{t\in[0,T]} \int_{\mathbb{R}^{2d}} |v|^2 f(t,x,v)\,dvdx \le M.$$

Then there exists a constant $C = C(M)$ *such that*

$$\begin{aligned} \|\rho\|_{L^\infty(0,T;L^p(\mathbb{R}^d))} &\le C, \quad \text{for every } p \in [1, \tfrac{d+2}{d}), \\ \|j\|_{L^\infty(0,T;L^p(\mathbb{R}^d))} &\le C, \quad \text{for every } p \in [1, \tfrac{d+2}{d+1}), \end{aligned} \tag{14}$$

where $\rho = \int f\,dv$ *and* $j = \int vf\,dv$.

2.1 The Velocity Averaging Lemma

When passing to the limit in (11), the main obstacle is to obtain compactness of the product $f\tilde{u}^r$. To achieve this, we will make use of the following version of the velocity averaging lemma due to Perthame and Souganidis [8]:

Proposition 2. *Let $\{f^n\}_n$ be bounded in $L^p_{loc}(\mathbb{R}^{2d+1})$ with $1 < p < \infty$, and $\{G^n\}_n$ be bounded in $L^p_{loc}(\mathbb{R}^{2d+1})$. If f^n and G^n satisfy*

$$f^n_t + v \cdot \nabla_x f^n = \nabla^k_v G^n, \qquad f^n|_{t=0} = f^0 \in L^p(\mathbb{R}^{2d}),$$

for some multi-index k, then for any $\varphi \in C^{|k|}_c(\mathbb{R}^{2d})$, the sequence

$$\left\{ \int_{\mathbb{R}^d} f^n \varphi(v)\, dv \right\}_n \tag{15}$$

is relatively compact in $L^p_{loc}(\mathbb{R}^{d+1})$.

The previous proposition cannot be directly applied to obtain the needed compactness. In fact, we will rely on the following lemma which can be seen as a corollary of the previous proposition. The proof can be found in [5].

Lemma 2. *Let $\{f^n\}_n$ and $\{G^n\}_n$ be as in Proposition 2 and assume that*

$$f^n \text{ is bounded in } L^\infty(\mathbb{R}^{2d+1}),$$

$$(|v|^2 + |x|^2) f^n \text{ is bounded in } L^\infty(0,T; L^1(\mathbb{R}^{2d+1})).$$

Then, for any $\varphi(v) \in C^\infty(\mathbb{R}^d)$ such that $|\varphi(v)| \leq c|v|$ and for any $q < \frac{d+2}{d+1}$, the sequence

$$\left\{ \int_{\mathbb{R}^d} f^n \varphi(v)\, dv \right\}_n \tag{16}$$

is relatively compact in $L^q((0,T) \times \mathbb{R}^d)$.

2.2 An Important Technical Lemma

In view of Lemmas 1 and 2, it is clear that in order to get convergence results for f^r and its moments, we will need to obtain some estimate on f^r that are uniform with respect to r. The main difficulty will be to show that the energy estimate (13) holds with constants independent of r (which does not obviously follow from the result of [5]). For this we will make use of the following technical lemma, which can be found in [5] (the proof is given below for completeness):

Lemma 3. *Let K satisfy* (5) *and assume there exist $0 < R_1 < R_2 < \infty$ such that*

$$K(x) > 0 \text{ for } |x| \leq R_1, \qquad K(x) = 0 \quad \text{for } |x| \geq R_2. \tag{17}$$

There exists a constant C depending only on

$$\frac{\sup_{B_{R_2}} K}{\inf_{B_{R_1}} K}\left(\frac{R_2}{R_1}\right)^d \tag{18}$$

such that

$$\int_{\mathbb{R}^d} K(x-y)\frac{\rho(x)}{\int_{\mathbb{R}^d} K(x-z)\rho(z)\,dz}\,dx \le C, \quad \forall y \in \mathbb{R}^d,$$

for all nonnegative functions $\rho \in L^1(\mathbb{R}^d)$.

The most important part of this lemma is the formula (18), which implies that if we replace the function $K(x)$ with $\alpha K(\beta x)$, for any $\alpha > 0$ and $\beta > 0$, then the same estimate holds with the same constant.

We deduce:

Corollary 1. *Assume that* K^r *is given by* (4) *where* K *satisfies* (5). *Then, there exists a constant* C *independent of* r *such that*

$$\int_{\mathbb{R}^d} K^r(x-y)\frac{\rho(x)}{\int_{\mathbb{R}^d} K^r(|x-z|)\rho(z)\,dz}\,dx \le C, \quad \forall y \in \mathbb{R}^d$$

for all nonnegative functions $\rho \in L^1(\mathbb{R}^d)$.

Proof of Lemma 3. We denote $\tilde\rho(x) = \int_{\mathbb{R}^d} K(x-z)\rho(z)\,dz$ and we observe that

$$\int_{\mathbb{R}^d} K(x-y)\frac{\rho(x)}{\tilde\rho(x)}\,dx \le (\sup K)\int_{B_{R_2}(y)} \frac{\rho(x)}{\tilde\rho(x)}\,dx.$$

Next, we cover $B_{R_2}(y)$ with balls of radius $R_1/2$: We can choose $(x_i)_{i=1}^N$ in such a way that

$$B_{R_2}(y) \subset \bigcup_{i=1}^N B_{R_1/2}(x_i)$$

with $N \sim (R_2/R_1)^d$. We can thus write

$$\int_{\mathbb{R}^d} K(x-y)\frac{\rho(x)}{\tilde\rho(x)}\,dx \le (\sup K)\sum_{i=1}^N \int_{B_{R_1/2}(x_i)} \frac{\rho(x)}{\tilde\rho(x)}\,dx.$$

Moreover, clearly,

$$\tilde\rho(x) = \int_{\mathbb{R}^d} K(x-z)\rho(z)\,dz \ge \int_{B_{R_1/2}(x_i)} K(x-z)\rho(z)\,dz.$$

By combining the two previous inequalities, we see that

$$\int_{\mathbb{R}^d} K(x-y)\frac{\rho(x)}{\tilde{\rho}(x)}\,dx \leq (\sup K)\sum_{i=1}^{N}\int_{B_{R_1/2}(x_i)}\frac{\rho(x)}{\int_{B_{R_1/2}(x_i)} K(x-z)\rho(z)\,dz}\,dx.$$

Now, using the fact that when $x, z \in B_{R_1/2}(x_i)$ we have $|x-z| \leq R_1$, we deduce

$$\begin{aligned}\int_{\mathbb{R}^d} K(x-y)\frac{\rho(x)}{\tilde{\rho}(x)}\,dx &\leq \frac{\sup K}{\inf_{B_{R_1}(0)} K}\sum_{i=1}^{N}\int_{B_{R_1/2}(x_i)}\frac{\rho(x)}{\int_{B_{R_1/2}(x_i)}\rho(z)\,dz}\,dx\\ &\leq \frac{\sup K}{\inf_{B_{R_1}(0)} K}N \leq C\frac{\sup K}{\inf_{B_{R_1}(0)} K}\left(\frac{R_2}{R_1}\right)^d\end{aligned}$$

and the proof is complete. □

2.3 A Priori Estimate

We can now conclude this preliminary section by proving that f^r satisfies some a priori estimates uniformly with respect to r. We recall that the energy functional is defined as

$$\mathcal{E}(t) = \int_{\mathbb{R}^{2d}}\left(\frac{|v|^2}{2}+\frac{|x|^2}{2}\right) f(t,x,v)\,dv\,dx. \tag{19}$$

We then prove:

Proposition 3 (Energy bound). *Let $0 \leq f_0 \in L^1(\mathbb{R}^{2d}) \cap L^\infty(\mathbb{R}^{2d})$ be given, let T be a finite final time, and let f be the corresponding weak solution of* (1). *There is a constant $C > 0$ independent of $r > 0$ such that for any $p \in [1,\infty]$, f satisfies*

$$\|f\|_{L^\infty(0,T;L^p(\mathbb{R}^{2d}))} \leq \|f_0\|_{L^p(\mathbb{R}^{2d})} e^{\frac{p-1}{p}CT}, \tag{20}$$

and

$$\begin{aligned}&\sup_{t\in(0,T)} \mathcal{E}(t) + \frac{1}{2}\int_{\mathbb{R}^{2d}} f|\tilde{u}^r - v|^2\,dv\,dx\\ &\quad + \frac{1}{2}\int_{\mathbb{R}^{4d}} \Phi(x-y) f(x,v) f(y,w)|w-v|^2\,dw\,dy\,dv\,dx \leq C(T)\mathcal{E}(0).\end{aligned} \tag{21}$$

The proof of Proposition 3 relies on two auxiliary results (Lemmas 4 and 5 below) that we prove first. We begin with the L^p estimate (20):

Lemma 4. *Let f be a weak solution of* (1)*. There is a constant C, independent of r, such that for any $p \in [1,\infty]$, f satisfies*

$$\sup_{t\in(0,T)} \|f\|_{L^p(\mathbb{R}^{2d})} \leq \|f_0\|_{L^p(\mathbb{R}^{2d})} e^{\frac{p-1}{p}CT}. \tag{22}$$

Proof. Let us first assume that f has sufficient regularity in x to make the following argument rigorous: Let B be a continuously differentiable function and define $b(f) = fB'(f) - B(f)$. By multiplying (1) by $B'(f)$ and integrating, we obtain

$$\begin{aligned}\frac{d}{dt}\int_{\mathbb{R}^{2d}} B(f)\,dv\,dx &= \int_{\mathbb{R}^{2d}} v\nabla_x b(f)\,dv\,dx \\ &\quad + \int_{\mathbb{R}^{2d}} (F(f) + L^r(f))\nabla_v b(f)\,dv\,dx \\ &= -\int_{\mathbb{R}^{2d}} b(f)\,(\mathrm{div}_v\,F(f) + \mathrm{div}_v\,L^r(f))\,dv\,dx.\end{aligned} \tag{23}$$

Next, using the definition of the alignment terms, we see that

$$\mathrm{div}_v\,F[f] = -d\int_{\mathbb{R}^{2d}} \Phi(x-y)f(y,w)\,dwitdy,$$
$$\mathrm{div}_v\,L^r[f] = -d\frac{\int_{\mathbb{R}^{2d}} K^r(x-y)f(y,w)\,dw\,dy}{\int_{R^{2d}} K^r(x-y)f(y,w)\,dw\,dy} = -d.$$

Substituting these identities into (23), we find that

$$\begin{aligned}\frac{d}{dt}\int_{\mathbb{R}^{2d}} B(f)\,dv\,dx &= \int_{\mathbb{R}^{2d}} b(f)\left(d + d\int_{\mathbb{R}^{2d}} \Phi(x-y)f(y,w)\,dw\,dy\right)dv\,dx \\ &\leq \int_{\mathbb{R}^{2d}} b(f)(d + dM\|\Phi\|_{L^\infty(\mathbb{R}^{2d})})\,dv\,dx,\end{aligned}$$

where M is the total mass. Next, we let $B(f) = f^p$, so that $b(f) = (p-1)f^p$. An application of the Gronwall inequality then provides the bound

$$\sup_{t\in(0,T)} \|f\|_{L^p(\mathbb{R}^{2d})} \leq \|f_0\|_{L^p(\mathbb{R}^{2d})} e^{\frac{p-1}{p}CT},$$

which is the desired inequality.

To make the previous calculation rigorous when f does not have sufficient regularity in x, one can convolve (1) with a standard mollifier η^ϵ in x to obtain an equation similar to (1) but with an additional commutator term. The corresponding calculation can then be carried out as above to obtain the desired inequality with an additional term on the right-hand side that converges to zero with ϵ. □

The main difficulty in proving the energy estimate (21) (even for $r > 0$) is to control the non-symmetric Motsch-Tadmor alignment term. This is the goal of the following Lemma, which relies on Lemma 3 and its Corollary 1:

Lemma 5. *There is a constant C, independent of r, such that*

$$\int_{\mathbb{R}^{2d}} f v L^r[f]\, dv\, dx \leq C\mathcal{E}(t) - \frac{1}{2}\int_{\mathbb{R}^{2d}} f|\tilde{u}^r - v|^2\, dv\, dx. \tag{24}$$

Proof. By definition of L^r, we have that

$$\begin{aligned} L^r[f] &= \frac{1}{\tilde{\rho}^r(x)}\int_{\mathbb{R}^{2d}} K^r(x-y) f(y,w)(w-v)\, dw\, dy \\ &= \frac{1}{\tilde{\rho}^r(x)}\int_{\mathbb{R}^d} K^r(x-y)(j(y)-\rho(y)v)\, dy = \frac{\tilde{j}^r}{\tilde{\rho}^r} - v := \tilde{u}^r - v. \end{aligned} \tag{25}$$

By adding and subtracting, we obtain

$$\begin{aligned} \int_{\mathbb{R}^{2d}} f v L^r[f]\, dv\, dx &= \int_{\mathbb{R}^{2d}} f(\tilde{u}^r - v)v\, dv\, dx \\ &= -\frac{1}{2}\int_{\mathbb{R}^{2d}} f(\tilde{u}^r - v)^2\, dv\, dx + \frac{1}{2}\int_{\mathbb{R}^{2d}} f(\tilde{u}^r)^2 - f v^2\, dv\, dx \\ &\leq -\frac{1}{2}\int_{\mathbb{R}^{2d}} f(\tilde{u}^r - v)^2\, dv\, dx + \frac{1}{2}\int_{\mathbb{R}^d} \rho(\tilde{u}^r)^2\, dx. \end{aligned} \tag{26}$$

From the Hölder inequality, we have that

$$\tilde{\rho}^r\tilde{u}^r := \int_{\mathbb{R}^{2d}} K^r(x-y) f(y,v)v\, dv\, dy \leq (\tilde{\rho}^r)^{\frac{1}{2}}\left(\int_{\mathbb{R}^{2d}} K^r(x-y) f(y,v)v^2\, dv\, dy\right)^{\frac{1}{2}}.$$

Hence, the following inequality holds

$$\tilde{\rho}^r(\tilde{u}^r)^2 \leq \int_{\mathbb{R}^{2d}} K^r(x-y) f(y,v)v^2\, dv\, dy,$$

from which we deduce

$$\begin{aligned} \int_{\mathbb{R}^d} \rho(\tilde{u}^r)^2\, dx = \int_{\mathbb{R}^d} \frac{\rho}{\tilde{\rho}^r}\tilde{\rho}^r(\tilde{u}^r)^2\, dx &\leq \int_{\mathbb{R}^{3d}} K^r(x-y)\frac{\rho(x)}{\tilde{\rho}^r(x)} f(y,v)v^2\, dy\, dv\, dx \\ &\leq \sup_y\left(\int_{\mathbb{R}} K^r(x-y)\frac{\rho(x)}{\tilde{\rho}^r(x)}\, dx\right) 2\mathcal{E}(t) \leq C\mathcal{E}(t), \end{aligned} \tag{27}$$

where the last inequality follows from Corollary 1. Inserting (27) into (26) concludes the proof. □

We have now gathered all the ingredients we need to prove Proposition 3.

Proof of Proposition 3. Only (21) remains to be proved. By direct calculation,

$$\begin{aligned}\frac{d}{dt}\int_{\mathbb{R}^{2d}} f\left(\frac{|v|^2}{2}+\frac{|x|^2}{2}\right) dv\,dx &= \int_{\mathbb{R}^{2d}} f_t\frac{|x|^2}{2} + f_t\frac{|v|^2}{2}\,dv\,dx \\ &= \int_{\mathbb{R}^{2d}} vfx + fvF[f] + fvL^r[f]\,dv\,dx \\ &\leq \mathcal{E}(t) + \int_{\mathbb{R}^{2d}} fvF[f] + fvL^r[f]\,dv\,dx,\end{aligned} \tag{28}$$

where the last inequality follows from the Cauchy-Schwarz inequality and Young's inequality with $\epsilon = 1/2$. Now, using the fact that $\Phi(-x) = \Phi(x)$, we write

$$\begin{aligned}\int_{\mathbb{R}^{2d}} fvF[f]\,dv\,dx &= \int_{\mathbb{R}^{4d}} \Phi(x-y)f(x,v)f(y,w)(w-v)v\,dw\,dy\,dv\,dx \\ &= \int_{\mathbb{R}^{4d}} \Phi(x-y)f(x,v)f(y,w)(v-w)w\,dw\,dy\,dv\,dx \\ &= -\frac{1}{2}\int_{\mathbb{R}^{4d}} \Phi(x-y)f(x,v)f(y,w)|w-v|^2\,dw\,dy\,dv\,dx.\end{aligned}$$

Then, we conclude the proof by applying this identity and Lemma 5 to (28) together with the Gronwall inequality. □

3 Convergence and Proof of Theorem 1

Equipped with the bounds of the previous section, we are ready to send r to 0 in (1) and thereby prove Theorem 1. For this purpose, we let $\{r^n\}_n$ be a sequence of positive numbers such that $r^n \to 0$ as $n \to \infty$ and consider corresponding solutions f^n of

$$f_t^n + \mathrm{div}_x(vf^n) + \mathrm{div}_v(f^nF[f^n]) + \mathrm{div}_v(f^n(\tilde{u}^n - v)) = 0, \tag{29}$$

where we recall the notation

$$\tilde{u}^n = \frac{\tilde{j}^n}{\tilde{\rho}^n} := \frac{\int_{\mathbb{R}^{2d}} K^{r^n}(x-y)f^n(y,w)w\,dw\,dy}{\int_{\mathbb{R}^{2d}} K^{r^n}(x-y)f^n(y,w)\,dw\,dy}.$$

Our starting point is the fact that Lemma 4, Proposition 3, together with Lemma 1, assert the existence of a function $0 \leq f \in C(0,T;L^1(\mathbb{R}^{2d})) \cap L^\infty(0,T;L^\infty(\mathbb{R}^{2d}))$, such that, as $n \to \infty$,

$$\begin{aligned} &f^n \overset{\star}{\rightharpoonup} f \quad \text{in } L^\infty((0,T);L^p(\mathbb{R}^{2d}))), \quad \text{for any } p \in (1,\infty), \\ &\rho^n \overset{\star}{\rightharpoonup} \rho \quad \text{in } L^\infty((0,T);L^p(\mathbb{R}^d)), \quad \text{for every } p \in \left(1, \frac{d+2}{d}\right), \\ &j^n \overset{\star}{\rightharpoonup} j \quad \text{in } L^\infty((0,T);L^p(\mathbb{R}^d)), \quad \text{for every } p \in \left(1, \frac{d+2}{d+1}\right). \end{aligned} \tag{30}$$

Moreover, the velocity averaging Lemma 2 is applicable. By setting $\varphi(v) = 1$ and $\varphi(v) = v$ in Lemma 2 we obtain respectively

$$\begin{aligned} &\rho^n \to \rho \quad \text{a.e. and } L^p((0,T)\times\mathbb{R}^d)\text{–strong}, \quad \text{for every } p \in \left(1, \frac{d+2}{d+1}\right), \\ &j^n \to j \quad \text{a.e. and } L^p((0,T)\times\mathbb{R}^d)\text{–strong}, \quad \text{for every } p \in \left(1, \frac{d+2}{d+1}\right), \end{aligned} \tag{31}$$

along some subsequence as $n \to \infty$. Furthermore, we can prove:

Lemma 6. *Up to another subsequence, we can also assume that*

$$\tilde{j}^n \to j, \quad \tilde{\rho}^n \to \rho, \quad \textit{a.e and } L^p((0,T)\times\mathbb{R}^d)\textit{–strong} \tag{32}$$

for all $p \in \left(1, \frac{d+2}{d+1}\right)$.

Proof. We begin by recalling the following classical result concerning mollifiers like $K^n = K^{r^n}$: For any $\epsilon > 0$, there is an m such that

$$\|K^n \star \rho - \rho\|_{L^p(\mathbb{R}^d)} < \epsilon, \quad \forall\, n \geq m.$$

Now, consider a subsequence n^k, where $n^k \geq m$, along which $\rho^n \to \rho$. By adding and subtracting, we obtain

$$\begin{aligned} \|\tilde{\rho}^n - \rho\|_{L^p(\mathbb{R}^d)} &\leq \|\tilde{\rho}^n - K^n \star \rho\|_{L^p(\mathbb{R}^d)} + \|K^n \star \rho - \rho\|_{L^p(\mathbb{R}^d)} \\ &= \|K^n \star (\rho^n - \rho)\|_{L^p(\mathbb{R}^d)} + \|K^n \star \rho - \rho\|_{L^p(\mathbb{R}^d)} \\ &\leq \|\rho^n - \rho\|_{L^p(\mathbb{R}^d)} + \epsilon = 2\epsilon, \end{aligned}$$

for any $p \in \left(1, \frac{d+2}{d+1}\right)$. The same argument can be applied to prove the convergence of $\tilde{j}^n$. □

Lemma 7. *From the convergences* (30) *to* (31), *it follows that*

$$f^n \tilde{u}^n \overset{\star}{\rightharpoonup} fu \text{ in } L^\infty((0,T); L^p(\mathbb{R}^{2d})) \quad \text{for every} \quad p \in \left(1, \frac{d+2}{d+1}\right).$$

Proof. For a given test function $\varphi(v)$, we let

$$\rho_\varphi^n = \int_{\mathbb{R}^d} f^n \varphi(v)\, dv, \qquad \tilde{m}_\varphi^n = \tilde{u}^n \rho_\varphi^n.$$

Consider now a test function $\psi(t,x,v) := \phi(t,x)\varphi(v)$, with $\phi \in C_c^\infty((0,T) \times \mathbb{R}^d)$ and $\varphi \in C_c^\infty(\mathbb{R}^d)$. We write

$$\begin{aligned} \int_0^T \int_{\mathbb{R}^{2d}} f^n \tilde{u}^n \psi \, dv\, dx\, dt &= \int_0^T \int_{\mathbb{R}^d} \tilde{u}^n(t,x)\phi(t,x) \left(\int_{\mathbb{R}^d} f^n(t,x,v)\varphi(v)\, dv \right) dx\, dt \\ &= \int_0^T \int_{\mathbb{R}^d} \tilde{u}^n(t,x)\rho_\varphi^n(t,x)\phi(t,x)\, dx\, dt \\ &= \int_0^T \int_{\mathbb{R}^d} \tilde{m}_\varphi^n(t,x)\phi(t,x)\, dx\, dt. \end{aligned} \tag{33}$$

Now, Hölder's inequality yields

$$\|\tilde{m}_\varphi^n\|_{L^p(\mathbb{R}^d)} \le \|\varphi\|_{L^\infty(\mathbb{R}^d)} \|\rho^n\|^{\frac{1}{2}}_{L^{\frac{p}{2-p}}(\mathbb{R}^d)} \|(\rho^n)^{\frac{1}{2}} \tilde{u}^n\|_{L^2(\mathbb{R}^d)}. \tag{34}$$

Using Proposition 3, (27), and Lemma 1, it is readily seen that the right-hand side of (34) is bounded provided

$$\frac{p}{2-p} \in \left(1, \frac{d+2}{d}\right)$$

which is equivalent to

$$p \in \left(1, \frac{d+2}{d+1}\right).$$

Hence, there exists a function $m \in L^\infty((0,T); L^p(\mathbb{R}^d))$ and a subsequence such that

$$\tilde{m}_\varphi^n \overset{\star}{\rightharpoonup} m \text{ in } L^\infty((0,T); L^p(\mathbb{R}^d)), \qquad \text{for every } p \in \left(1, \frac{d+2}{d+1}\right),$$

and we have to prove that

$$m = u\rho_\varphi, \quad \text{where } u \text{ is such that} \quad j = \rho u.$$

Let us first verify the existence of such a function u. Consider the set

$$A_R = \{(t,x) \in B_R(0) \times (0,T); \rho(t,x) = 0\},$$

where $B_R(0)$ is the ball of radius R centered at 0. By direct calculation,

$$\begin{aligned}\int_{A_R} |j_n| \, dx \, dt &\leq \left(\int_{A_R} \rho^n |u^n|^2 \, dx \, dt \right)^{\frac{1}{2}} \left(\int_{A^R} \rho^n \, dx \, dt \right) \\ &\leq CT \left(\int_{A_R} \rho^n \, dx \, dt \right) \overset{n \to \infty}{\longrightarrow} 0,\end{aligned}$$

and hence we have that $j = 0$ a.e in A^R. If we define the function u as

$$u(t,x) = \begin{cases} \frac{j(t,x)}{\rho(t,x)}, & \text{if } \rho(t,x) \neq 0, \\ 0, & \text{if } \rho(t,x) = 0, \end{cases} \tag{35}$$

we have that $j = \rho u$ and it remains to prove that $m = \rho_\varphi u$. We first observe that we can show as in (34) that

$$\|m_\varphi^n\|_{L^p(A_R)} \leq C \|\rho^n\|_{L^p(A_R)}^{\frac{1}{2}} \overset{n \to \infty}{\longrightarrow} 0,$$

and hence it suffices to check that

$$m(t,x) = u(t,x)\rho_\varphi(t,x), \quad \text{whenever } \rho(t,x) \neq 0.$$

For this purpose, we consider the set

$$B_R^\epsilon = \{(t,x) \in B_R(0) \times (0,T); \rho(t,x) > \epsilon\}.$$

From Egorov's theorem and the compactness of ρ^n and $\tilde{\rho}^n$ (Lemma 6), we have the existence of a set $C_\eta \subset B_R^\epsilon$ with measure $|B_R^\epsilon \setminus C_\eta| < \eta$ on which $\tilde{\rho}^n$ and ρ^n converge uniformly to ρ. Then, for n sufficiently large,

$$\tilde{\rho}^n \geq \epsilon/2 \quad \text{in } C_\eta,$$

and since

$$m_\varphi^n = \tilde{u}^n \rho_\varphi^n = \frac{\tilde{j}^n}{\tilde{\rho}^n} \rho_\varphi^n,$$

we can pass to the a.e limit on C_η to deduce

$$m = \frac{j}{\rho} \rho_\varphi = u \rho_\varphi \quad \text{in } C_\eta.$$

Since this holds for all $\eta > 0$, we can conclude

$$m = u\rho_\varphi \quad \text{in } B_R^\epsilon,$$

for every R and ϵ. We conclude that,

$$m = u\rho_\varphi \quad \text{on } \{\rho > 0\}.$$

We have thus shown that

$$\int_0^T \int_{\mathbb{R}^{2d}} f^n \tilde{u}^n \psi \, dv\, dx\, dt \longrightarrow \int_0^T \int_{\mathbb{R}^d} u\rho_\varphi \phi \, dx\, dt = \int_0^T \int_{\mathbb{R}^{2d}} f u \psi \, dv\, dx\, dt$$

for all test functions of the form $\psi(t,x,v) := \phi(t,x)\varphi(v)$. Finally, the density of the vector space generated by $C_c^\infty((0,T)\times\mathbb{R}^d)\times C_c^\infty(\mathbb{R}^d)$ in $L^1((0,T); L^{p'}(\mathbb{R}^{2d}))$ yields the result. □

Proof of Theorem 1:. The weak formulation of (29) reads

$$\begin{aligned}\int_0^T \int_{\mathbb{R}^{2d}} & f^n(\psi_t + v\cdot\nabla_x\psi)\, dv\, dx\, dt \\ &= I_1^n + I_2^n - \int_{\mathbb{R}^2 d} f_0^n \psi(0,\cdot)\, dv\, dx, \quad \forall \psi \in C_c^\infty((0,T)\times\mathbb{R}^{2d}),\end{aligned} \tag{36}$$

where we have introduced the quantities

$$I_1^n := -\int_0^T \int_{\mathbb{R}^{4d}} \Phi(x-y) f^n(x,v) f^n(y,w)(w-v)\nabla_v\psi(x,v)\, dw\, dy\, dv\, dx\, dt,$$

$$I_2^n := -\int_0^T \int_{\mathbb{R}^{2d}} f^n(\tilde{u}^n - v)\nabla_v\psi \, dv\, dx\, dt.$$

Since I_1^n contains only integrated quantities, we can apply (30) to pass to the limit in (36) and conclude

$$\begin{aligned}\int_0^T \int_{\mathbb{R}^{2d}} & f(\psi_t + v\cdot\nabla_x\psi)\, dv\, dx\, dt \\ &= I_1 + \lim_{n\to\infty} I_2^n - \int_{\mathbb{R}^2 d} f_0 \psi(0,\cdot)\, dv\, dx,\end{aligned} \tag{37}$$

where $I_1 := -\int_0^T \int_{\mathbb{R}^{4d}} \Phi(x-y) f(x,v) f(y,w)(w-v)\nabla_v\psi(x,v)\, dw\, dy\, dv\, dx\, dt$.

From Lemma 7, we have that $f^n\tilde{u}^n \overset{\star}{\rightharpoonup} fu$ in $L^\infty((0,T); L^p(\mathbb{R}^{2d}))$, for any $p \in \left(1, \frac{d+2}{d+1}\right)$. Hence, we can pass to the limit in I_2^n and conclude

$$\lim_{n\to\infty} I_2^n = -\lim_{n\to\infty} \int_0^T \int_{\mathbb{R}^{2d}} f^n(\tilde{u}^n - v)\nabla_v\psi\, dv\, dx\, dt = -\int_0^T \int_{\mathbb{R}^{2d}} f(u-v)\nabla_v\psi\, dv\, dx\, dt.$$

In view of (37), we can then conclude that the limit f is a weak solution to

$$f_t + \mathrm{div}_x(vf) + \mathrm{div}_v(fF[f]) + \mathrm{div}_v(f(u-v)) = 0.$$

This concludes the proof of Theorem 1. □

Acknowledgements The work of Trygve K. Karper was supported by the Research Council of Norway through the project 205738. The work of Antoine Mellet was supported by the National Science Foundation under the Grant DMS-0901340. The work of Konstantina Trivisa was supported by the National Science Foundation under the Grant DMS-1109397.

References

1. J.A. Carrillo, M. Fornasier, J. Rosado, G. Toscani, Asymptotic flocking dynamics for the kinetic Cucker-Smale model. SIAM J. Math. Anal. **42**(1), 218–236 (2010)
2. F. Cucker, S. Smale, Emergent behavior in flocks. IEEE Trans. Autom. Control **52**(5), 852–862 (2007)
3. F. Cucker, S. Smale, On the mathematics of emergence. Jpn. J. Math. **2**(1), 197–227 (2007)
4. S.-Y. Ha, E. Tadmor, From particle to kinetic and hydrodynamic descriptions of flocking. Kinet. Relat. Models **1**(30), 415–435 (2008)
5. T. Karper, A. Mellet, K. Trivisa, Existence of weak solutions to kinetic flocking models. SIAM J. Math. Anal. **45**(1), 215–243 (2013)
6. T. Karper, A. Mellet, K. Trivisa, Hydrodynamic limit of the kinetic Cucker-Smale flocking with strong local alignment (2012, Preprint)
7. S. Motsch, E. Tadmor, A new model for self-organized dynamics and its flocking behavior. J. Stat. Phys. **141**(5), 923–947 (2011)
8. B. Perthame, P.E. Souganidis, A limiting case for velocity averaging. Ann. Sci. École Norm. Sup. **31**(4), 591–598 (1998)

Multi-dimensional Systems of Conservation Laws: An Introductory Lecture

Denis Serre

Abstract These notes are written after the crash course given at the ICMS conference on Hyperbolic conservation laws. We intend to review several aspects of the theory of the Cauchy problem and the Initial-boundary value problem (IBVP). On the one hand, we give a thorough account of the theory for linear, constant coefficient operators, following Gårding, Hersch, Kreiss and others. Hyperbolicity raises interesting questions in real algebraic geometry, a topic to which Petrowski's school (in particular Oleĭnik) contributed. Next, we turn towards quasilinear systems and recall the interplay between entropies and symmetrizability. This leads us to the local existence of a classical solution. The global-in-time Cauchy problem necessitates weak solutions; these must be selected by admissibility criteria. We give a review of the various criteria that have been elaborated so far. Some of them lead us to the 'viscous' approximation of hyperbolic systems. We review the structural properties of these models, whose paradigm is the Navier-Stokes-Fourier (NSF) system of gas dynamics. This is more or less Kawashima's theory, in the simplified description that we have given in recent papers. We end with results about singular limits, such as the convergence of NSF towards Euler-Fourier when Newtonian viscosity tends to zero, and the analysis of the principal sub-systems introduced by Boillat and Ruggeri.

Despite the length of these notes, they contain only very few proofs. We focus instead on the concepts and the theorems of the theory.

2010 Mathematics Subject Classification Primary: 35L04, 35L60, 35L65, 35L67, 35K40, 35M30; Secondary: 76L05, 76N17

D. Serre (✉)
UMPA, UMR 5669 CNRS, École Normale Supérieure de Lyon, Lyon, France
e-mail: denis.serre@ens-lyon.fr

G.-Q.G. Chen et al. (eds.), *Hyperbolic Conservation Laws and Related Analysis with Applications*, Springer Proceedings in Mathematics & Statistics 49,
DOI 10.1007/978-3-642-39007-4_12,

Notations

The most general form of a first-order system of conservation laws is

$$\frac{\partial u_i}{\partial t} + \sum_{\alpha=1}^{d} \frac{\partial f_i^\alpha(u)}{\partial x_\alpha} = 0, \qquad \forall i = 1, \dots, n, \tag{1}$$

where $u(x,t) \in \mathbb{R}^n$ is the unkown and $x \in \mathbb{R}^d$ is the space variable. The unknown takes values in some phase space $\mathcal{U}$, an open subset of $\mathbb{R}^n$. The vector fields $f^\alpha : \mathcal{U} \to \mathbb{R}^n$ are given smooth functions that describe the underlying physics; we call them the *constitutive fluxes*.

In general, an initial data is given at time $t = 0$:

$$u(x,0) = a(x). \tag{2}$$

In the Cauchy problem, the physical domain is the whole space $\mathbb{R}^d$, whereas in an IBVP, it is an open subset $\Omega \subset \mathbb{R}^d$. Then boundary conditions have to be imposed; the number and the nature of the boundary conditions is a difficult topic, which is discussed in Sect. 2.

Instead of the fully developed form (1), we may opt for a more compact expression, either by using operators ∂:

$$\partial_t u_i + \sum_{\alpha=1}^{d} \partial_\alpha f_i^\alpha(u) = 0, \qquad \forall i = 1, \dots, n,$$

or a vector form

$$\partial_t u + \sum_\alpha \partial_\alpha f^\alpha(u) = 0,$$

or a component-wise form

$$\partial_t u_i + \operatorname{div} f_i(u) = 0, \qquad i = 1, \dots, n.$$

The more compact form is of course

$$\partial_t u + \operatorname{Div} F(u) = 0,$$

where the capital in the divergence operator indicates that the argument is a tensor (an $n \times d$ one), and the divergence has to be taken row-wise.

If $A \in \mathbf{M}_n(\mathbb{C})$, the ambient space $\mathbb{C}^n$ splits in a unique manner as $U(A) \oplus C(A) \oplus S(A)$, where each factor is an invariant subpace (that is $A(E) \subset E$), and the spectrum of the restriction of A to $U(A)$ (resp. $C(A)$, resp. $S(A)$) has positive (resp. null, resp. negative) real part. These spaces are respectively called the unstable, central and stable invariant subspaces.

1 Hyperbolicity

When studying the stability of a classical solution of an evolution PDE under some initial disturbance, we are led to the analysis of the linearized PDE. Linearizing (1) about u yields a system

$$\partial_t v + \mathrm{Div}(\mathcal{A}(x,t)v) = 0,$$

where $\mathcal{A} := \mathrm{d}F(u)$. Now a rescaling $(x,t) \mapsto (\mu(x-x_0), \mu(t-t_0))$ with $\mu \to +\infty$ a constant suggests to begin the analysis by freezing (x,t) to any value (x_0,t_0) and retaining only the principal part, which consists of the derivatives of order one. These are the reasons why we are interested in *linear operators with constant coefficients*.

Let us point out that rescaling at points interior to a domain Ω leads us to the Cauchy problem: $\Omega \ni 0$ rescales as $\mu\Omega$, which tends to $\mathbb{R}^d$ as $\mu \to +\infty$. Instead, at boundary points the limit domain is a half-space and we face an IBVP. The latter situation will be studied in Sect. 2. We concentrate for the moment of the pure Cauchy problem. Most of the topics listed below may be found in the first chapter of the monograph [1].

We therefore consider a linear equation $Lu = f$ in $\mathbb{R}^d \times (0,T)$, where

$$L := \partial_t + \sum_\alpha A^\alpha \partial_\alpha, \qquad A^\alpha \in \mathbf{M}_n(\mathbb{R}). \tag{3}$$

Because we intend to employ the Duhamel's principle in order to treat the right-hand side, we may reduce our analysis to the problem

$$Lu = 0 \qquad \text{in } \mathbb{R}^d \times (0,T), \qquad u(\cdot,0) = a. \tag{4}$$

The operator L is said to be *hyperbolic* if the problem (4) is well-posed in the space[1] $L^2(\mathbb{R}^d)^n$. This means that for every initial data $a \in L^2$, there exists a unique solution $u \in C(0,T;L^2)$, and the linear map $S_t : a \mapsto u(t)$ is L^2-bounded:

$$\|u(t)\|_{L^2} \leq c(t)\|a\|_{L^2}.$$

Then $(S_t)_{t\geq 0}$ forms a continuous semi-group over L^2. Of course, the solution is understood in the distributional sense. Because the Fourier transform is an isometry over L^2, this is equivalent to saying that the Cauchy problem for the operator

$$\mathcal{L} := \mathcal{F}L\mathcal{F}^{-1} = \partial_t + i\sum_\alpha \xi_\alpha A^\alpha$$

[1]Gårding's original definition involves well-posedness in C^∞, a weaker notion than the one used here. In particular it is not tailored to apply Duhamel's principle. Our stronger version of hyperbolicity used to be called *strong* hyperbolicity. It is more practical, at least when we have quasi-linear Cauchy problems in mind.

is well-posed over L^2. Here ξ is the independent variable in Fourier space and the new unknown is a function $v(\xi, t)$. If the initial data is $b(\xi)$, the solution is given explicitly by the formula

$$v(\xi, t) = \exp(-itA(\xi))b(\xi),$$

where

$$A(\xi) := \sum_\alpha \xi_\alpha A^\alpha$$

is the symbol of the spatial derivatives. Because the map $b \mapsto v(t)$ is a pointwise multiplication (by a matrix depending upon the variable), its L^2-norm is

$$\sup_{\xi \in \mathbb{R}^d} \| \exp(-itA(\xi))\|.$$

Thanks to the linearity $-tA(\xi) = A(-t\xi)$, this expression does not depend upon $t \neq 0$. We therefore have

Proposition 1. *The operator L is hyperbolic if and only if the family of matrices*

$$\{\exp(iA(\xi)) \mid \xi \in \mathbb{R}^d\}$$

is bounded in $\mathbf{M}_n(\mathbb{C})$.

In particular, this notion does not depend upon the time interval, or upon the time arrow: if the Cauchy problem is well-posed forward, it is well-posed backward. This property will be lost either in quasi-linear problems or in IBVPs. We notice that because $\mathbf{M}_n(\mathbb{C})$ is a finite-dimensional vector space, there is no need to specify the norm when speaking of boundedness.

If (λ, r) is an eigenpair of $A(\xi)$, that is $r \neq 0$ and $A(\xi)r = \lambda r$, we have $\exp(iA(\xi))r = e^{i\lambda}r$. Hence the supremum above is at least $|e^{i\lambda}|$, and even $\sup_{s\in\mathbb{R}} |e^{is\lambda}|$ if we consider $A(s\xi)$ instead. Hence the necessary condition

Proposition 2. *If L is hyperbolic, the eigenvalues of the symbol $A(\xi)$ are real, for every $\xi \in \mathbb{R}^d$.*

They actually are semi-simple.

The second part of the propotion makes use of the fact that if $Ar = \lambda r$ and $Av = \lambda v + r$ (λ being non semi-simple), then $\exp(isA)v = e^{is\lambda}(v + isr)$.

The condition of Proposition 2 is however not sufficient to guarantee hyperbolicity. A necessary and sufficient condition is

Theorem 1 (Kreiss). *The operator L is hyperbolic if and only if the symbol $A(\xi)$ is diagonalizable with real eigenvalues, uniformly in ξ:*

$$A(\xi) = P(\xi)\Lambda(\xi)P(\xi)^{-1}, \qquad \forall \xi \neq 0,$$

where $\Lambda(\xi)$ is real diagonal and $P(\xi)$ is well-conditioned:

$$\sup_{\xi\neq 0} \|P(\xi)\| \cdot \|P(\xi)^{-1}\| < +\infty.$$

One part of the proof is easy because if $\Lambda(\xi)$ is real diagonal, then

$$\|\exp(iA(\xi))\| \leq \|P(\xi)\| \cdot \|\exp(i\Lambda(\xi))\| \cdot \|P(\xi)^{-1}\| = \|P(\xi)\| \cdot \|P(\xi)^{-1}\|.$$

The converse is more involved, even if we already know the diagonalizability within the reals (Proposition 2). The necessity of the well-conditioning is really a delicate issue.

Kreiss' theorem is actually more general and deals with arbitrary evolution operators with constant coefficients. It is not always practical, because it necessitates the calculation of the eigenvalues and an eigenbasis of the symbol $A(\xi)$. For a rather general operator, this cannot be done, or it could simply be too complicated to carry out. However, most of the interesting hyperbolic operators fall in either of two classes: strictly hyperbolic (in a generalized sense) or Friedrichs symmetric.

Strict hyperbolicity: The operator L is *strictly hyperbolic* if $A(\xi)$ is diagonalizable with real and *simple* eigenvalues. Then the eigen-elements can be chosen continuously on every hemisphere of $\mathbf{S}^{d-1}$. Whence the well-conditioning.

Constant rank hyperbolicity: Same as above, but the eigenvalues have multiplicities that do not depend upon $\xi \neq 0$. Not all eigenvalues have the same multiplicity, but the list of multiplicities (for instance $(1, d, 1)$ in linearized gas dynamics) don't depend on ξ. Strict hyperbolicity is the special case where this list is $(1, \dots, 1)$.

Symmetric hyperbolicity: Here, one assumes that there exists a positive definite symmetric matrix S^0 such that every $S^\alpha := S^0 A^\alpha$ is symmetric. Then $P(\xi) = (S^0)^{-1/2} U(\xi)$ where $U(\xi) \in \mathbf{SO}_n$. Hence the well-conditioning, because $\|P(\xi)\| \cdot \|P(\xi)^{-1}\|$ does not depend on ξ at all. Symmetric hyperbolicity is associated with a supplementary conservation law

$$\partial_t (S^0 u, u) + \sum_\alpha \partial_\alpha (S^\alpha u, u) = 0,$$

which yields the a priori estimate

$$\int_{\mathbb{R}^d} (S^0 u(x,t), u(x,t))\, dx = \int_{\mathbb{R}^d} (S^0 a(x), a(x))\, dx, \qquad \forall t \in \mathbb{R}.$$

It comes naturally in the linearization of a system endowed with a *convex entropy* (entropies are studied in Sect. 3).

All these classes are subclasses of the set of hyperbolic operators, as suggested by the terminology.

1.1 Wave Velocities

Eigenvalues of $A(\xi)$ are associated with planar waves, which are special solutions of $Lu = 0$, of the form

$$u(x,t) = \phi(x \cdot \xi - \lambda t)r.$$

Hereabove, r is an eigenvector associated with λ. Such a solution moves in the direction ξ, at velocity $\lambda/|\xi|$. Because the eigenvalues are homogeneous of degree one in ξ, we may say that $\lambda(\xi)$ is a wave velocity in direction ξ whenever $\xi \in \mathbf{S}^{d-1}$. Traditionally, one lists the eigenvalues of $A(\xi)$ in increasing order:

$$\lambda_1(\xi) \leq \cdots \leq \lambda_n(\xi).$$

If $\xi \mapsto \lambda(\xi)$ is a smoothly varying eigenvalue,[2] one may construct L^2-solutions of $Lu = 0$ by modulation in ξ:

$$u(x,t) = \int_{\mathbb{R}^d \setminus \{0\}} \phi(x \cdot \xi - \lambda t; \xi) r(\xi)\, d\xi,$$

where $\xi \mapsto \phi(\cdot;\xi)$ is smooth with compact support. This construction provides waves moving at *group velocity* $\mathrm{grad}_\xi \lambda$. The notion of wave velocity is altogether one of the most natural, although impossible to define in closed form.

1.2 Inequalities Involving the Eigenvalues

The symbol of the full operator L is the matrix-valued $\tau I_n + A(\xi)$. When L is hyperbolic, the determinant $P_L(\tau,\xi) := \det(\tau I_n + A(\xi))$ is said to be a *hyperbolic polynomial*. More generally, a homogeneous polynomial P is *hyperbolic* (Gårding) in a direction $V \neq 0$ if on the one hand $P(V) \neq 0$, and on the other hand, the roots of the univariate polynomial $s \mapsto P(\xi + sV)$ are real, for every $\xi \in \mathbb{R}^d$. In our case, we may take $V = (1,0)$. The *characteristic cone* of P is the set $\{\xi \in \mathbb{R}^d \mid P(\xi) = 0\}$.

We shall not develop the theory of hyperbolic polynomials here, because it is reminiscent of the Gårding's weaker notion of hyperbolic operators. However, the following results have a counterpart in this generality.

Proposition 3 (Gårding). *The eigenvalue λ_n is a convex function of ξ.*

[2] This is always true in the constant rank hyperbolic case.

Likewise, $\lambda_1(\xi) = -\lambda_n(-\xi)$ is a concave function. Therefore, the connected component of $(1,0)$ in the complement of the characteristic cone is an open convex cone

$$\Gamma := \{(\tau,\xi) \mid \tau > -\lambda_1(\xi)\}.$$

It is called the *cone of hyperbolicity* for the following reason. If we take a non-zero vector $V \in \mathbb{R}^{d+1}$, we may perform a change of variables $(x,t) \mapsto (x',t')$, such that the hyperplane $t' = 0$ has equation $V \cdot (x,t) = 0$. The choice of the variables x' is irrelevant, provided that $(x,t) \mapsto (x',t')$ is a change of variables. After the change of variables, the operator L becomes $L' = \partial_{t'} + \sum_\alpha B^\alpha \partial_{x'_\alpha}$. It is natural to ask whether L' is still hyperbolic. This amounts to asking whether the Cauchy problem for L, with data on the hyperplane $V \cdot (x,t) = 0$, is well-posed in L^2. It happens that if $V \in \Gamma$, then L' is hyperbolic (Gårding). This phenomenon is reminiscent to the theory of special relativity: unless human constraints impose you a reference frame (often called the *laboratory frame*), there is no prefered time variable. A time variable has only to be such that its level sets are *space-like*, meaning that their normals belong to Γ.

The convexity of λ_n is not the end of the story. It turns out that the eigenvalues satisfy a lot of other inequalities, for instance

Weyl inequalities: $\lambda_k(\xi+\eta) \le \lambda_i(\xi) + \lambda_j(\eta)$, if $k+n = i+j$ and $\xi,\eta \in \mathbb{R}^d$. Of course this is true if $k+n \le i+j$ as well.

Lidskiĭ–Wielandt inequalities: If $1 \le r \le n$, $i_1 < \cdots < i_r \le n-r$ and $k_j = i_j + j$, then

$$\lambda_{k_1}(\xi+\eta) + \cdots + \lambda_{k_r}(\xi+\eta) \ge \lambda_{i_1}(\xi) + \cdots + \lambda_{i_r}(\xi) + \lambda_1(\eta) + \cdots + \lambda_r(\eta).$$

For instance, $\xi \mapsto \lambda_1(\xi) + \cdots + \lambda_r(\xi)$ is a concave function (Ky Fan).

More generally, the eigenvalues satisfy all the inequalities that the eigenvalues of Hermitian matrices satisfy (A. Horn's conjecture, eventually proved by Knutson and Tao [18]). This is due to the proof by Helton and Vinnikov [14] of Lax's conjecture [21] that every homogeneous hyperbolic polynomial in three variables is a P_L for some symmetric hyperbolic operator L. See [33] for a review of this topic.

Finally, let us mention an inequality of a different type,

Proposition 4 (Gårding). *Let the homogeneous polynomial P of degree n be hyperbolic in the direction $(1,0)$. Let Γ be its cone of hyperbolicity (the connected component of $(1,0)$ in $\{P \neq 0\}$).*

Then the map $\xi \mapsto P(\xi)^{1/n}$, which is positive and homogeneous of degree one, is concave over Γ.

A typical example is the concavity of $H \mapsto \det^{1/n} H$ over the cone of positive definite Hermitian matrices ($H \mapsto \det H$ is hyperbolic in the direction of I_n).

This well-know fact is reminiscent of the Brunn–Minkowski theorem that the function $A \mapsto \mathcal{L}(A)^{1/n}$ ($\mathcal{L}$ the Lebesgue measure on $\mathbb{R}^n$) is concave, in the sense that

$$(\mathcal{L}(A+B))^{1/n} \geq (\mathcal{L}(A))^{1/n} + (\mathcal{L}(B))^{1/n}$$

for every measurable subsets A and B.

Gårding also proved the following related result.

Proposition 5. *Let Φ be the symmetric n-linear form such that $\Phi(\xi,\dots,\xi) \equiv P(\xi)$. For every $\xi^1,\dots,\xi^n \in \Gamma$, one has*

$$P(\xi^1)^{1/n} \cdots P(\xi^n)^{1/n} \leq \Phi(\xi^1,\dots,\xi^n).$$

The case $n = 2$ is a *reverse Cauchy–Schwarz* inequality: if Q is a hyperbolic quadratic form, then for every $\xi, \eta \in \Gamma_Q$, one has

$$\sqrt{Q(\xi)Q(\eta)} \leq \Phi(\xi,\eta).$$

There is a wide literature on hyperbolic polynomials, which has connections with geometric inequalities (as above), combinatorics (van der Warden conjecture for the permanent of bistochastic matrices, now a theorem), optimization (quadratic programing) and many other topics.

1.3 L^p-Theory of the Cauchy Problem

When $p \neq 2$, the situation becomes completely different. Although the Cauchy problem may be well-posed in $L^p(\mathbb{R}^d)^n$ in very special situations, namely

- Scalar equations ($n = 1$), because it is pure transport ($(S_t)_{t\geq 0}$ is a translation semi-group),
- One-dimensional systems ($d = 1$), because they decouple into independent transport equations,

it is *not* well-posed for general systems when $p \neq 2$. The precise statement reads as follows.

Theorem 2 (Brenner). *Suppose that L is a hyperbolic operator. Then the following properties are equivalent to each other.*

1. *The Cauchy problem for $Lu = 0$ is well-posed for some $p \neq 2$.*
2. *The Cauchy problem for $Lu = 0$ is well-posed for every $1 \leq p \leq \infty$.*
3. *The matrices A^α pairwise commute, that is $A^\alpha A^\beta = A^\beta A^\alpha$ for every $\alpha \neq \beta$.*

Now, we understand why the particular cases above are well-posed: – if $n = 1$, the matrices are scalar, thus commute, – if $d = 1$, there is only one matrix, which commutes with itself, of course. But problem (4) for an operator as simple as

$$\partial_t + \begin{pmatrix} 1 & 0 \\ 0 & -1 \end{pmatrix} \partial_1 + \begin{pmatrix} 0 & 1 \\ 1 & 0 \end{pmatrix} \partial_2 \tag{5}$$

is ill-posed for $p \neq 2$.

To resolve this obstruction, we might be tempted to relax the notion of well-posedness, by asking for inequalities $\|u\| \leq c\|a\|'$ in different norms. This is the strategy used in *Strichartz estimates*. We point out that here the solution is evaluated globally in space and time, instead of simply evaluating the trace of u at a given time t. Let us provide an example with the wave equation

$$\partial_t^2 \phi = \Delta_x \phi.$$

When $d = 2$, this second-order equation can be recast as a first-order system with operator given by (5). More generally, it is related to Dirac operators, for which $(A^\alpha)^2 = I_n$ and $A^\alpha A^\beta + A^\beta A^\alpha = 0_n$ instead. If the exponents p and q satisfy the constraints

$$\frac{1}{p} + \frac{d}{q} = \frac{d}{2} - 1, \quad \frac{2}{p} + \frac{d-1}{p} \leq \frac{d-1}{2}, \quad 2 \leq p, q \leq \infty,$$

and if $d \geq 2$, then there exists a finite constant $c(p, q, d)$ such that every solution with $\nabla_{x,t}\phi|_{t=0} \in L^2$ satisfies

$$\|\phi\|_{L_t^p(L_x^q)} \leq c\|\nabla_{x,t}\phi|_{t=0}\|_{L^2}. \tag{6}$$

Strichartz estimates are inequalities for distributions whose Fourier transforms are supported by hypersurfaces. This theory is therefore ubiquitous in linear, constant coefficient PDEs: if $P(\partial)u = 0$, then $\hat{u}$ is supported by the characteristic set of P. However, it is not always efficient; for instance the PDE $\partial_t u = 0$ does not yield any interesting inequalities. The reason is the crucial assumption in Strichartz theory that $\{P = 0\}$ has a non-vanishing curvature (except along rays when P is homogeneous). In particular, in order that a hyperbolic system $Lu = 0$ obey a Strichartz estimate, it is necessary that the component $\tau > -\lambda_1(\xi)$ not only be convex as stated in Proposition 3, but be strictly convex in the direction not passing through the origin. This is precisely saying that the matrices A^α do not commute!

We thus envision a deep relation between Brenner's theorem and Strichartz estimates. It turns out that one can prove that if a hyperbolic system $Lu = 0$ satisfies a Strichartz estimate, then the Cauchy problem is ill-posed in L^p for every $p \neq 2$.

The importance of Brenner's result appears when we ask ourselves which functional space could be adapted to the analysis of quasi-linear systems of conservation laws in several space dimensions. When $d = 1$, the only functional framework that has been succesful so far is that of BV-functions (Glimm's theory, uniqueness and well-posedness by Bressan and coll.) But the BV-space is closely related to L^1, and the fact that the linearized systems yield ill-posed Cauchy problems in L^1 lead Rauch to claim that BV is likely a bad space in several space dimensions. So far, nobody has suggested a promising functional space for systems in dimension $d \geq 2$. As long as this holy grail remains hidden, the theory of the multi-dimensional Cauchy problem for quasi-linear systems will necessarily be restricted to special cases: – decoupling systems, – piecewise smooth solutions, – the Riemann problem, and so on.

2 The Initial-Boundary Value Problem

Continuing our analysis of 'simple' problems, we keep a linear operator L with constant coefficients and we focus on the problem in a half-space domain:

$$Lu = 0 \quad \text{in } x_d, t > 0, \quad u(\cdot, 0) = a \quad \text{in } x_d > 0, \quad Bu = g \quad \text{on } x_d = 0, t > 0. \tag{7}$$

We denote by $H = \mathbb{R}^{d-1} \times (0, +\infty)$ the physical domain. The boundary condition involves a matrix $B \in \mathbf{M}_{p \times n}(\mathbb{R})$ with $p \leq n$. Because $Bu = g$ can be rewritten as $B'u = g'$, with $B' = PB$ and $g' = Pg$, whenever $P \in \mathbf{GL}_p(\mathbb{R})$, we see that only the right coset $\mathbf{GL}_n \cdot B$ matters; in other words, only $\ker B$ matters.

2.1 *The Number of Boundary Conditions*

Let us observe that p is often strictly smaller than n. The number of boundary conditions must be, roughly speaking, equal to the number of waves that are *incoming* at the boundary. To see this, let us begin with the scalar case

$$L = \partial_t + \vec{v} \cdot \nabla_x.$$

Any solution has the form $u(x,t) = \phi(x - t\vec{v})$. In order to determine ϕ from the data, there are two cases:

- Either $v_d \leq 0$, then $\phi \equiv a$ determines u completely, because for every $x \in H$ and $t > 0$, we have $x - t\vec{v} \in H$.
- Or $v_d > 0$. Then if $0 < x_d < tv_d$, the determination of $u(x,t)$ requires boundary data at the point $(x - \frac{x_d}{v_d}\vec{v}, \frac{x_d}{v_d})$. Here, the well-posedness requires a Dirichlet boundary condition $u = g$ on $\partial H \times (0, T)$.

If instead $d = 1$ and $n \geq 1$, the domain is $H = (0, +\infty)$ and $L = \partial_t + A\partial_x$. Because of hyperbolicity, we may assume that A is diagonal, with eigenvalues $a_1 \leq \cdots \leq a_n$; say that $a_q \leq 0 < a_{q+1}$. We therefore have a list of transport equations $(\partial_t + a_j \partial_x)u_j = 0$. The analysis above tells us that the boundary values $u_j(0,t)$

must not be imposed if $j \leq q$, but those with $j \geq q+1$ must be prescribed. Hence the number $p = n - q$ of boundary data. But now that we have several unknowns, we may introduce a coupling by means of the boundary condition: we may prescribe the entering modes ($j > q$) in terms of the outgoing ones ($j \leq q$). Denoting by u_- (resp. u_+) the list of outgoing (resp. incoming) modes, an appropriate boundary condition has the form

$$u_+(0,t) = Mu_-(0,t) + h(t),$$

where M is some $p \times q$-matrix. This is equivalent to $Bu = g$, where

$$B = \begin{pmatrix} M \\ I_p \end{pmatrix}.$$

Thus the IBVP (7) is well-posed if and only if

$$\mathbb{R}^n = \ker B \oplus U(A),$$

where $U(A)$ is the *unstable* invariant subspace of A.

We now consider the general situation where $d, n \geq 1$. Because the multi-dimensional IBVP contains the particular case of data and solutions depending only upon (x_d, t), we see that p must be equal to the number of positive eigenvalues of A^d, and a necessary condition for well-posedness is

$$\mathbb{R}^n = \ker B \oplus U(A^d). \tag{8}$$

We shall see however that (8) does not imply well-posedness when $d \geq 2$.

2.2 The Symmetric Dissipative Case

Let us assume for the moment that L is symmetric hyperbolic: the matrices A^α are symmetric. Then we have the identity

$$\partial_t |u|^2 + \sum_\alpha \partial_\alpha (A^\alpha u, u) = 0,$$

from which we derive formally

$$\frac{d}{dt} \int_H |u|^2 dx = \int_{\partial H} (A^d u, u)\, dy.$$

Hereabove $y = (x_1, \dots, x_{d-1})$. If it happens that the restriction of A^d to $\ker B$ is negative semi-definite, then the homogeneous (by this, we mean $g \equiv 0$, thus a boundary condition $Bu = 0$) IBVP satisfies

$$\int_{\partial H} (A^d u, u)\, dy \leq 0$$

and we have an estimate $\|u(t)\|_{L^2} \leq \|a\|_{L^2}$, which implies well-posedness. In the present dissipative case, the homogeneous IBVP generates a *contraction* semi-group in L^2.

We point out that the non-homogeneous IBVP, unlike the non-homogeneous Cauchy problem, cannot be treated by semi-group tools. As a matter of fact, the trick above fails by a little amount in the non-homogeneous case. But if we assume that the restriction of A^d to $\ker B$ is *negative definite*, then there exist constants $\epsilon > 0$ and $C < \infty$ such that $(A^d v, v) \leq C|Bv|^2 - \epsilon|v|^2$ for all $v \in \mathbb{R}^n$, and therefore we have an estimate

$$\frac{d}{dt}\int_H |u|^2 dx + \epsilon \int_{\partial H} |u|^2 dy \leq C \int_{\partial H} |g|^2 dy.$$

We then say that the boundary condition is *strongly dissipative*. We point out that because $\ker B$ is of dimension $n - p$, it is maximal for this dissipative property.

In order to achieve explicit bounds on the solution, we may integrate the inequality above from 0 to T, but in view of the relevance of the Laplace transform in the general theory, it is even better to pre-multiply by the exponential weight $\exp(-2\gamma t)$, where $\gamma > 0$ is a parameter. We obtain easily

$$\begin{aligned} e^{-2\gamma T}\|u(T)\|_{L^2}^2 + \gamma \int_{H\times(0,T)} e^{-2\gamma t}|u|^2 dx\, dt \\ +\epsilon \int_{\partial H\times(0,T)} e^{-2\gamma t}|u|^2 dy\, dt \leq \|a\|_{L^2}^2 + C \int_{\partial H\times(0,T)} e^{-2\gamma t}|g|^2 dy\, dt. \end{aligned} \tag{9}$$

What is appealing in (9) is that the unknown is estimated in the same spaces as those where the data are given:

- The initial data a and the final (as well as intermediate) state $u(T)$ in $L^2(H)$.
- The boundary data g and the trace of u in $L^2(\partial H \times (0,T))$.
- The interior data f (not present in our calculation for the sake of simplicity) and the solution in $L^2(H \times (0,T))$.

We point out that in a strongly dissipative IBVP, the matrix A^d has a negative subspace $\ker B$ of dimension $n - p$ and a positive subspace $U(A^d)$ of dimension p. Therefore there is no room for a null eigenvalue: A^d is non-singular. It turns out that the general IBVP is significantly more involved when the matrix A^d is singular. More generally, if the physical domain Ω has a smooth boundary, then it is important to know whether $A(\nu)$ is singular or not, where ν is the normal to the boundary. One says that the boundary is *characteristic* if $A(\nu)$ is singular at some boundary point, and *non-characteristic* otherwise. In these notes, we shall restrict our attention to the non-characteristic case. An (incomplete) analysis of characteristic IBVPs can be found in [1].

2.3 Maximal Estimates

There are two reasons why the above theory is not completely satisfactory. On the one hand, not all linear hyperbolic operators can be symmetrized. And on the other hand, even if L is symmetric, strong dissipativity is only a sufficient condition for well-posedness. Thus the theory of dissipative symmetric IBVPs does not cover all the interesting problems. For instance, the analysis of multi-dimensional shock waves by Majda requires a more complete treatment of the linear IBVP. This linear theory was developed by Kreiss [19] in the case of first-order systems, and by Sakamoto [29] for scalar higher-order operators.

As in the Cauchy problem, we will not be content with a C^∞-well-posedness, in which a priori estimates suffer a loss of derivatives. It is hard to exploit such estimates when dealing with quasi-linear systems, even though this has been done succesfully in some instances by Coulombel and Secchi. Also, the present notes have a finite length and we may only focus on the nicest situation.

The best framework is that of *maximal estimates*. They mimic those already obtained in the symmetric, strongly dissipative case. We say that our IBVP satisfies a maximal estimate whenever there exists a finite constant C such that, for every smooth u, the following inequality holds true for every $\gamma, T > 0$:

$$e^{-2\gamma T}\|u(T)\|^2_{L^2(H)} + \gamma \int_{H\times(0,T)} e^{-2\gamma t}|u|^2 dx\,dt + \int_{\partial H\times(0,T)} e^{-2\gamma t}|u|^2 dy\,dt \le C\left(\|u(0)\|^2_{L^2(H)} + \frac{1}{\gamma}\int_{H\times(0,T)} e^{-2\gamma t}|Lu|^2 dx\,dt + \int_{\partial H\times(0,T)} e^{-2\gamma t}|Bu|^2 dy\,dt\right). \qquad (10)$$

Traces. Because we presented the maximal estimates in the context of smooth fields, we did not worry about the existence of the trace of u at the boundary $x_d = 0$. However, the theory will culminate with existence results in spaces of limited regularity (see Theorems 4 and 5). When the system is linear, we are interested in fields that are only square-integrable in $H \times (0,T)$. In such a case, we have to prove that the trace is well-defined. To this end, we use a classical lemma, which says that if $\vec{q}$ and div$\vec{q}$ are square integrable in Ω, then the *normal* trace of $\vec{q}$ at the boundary exists in $H^{-1/2}(\partial\Omega)$. With both u and Lu square integrable, this tells us that $A^d u$ has a well-defined trace at $x_d = 0$. Recalling that A^d is non-singular, we obtain that the restriction of u to $x_d = 0$ makes sense, at least in $H^{-1/2}(\partial H \times (0,T))$. Therefore the maximal estimate, plus a density argument, tells us that this restriction is actually in $L^2(\partial H \times (0,T))$.

We point out that in the characteristic case, u may not have a well-defined trace. Only $A^d u$ is well-defined. A maximal estimate in the spirit of (10) will involve instead a boundary term

$$\int_{\partial H \times (0,T)} e^{-2\gamma t} |A^d u|^2 dy\, dt$$

in the left-hand side. Likewise, the boundary data Bu has to be defined in terms of $A^d u$ only; in other words, we must have $\ker A^d \subset \ker B$.

2.4 Necessary Condition: The Kreiss–Lopatinskiĭ Condition

When looking for a condition that is necessary for a maximal estimate, we begin by restricting to fields u that satisfy $Lu \equiv 0$ in $H \times \mathbb{R}$. Then we discard the final and interior norms of u. Thus it remains to treat the inequality

$$\int_{\partial H \times (0,T)} e^{-2\gamma t} |u|^2 dy\, dt \le C \left(\|u(0)\|_{L^2}^2 + \int_{\partial H \times (0,T)} e^{-2\gamma t} |Bu|^2 dy\, dt \right),$$

whenever $Lu = 0$. On the other hand, the translational invariance in the spatial directions tangential to the boundary suggests to apply the Fourier transform $\mathcal{F} = \mathcal{F}_y$ with respect to the tangential variable. Because $\mathcal{F}$ is an isometry of L^2, the above estimate amounts to

$$\int_{\partial H \times (0,T)} e^{-2\gamma t} |v|^2 d\eta\, dt \le \left(C \|v(0)\|_{L^2}^2 + \int_{\partial H \times (0,T)} e^{-2\gamma t} |Bv|^2 d\eta\, dt \right), \qquad (11)$$

whenever $\partial_t v + iA(\eta)v + A^d \partial_d v = 0$. Hereabove, η denotes the frequency associated with the tangential variables y.

The inequality (11) decouples into

$$\int_0^T e^{-2\gamma t} |v(\eta,0,t)|^2 dt \le C \left(\int_0^\infty |v(\eta,x_d,0)|^2 dx_d + \int_0^T e^{-2\gamma t} |Bv(\eta,0,t)|^2 dt \right), \quad \forall \eta \in \mathbb{R}^{d-1}. \qquad (12)$$

We test this criterion on 'normal modes'. These are fields of the form $v = e^{\tau t} w(x_d)$, with $\Re\tau > 0$ and $w \in L^2(0,+\infty)$. Such growing modes are those that could cause a Hadamard instability. Their treatment explains why the Laplace transform in the time variable is so important in the analysis of the linear evolution equation.

The differential constraint tells us that w is a solution of the ODE

$$A^d w' + (\tau I_n + iA(\eta))w = 0. \tag{13}$$

This is where the assumption that the boundary is non-characteristic becomes crucial: (13) can be recast as

$$w' = \mathcal{A}(\tau,\eta)w, \qquad \mathcal{A}(\tau,\eta) := -(A^d)^{-1}(\tau I_n + iA(\eta)).$$

The fact that $w \in L^2(0,+\infty)$ when w solves a linear ODE $w' = \mathcal{A}w$ is equivalent to w having an exponential decay. It amounts to saying that $w(0)$ belongs to the stable invariant subspace of the matrix $\mathcal{A}$. The following statement tells us that the dimension of this subspace does not depend upon the parameters (τ,η).

Lemma 1 (Hersch [15]). *Let us assume that L is hyperbolic and that A^d is non-singular (non-characteristic boundary). Then for every (τ,η) with $\Re\tau > 0$ and $\eta \in \mathbb{R}^{d-1}$, the eigenvalues of the matrix $\mathcal{A}(\tau,\eta)$ have a non-vanishing real part (we say that these matrices are hyperbolic in the terminology of dynamical systems theory). Its stable invariant subspace $S(\tau,\eta)$ has dimension p, that of the unstable subspace of A^d. Finally, the map $(\tau,\eta) \mapsto S(\tau,\eta)$ is analytic in η and holomorphic in τ.*

For such modes, (12) becomes

$$|w(0)|^2 \le C\left(\frac{1}{I(\gamma)}\|w\|_{L^2}^2 + |Bw(0)|^2\right), \qquad I(\gamma) := \int_0^T e^{2(\Re\tau-\gamma)t}\,dt.$$

By letting γ tend to $\Re\tau$ by above, we obtain the necessary condition that

$$|W| \le C\cdot|BW|, \qquad \forall W \in S(\tau,\eta). \tag{14}$$

This tells us two important things:

- For every (τ,η) with $\Re\tau > 0$ and $\eta \in \mathbb{R}^{d-1}$, the map $B : S(\tau,\eta) \to \mathbb{C}^p$ is an isomorphism. This is called the *Kreiss–Lopatinskiĭ* condition (KL).
- The norm of its reciprocal $B^{-1} : \mathbb{C}^p \to S(\tau,\eta)$ is bounded independently of (τ,η). This is the *uniform* KL condition. We write it as (UKL).

When (KL) fails, the IBVP experiences a Hadamard instability. To see this, choose a pair (τ_0,η_0) with $\Re\tau_0 > 0$, at which B is not one-to-one over $S(\tau_0,\eta_0)$. If η is close to η_0, then because of Rouché's theorem and the τ-holomorphy, there exists a τ close to τ_0 such that B is still not one-to-one over $S(\tau,\eta)$. Up to a small change of η_0, we may assume that this τ depends analytically on η. Likewise, we can choose analytically a non-zero vector $r(\eta) \in \ker B \cap S(\tau(\eta),\eta)$. Choosing now any non-trivial test function $\phi(\eta)$, we form a solution of $Lu = 0$ by the formula

$$u(x,t) := \int \phi(\eta)e^{\tau(\eta)t+i\eta\cdot y}\left(\exp(x_d\mathcal{A}(\tau(\eta),\eta))\right)r(\eta)\,d\eta.$$

Letting the support of ϕ shrink to τ_0, we obtain that the constant C in (10) must be larger than or equal to $\exp(2(\Re\tau_0 - \gamma)T)$. Replacing now the pair (τ_0, η_0) by $(N\tau_0, N\eta_0)$ (this does not change the stable subspace, thus (KL) still fails at such points), we see that C cannot be finite. Hence a maximal estimate does not hold. As a matter of fact, this construction shows also that even a non-maximal estimate, in which we allow a loss of finitely many derivatives, does not hold either.

When (KL) is satisfied but (UKL) fails, the situation is far more tricky. We do not have a maximal estimate, but there is an estimate with loss of finitely many derivatives. We might think that this is a borderline case, occuring at the interface between (UKL) and (nonKL); this was claimed by Kreiss in [19], but the situation is more subtle. It has been shown that within the class of IBVPs satisfying (KL) in a non-uniform way, there is a subclass which is structurally stable in the sense that every IBVP in a small neighbourhood (L and B being slightly perturbed) retains the same property. See [3] or Sect. 8.3 in [1].

2.5 *The Estimate Under (UKL)*

It remains to see whether (UKL) is sufficient to imply the maximal estimates. There are two difficulties here. On the one hand, we discarded several important terms to derive our necessary condition. Would these other terms add new conditions? On the other hand, an IBVP does not belong to the category of semi-group problems; therefore the boundary condition, the initial data and the inner data f play different but coupled roles. The strategy consisting in decoupling a full IBVP into three sub-problems in which only one of the three data is non-zero, is not really efficient.

The strategy adopted by Kreiss was to mimic the symmetric dissipative case by looking for a *dissipative symmetrizer* $\mathcal{K}$. But instead of having $\mathcal{K} \equiv \mathrm{id}$, we must search for a pseudo-differential symmetrizer, with symbol $K(\tau, \eta)$. The properties that are required are the following:

1. $(\tau, \eta) \mapsto K(\tau, \eta)$ is bounded and homogeneous of degree zero.
2. $\Sigma(\tau, \eta) := K(\tau, \eta)A^d$ is Hermitian.
3. There exists a $c_0 > 0$ such that, for every (τ, η), we have
$$w^*\Sigma(\tau, \eta)w \le -c_0|w|^2, \qquad \forall w \in \ker B.$$
4. There exists a $c_0 > 0$ such that, for every (τ, η), we have
$$\Re(v^*M(\tau, \eta)v) \ge c_0(\Re\tau)|v|^2, \qquad \forall v \in \mathbb{C}^n,$$
where we have used
$$M(\tau, \eta) := K(\tau, \eta)(\tau I_n + iA(\eta)).$$

In practice, $(\tau, \eta) \mapsto K$ is positively homogenenous of degree zero. Note the uniformity of the third and fourth conditions, in terms of (τ, η). In the symmetric, strongly dissipative case, we may choose $K \equiv I_n$; then the third property is dissipativeness while the others follow from the symmetry. Note also that because $\Re(\Sigma\mathcal{A}) = -\Re M$ is negative definite, Σ is non-singular, with inertia[3] $(n-p, p)$. Therefore B is maximal for the third property.

If the IBVP (7) admits a Kreiss symmetrizer, it is straightforward to establish a maximal estimate for the *pure BVP* (the time runs over $\mathbb{R}$ instead of $(0, T)$)

$$Lu = f \qquad \text{in } H \times (-\infty, +\infty), \qquad Bu = g \qquad \text{in } \partial H \times (-\infty, +\infty),$$

provided u has a compact support. The estimate is then

$$\gamma \int_{H\times\mathbb{R}} e^{-2\gamma t}|u|^2 dx\,dt + \int_{\partial H\times\mathbb{R}} e^{-2\gamma t}|u|^2 dy\,dt \leq C\left(\frac{1}{\gamma}\int_{H\times\mathbb{R}} e^{-2\gamma t}|Lu|^2 dx\,dt \right.$$
$$\left. + \int_{\partial H\times\mathbb{R}} e^{-2\gamma t}|Bu|^2 dy\,dt\right). \tag{15}$$

This estimate, thanks to its dependence upon the parameters, implies that the restriction of a solution to times $t < T$ depends only upon the restriction of the data to $(-\infty, T)$. In other words, the fact that $Lu \equiv 0$ in $H \times (-\infty, T)$ and $Bu \equiv 0$ on $\partial H \times (-\infty, T)$ implies that $u \equiv 0$ in $H \times (-\infty, T)$. Therefore an IBVP satisfying (UKL) yields a genuine evolution problem.

Weighted norms. The form of the estimate (15) suggests working in weighted L^2-spaces $e^{\gamma t}L^2$, which we denote by L^2_γ. The norm in this space is

$$\|h\|_\gamma := \|e^{-\gamma t}h\|_{L^2}.$$

This notation applies either to the domain $H \times \mathbb{R}$, or to $\partial H \times \mathbb{R}$. Estimate (15) thus reads as

$$\gamma\|u\|_\gamma^2 + \|\gamma_0 u\|_\gamma^2 \leq C\left(\frac{1}{\gamma}\|Lu\|_\gamma^2 + \|\gamma_0 Bu\|_\gamma^2\right),$$

where γ_0 denotes the restriction to the boundary $\partial H \times \mathbb{R}$ and is called a *trace operator*. This notation is slightly incorrect, because the ambient space is not contained in $H^1(H \times \mathbb{R})$, but we have seen above how the control of Lu helps us.

[3] A classical result for Lyapunov equations $\Sigma X + X^*\Sigma = S$ where $S \in \mathbf{H}_n$ is given and $\Sigma \in \mathbf{H}_n$ is the unknown.

2.6 The Boundary of the Frequency Domain

The construction of a Kreiss symmetrizer is a technical process. Because of homogeneity, it is enough to restrict to the open half-sphere $\mathbf{S}^+$:

$$\Re\tau > 0, \qquad |\tau|^2 + |\eta|^2 = 1.$$

The construction of K at a given point (τ, η) is not too difficult under (KL). By a covering argument, everything is fine so long as (τ, η) remains in a compact domain of $\mathbf{S}^+$. The difficulty comes when one approaches from the boundary of the frequency space, that is when $\Re\tau \to 0^+$. A general solution has not been provided so far, and it is suspected that a Kreiss symmetrizer does not always exist under (UKL) only. This is why Kreiss and Majda introduced the so-called *block structure condition*, which concerns only the operator L. Instead of describing it, let us say that this condition is satisfied whenever L has characteristics of constant multiplicities; Kreiss proved this fact in the strictly hyperbolic case, and Métivier [24] extended it to the case of multiple characteristics. We thus have

Theorem 3 (Kreiss, Métivier). *Assume that L is a hyperbolic operator whose characteristic fields have constant multiplicities. If the pair (L, B) satisfies (UKL), then there exists a Kreiss symmetrizer.*

Later on, Métivier and Zumbrun [25] extended this result by showing that if L is symmetric and if *at most two* of the eigenvalues of the symbol $A(\xi)$ cross at some points, then the block structure is still satisfied and therefore the symmetrizer exists. They applied this to the linearized MHD.

In the course of the proof, Kreiss and Métivier prove actually that if $(\rho_0, \eta_0) \in S^{d-1}$, then even if $\mathcal{A}(i\rho_0, \eta_0)$ is not a hyperbolic matrix (Hersch's lemma tells us nothing at such points), the stable subspace of $\mathcal{A}(\tau, \eta)$ has a limit when $\Re\tau > 0$, as $(\tau, \eta) \to (i\rho_0, \eta_0)$. We still denote this limit by $S(i\rho_0, \eta_0)$, even though it is no longer a stable subspace. Note however that it is included in the center-stable subspace of $\mathcal{A}(i\rho_0, \eta_0)$, by continuity.

The existence of this limit is particularly interesting because it yields a characterization of property (UKL). We are now in the situation where the map $(\tau, \eta) \mapsto S(\tau, \eta)$ is continuous over the *closed* half-sphere

$$\Re\tau \geq 0, \qquad |\tau|^2 + |\eta|^2 = 1, \tag{16}$$

a compact space. Therefore (UKL) is equivalent to the fact that (KL) be satisfied at every point, *including those at the boundary*.

Proposition 6. *Assume that L is a hyperbolic operator whose characteristic fields have constant multiplicities. Then the pair (L, B) satisfies (UKL) if and only if the maps $B : S(\tau, \eta) \to \mathbb{C}^n$ are one-to-one for every (τ, η) in the closed half-sphere defined by (16).*

We warn the reader that it may be cumbersome to verify (KL) pointwise at every such point; but there is no better way to proceed, unless the system is obviously symmetric dissipative. It is particularly delicate to work at boundary points, where a careless calculation may lead to a wrong description of $S(i\rho_0, \eta_0)$. People who are fond of complex analysis may prefer to find a basis bundle $(\tau, \eta) \mapsto (V^1, \dots, V^p)$ and form a *Lopatinskiĭ determinant*

$$\Delta(\tau, \eta) := \det(BV^1(\tau, \eta), \dots, BV^p(\tau, \eta)).$$

This object is not canonical (because there is no prefered choice of the bundle), but it may be chosen holomorphically in τ, analytically in η, and continuous up to the boundary $\Re\tau = 0$ under the assumption of constant multiplicities. It vanishes precisely at points where (KL) fails. Therefore the characterization given in Proposition 6 can be written as $\Delta(\tau, \eta) \neq 0$ for every pair in the closed half-sphere.

2.7 How the A Priori Estimate Implies Well-Posedness of the BVP

We now assume that L is hyperbolic and that the pair (L, B) admits a Kreiss symmetrizer. So far we have been able to derive an estimate (15) for smooth functions u. Let us show that it remains true for the class of fields u such that $e^{-\gamma t} u$, $e^{-\gamma t} Lu$ and $e^{-\gamma t} Bu$ are square integrable.

Thus let $u \in L^2_\gamma$ be such that $Lu \in L^2_\gamma$. We first observe that the j-th line of Lu is a divergence of some vector field $\vec{q}^j$. By assumption, $\vec{q}^j$ and its divergence are in L^2, and therefore the normal component q^j_d has a well-defined trace in $H^{-1/2}_\gamma(\partial H \times \mathbb{R})$. This amounts to saying that $A^d u$ has such a trace, but since A^d is non-singular, we find that u has a trace in $H^{-1/2}_\gamma$. Therefore it makes sense to assume that $Bu|_{\partial H \times (0,T)} \in L^2_\gamma$; this simply means that the trace, which is known to be $H^{-1/2}_\gamma$, actually belongs to the subspace L^2_γ.

We next convolve u with a test function

$$\phi_\epsilon(y, t) := \frac{1}{\epsilon}\phi\left(\frac{y}{\epsilon}, \frac{t}{\epsilon}\right)$$

(note that the convolution does not act on the last variable x_d). We find that the sequence $u^\epsilon := u *_{y,t} \phi_\epsilon$ is bounded in L^2_γ, as well as Lu^ϵ and the trace of Bu^ϵ. In addition u^ϵ is smooth in (y, t), in particular the derivatives $\nabla_{y,t} u^\epsilon$ are in L^2_γ. This, together with $Lu^\epsilon \in L^2_\gamma$ and the fact that A^d is non-singular, imply that $\partial_d u^\epsilon$ is in L^2_γ. Then it is not difficult to validate the estimate (15) for u^ϵ. Finally, we pass to the limit as $\epsilon \to 0$. The right-hand side being bounded, we find that the limit u satisfies (15) too. In particular, this forces the trace of Bu to be in L^2_γ.

We now turn towards existence, for which we use duality arguments. We therefore introduce the *adjoint* IBVP, characterized by a pair (L^*, C) where

$$L^* = -\partial_t - \sum_\alpha (A^\alpha)^T \partial_\alpha$$

and $C \in \mathbf{M}_{(n-p)\times n}(\mathbb{R})$ is such that $\ker C = (A^d \ker B)^\perp$. It has the property that an identity holds for smooth fields:

$$\int_{H\times\mathbb{R}} ((Lu, v) - (u, L^*v))dx\,dt + \int_{\partial H\times\mathbb{R}} ((Nu, Cv) + (Bu, Mv))dy\,dt = 0, \tag{17}$$

for some matrices $N \in \mathbf{M}_{(n-p)\times n}(\mathbb{R})$ and $M \in \mathbf{M}_{p\times n}(\mathbb{R})$. The latter are determined by the equation $A^d = C^T N + M^T B$.

Let us point out that in the duality method, we are interested in a priori estimates for the *backward* adjoint BVP. In particular, the relevant time frequencies will now have *negative* real part.

It is not too difficult to verify that if K is a Kreiss symmetrizer for the forward problem attached to (L, B), then

$$K'(\tau, \eta) := K(-\bar{\tau}, \eta)^{-1}$$

is a Kreiss symmetrizer attached to the backward adjoint problem.[4] This requires only linear algebra, the most involved point being that if a non-singular Hermitian matrix S is negative definite over a subspace E, and if E is maximal for this property, then S^{-1} is positive definite over $E^\perp$. We warn the reader that, because the adjoint BVP is backward, the dissipative inequality is that $\Sigma'(\tau, \eta)$ is *positive* definite over $\ker C$.

All this allows us to establish a similar estimate for the adjoint BVP, namely that if $\gamma > 0$ and $v \in L^2_{-\gamma}$ is such that $L^*v \in L^2_{-\gamma}$ and $Cv \in L^2_{-\gamma}$ on the boundary, then

$$\gamma \|v\|^2_{-\gamma,H\times(0,T)} + \|v\|^2_{-\gamma,\partial H\times(0,T)} \le C\left(\frac{1}{\gamma}\|L^*v\|^2_{-\gamma,H\times(0,T)} + \|Cv\|^2_{-\gamma,\partial H\times(0,T)}\right). \tag{18}$$

Thanks to (18), we know that the adjoint BVP has the uniqueness property. If f and g are L^2_γ, we may therefore define a linear form ℓ by

$$\ell(L^*v) := \int_{H\times\mathbb{R}} (v, f)\,dx\,dt + \int_{\partial H\times\mathbb{R}} (g, Mv)\,dy\,dt.$$

[4] In [1], we missed this easy argument.

The domain of ℓ consists of those $f' \in L^2_{-\gamma}$ for which there exists a $v \in L^2_{-\gamma}$ such that $L^* v = f'$ and $Cv \equiv 0$ on the boundary; this forms a subspace X in $L^2_{-\gamma}$. The adjoint estimate guarantees that ℓ is bounded over X for the $L^2_{-\gamma}$-norm. By Hahn–Banach, it admits a bounded extension to $L^2_{-\gamma}$. Because the dual of $L^2_{-\gamma}$ is L^2_γ we deduce that there exists a $u \in L^2_\gamma$ such that

$$\ell(f') = \int_{H\times\mathbb{R}} (f', u)\, dx\, dt, \qquad \forall f' \in X.$$

We thus obtain

$$\int_{H\times\mathbb{R}} (u, L^* v)\, dx\, dt = \int_{H\times\mathbb{R}} (v, f)\, dx\, dt + \int_{\partial H\times\mathbb{R}} (g, Mv)\, dy\, dt,$$

for every smooth v such that Cv vanishes on $\partial H \times \mathbb{R}$. One easily verifies that this is the variational formulation of our BVP.

2.8 From the BVP to the IBVP

So far, we have treated only the BVP: given $f \in L^2_\gamma(H\times\mathbb{R})$ and $g \in L^2_\gamma(\partial H\times\mathbb{R})$, find a solution $u \in L^2_\gamma(H\times\mathbb{R})$ such that $u|_{\partial H\times(0,T)} \in L^2_\gamma(\partial H\times\mathbb{R})$. We explained above that the existence and uniqueness follow from the existence of a Kreiss symmetrizer, the latter being ensured if L has characteristics of constant multiplicities and (L, B) satisfies (UKL). We also know that if $f, g \equiv 0$ for $t < T$, then $u \equiv 0$ as well for $t < T$. In other words, we know how to solve the IBVP whenever the initial data is $a \equiv 0$. It remains to treat the case of a general data $a \in L^2(H)$.

Toward this goal, we first establish an additional estimate, namely

$$e^{-2\gamma T}\int_H |u(T)|^2 dx \le C\left(\frac{1}{\gamma}\int_{H\times\mathbb{R}} e^{-2\gamma t}|Lu|^2 dx\, dt + \int_{\partial H\times\mathbb{R}} e^{-2\gamma t}|Bu|^2 dx\, dt\right). \tag{19}$$

In his article, Rauch [28] showed that (19) follows, in a non trivial way, directly from (15). Before that, he had described his PhD thesis a clever and much simpler proof in the symmetric (not necessarily dissipative) case. We present below the latter.

We may always assume that $Lu, Bu \equiv 0$ for $t > T$, because the data for $t > T$ do not influence the values of u at time T. Therefore the integrals may be restricted to $(\partial)H \times (-\infty, T)$. Then the energy identity

$$\partial_t(e^{-2\gamma t}|u|^2) + \sum_\alpha \partial_\alpha(e^{-2\gamma t} u^* A^\alpha u) = 2e^{-2\gamma t}((Lu, u) - \gamma|u|^2)$$

yields (19) after integrating over $H \times (-\infty, T)$, using Cauchy–Schwarz and finally using (15).

Thanks to (19), we establish that the solution of the BVP is in $C(\mathbb{R}; L^2(H))$. Of course, the same property holds true for the adjoint BVP. Then we may prove the existence of the general IBVP by the same duality argument as above, using the linear form m defined by

$$m(L^*v) := \int_H (a, v(0))\, dx + \int_{H\times\mathbb{R}} (v, f)\, dx\, dt + \int_{\partial H\times\mathbb{R}} (g, Mv)\, dy\, dt.$$

We find again that if $f, g \in L^2_\gamma$ and $a \in L^2$, then m can be extended as a bounded linear form over $L^2_{-\gamma}$, and therefore there exists a $u \in L^2_\gamma$ such that $m(f') \equiv \int_{H\times\mathbb{R}} (u, f')\, dx\, dt$. This tells us that u solves the variational formulation of the full IBVP. Hence the fundamental statement:

Theorem 4. *Let L, given by (3), be a hyperbolic operator. We assume that the boundary $x_d = 0$ is not characteristic, and that the pair (L, B) admits a Kreiss symmetrizer.*

Then the IBVP (7) is well-posed in L^2_γ: for every $f \in L^2_\gamma(H \times (0,T))$, $g \in L^2_\gamma(\partial H \times (0,T))$ and $a \in L^2(H)$, there exists a unique solution $u \in L^2_\gamma(H \times (0,T)) \cap C([0,T]; L^2(H))$, such that (7) holds true in the sense of distributions. In addition, u satisfies the maximal estimates (10), where the constant C neither depends upon the data, nor upon $\gamma > 0$ and $T > 0$.

From this and Theorem 3, we conclude:

Theorem 5. *Let L, given by (3), be a hyperbolic operator with characteristic fields of constant multiplicities. If the pair (L, B) satisfies (UKL), then the IBVP (7) is well-posed in the sense described in Theorem 4.*

2.9 An Example: The Wave Equation

Higher-order hyperbolic equations or systems can be treated in the same way as we described in the first-order case. It is enlighting to consider the wave equation

$$\partial_t^2 u = c^2 \Delta_x u, \tag{20}$$

where u is a scalar unknown. Because the equation is second-order, we look for a solution such that $\nabla_{x,t} u$ is square integrable. The initial data is

$$u(x,0) = a_0(x), \qquad \partial_t u(x,0) = a_1(x),$$

with $\nabla a_0 \in L^2(H)$ and $a_1 \in L^2(H)$.

The wave operator is strictly hyperbolic, with velocities $\pm c$ in every direction $\xi \in \mathbf{S}^{d-1}$. The boundary is non-characteristic (equivalently, when ξ is normal to the boundary, the velocities don't vanish) and there is one incoming characteristic.

Therefore we have to impose one boundary condition. We choose a condition of order one, because it would be of the form $BU = G$ when we rewrite the equation as a first-order system in $U = \nabla_{x,t} u$:

$$\frac{\partial u}{\partial \nu} = \kappa \frac{\partial u}{\partial t} + g, \tag{21}$$

where $g \in L^2(\partial H \times (0, T))$ is a data. Hereabove, ν is the unit outward normal and $\kappa \in \mathbb{R}$ is a parameter. Equivalently, we have

$$\frac{\partial u}{\partial x_d} + \kappa \frac{\partial u}{\partial t} + g = 0.$$

When $\kappa = 0$, this is the (non-homogeneous) Neumann condition, whereas the Dirichlet boundary condition is obtained formally in the limit $\kappa \to \infty$.

If $d = 1$, the IBVP is very simple. It is well-posed if and only if $\kappa \neq \frac{1}{c}$; the forbidden case corresponds as expected to the situation where $\ker B$ is not transversal to $U(A^d)$, in the terminology of Sect. 2.1. The situation is totally different when $d \geq 2$:

If $\kappa < 0$, then the IBVP is well-posed. Actually, this is a symmetric, strongly dissipative situation.
If $0 \leq \kappa < \frac{1}{c}$, the IBVP satisfies (KL) in a non-uniform way. Hence the IBVP is well-posed in C^∞, but the estimates are not the maximal ones.
If $\frac{1}{c} \leq \kappa$, the IBVP does not satisfy (KL). It is ill-posed in the Hadamard sense.

The reader may be surprised that the pure Neumann boundary condition

$$\frac{\partial u}{\partial \nu} = g$$

does not yield a strongly well-posed IBVP. This seems to be in contradiction to what we learn throughout our graduate studies. It turns out that the Neumann IBVP, with homogenenous boundary condition (that is $g \equiv 0$), has maximal estimates provided we do not ask for a trace estimate. If

$$\frac{\partial u}{\partial \nu} = 0,$$

then the solution satisfies

$$\gamma \int_{H\times(0,T)} e^{-2\gamma t} |\nabla_{x,t} u|^2 dx\,dt + e^{-2\gamma T} \int_H |\nabla_{x,t} u(x,T)|^2 dx$$
$$\leq C \left(\frac{1}{\gamma} \int_{H\times(0,T)} e^{-2\gamma t} |Lu|^2 dx\,dt + \int_H (|\nabla a_0|^2 + |a_1|^2)\, dx \right).$$

However, when g is arbitrary, the fact that $g \in L^2_\gamma$ does not ensure that there exists a solution $u \in L^2_\gamma$. We need a more regular g in order to derive this conclusion.

Calculations. This is quite a simple situation because the space of modes at frequency (τ, η) is a line. We have to test whether the boundary operator is one-to-one on this line, uniformly in (τ, η).

The modes solve the ODE

$$\tau^2 w = c^2 \left(\frac{d^2 w}{dx_d^2} - |\eta|^2 w \right),$$

with $w(+\infty) = 0$. The solutions of the ODE are linear combinations of the exponentials $e^{\pm\omega(\tau,\eta)x_d}$, where ω is the square root of $\frac{\tau^2}{c^2} + |\eta|^2$ whose real part is positive (note that $\tau^2 + |\eta|^2$ is not a negative real; this is the essence of Hersch's lemma). The decay at $+\infty$ tells us that

$$w(x_d) = w(0)e^{-\omega(\tau,\eta)x_d}.$$

This yields the relations

$$\frac{\partial u}{\partial x_d} = -\omega(\tau, \eta)w(0), \qquad \frac{\partial u}{\partial t} = \tau w(0).$$

The Lopatinskiĭ condition is thus satisfied at (τ, η) if and only if

$$\Delta(\tau, \eta) := \kappa\tau - \omega(\tau, \eta) \neq 0.$$

This Δ is our Lopatinskiĭ determinant.

Let us look at the zeroes of Δ within $\Re\tau > 0$. If $\Delta = 0$, then $\kappa\tau = \omega$. If $\kappa \leq 0$, this is impossible because $\Re\tau, \Re\omega > 0$. Thus let us assume that $\kappa > 0$. We must have $(\kappa\tau)^2 = \frac{\tau^2}{c^2} + |\eta|^2$, that is

$$(\kappa^2 - c^{-2})\tau^2 = |\eta|^2.$$

If $0 < \kappa < \frac{1}{c}$, this is again impossible, because τ^2 is not a negative real number. Now, if instead $\kappa \geq \frac{1}{c}$, then Δ vanishes at $\tau = 1$ and $|\eta| = \sqrt{\kappa^2 - c^{-2}}$. Therefore (KL) is satisfied if and only if $\kappa < \frac{1}{c}$.

We now turn to (UKL). As (KL) fails if $\kappa \geq \frac{1}{c}$, we may restrict to the case $\kappa < \frac{1}{c}$. Because L is strictly hyperbolic, we know from the general version of Kreiss' theorem that ω admits a continuous extension to boundary frequencies $(i\rho, \eta)$, where ρ is real (we could prove this directly of course). Then (KL) will be uniformly satisfied if in addition Δ does not vanish at boundary points (except at $(0, 0)$, which is irrelevant). So, let $(i\rho, \eta)$ be a zero of Δ. We have $i\kappa\rho = \omega$. Therefore ω must be pure imaginary, that is

$$|\eta| \leq \frac{|\rho|}{c}.$$

We then have

$$\omega = \epsilon i \sqrt{\frac{\rho^2}{c^2} - |\eta|^2}$$

where $\epsilon = \pm 1$ must be determined so that $\Re\omega$ increases when $\Re\tau$ does. With

$$\frac{\partial \omega}{\partial \tau} = \frac{\tau}{c\omega},$$

we find that ϵ is the sign of ρ, hence

$$\omega(i\rho, \eta) = i\rho \sqrt{\frac{1}{c^2} - \frac{|\eta|^2}{\rho^2}}.$$

Thus the equation $\Delta = 0$ at the boundary amounts to

$$\kappa = \sqrt{\frac{1}{c^2} - \frac{|\eta|^2}{\rho^2}},$$

which has a root η if and only if $\kappa \in [0, \frac{1}{c})$.

3 The Quasi-linear Cauchy Problem

We now turn to quasi-linear systems, which are of the form

$$\partial_t u + \sum_\alpha A^\alpha(u) \partial_\alpha u = 0. \tag{22}$$

Hereabove, $u \mapsto A^\alpha(u)$ is a field of $n \times n$ matrices with real entries. The latter are sufficiently smooth functions over a phase space $\mathcal{U}$. The symbol now also depends on the unknown:

$$A(u; \xi) := \sum_\alpha \xi_\alpha A^\alpha(u).$$

We say that (22) is hyperbolic if every linear system of the form

$$\partial_t u + \sum_\alpha A^\alpha(\bar{u}) \partial_\alpha u = 0$$

is hyperbolic. In addition, we have the notions of strict (resp. constant rank) hyperbolicity and symmetric hyperbolicity. We denote the eigenvalues of $A(u; \xi)$

by

$$\lambda_1(u;\xi) \leq \cdots \leq \lambda_n(u;\xi).$$

A particular, extremely important case is that of *systems of conservation laws*

$$\partial_t u + \sum_\alpha \partial_\alpha f^\alpha(u) = 0, \tag{23}$$

where the matrices A^α are the Jacobians of the f^α's. We say then that $u_1, \dots, u_n$ are the *conserved quantities*.

3.1 Entropies

Definition 1. For a system (23), an *entropy-flux pair* is a smooth map

$$u \mapsto (\eta(u), \vec{q}(u)) \in \mathbb{R} \times \mathbb{R}^d,$$

such that (23) implies formally the additional conservation law

$$\partial_t \eta(u) + \mathrm{div}\vec{q}(u) = 0. \tag{24}$$

In other words, an entropy-flux pair is a solution of the linear system of PDEs

$$\mathrm{d}q^\alpha = \mathrm{d}\eta \mathrm{d}f^\alpha, \qquad \alpha = 1, \dots, d, \quad u \in \mathcal{U}, \tag{25}$$

or in full detail

$$\frac{\partial q^\alpha}{\partial u_i} = \sum_{j=1}^n \frac{\partial \eta}{\partial u_j} \frac{\partial f_j^\alpha}{\partial u_i}, \qquad \forall i = 1, \dots, n, \ \forall \alpha = 1, \dots, d, \ \forall u \in \mathcal{U}.$$

We warn the reader that this system comprises dn equations in $d + 1$ unknowns. It is overdetermined whenever $dn \geq d + 2$. Therefore, we do not expect non-trivial entropy-flux pairs for general systems unless either $n = 1$ (scalar equations) or $(n, d) = (2, 1)$ (so-called 2×2 systems). Trivial entropies are affine: to an entropy $\eta(u) = \ell \cdot u + c$, there corresponds the flux $q^\alpha(u) = \ell \cdot f^\alpha(u)$, and then (24) is nothing but a linear combination of the rows of (23). In the scalar case, every function $u \mapsto \eta$ is an entropy. The 2×2 case is in between; not all functions of u are entropies, but the vector space of entropies is infinite-dimensional, parametrized by two arbitrary functions of one variable; see [31] or Chap. 9 of [32].

It is worth mentioning that most of the physically relevant systems do admit a non-trivial (= non-affine) entropy; this is because they have an underlying thermodynamical formalism. We shall give some examples soon. The fact that η is not affine can be translated in terms of properties of the Hessian $\mathrm{D}^2\eta(u)$. It will be non-degenerate, and often have some positivity properties. In the most favorable case, we shall have $\mathrm{D}^2\eta(u) > 0_n$. We say in this case that η is *strongly convex*.

If we eliminate $\vec{q}$ from (25), we obtain the property that the matrix $\mathrm{D}^2\eta\,\mathrm{d}f^\alpha$ is symmetric. More generally, $(\mathrm{D}^2\eta)A(u;\xi)$ is symmetric for every $\xi \in \mathbb{R}^d$. This implies that $\mathrm{D}^2\eta$ is diagonal in the eigenbasis of A:

Proposition 7. *Suppose that the eigenvalues of $A(u;\xi)$ are simple (strict hyperbolicity). Then its eigenbasis is orthogonal with respect to the scalar product induced by $\mathrm{D}^2\eta(u)$. If $\ell(u)$, $r(u)$ denote left- and right-eigenvectors, then*

$$\mathrm{D}^2\eta(r, X) = \alpha\ell \cdot X, \qquad \forall X \in \mathbb{R}^n, \tag{26}$$

with $\alpha := \mathrm{D}^2\eta(r,r)/\ell \cdot r$.

Euclidian vs. Riemannian structure. Even if one reads now and then that $\mathbb{R}^n$ is endowed with the natural, or even canonical scalar product, and the eigenvectors are chosen of unit length, this is nonsense. The only structure (over $\mathbb{R}^n$) associated with hyperbolic systems of conservation laws is the affine geometry, because a system like (23) can be transformed into another one, without changing the notion of weak solution, by applying an affine transformation $u \mapsto v = Mu + b$ with $M \in \mathbf{GL}_n(\mathbb{R})$ and $b \in \mathbb{R}^n$. Then f^α is replaced by $g : v \mapsto Mf^\alpha(M^{-1}(v-b))$. When our system admits a strongly convex entropy η, we may superimpose a Riemannian structure over the phase space $\mathcal{U}$, where $\mathrm{D}^2\eta$ plays the role of the metric. Then it makes sense to normalize the eigenvectors by $\mathrm{D}^2\eta(r,r) = 1$. So far, this geometrical aspect has not been exploited in the analysis of the Cauchy problem. For instance, it is an open question whether or not the sign of the curvature plays a role in the dynamics.

3.2 An Example: Gas Dynamics

Gas dynamics is governed by the Euler system

$$\begin{aligned}
\partial_t \rho + \mathrm{div}(\rho\mathbf{v}) &= 0,\\
\partial_t(\rho\mathbf{v}) + \mathrm{Div}(\rho\mathbf{v}\otimes\mathbf{v}) + \nabla p &= 0,\\
\partial_t\left(\frac{1}{2}\rho|\mathbf{v}|^2 + \rho e\right) + \mathrm{div}\Big(\left(\frac{1}{2}\rho|\mathbf{v}|^2 + \rho e + p\right)\mathbf{v}\Big) &= 0,
\end{aligned}$$

where ρ is the density, p the pressure, e the specific internal energy and $\mathbf{v}$ the velocity. The second line above is itself a system and $\mathbf{v}\otimes\mathbf{v}$ stands for the matrix with entries v_iv_j. An *equation of state* relates ρ, e and p. The conserved quantities are $u := (\rho, \rho\mathbf{v}, \frac{1}{2}\rho|\mathbf{v}|^2 + \rho e)$. The system is compatible with

$$\partial_t(-\rho s) + \mathrm{div}(-\rho s \mathbf{v}) = 0,$$

where $s = s(\rho, e)$ is the entropy of physicists. Thus $\eta = -\rho s$ and $\vec{q} = \eta \mathbf{v}$. The fact that η is an entropy is equivalent to the fact that the differential $\mathrm{d}s$ is proportional to $\mathrm{d}e + p\mathrm{d}\frac{1}{\rho}$. In other words, we have the *Gibbs' relation* for fluids in classical thermostatics

$$\theta \mathrm{d}s = \mathrm{d}e + p\mathrm{d}\frac{1}{\rho}, \tag{27}$$

for some function $\theta(\rho, e)$. The latter, which turns out to be positive, is known as the *absolute temperature*.

3.3 *The Strongly Convex Case*

We now assume that η is strongly convex: $\mathrm{D}^2\eta > 0_n$. The main consequence is

Theorem 6 (Godunov [13], Las–Friedrichs [11], and Boillat [4]). *If (23) admits a strongly convex entropy, then it is symmetrizable, hence hyperbolic.*

There are several proofs of this result, because there are several symmetric forms of the system. Actually, it can be symmetrized for any choice of primary variables $v = \phi(u)$ (ϕ a diffeomorphism). Say that (23) is formally equivalent to $S_0(v)\partial_t v + \sum_\alpha S^\alpha(v)\partial_\alpha v = 0$, where S_0 and S^α symmetric, and S_0 is positive definite. Then, if $v = \psi(w)$, this system can be transformed into the are symmetric form $\Sigma_0(w)\partial_t w + \sum_\alpha \Sigma^\alpha(w)\partial_\alpha w = 0$, with

$$\Sigma_0 = (\nabla\psi)^T S_0 \nabla\psi, \quad \Sigma^\alpha = (\nabla\psi)^T S^\alpha \nabla\psi.$$

Godunov first symmetrized (23) in terms of $v := \nabla\eta(u)$, which Boillat later called the *main field*. Their symmetrizer is $S_0 = (\mathrm{D}^2\eta)^{-1}$, the Hessian of the Legendre transform η^*. Instead, Friedrichs and Lax symmetrized directly in terms of u, with $S_0 = \mathrm{D}^2\eta$.

The importance of symmetrization is given by the following existence result for the Cauchy problem. We give it in Dafermos' formulation [6].

Theorem 7. *Assume that the system (23) is endowed with a C^3 entropy η, which is strongly convex. Suppose that the initial data $a \in C^1(\mathbb{R}^d)$ takes values in some compact set of $\mathcal{U}$ and $\nabla a \in H^\ell(\mathbb{R}^d)$ for some $\ell > \frac{d}{2}$. Then there exists a $T_\infty \in (0, +\infty]$, and a unique continuously differentiable function $u : \mathbb{R}^d \times [0, T_\infty) \to \mathcal{U}$, which is a classical solution of the Cauchy problem (23, 2). Furthermore,*

$$\nabla u \in \bigcap_{k=0}^{\ell} C^k([0, T_\infty); H^{\ell-k}(\mathbb{R}^d)).$$

The time interval is maximal in that if $T_\infty < \infty$, then

$$\int_0^{T_\infty} \|\nabla u(\cdot,t)\|_{L^\infty} dt = \infty$$

and/or the range of $u(\cdot,t)$ escapes from every compact subset of $\mathcal{U}$ as $t \to T_\infty$.

We emphasize that because $H^\ell(\mathbb{R}^d)$ is included in $C^0(\mathbb{R}^d)$, such solutions are continuously differentiable over $[0,T)\times\mathbb{R}^d$. They are therefore *classical solutions*.

3.4 Limitations of the Classical Theory

The local well-posedness stated in Theorem 7 is accurate in many directions. On the one hand, it promisses solutions whose regularity is that of the initial data, but not more. This cannot be improved in general, because the class of systems of conservation laws with a convex entropy is reversible in time, as long as classical solutions are concerned. Therefore, because we may write $a = S_{-t} \circ S_t(a)$ in terms of the local semi-group, there cannot be any kind of regularization effect.

On the other hand, the solution is not global in general, because of the non-linearity. The derivatives $v^\alpha := \partial_\alpha u$ satisfy a system

$$\partial_t v^\alpha + \sum_\beta A^\beta(u)\partial_\beta v^\alpha + \sum_\beta (\mathrm{d}A^\beta(u)\cdot v^\alpha)v^\beta = 0, \qquad \alpha = 1,\dots,d,$$

which looks like a Ricatti equation. Thus we anticipate a blow-up of the first derivatives in finite time, for rather general and smooth initial data. This phenomenon is particularly easy to analyze in the scalar case, say for the Burgers equation

$$\partial_t u + \partial_x \frac{1}{2}u^2 = 0. \tag{28}$$

Then $v := u_x$ satisfies $(\partial_t + u\partial_x)v + v^2 = 0$, which is a Ricatti equation $v' + v^2 = 0$ along the *characteristic lines* defined by

$$\frac{dX}{dt} = u(X,t).$$

If $a \in C^1(\mathbb{R})$ is not an increasing function, we may choose a point x_0 at which a_x is negative. Then, along the characteristic line originated at x_0 ($X(0) = x_0$ above), u_x has to blow up in finite time $T = -1/a_x(x_0)$, unless the solution ceases to exist sooner for some other reason.

3.5 Weak Solutions

We therefore don't expect global classical solutions for general, even smooth, initial data. This suggests that the actual solutions, which have to represent some physical process, would be less regular. This view is supported by physical experiments in gas dynamics, which provide evidence that shock waves develop in finite time. Shock waves are discontinuities of the state (density, temperature, velocity).

In mathematical analysis, we have known for decades that the physical relevance of PDEs can be expressed in terms of distributions. If we think that the conservation laws in a system (23) have a physical meaning and must be satisfied accurately, then it is natural to consider the following notion of solutions. A full justification in the case of continuum thermomechanics is given in Dafermos' book [6].

Definition 2. Let $a \in L^\infty(\mathbb{R}^d)^n$ be given. A field $u \in L^\infty((0,T)\times\mathbb{R}^d)^n$ is a *weak solution* of the Cauchy problem (23) with initial data $u(\cdot,0) = a$ if, for every test function $\phi \in \mathcal{D}(\mathbb{R}^{d+1})^n$, we have

$$\int_{(0,T)\times\mathbb{R}^d} (\phi_t \cdot u + \nabla_x\phi : f(u))dx\,dt + \int_{\mathbb{R}^d} \phi(x,0)\cdot a(x)\,dx = 0. \tag{29}$$

In (29), the first integral affords for the PDEs in $(0,T)\times\mathbb{R}^d$, whereas the second takes into account the initial data. As usual, the notion of weak solution is an extension of the classical notion, in the sense that if u is a classical solution, then it is a weak solution. The converse is of course false. For instance, $u(x,t) = a(x-ct)$ is a weak solution of the transport equation $\partial_t u + c\partial_x u = 0$, regardless of whether a is smooth or not. As a matter of fact, the notion of weak solutions allows us to consider non-smooth initial data a that are only bounded measurable.

It is instructive to characterize those weak solutions that are piecewise smooth, because this gives a description of the physical discontinuities mentioned above. To this end, we ignore the initial data and just say that u is a weak solution of (23) in some open set $\omega \subset \mathbb{R}^{d+1}$ if, for every test function $\phi \in \mathcal{D}(\omega)^n$, we have

$$\int_{(0,T)\times\omega} (\phi_t \cdot u + \nabla_x\phi : f(u))dx\,dt = 0.$$

Proposition 8. *Let u be piecewise smooth in ω, in the following sense: There exists a smooth hypersurface Σ that separates ω into two pieces $\omega_\pm$ in which $u \in C^1(\omega_- \cup \omega_+)$, and u has limit $u_\pm$ along Σ, from each side.*

Then u is a weak solution of (23) in ω if and only if, on the one hand it is a classical solution in $\omega_- \cup \omega_+$, and on the other hand, it satisfies the Rankine–Hugoniot relation across Σ:

$$\nu_t(u_+ - u_-) + \sum_\alpha \nu_\alpha(f^\alpha(u_+) - f^\alpha(u_-)) = 0. \tag{30}$$

Hereabove, ν is the normal to Σ.

Observe that because f is smooth, and thus Lipschitz on bounded sets, we have

$$\left|\sum_\alpha \nu_\alpha (f^\alpha(u_+) - f^\alpha(u_-))\right| \leq M\,|u_+ - u_-| \max_\alpha |\nu_\alpha|,$$

which implies $|\nu_t| \leq M \max_\alpha |\nu_\alpha|$. This shows that $(\nu_1, \dots, \nu_d)$ may not vanish. In other words, Σ is time-like: it does not have a horizontal tangent space. Then, normalizing ν by $\nu = (\xi, -V)$ with $|\xi| = 1$, we call ξ the *direction of propagation* and V the *normal velocity* of the discontinuity. We warn the reader that there remains an ambiguity in that we may always choose $-\nu$ instead of ν.

In one space dimension, the ambiguity is removed by adopting the convention that $\xi = 1$. Then $V > 0$ (resp. $V < 0$) amounts to saying that the discontinuity $X(t)$ propagates to the right (resp. left); the velocity is nothing but $X'(t)$. The Rankine–Hugoniot condition then reads

$$f(u_+) - f(u_-) = V(u_+ - u_-). \tag{31}$$

For instance, the velocity of a discontinuity in the Burgers equation (28) is given by

$$V = \frac{1}{2}(u_+ + u_-).$$

Notations. The *jump* $h_+ - h_-$ of a quantity h from ω_- to ω_+ is denoted by $[h]$. Its arithmetic mean $\frac{1}{2}(h_- + h_+)$ is written as $\langle h \rangle$. The Rankine–Hugoniot condition can thus be condensed to

$$\nu_t[u] + \sum_\alpha \nu_\alpha [f^\alpha(u)] = 0.$$

The Hugoniot locus. The analysis of equation (31) was first made by Lax [20], by means of bifurcation analysis. If we impose u_- and search for solutions (u_+, V), we first have the trivial solution $u_+ = u_-$, with $V \in \mathbb{R}$ arbitrary. This solution is useless because it does not describe a genuine discontinuity. The non-trivial solutions form the *Hugoniot locus* $\mathcal{H}(u_-)$. To determine this set, we may bifurcate from points (u_-, V) at which the differential of $u \mapsto f(u) - f(u_-) - V(u - u_-)$ is singular. This happens precisely when $V = \lambda_k(u_-)$ for some index k. Lax proved the following result.

Theorem 8 (Lax [20]). *We assume that the system*

$$\partial_t u + \partial_x f(u) = 0 \tag{32}$$

is strictly hyperbolic at u_-. Let us choose an index $1 \leq k \leq n$. Then, in a neighbourhood of $(u_-, \lambda_k(u_-))$, the equation (31) defines the union of the line

$\{u_-\} \times \mathbb{R}$ and of a non-trivial curve $s \mapsto (u_+(s), V(s))$ (say that $u_+(0) = u_-$). The derivative

$$r := \left.\frac{du_+}{ds}\right|_{s=0}$$

is an eigenvector:

$$\mathrm{d}f(u_-)r = \lambda_k(u_-)r.$$

Actually, the curve $s \mapsto u_+(s)$ is tangent at u_- to the integral curve of the eigenfield $r_k(u)$, to the order two. In addition, we have

$$V(s) = \frac{1}{2}(\lambda_k(u_-) + \lambda_k(u_+(s))) + O(s^2).$$

For a proof, see for instance Chap. 4 of [32]. We denote by $\mathcal{H}_k(u_-)$ the curve $s \mapsto u_+(s)$ described above. A discontinuity $(u_-, u_+; V)$ with $u_+ \in \mathcal{H}_k(u_-)$ is called a *k-discontinuity*.

3.6 Entropy Admissible Solutions

In thermodynamics, the second fundamental principle tells us that not all mathematically possible discontinuities can be observed. More precisely, many discontinuities are irreversible, in the sense that if $U(x,t)$ is a flow (hence, a weak solution of the Euler system) with a shock wave, then $\tilde{U}(x,t) = U(-x,-t)$ is not a flow. This is surprising at a first glance, because the Euler system, as well as every system of the form (23), is formally reversible; therefore $\tilde{U}$ is a weak solution. Thus not all weak solutions are physically relevant.

There are several mathematical motivations to the second fundamental principle. At the beginning, there is the observation that a Cauchy problem for a nonlinear system (23) has far too many weak solutions. Thus passing from classical solutions to weak ones, we left Scylla (lack of global solutions) for Charybdis (high non-uniqueness). There is therefore a need to select what will be called *admissible* solutions.

Evidence of non-uniqueness is provided again by the Burgers equation (28). In the best of the worlds, the initial data $a \equiv 0$ should yield the trivial solution $u \equiv 0$ and only this one. But we do have at least a continuum of solutions, parametrized by $p > 0$:

$$u^p(x,t) = \begin{cases} 0, & x < -pt, \\ -2p, & -pt < x < 0, \\ 2p, & 0 < x < pt, \\ 0, & x > pt. \end{cases}$$

To verify that u^p is a weak solution, one remarks that constants are classical solutions, and that (30) is satisfied across discontinuities.

The selection of an admissible solution is the second place where a convex entropy may play a role. Firstly, we remark that a weak solution does not necessarily solve (24) in the distributional sense, because we cannot use the chain rule to calculate $\partial_t \eta(u)$ or $\partial_\alpha q^\alpha(u)$. Secondly, the jump relation for (24) does not follow from the Rankine–Hugoniot relations.

Definition 3. We assume that the system (23) admits a strongly convex entropy η, associated with an entropy flux $\vec{q}$.

Let $a \in L^\infty(\mathbb{R}^d)^n$ be given. A weak solution of the Cauchy problem $u \in L^\infty((0,T) \times \mathbb{R}^d)^n$ is *entropy admissible* if, for every non-negative test function $\phi \in \mathcal{D}((-\infty, T) \times \mathbb{R}^d)$, we have

$$\int_{(0,T)\times\mathbb{R}^d} (\phi_t \eta(u) + \nabla_x \phi \cdot \vec{q}(u)) dx\, dt + \int_{\mathbb{R}^d} \phi(x,0)\eta(a(x))\, dx \geq 0. \tag{33}$$

We emphasize that the definition above involves two inequalities: $\phi \geq 0$ and the integrals sum up to a non-negative number. This is where our notion of solution becomes irreversible: if u is an admissible solution, then $\tilde{u}(x,t) := u(-x,-t)$ need not be. Yet, $\tilde{u}$ was a weak solution.

When restricting to test functions with compact support in $(0,T) \times \mathbb{R}^d$, the admissibility amounts to the Lax *entropy inequality*

$$\partial_t \eta(u) + \mathrm{div}\vec{q}(u) \leq 0, \tag{34}$$

in the distributional sense.

When u is a piecewise smooth flow, entropy admissibility can be interpreted in the following way.

Proposition 9. *Let u be piecewise smooth in ω, as described in Proposition 8.*

Then u is an entropy admissible solution if and only if on the one hand it is a weak solution of (23) in ω, and on the other hand, it satisfies the following inequality across Σ:

$$\nu_t(\eta(u_+) - \eta(u_-)) + \sum_\alpha \nu_\alpha (q^\alpha(u_+) - q^\alpha(u_-)) \leq 0. \tag{35}$$

Hereabove, ν is oriented from ω_- to ω_+.

Notice that the inequality does not depend on which side is labelled with a $+$ or a $-$: if we switch the labels, the signs of the jumps $[\eta]$ or $[q^\alpha]$ are changed, but the normal ν is flipped and therefore the products remain the same.

Example. Let us consider the Burgers equation. If a discontinuity propagates at velocity V, we have

$$-V[\eta] + [q] = q(u_+) - q(u_-) - \langle u\rangle[\eta] = \frac{1}{2}\int_{u_-}^{u_+} (b-s)(s-a)\eta''(s)\,ds.$$

Therefore the discontinuity is admissible if and only if $u_+ \leq u_-$. We point out that in this favorable case, the admissibility does not depend on the choice of the strongly convex entropy. This remains true for every scalar conservation law in one space dimension, provided the flux f is itself a convex function (or if it is concave). When f is arbitrary, it becomes important to strengthen the admissible criterion by imposing (33) for *every* strongly convex η. We then obtain the celebrated *Oleinik's criterion*:

A scalar discontinuity $(u_-, u_+; V)$ in one space dimension is admissible if and only if

- either $u_- \leq u_+$ and the graph of f over (u_-, u_+) lies above the chord between the points $(u_\pm, f(u_\pm))$,
- or $u_+ \leq u_-$ and the graph of f over (u_+, u_-) lies below the chord.

3.7 Weak-Strong Uniqueness

It happens frequently in the theory of nonlinear PDEs that we are not able to prove the regularity and the uniqueness of the solutions that have been proven to exist. This is for instance the case for the 3-dimensional Navier-Stokes equations for an incompressible fluid, where this open problem is worth one million dollars. However, we often are able to prove that if a solution is smooth enough, then it is unique *among the weak solutions*. This is the case here. We borrow the following statement from Chap. 5 of Dafermos' book. It states a slightly stronger stability result.

Theorem 9. *Assume that the system (23) is endowed with an entropy-flux pair* $(\eta, \vec{q})$*, where* η *is a strongly convex entropy over* $\mathcal{U}$*. Suppose that* $\bar{U}$ *is a classical solution of (23) in* $\mathbb{R}^d \times (0,T)$*, taking values in a compact subset* $\mathcal{K}$ *of* $\mathcal{U}$*, with initial data* $\bar{U}_0$*. Let* U *be any entropy admissible solution of the Cauchy problem for (23), taking values in* $\mathcal{K}$*, with initial data* U_0*. Then*

$$\int_{|x|<r} |U(x,t) - \bar{U}(x,t)|^2 dx \leq a e^{bt} \int_{|x|<r+vt} |U_0(x) - \bar{U}_0(x)|^2 dx \qquad (36)$$

holds true for any $r > 0$ *and* $t \in [0,T)$*, with positive constant* a, v *depending solely on* $\mathcal{K}$*, and* b *depending also on the Lipschitz constant of* $\bar{U}$*.*

In particular, $\bar{U}$ *is the unique entropy admissible solution with initial data* $\bar{U}_0$*.*

4 Improved Admissibility Criteria

Propositions 8 and 9 suggest that, even if admissibility is stated in terms of differential inequalities, it concerns only the triplets $(u_-, u_+; \nu)$ that solve the Rankine–Hugoniot relation. This is certainly true if we deal with piecewise smooth solutions, even if discontinuities meet, provided their intersection is a codimension-2 subset. This is because of the following obvious fact:

Proposition 10. *Let $\vec{Q}$ be a bounded measurable vector field in Ω, an open subset of $\mathbb{R}^D$. Let Γ be a closed subset contained in the denumerable union of codimension-2 submanifolds of class C^1. If the inequality* div $\vec{Q} \leq 0$ *is satisfied in $\Omega \setminus \Gamma$ in the distributional sense, then it is satisfied in Ω.*

Of course, multi-dimensional flows are not that smooth in general, and the evidence of turbulence tells us that there must be wilder flows,[5] for which the chain rule does not apply and therefore the entropy inequality might tell us something non-trivial. Nevertheless, researchers are convinced that classifying the triples and selecting among them the admissible ones is a sensible goal, and they faint to believe that there is nothing more subtle regarding admissibility. Whether this attitude is right or not will be known only after the theory has developed far enough, perhaps only within decades. We therefore adopt this simplistic point of view. We warn the reader however that this approach is useless when the solutions are so wild that we are unable to distinguish well-drawn discontinuities.

There are two approaches in the search for admissibility criteria for discontinuities. On the one hand, we may view a discontinuity as a free boundary problem (FBP), because the hypersurface of discontinuity is not known a priori. By this, we mean that in general we cannot identify a discontinuity *before* constructing the solution. Say that we have only one such surface, which we may parametrize by some coordinate $x_d = \psi(y,t)$, with $y = (x_1, \dots, x_{d-1})$. Then the strategy is to make the change of independent variables $(x,t) \mapsto (y,z,t)$ where $z := x_d - \psi(y,t)$. This fixes the discontinuity to the hyperplane $z = 0$. The price to pay is the introduction of an additional unknown function ψ. If we fold the left half-plane $z < 0$ onto the right half-plane $z > 0$ by flipping $z \mapsto -z$, then we obtain an IBVP coupling a field $\tilde{u}(y,z,t) \in \mathcal{U}^2 \subset \mathbb{R}^{2n}$ with $\psi(y,t) \in \mathbb{R}$. This strategy was developed by Majda in his celebrated memoirs [22]. For a modern and more complete analysis, the reader may refer to [1]. Of course, a physically relevant discontinuity must be somehow stable under small disturbances.

The second approach follows from the following claim: a first-order system of conservation laws (1) is usually not physically correct at small scales. For instance, the description of fluid flows should involve the effects of Newtonian viscosity

[5] But it does not tell us whether these flows solve the Euler equation, that is if the Euler system is a good physical model.

and heat conduction.[6] These phenomena are represented by additional terms in the PDEs; these terms are of higher order (two or more) and they contain small factors which let them act significantly at short scales, while being innocuous at large scales. If a discontinuity of (23) is to be physically relevant, it must correspond to an actual wave of the higher-order system, a travelling wave. The existence of this wave, and its stability properties, thus provides meaningful information.

4.1 The FBP Approach: Number of Boundary Conditions

Let us follow the approach of the FBP. It comprises $2n$ equations in the domain $z > 0$, plus a scalar equation along $z = 0$. Let us choose a point $\bar{P} \in \Sigma$ where the normal is $(-V, \xi)$ with $|\xi| = 1$ and $V \in \mathbb{R}$ is the normal velocity of Σ at $\bar{P}$. The state at both sides $\pm z > 0$ of $\bar{P}$ are denoted $u_\pm$. We know that $(u_-, u_+; V)$ satisfies the Rankine–Hugoniot condition in the direction ξ. We shall denote by $\lambda_j(v;\zeta)$ the eigenvalues of the Jacobian

$$\mathrm{d}f(v;\zeta) := \sum_\alpha \zeta_\alpha \mathrm{d}f^\alpha(v).$$

It is not difficult to see that the boundary $z = 0$ is characteristic if and only if one of the eigenvalues $\lambda_j(u_\pm;\xi)$ equals V. We suppose here that this is not the case: the discontinuity is not characteristic. Then there are two indices $1 \le j, k \le n$ such that

$$\lambda_j(u_-;\xi) < V < \lambda_{j+1}(u_-;\xi), \qquad \lambda_k(u_+;\xi) < V < \lambda_{k+1}(u_+;\xi). \tag{37}$$

Related to the previous observation, we can see that there are $n - k + j$ incoming characteristics. Thus a necessary condition for the well-posedness of the IBVP is that we are given $n-k+j$ boundary conditions. These are provided by the Rankine–Hugoniot relation, which is vectorial and comprises n equations relating $u_\pm$ and ψ. Because we need precisely one evolution equation for ψ, we really have only $n - 1$ boundary conditions for $u_\pm$. Thus we must have $n-k+j = n-1$, that is $j = k-1$. This gives us an admissibility condition, necessary but not sufficient,

$$\exists k \text{ such that } \lambda_{k-1}(u_-;\xi) < V < \lambda_k(u_-;\xi), \qquad \lambda_k(u_+;\xi) < V < \lambda_{k+1}(u_+;\xi). \tag{38}$$

This is called the *Lax shock condition*. A discontinuity satisfying it is called a Lax k-shock. If $k = 1$, we discard $\lambda_{k-1}(u_-;\xi)$, and likewise $\lambda_{k+1}(u_+;\xi)$ if $k = n$.

For a scalar equation in one space dimension, the condition (38) reduces to $f'(u_+) < V < f'(u_-)$, that is

[6] Such perturbative effects are treated in Sect. 5.

$$f'(u_+) < \frac{f(u_+) - f(u_-)}{u_+ - u_-} < f'(u_-).$$

We point out that this does not ensure the entropy inequality

$$[q(u)] \le V[\eta(u)], \tag{39}$$

and it is not implied by it either. Observe however that a weak Lax shock inequality

$$f'(u_+) \le \frac{f(u_+) - f(u_-)}{u_+ - u_-} \le f'(u_-)$$

is implied by the Oleinik condition. Notice that the derivation of the latter uses the fact that all functions $\eta(u)$ are entropies. Because this property is false for systems, one does not see a clear counterpart of the Oleinik condition if $n \ge 2$. It is even less clear when $n \ge 3$, because then the space of entropies is usually finite dimensional, of dimension $n + 2$ or $n + 3$, including the worthless subspace of affine functions. This leaves us with only a few independent inequalities of the form (39), instead of an infinity. Thus the entropy admissibility is unlikely, in many cases, to ensure the Lax shock inequality, even in a weak form.

We may now give a complement to Theorem 8. The derivative of $s \mapsto \lambda_k(u_+(s))$ at $s = 0$ equals

$$d\lambda_k(u_-) \cdot r_k(u_-), \qquad r_k(u_-) := \left.\frac{du_+}{ds}\right|_{s=0}.$$

Let us assume that $d\lambda_k(u_-) \cdot r_k(u_-) \neq 0$; we say that the k-th characteristic field is *genuinely nonlinear* at u_-. Then $s \mapsto \lambda_k(u(s))$ is strictly monotonous near $s = 0$. We may always normalize $r_k(u_-)$ by

$$d\lambda_k(u_-) \cdot r_k(u_-) = 1,$$

so that $\lambda_k(u_+) = \lambda_k(u_-) + s + O(s^2)$. Lax [20] made the important observation that we also have

$$V = \lambda_k(u_-) + \frac{s}{2} + O(s^2),$$

from which it follows that the triple $(u_-, u_+(s); V)$ is a Lax shock if and only if $s < 0$. Therefore the Lax shock condition eliminates locally half of the Hugoniot curve $\mathcal{H}_k(u_-)$.

For a scalar equation, genuine nonlinearity amounts to saying that the second derivative of the flux f does not vanish.

4.2 The Role of Lax Shocks in the Riemann Problem

An interesting case for our uniqueness/stability problem is the *Riemann Problem.* In one space dimension, it consists of searching for self-similar solutions:

$$u(x,t) = U\left(\frac{x}{t}\right).$$

The data of the Riemann problem is given in terms of a pair $(u_\ell, u_r) \in \mathcal{U}^2$,

$$a(x) = \begin{cases} u_\ell, & \text{if } x < 0, \\ u_r, & \text{if } x > 0. \end{cases}$$

For a self-similar function, (32) becomes

$$f(U)' = yU', \tag{40}$$

where y stands for the variable x/t.

When $|u_\ell - u_r|$ is small, we search for a solution U of (40) that remains in a small neighbourhood of u_ℓ. The general analysis was begun by Lax [20], who observed that the solution mimics that in the linear case: if the system is strictly hyperbolic, U is made of $n+1$ constant states $u_0 = u_\ell, u_1, \dots, u_n = u_r$, separated by simple waves. The wave between u_{k-1} and u_k travels at a velocity that remains close to $\lambda_k(u_\ell)$.

It turns out that in this picture, the discontinuities associated with the Hugoniot locus $\mathcal{H}_k(u_k)$ come into competition with the so-called *rarefaction waves*. The latter are smooth self-similar solutions; they therefore satisfy $(\mathrm{d}f(U) - y)U' = 0$, which tells us that there exists an index $k = 1, \dots, n$ such that

$$y = \lambda_k(U(y)), \qquad U'(y) \parallel r_k(U(y)). \tag{41}$$

We then speak of a k-rarefaction. If we ignore the rarefaction waves, it is always possible to go from u_ℓ to u_r by a succession of k-discontinuities from $k = 1$ to $k = n$. This is a simple consequence of the Implicit Function Theorem, plus the fact that the tangent space of $\mathcal{H}_k(u_\ell)$ at u_ℓ is the k-th eigenspace of $\mathrm{d}f(u_\ell)$, and $\mathbb{R}^n$ is the direct sum of these eigenspaces. This construction has two flaws. On the one hand, if a rarefaction wave $R(x/t)$ between two states $v_{\ell,r}$ is given, we find an alternate piecewise constant solution $U(x/t)$ with the same initial data v_ℓ / v_r, thus yielding to non-uniqueness. On the other hand, some of the discontinuities involved in the construction might not satisfy the Lax shock condition.

It is a general principle in thermodynamics that smooth flows (often called *quasi-reversible*) are physically relevant. In the example above, this means that R is acceptable, and thus U must not be, if we believe to uniqueness. This implies again that we have to get rid of discontinuities, at least of some of them. The reason why we need to accept some discontinuities is that rarefaction waves are not reversible in the following sense: if we can join a left state u_- to a right state u_+ by a k-rarefaction, then we have $\lambda_k(u_+) > \lambda_k(u_-)$ because of (41).

In the rather simple situation of a genuinely nonlinear characteristic field, where we have $d\lambda_k(u) \cdot r_k(u) \equiv 1$ after normalization, we see that λ_k is strictly monotone along the integral curve of r_k. If a, b belong to some integral curve of r_k, we may join a to b by a k-rarefaction if and only if $\lambda_k(a) \leq \lambda_k(b)$, but the converse is impossible. For this reason, it is not always possible to solve a Riemann problem by gluing only rarefaction waves. But the global picture leaves us optimistic, because k-rarefactions can be used when $\lambda_k(u_+) > \lambda_k(u_-)$, whereas k-shocks can be used when instead $\lambda_k(u_+) < \lambda_k(u_-)$.

Linearly degenerate fields. Before giving Lax' result concerning the Riemann problem for small data, we need to mention the type that stands at the tip opposite to genuine nonlinearity: the k-th characteristic field is *linearly degenerate* (LD) if $d\lambda_k \cdot r_k \equiv 0$. Linear degeneracy means that λ_k is constant along the integral curves of r_k. For a linearly degenerate field, (41) implies $U' = 0$ and therefore there are no k-rarefactions. Instead, we have

Proposition 11. *If the k-th characteristic field is linearly degenerate, then for any two points a and b on the same integral curve of r_k, the triple $(a, b; V = \lambda_k(a) = \lambda_k(b))$ satisfies the Rankine–Hugoniot relation, and the entropy equality $[q(u)] = V[\eta(u)]$.*

Such triplets are called *contact discontinuities*. They are generally accepted as physically relevant, at least in the one-dimensional case. We point out that such discontinuities are reversible, and that they satisfy the equalities instead of inequalities in (38).

According to Proposition 11, the integral curve of r_k and the Hugoniot locus $\mathcal{H}_k(u_-)$ coincide when the field is linearly degenerate. This is false in general for other characteristic fields. This coincidence was studied in 2×2 systems by Temple [38], where he found another particular case, which is now known as that of the "Temple field". This notion has been extended to $n \times n$ systems and appears to be a natural extension of the scalar situation. See volume II of [32] for a complete presentation. The k-th characteristic field is Temple if the left eigenfield $u \mapsto \ell_k(u)$ is normal to a foliation of $\mathcal{U}$ into hyperplanes. In other words, classical solutions u obey to a transport equation $(\partial_t + \lambda_k(u)\partial_x)w(u) = 0$ where w is a function whose level sets are affine hyperplanes.

When a characteristic field has constant multiplicity $m \geq 2$, it is always linearly degenerate, and even more:

Theorem 10 (Boillat). *Let us assume that the system (32) is hyperbolic in a neighbourhood $\mathcal{V}$ of u_-, and has a characteristic field of constant multiplicity $m \geq 2$, associated with the eigenvalue $\lambda(u)$. Then the field of affine subspaces*

$$u \mapsto u + \ker(df(u) - \lambda(u)I_n)$$

is integrable in the sense of Liouville: it is the tangent bundle to a foliation of $\mathcal{V}$ by smooth manifolds of dimension m. In addition, λ is constant on each leaf (linear degeneracy, see above).

It is not too hard to deduce from Theorem 10 the following description of the Hugoniot locus: $\mathcal{H}(u_-)$ is locally the union of the line $\{u_-\} \times \mathbb{R}$ and of $\mathcal{F}(u_-) \times \{\lambda(u_\pm)\}$, where $\mathcal{F}$ is the leaf passing through u_-.

A kind of converse to Theorem 10 turns out to be true: if a characteristic field of (32) is linearly degenerate, it is possible to embed the system into a larger one ($n + 1$ unknowns and equations) in which the corresponding eigenvalue has multiplicity ≥ 2.

Lax's treatment of the Riemann Problem. The first general theorem on the Riemann problem is due to Lax. It concerns the case where every characteristic field is either genuinely nonlinear or linearly degenerate.

Theorem 11 (Lax [20]). *We assume that (32) is strictly hyperbolic. Let $\bar{u}$ be a state at which all the characteristic fields are either genuinely nonlinear or linearly degenerate. Then for a small enough neighbourhood $\mathcal{O}_1$ of $\bar{u}$, there exists a neighbourhood $\mathcal{O}_0$ of $\bar{u}$ such that if $u_\ell, u_r \in \mathcal{O}_0$, then the Riemann problem from u_ℓ to u_r admits a unique solution with the following properties:*

- *U takes values in $\mathcal{O}_1$,*
- *U is made of $n+1$ constant states $u_0 = u_\ell, u_1, \dots, u_n = u_r$ separated by simple waves,*
- *If the k-field is genuinely nonlinear (GNL), the wave from u_{k-1} to u_k is either a k-rarefaction or a k-shock,*
- *If the k-field is linearly degenerate (LD), the wave from u_{k-1} to u_k is a contact discontinuity.*

Lax's Theorem is subtle. When the Hugoniot curves are not integral curves of r_k, the Riemann problem cannot be resolved by first applying a non-linear change of coordinates and then following lines parallel to the coordinate axes.

4.3 The FBP Approach (II)

In [22], Majda consider the linearized stability of a discontinuous wave. As explained above, the linearization, carried out after Σ has been sent to a fixed boundary $x_d = 0$, looks very much like a linear IBVP, with variable coefficients. Its stability amounts to that of each IBVP with constant coefficients, obtained by freezing the background state at some point $\bar{P} \in \Sigma$. Thus it is encoded into a uniform Kreiss–Lopatinskiĭ condition; because the latter involves an extra term which accounts for the disturbances of the front, it bears the name of the uniform Majda–Kreiss–Lopatinskiĭ condition (MKL). It is parametrized by frequency pairs

(η, τ), where $\eta \in \xi^\perp$ (ξ is the direction in which the shock propagates), $\Re\tau \geq 0$ and $|\eta|^2 + |\tau|^2 = 1$. Without loss of generality, we identify $\xi^\perp$ with $\mathbb{R}^{d-1}$.

Following Sect. 4.1, we assume that the discontinuity is a Lax shock. In particular, the linearized FBP is non-characteristic. Because $n - 1$ characteristics enter the domain (they leave the shock), we should expect that the Lopatinskiĭ condition expresses as the non-vanishing of a determinant of size $n - 1$. But because the FBP involves also a scalar unknown ψ along the boundary, we actually get an $n \times n$ determinant, the *Lopatinskiĭ determinant*.

As in the analysis of a linear IBVP, the Lopatinskiĭ determinant is a function $\Delta(\eta, \tau)$ that can be taken analytic in η and holomorphic in τ. When our system (23) is strictly hyperbolic, or more generally has characteristic fields of constant multiplicity, the Kreiss block structure allows us to extend Δ by continuity to the boundary of the half-sphere.

4.3.1 The Liu–Majda Condition

The case $(\eta, \tau) = (0, 1)$ gives the Liu–Majda condition

$$\Delta(0,1) = \det(r_1(u_+;\xi), \dots, r_{k-1}(u_+;\xi), u_+ - u_-, r_{k+1}(u_-;\xi), \dots, r_n(u_-;\xi)) \neq 0. \tag{42}$$

When $u_\pm$ are close to some state $\bar{u}$, then $r_j(u_\pm;\xi) \sim r_j(\bar{u};\xi)$, whereas $u_+ - u_-$ is approximately colinear to $r_k(\bar{u};\xi)$. Thus (42) is satisfied, because of the linear independence of the eigenvectors of $\mathrm{d}f(\bar{u};\xi)$.

In one space dimension, the Liu–Majda condition is nothing but the transversality criterion (8). Therefore a one-dimensional, non-characteristic discontinuity is linearly strongly stable whenever it satisfies the Lax shock condition for some $k = 1, \dots, n$, plus (42). In particular, every Lax shock of moderate amplitude is linearly strongly stable. Strong linear stability turns out to imply nonlinear stability in spaces of differentiable functions, but we shall not address this question in these notes; the interested reader is refered to [22] or to [1].

We remark that (42) is not an admissibility criterion by itself. It comes only as a complement of the Lax shock inequalities (38). An admissibility condition usually rejects 'half' of the discontinuities satisfying the Rankine–Hugoniot condition, even small ones, whereas (42) rejects only exceptional ones; in addition, if the characteristic fields are genuinely non-linear, only large shocks may be thrown out by (42).

4.3.2 The Majda–Kreiss–Lopatinskiĭ Condition

In several space dimensions, the MKL condition is again a complement to (38). We have now non-zero frequencies $\eta \in \mathbb{R}^{d-1}$. The non-uniform version of MKL is that for every pair (η, τ) with $|\eta|^2 + |\tau|^2 = 1$ and $\Re\tau > 0$,

$$\Delta(\eta, \tau) \neq 0. \tag{43}$$

Assuming strict hyperbolicity, or more generally that characteristic fields have constant multiplicities, we know that Δ extends by continuity to the boundary of the half-sphere. Then uniform MKL is equivalent to the validity of (43) up to this boundary.

The Majda–Kreiss–Lopatinskiĭ condition is always satisfied for scalar shocks ($n = 1$), but only in a non-uniform way if $d \geq 2$. This weak stability is consistent with the fact that a scalar equation generates an L^1-contraction semi-group. Of course, these two facts are difficult to relate rigourously, because on the one hand the MKL condition refers to the L^2-stability instead of that in L^1, and on the other hand, we do not expect fast decay rates for the disturbance.

The MKL condition has been studied in detail for gas dynamics, and a wide range of shock data have been found to be strongly stable. For real equations of state, Lax shocks of large amplitude may be unstable in the Hadamard sense (MKL is violated at some pair (η, τ) with $\Re\tau > 0$); shocks of intermediate strength may be non-uniformly stable (MKL is violated at some boundary frequency (η, τ)). Majda's calculations of [22] have been simplified by Jenssen and Lyng [16]. A complete treatment can be found in Sect. 15.2 of [1].

4.3.3 How Non-characteristic Are Small Shocks?

Let us come back to the fact that a discontinuity is non-characteristic if and only if $V \neq \lambda_j(u_\pm; \xi)$ for every $j = 1, \dots, n$. We have seen above that this turns out to be true under the conditions that $u_\pm$ are close to each other, u_+ being on the k-th branch of the Hugoniot locus $\mathcal{H}(u_-)$, and the k-th characteristic field is genuinely nonlinear. However, the difference between V and $\lambda_k(u_\pm)$ is very small, typically

$$V - \lambda_k \left(\frac{u_- + u_+}{2} \right) = O\left(|u_+ - u_-|^2 \right).$$

A small shock is therefore *almost* characteristic. This raises a technical difficulty, as well as an important question. If the shock satisfies MKL, the maximal estimates for the linearized FBP are likely to involve large constants when $V - \lambda_k(u_\pm)$ is small. Thus it may be difficult to handle the nonlinear terms in a fixed point argument where one uses a kind of Duhamel formula. For this reason, the existence time found by Majda shrank to zero with the shock strength. This is a bit puzzling, because the zero shock strength corresponds to the so-called *weak shock*, which is a Lipschitz solution with a discontinuity of the gradient, a situation in which we are able to establish an existence result on a positive time interval. Majda raised therefore the problem to prove a uniform lower bound for the existence time as the shock strength tends to zero. This was eventually achieved by Francheteau and Métivier in [7].

4.4 Liu's E-Condition

The accurate theory for scalar conservation laws includes the Oleinik condition. The latter suggests that when a characteristic field is neither GNL nor LD (for instance $f : \mathbb{R} \to \mathbb{R}$ has an inflexion point), the Lax shock condition (38) is not sufficient to make a good selection. Oleinik's condition has been generalized for strictly hyperbolic systems by T.-P. Liu in the following way, called the *E-condition*.

A discontinuity $(u_-, u_+; V)$ satisfies the E-condition if

- on the one hand, the Hugoniot locus $\mathcal{H}(u_-)$ contains an arc γ from u_- to u_+; if $u \in \gamma$ we denote by $v_-(u)$ the velocity of the discontinuity from u_- to u,
- on the other hand, for every $u \in \gamma$, we have $V \leq v_-(u)$.

The definition above looks odd as first glance: we could as well ask for the dual condition that $\mathcal{H}(u_+)$ contains an arc δ from u_- to u_+, on which $v_+(u) \leq V$. It is not immediate that both conditions are equivalent, and it could be that they are not when $u_\pm$ are far apart. It is easy to see that they always are in the scalar case. For systems, Liu found that when $|u_+ - u_-|$ is not too large, then each of them is equivalent to the existence of a viscous profile, a notion that we consider below. In this case, they are therefore equivalent to each other.

Liu's E-condition turns out to contain the Lax shock condition (38). This is shown by considering the limit of the inequalities above as u tends to the base point of the Hugoniot locus. For instance, $v_-(u) \geq V$ gives $\lambda_k(u_-) \geq V$ when $u \to u_-$ along γ. The index k is given by the point $(u_-, \lambda_k(u_-))$ at which the arc γ bifurcates from the line $u \equiv u_-$.

One sometimes says that a triple satisfying Liu's E-condition is a *Liu shock*.

4.5 Existence of a Viscous Profile

We now turn to the second side of the admissibility theory, that given by the study of travelling waves for systems that are higher-order completions of (23).

Let us consider first what is called "artificial viscosity": we consider the slightly modified system

$$\partial_t u + \partial_x f(u) = \epsilon \partial_x^2 u, \tag{44}$$

where $0 < \epsilon \ll 1$. This is a parabolic system, for which the Cauchy problem is well-posed in classes of smooth functions under very general assumptions. Contrary to (32), this system is not scaling invariant. On the contrary, the effect of a space-time dilation is to change the parameter ϵ, because the latter has dimension L^2T^{-1}. This suggests that we look for travelling waves of the form

$$u^\epsilon(x,t) = U\left(\frac{x - Vt}{\epsilon}\right). \tag{45}$$

We ask the *viscous profile* U to have limits $u_\pm$ at $\pm\infty$. This amounts to saying that

$$u^0(x,t) := \lim_{\epsilon\to 0+} u^\epsilon(x,t)$$

exists almost everywhere; this limit is then equal to u_- for $x < Vt$ and u_+ for $x > Vt$. The profile satisfies an ODE

$$U'' = (f(U))' - VU', \tag{46}$$

which can be integrated as

$$U' = f(U) - VU - z, \tag{47}$$

where $z \in \mathbb{R}^n$ is a constant, to be determined. If the limits $u_\pm$ exist, they must satisfy $f(u_\pm) - Vu_\pm = z$, and it follows that the triplet $(u_-, u_+; V)$ satisfies the Rankine–Hugoniot condition. In other words, u^0 is a weak solution of (32). We remark in passing that V is uniquely determined by the pair (u_-, u_+), and therefore there was no need to put $V = V_\epsilon$ in (45).

Going beyond that, we have for every entropy η with flux q

$$(\mathrm{d}\eta(U)U' - q(U) + V\eta(U))' = \mathrm{D}^2\eta(U', U').$$

If η is convex, then $y \mapsto \mathrm{d}\eta(U)U' - q(U) + V\eta(U)$ is non-decreasing. By looking at its limits as $y \to \pm\infty$, we deduce that our triplet also satisfies (39).

Of course, the existence of a viscous profile is a much more complex statement than just an algebraic equality plus an inequality. Therefore, we expect that the converse does not hold in general: (31, 39) do not imply that such a profile exist. For instance, it is a simple exercise to verify that for a scalar equation, the existence of a viscous profile is equivalent to the Oleinik condition, written in a strict sense ("above/below" have to be understood as "strictly above/below"). For strictly hyperbolic systems, one does not know a simple necessary and sufficient criterion for this existence, unless $|u_+ - u_-|$ is small enough:

Theorem 12 (Majda–Pego [23]). *Let us assume that (32) is strictly hyperbolic. Let $\bar{u} \in \mathcal{U}$ and $1 \le k \le n$ be given. Then there is a neighbourhood $\mathcal{O} \ni \bar{u}$ such that, for every $u_\pm \in \mathcal{O}$, a viscous profile from u_- to u_+, taking values in $\mathcal{O}$, exists if and only if $u_+ \in \mathcal{H}_k(u_-)$ and Liu's E-condition is satisfied in the strict sense: for every $u \in \gamma \setminus u_+$, we have $V < v_-(u)$.*

This theorem shows that Liu's E-condition is the appropriate generalization of Oleinik's admissibility criterion to systems. Majda and Pego actually proved their theorem in the much more general context of general viscous systems of conservation laws

$$\partial_t u + \partial_x f(u) = \partial_x (B(u)\partial_x u),$$

under the condition that the matrix B satisfies

$$\ell_k(\bar{u})B(\bar{u})r_k(\bar{u}) > 0, \tag{48}$$

where $\ell_k(u)$ and $r_k(u)$ are the left- and right-eigenvectors of $\mathrm{d}f(u)$, normalized by $\ell_k \cdot r_k = 1$. A viscous profile is a heteroclinic solution of the differential equation

$$B(u)u' = f(u) - f(u_-) - V(u - u_-), \tag{49}$$

called the *profile equation*. The proof proceeds by adding to the system (49) the trivial ODE $V' = 0$. The resulting system is autonomous and one reduces its flow to that on the *center manifold* at $(u_-, \lambda_k(u_-))$.

We do not even need that $B(u)$ be non-singular; the equation (49) might be a *differential-algebraic system*. We only need (48). This situation was investigated by Pego [27]. Remarkably enough, Theorem 12 tells us that the existence of a viscous profile does not really depend on which 'reasonable' tensor B we deal with, as long as the strength of the discontinuity is small enough. We warn the reader that this is no longer true for large discontinuities; it is possible to construct examples for which this existence does depend on the choice of B.

It is not surprising that the existence of a viscous profile has something to do with other admissibility criteria, for instance with the Lax shock condition. A viscous profile is a heteroclinic orbit of (49) between two equilibria $u_\pm$. In the case of artificial viscosity, the number of positive numbers $\lambda_j(u_-) - V$ equals the dimension of the unstable manifold $W^u(u_-)$ of (49), and the number of negative numbers $\lambda_j(u_+) - V$ equals the dimension of the stable manifold $W^s(u_+)$. Thus the Lax shock condition tells us that the sum of these dimensions equals $n + 1$; this amounts to saying that the profile is structurally stable, which roughly means that it persists under small disturbances of either the data $(u_-, u_+; V)$ within the Hogoniot locus, or the profile equation itself. This confirms the comment made above, that the existence of a viscous profile does not depend much upon the tensor B.

4.6 Stable Viscous Profile

Even if the existence of a viscous profile is a significant improvement of the entropy inequality or of the Lax shock inequalities, it is not the end of the story. A shock profile is associated with explicit solutions u^ϵ through (45). These are travelling waves which, up to the choice of a moving frame, may be assumed stationary ($V = 0$). Their dynamical stability is encoded at the leading order into a second-order linearized operator L_ϵ. In the case of artificial viscosity, we have

$$L_\epsilon z := \frac{d}{dx}\left(\frac{dz}{dx} - \mathrm{d}f(u^\epsilon(x))z\right) = \frac{d}{dx}\left(\frac{dz}{dx} - \mathrm{d}f \circ U\left(\frac{x}{\epsilon}\right)z\right).$$

It is not hard to see that L_ϵ is conjugated, through the rescaling $x \mapsto \epsilon x$, to $\frac{1}{\epsilon}L_1$. Its spectrum is therefore ϵ^{-1} times that of L_1. It is possible to show, under very reasonable assumptions, that the continuous part of the L^2-spectrum of L_1 belongs to the left half-plane $\Re z \leq 0$. Besides, $\lambda = 0$ is an eigenvalue of L_1, because of the identity

$$L_1 \frac{dU}{dx} = 0.$$

But if L_1 admits an eigenvalue λ of positive real part, then L_ϵ has the eigenvalue λ/ϵ; this yields an amplification of small initial disturbances by a factor

$$\exp \frac{t\Re\lambda}{\epsilon},$$

which becomes enormous at fixed time $t > 0$ when $\epsilon \to 0^+$. In practice, the travelling wave cannot be observed because the disturbance is so big that it overcomes the profile itself in time t of order ϵ.

For this reason, we declare that a viscous profile is admissible only when the spectrum of L_1 is contained in the left half $\Re z \leq 0$ of the complex plane. For technical reasons, we even prefer to say that a profile is *strongly stable* if this spectrum is contained to the left of some parabola $\mathcal{P}$ whose tip is at the origin, and the zero eigenvalue is simple. When dealing with moving shocks ($V \neq 0$), the tip is shifted to $-iV$, which is an eigenvalue. With Fredholm theory, one can show that $(z - L_1)^{-1}$ is a bounded operator over L^2 for every z to the right of the parabola $\mathcal{P}$, and the map $z \mapsto (z - L_1)^{-1}$ is holomorphic. This is at the basis of the stability analysis, initiated by Sattinger [30] and culminating in a list of papers by Zumbrun and coll., see for instance [40] and the references herein.

It turns out that small amplitude viscous profiles found in Theorem 12 (see [10]) are strongly stable. The same is true for profiles of scalar equations, because of the following simple arguments (wlog, we assume that $V = 0$):

- L_1 is a Sturm–Liouville operator, hence its eigenvalues are real, and the largest one is associated with the unique positive eigenfunction,
- As mentioned above, one has $L_1 U' = 0$ and $U'(\pm\infty) = 0$, thus 0 is an eigenvalue of L_1,
- The eigenfunction U' has a constant sign, because the solutions of a scalar equation (47) are monotonous,
- Hence $\lambda = 0$ is the largest eigenvalue of L_1.

In conclusion, instability happens only for systems ($n \geq 2$), and only if the amplitude of the profile is large enough. We point out that $\lambda = 0$ is a generalized eigenvalue of the adjoint operator L_1^*, and the constants form an n-dimensional generalized eigenspace.[7] But there is nothing analogous to the Sturm–Liouville theory for systems of second-order differential operators. Therefore we cannot draw a general conclusion about the spectrum of L_1.

In a celebrated paper [12], Gardner and Zumbrun showed that the simplicity of the eigenvalue $\lambda = 0$ ($= -iV$ in the general case) implies the Liu–Majda condition (42). Therefore the latter is a necessary condition for strong stability. When the shock is extreme, that is when it is a 1-shock or an n-shock, they actually proved a stronger result, which we describe now. By symmetry, we may consider an n-shock. Let us assume (48). Then the unstable manifold of equation (49) at u_- has dimension one, and the normalized derivative $|U'(s)|^{-1}U'(s)$ admits a limit s_n^- as $s \to -\infty$. The main result of [12] when $n = 2$, extended to general values of n in [2], is that the parity of the number of unstable eigenvalues (those of L_1 with positive real part) is given by the sign of the "stability index"

$$\iota := (\ell_n(u_-)(u_+ - u_-)) \times \left(\ell_n(u_-)s_n^-\right).$$

More precisely, the number of unstable eigenvalues is odd if and only if $\iota < 0$. We point out that the first factor $\ell_n(u_-)(u_+ - u_-)$ equals Liu–Majda's determinant $\Delta(0,1)$. Therefore, a necessary condition for strong stability is that $\Delta(0,1)$ has the same sign as $\ell_n(u_-)s_n^-$. In geometric terms, this condition is that the profile U crosses the affine hyperplane

$$u_- + \mathrm{Span}\{r_1(u_-), \dots, r_{n-1}(u_-)\}$$

in an even number of points. For small shocks, the profile given by Theorem 12 does not cross this hyperplane at all.

4.7 Multi-dimensional Stability of Viscous Profiles

Let us now consider a viscous shock profile $U((x \cdot \xi - Vt)/\epsilon)$ of a system (23) in d space variables. The physically relevant viscosity tensors will be described in Sect. 5, and there is a rather general stability theory to treat them. But for the sake of clarity, we content ourselves with the simplest situation of artificial viscosity

$$\partial_t u + \mathrm{Div} F(u) = \epsilon \Delta U. \tag{50}$$

Up to the choice of a Galilean frame, we may assume that $\xi = \vec{e}_1$ and $V = 0$. Therefore $u^\epsilon(x,t) = U(\epsilon^{-1}x_1)$ is a steady solution of (50). The linearized operator L_ϵ at u^ϵ is again conjugated to $\frac{1}{\epsilon}L_1$ and therefore the stability of u^ϵ requires that the spectrum of L_1 be contained in the left half of the complex plane. We have

[7] This is associated with the conservation of mass.

$$L_1 z = \Delta z - \mathrm{Div}(\mathrm{d}F \circ U(x_1)z).$$

Because the coefficients of this differential operator depend only on x_1, a partial decoupling is obtained by conjugating L_1 by the Fourier transform in $y = (x_2, \ldots, x_d)$. If $\eta \in \mathbb{R}^{d-1}$ is the frequency variable, we obtain a family of operators $L_{1,\eta}$, which operate over functions of x_1 by

$$L_{1,\eta} h = h'' - |\eta|^2 h - ((\mathrm{d}f^1 \circ U)h)' - i \left(\sum_{\alpha=2}^{d} \eta_\alpha \mathrm{d}f^\alpha \circ U \right) h.$$

The spectrum of L_1 is the union of the spectra of $L_{1,\eta}$ as η varies. We point out that $L_{1,0}$ is nothing but the linearized operator about U of the one-dimensional system $\partial_t u + \partial_1 f^1(u) = \partial_1^2 u$; therefore the multi-dimensional stability of U necessitates its one-dimensional stability, and a little (a lot?) more.

We lack time and space to develop this stability theory here. But let us mention that it is tied to the FBP approach:

Theorem 13 (Zumbrun and Serre [41]). *Under rather natural assumptions on the structure of the system (54), let us consider a viscous profile U for a Lax shock. Then the Lopatinskiĭ condition is the small frequency asymptotics of the spectral stability of U, in the following sense:*

If $\Delta(\eta, \tau)$ vanishes for some pair $\eta \in \mathbb{R}^{d-1}$ and $\Re\tau > 0$, then for $0 < s \ll 1$, the operator $L_{1,s\eta}$ admits an eigenvalue $\lambda(s\eta) \sim s\tau$. In particular, $\Re\lambda(s\eta)$ is positive and the profile is linearly unstable.

Corollary 1. *As an admissibility condition, the multi-dimensional stability of the viscous profile is stronger than or equal to the Lopatinskiĭ condition.*

4.8 Conclusion

An overview of the various admissibility condition for shocks is given in Fig. 1. Points represent pairs (system, shock). When a set A is contained in B, the corresponding criterion C_A implies C_B.

Our presentation above might lead the reader to think that there is no room for those shocks that do not satisfy the Lax inequalities (38). This is a drawback of our choice to present the FBP side of the theory. But there are circumstances where some non-Lax shocks must be accepted in order to have a well-posed Cauchy Problem. This happens in systems whose strict hyperbolicity is lost somewhere; that could be because hyperbolicity is lost, or just two or more eigenvalues cross at some states. We may distinguish under-compressive discontinuities ($j \geq k$ in (37)) or over-compressive ones ($j \leq k - 2$).

Under-compressive shocks may be treated with the same tools as Lax shocks, by studying the well-posedness of some Free-Boundary value problem, but where

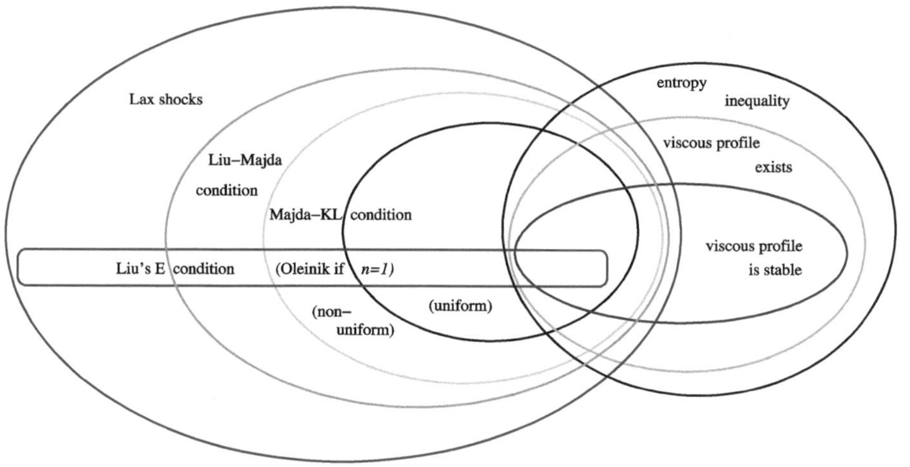

Fig. 1 How the admissibility conditions interact; the shock strength is arbitrary

$j - k + 1$ algebraic jump conditions are added to the Rankine–Hugoniot relations. This method was developed by Freistühler [8]. The additional jump condition might well express the existence of a viscous profile; as a matter of fact, such profiles are not structurally stable and therefore exist only for triples $(u_-, u_+; s)$ in a submanifold of codimension $j - k + 1$ of the Hugoniot locus. We warn the reader that in contrast with the case of Lax shocks, the existence of a profile for an under-compressive shock depends crucially upon the viscosity tensor. The profile may or may not be spectrally stable; the calculation of the Evans function in the case of a 2×2 system was carried out by Gardner and Zumbrun [12].

Over-compressive shocks are subtle in another way. When they admit a viscous profile, it is far from being unique up to a space shift. Typically, the union of the trajectories from u_- to u_+ forms a manifold of dimension $k - j$. They may or may not be spectrally stable (again, see [12]) but a strange phenomenon may happen: Freistühler and Liu [9] observed that if the viscosity tensor is $B_\epsilon = \epsilon B$, then their nonlinear stability vanishes as $\epsilon \to 0$.

Small shocks. When the shock amplitude $|u_+ - u_-|$ is small, we know that Liu's E condition is equivalent to the existence of a viscous profile. We cannot say more, unless we assume that the characteristic field to which the discontinuity is associated is genuinely non-linear. When it is so, it is expected that all the admissibility conditions will be equivalent to each other. When $B \equiv I_d \otimes I_n$ (artificial viscosity: the diffusion is induced by Δu), Freistühler and Szmolyan [10] proved that extreme shocks (1-shocks or n-shocks) satisfy the uniform (MKL). On the other hand, it is classical that for a genuinely non-linear field, (38) is equivalent to the entropy inequality in the small.

5 Viscosity and Dissipativity

As mentionned above, many among the first-order systems of conservation laws are not accurate,[8] because they are only approximations of higher-order models which take into account dissipative physical processes. For instance, the Euler system (see Sect. 3.2) is a simplification of the Navier-Stokes–Fourier system

$$\partial_t \rho + \operatorname{div}(\rho \mathbf{v}) = 0, \tag{51}$$

$$\partial_t (\rho \mathbf{v}) + \operatorname{Div}(\rho \mathbf{v} \otimes \mathbf{v}) + \nabla p = \operatorname{Div} T, \tag{52}$$

$$\partial_t \left(\frac{1}{2} \rho |\mathbf{v}|^2 + \rho e \right) + \operatorname{div}\left(\left(\frac{1}{2} \rho |\mathbf{v}|^2 + \rho e + p \right) \mathbf{v} \right) = \operatorname{div}(T\mathbf{v} + \kappa \nabla \theta). \tag{53}$$

The equations above involve two additional objects: a symmetric tensor accounting for Newtonian viscosity, given by

$$T := \mu(\nabla \mathbf{v} + \nabla \mathbf{v}^T) + \zeta(\operatorname{div} \mathbf{v}) I_d$$

and the temperature $\theta = \theta(\rho, e)$. The dissipation coefficients μ, ζ and κ are positive and may depend upon ρ and θ. Although we do not assume their smallness, we may derive the Euler system as an approximation, by rescaling (51–53) *via* $(x,t) \mapsto (x/\epsilon, t/\epsilon)$ and letting $\epsilon \to 0$. In other words, we expect that the Euler system will be accurate on large scales.

More generally, given a general first-order system (23) in conservation form, we consider systems that are obtained by introducing first-order terms in the spatial flux, that is by replacing $f^\alpha(u)$ by

$$f^\alpha(u) - \sum_\beta B^{\alpha\beta}(u) \partial_\beta u, \qquad B^{\alpha\beta} : \mathcal{U} \to \mathbf{M}_n(\mathbb{R}).$$

Such systems can be written in the abstract form

$$\partial_t u + \operatorname{Div} F(u) = \operatorname{Div}(B(u) \nabla u), \tag{54}$$

where $B(u)$ must be seen as a linear operator within the space $\mathbf{M}_{n \times d}(\mathbb{R})$, acting by

$$G \mapsto H = B(u)G, \qquad h_{i\alpha} := \sum_{\beta,j} B_{ij}^{\alpha\beta}(u) g_{j\beta}.$$

[8]There is at least one notable exception, namely the Maxwell system governing the electromagnetic field in the vacuum. If it was dissipative, our world would be completely dark after billions of years. There are also the Einstein equations of the gravitational field and more generally all models dealing with fundamental forces.

Again, a space-time dilation followed by a limit towards the small frequencies gives back (23).

When dealing with the class of models (54), we face a few natural questions:

Q.1. Suppose that (23) is endowed with an entropy-flux pair with a strongly convex entropy. What can be the most general form of the dissipation tensor B so that $\int \eta(u)\,dx$ is a Lyapunov function? Here, we assume that u tends to a constant state u_∞ at infinity, and η is normalized by $\eta(u_\infty) = 0$ and $\eta \geq 0$.

Q.2. If the system admits many independent entropies, is there a canonical entropy from the point of view of (54)? For instance, is there a canonical entropy-temperature pair in gas dynamics?

Q.3. Is there a symmetrization similar to that of Godunov, or that of Lax–Friedrichs? Can it be used to treat the Cauchy problem?

Q.4. When the Cauchy problem is well-posed, but the dissipative tensor depends on parameters, how does the solution depend on them, especially when the rank of B drops in the limit? Such situations are called singular limits.

Q.5. Can dissipation ensure global existence and regularity, at least when the initial data is small and smooth?

We point out that there is absolutely no reason why $B(u)$ would be a symmetric operator with respect to the scalar product

$$\langle G, H\rangle = \mathrm{Tr}(G^T H) = \sum_{\alpha,i} g_{\alpha i} h_{\alpha i}.$$

We shall see instead that a natural symmetry occurs when rewriting the system in a different set of variables. This is reminiscent of *Onsager's reciprocity relations*, and contributes to a positive answer to Q.4.

Another important remark is that the action of the viscous tensor concerns only certain equations of the system, not all of them. For instance, it is absent from the conservation of mass (51). Therefore the resulting system (54) is not fully parabolic. Because it is obviously not hyperbolic, we anticipate that it mixes both hyperbolic and parabolic features. This composite aspect is involved in the answer to Q.2.

5.1 *Strong Versus Weak Dissipativeness*

We assume from now on that the first-order part (the left-hand side) of the 'viscous' system (54) admits a strongly convex entropy η, whose flux is $\vec{q}$. We may always assume that $\eta \geq 0$ everywhere, with $\eta(u_\infty) = 0$ at some point. Then we are interested in solutions that tend towards u_∞ as $|x| \to \infty$.

For a classical solution, the multiplication of (54) by $\mathrm{d}\eta(u)$, and the chain rule yield

$$\partial_t \eta(u) + \mathrm{div}\vec{q}(u) + \sum_{\alpha,\beta} \mathrm{D}^2_u \eta(B^{\alpha\beta}(u)\partial_\beta u, \partial_\alpha u) = \sum_{\alpha,\beta} \partial_\alpha(\mathrm{d}\eta(u) \cdot B^{\alpha\beta}(u)\partial_\beta u).$$

Roughly speaking, the system *dissipates the entropy* if

$$\mathcal{D}[u] := \int_{\mathbb{R}^d} \sum_{\alpha,\beta} \mathrm{D}_u^2\eta(B^{\alpha\beta}(u)\partial_\beta u, \partial_\alpha u)\, dx$$

is non-negative; this quantity is called the *dissipation rate*. Then, assuming that $\eta(u), \vec{q}(u)$ as well as ∇u decay fast enough at infinity, we obtain the property that

$$t \mapsto \int_{\mathbb{R}^d} \eta(u(x,t))\, dx$$

is non-increasing, and therefore

$$\int_{\mathbb{R}^d} \eta(u(x,t))\, dx \leq \int_{\mathbb{R}^d} \eta(u(x,0))dx.$$

This is the first a priori estimate on the way of the well-posedness theory. For a reason that will become clear in a moment, it is far from sufficient, and we also want to estimate the flux $B(u)\nabla u$. This can be done only by using the dissipation rate in order. Hence the definition:

Definition 4. The entropy η of (SCL) is *strongly dissipated* by the viscous tensor B if, for every constant state $\bar{u}$, there exists an $\omega(\bar{u}) > 0$ such that

$$\sum_{\alpha,\beta} \mathrm{D}_{\bar{u}}^2\eta(B^{\alpha\beta}(\bar{u})X_\beta, X_\alpha) \geq \omega(\bar{u}) \sum_\alpha |\sum_\beta B^{\alpha\beta}(\bar{u})X_\beta|^2, \qquad \forall X_1,\ldots,X_d \in \mathbb{R}^n. \tag{55}$$

The sum on the right-hand side can be written as $|B(\bar{u})\mathbb{X}|^2$, where $\mathbb{X} = (X_1,\ldots,X_d) \in \mathbf{M}_{n\times d}(\mathbb{R})$.

Example: If $\eta(u) = \frac{1}{2}|u|^2$, this precisely means that $B(u)$ satisfies $\langle B(u)G, G\rangle \geq \omega(u)|B(u)G|^2$ for every $G \in \mathbf{M}_{n\times d}(\mathbb{R})$. This is certainly true if $B(u)$ is symmetric and semi-definite positive, but the symmetry is not absolutely necessary.

Strong dissipation gives immediately the inequality

$$\partial_t\eta(u) + \mathrm{div}\vec{q}(u) + \omega(u)|B(u)\nabla u|^2 \leq \sum_{\alpha,\beta} \partial_\alpha(\mathrm{d}\eta(u)\cdot B^{\alpha\beta}(u)\partial_\beta u), \tag{56}$$

which gives a little more than the decay of the total entropy:

$$\frac{d}{dt}\int_{\mathbb{R}^d} \eta(u)dx + \int_{\mathbb{R}^d} \omega(u)|B(u)\nabla u|^2 dx \leq 0. \tag{57}$$

From this, we deduce a second a priori estimate

$$\int_0^T \int_{\mathbb{R}^d} \omega(u)|B(u)\nabla u|^2 dx\, dt \le \int_{\mathbb{R}^d} \eta(u(x,0))dx. \tag{58}$$

We emphasize that strong dissipation, or even its weakest form

$$\sum_{\alpha,\beta} \mathrm{D}^2_{\bar{u}}\eta(B^{\alpha\beta}(\bar{u})X_\beta, X_\alpha) \ge 0, \qquad \forall X_1,\dots,X_d \in \mathbb{R}^n$$

is only a sufficient condition for having $\mathcal{D}[u] \ge 0$, but it is not necessary. Cancellations may occur because of integration, even when the left-hand side of (55) is not positive semi-definite over $\mathbf{M}_{n\times d}(\mathbb{R})$. Again, because we wish to control the additional flux $B(u)\nabla u$, we actually ask that

$$\mathcal{D}[u] \ge \int_{\mathbb{R}^d} \omega(u)|B(u)\nabla u|^2 dx \tag{59}$$

for some positive ω. It is not easy to characterize the latter property in general, because of the dependence of $\mathrm{D}^2\eta$ and B upon u. But if $\eta(u) = \frac{1}{2}u^T Su$ is quadratic and B is constant, then $\mathcal{D}$ is just a quadratic form

$$u \mapsto \int_{\mathbb{R}^d} \sum_{\alpha,\beta} (SB^{\alpha\beta}\partial_\beta u, \partial_\alpha u)\, dx.$$

Its non-negativity is equivalent to the Legendre–Hadamard condition that $(SB(\xi)X, X) \ge 0$ for every $X \in \mathbb{R}^n$ and $\xi \in \mathbb{R}^d$, where the symbol $B(\xi)$ is defined by

$$B(\xi) = \sum_{\alpha,\beta} \xi_\alpha \xi_\beta B^{\alpha\beta}.$$

The slightly stronger condition (59) is instead $(SB(\xi)X, X) \ge \omega \sum_\alpha |B^\alpha(\xi)X|^2$, where

$$B^\alpha(\xi) = \sum_\beta \xi_\beta B^{\alpha\beta}(\bar{u}).$$

This yields our next definition, which is now a necessary condition for (59):

Definition 5. The entropy η of (SCL) is *weakly dissipated* by the viscous tensor B if, for every constant state $\bar{u}$, there exists an $\omega(\bar{u}) > 0$ such that

$$\mathrm{D}^2_{\bar{u}}\eta(B(\bar{u};\xi)X, X) \ge \omega(\bar{u}) \sum_\alpha |B^\alpha(\bar{u};\xi)X|^2, \qquad \forall X \in \mathbb{R}^n,\ \forall \xi \in \mathbb{R}^d.$$

In this inequality, the symbols $B(u;\xi)$ and $B^\alpha(u;\xi)$ are defined as above, with $B = B(u)$.

In one space dimension, both strong and weak dissipativeness coincide, and mean

$$\mathrm{D}^2_{\bar{u}}\eta(B(\bar{u})X, X) \geq \omega(\bar{u})|B(\bar{u})X|^2, \qquad \forall X \in \mathbb{R}^n. \tag{60}$$

Because of Proposition 7, the case $X = r_k(\bar{u})$ in (60) gives

$$\ell_k(\bar{u})B(\bar{u})r_k(\bar{u}) \geq 0.$$

If moreover

$$Br_k(\bar{u}) \neq 0, \tag{61}$$

we even have the strict inequality, which is nothing but the assumption (48) under which the existence of viscous profiles was proved for shocks of small amplitude. The generic condition (61) is the so-called *Kawashima condition*, which says that the kernel of $B(u)$ does not contain the eigenvectors of $A(u)$. It can be seen as a property of genuine coupling between the hyperbolic and the parabolic part of the system (54), at least in one space dimension. It has deep consequences, for instance in terms of decay to equilibrium, but we shall not develop these aspects here.

In several space dimensions, strong dissipativeness implies weak dissipativeness; we only need to choose $X_\alpha = \xi_\alpha X$, that is to restrict to rank-one tensors $\mathbb{X} = X \otimes \xi$.

When (59) is true, we obtain the a priori estimates

$$\int_{\mathbb{R}^d} \eta(u(t,x))dx \leq \int_{\mathbb{R}^d} \eta(a(x))dx,$$

$$\int_0^T \int_{\mathbb{R}^d} \omega(u)|B(u)\nabla u|^2 dx\,dt \leq \int_{\mathbb{R}^d} \eta(a(x))dx.$$

We point out that because $B(\bar{u};\xi)$ may be singular, ∇u is *not* controlled in $L^2_{x,t}$. Only $B(u)\nabla u$ is controlled.

Vanishing viscosity limit. We now assume that the dissipation process is of small intensity ϵ ($0 < \epsilon \ll 1$):

$$\partial_t u^\epsilon + \mathrm{Div}\, F(u^\epsilon) = \epsilon \sum_{\alpha,\beta} \partial_\alpha (B^{\alpha\beta}(u^\epsilon)\partial_\beta u^\epsilon).$$

With a fixed initial data $a(x)$, we consider the limit $\epsilon \to 0^+$. One has

$$\epsilon \int_0^T \int_{\mathbb{R}^d} \omega(u^\epsilon)|B(u^\epsilon)\nabla u^\epsilon|^2 dx\,dt \leq \int_{\mathbb{R}^d} \eta(a(x))dx,$$

whence

$$\epsilon \mathrm{Div}(B(u^\epsilon)\nabla u^\epsilon) = \epsilon^{1/2}\mathrm{Div}(\underbrace{\epsilon^{1/2}B(u^\epsilon)\nabla u^\epsilon}_{\text{bounded in } L^2_{x,t}}) \to 0 \quad \text{in } H^{-1}_{x,t}.$$

If $u^\epsilon(t,x) \to u(t,x)$ boundedly a.e.,[9] we may pass to the limit in (54) and obtain

$$\partial_t u + \mathrm{Div} F(u) = 0.$$

This tells us that u is a weak solution. In addition strong dissipativeness tells us that

$$\partial_t \eta(u^\epsilon) + \mathrm{div}\vec{q}(u^\epsilon) \leq \epsilon \sum_{\alpha,\beta} \partial_\alpha(\mathrm{d}\eta(u^\epsilon)B^{\alpha\beta}(u^\epsilon)\partial_\beta u^\epsilon) \to 0 \quad \text{in } H^{-1}_{x,t},$$

whence in the limit

$$\partial_t \eta(u) + \mathrm{div}\vec{q}(u) \leq 0.$$

5.2 *Algebraic Facts*

We now make a natural assumption, which is met by all the physical examples we have at hand:

(H1) The range of $B(\bar{u};\xi)$ is independent of $(\bar{u}, \xi \neq 0)$.

For instance, in the Navier-Stokes–Fourier system (51–53), we have $R(B(\bar{u};\xi)) \equiv \{0\} \times \mathbb{R}^{d+1}$, while in the Euler–Fourier system (no Newtonian viscosity, that is $T = 0$, but with heat diffusion, that is $\kappa > 0$), we have $R(B(\bar{u};\xi)) \equiv \{0\}^{d+1} \times \mathbb{R}$. Up to a linear change of the unknowns, we may always assume that the above mentioned range is of the form $\{0\}^p \times \mathbb{R}^{n-p}$.

Theorem 14 ([34]). *Let us assume either strong or weak dissipativeness and the block structure (H1). Without loss of generality, assume that $R(B(u;\xi)) \equiv \{0\}^p \times \mathbb{R}^{n-p}$ for every $u \in \mathcal{U}$ and $\xi \neq 0$:*

$$B(\bar{u};\xi) = \begin{pmatrix} 0_{n\times p} \\ b(\bar{u};\xi) \end{pmatrix} \qquad (\textit{hence } b(\bar{u};\xi) \in \mathbf{M}_{p\times(n-p)}(\mathbb{R}) \textit{ is onto}).$$

Set $z_j := \frac{\partial \eta}{\partial u_j}$ for $j = p+1, \ldots, n$. Then

1. $u \mapsto (u_1, \ldots, u_p, z_{p+1}, \ldots, z_n)$ is a change of variables.

[9]The big open problem!

2. *The tensor $b(u)\nabla u$ can be rewritten as $R(u)\nabla z$.*
3. *Weak dissipation is that R satisfies the Legendre–Hadamard condition:*

$$\langle R(u;\xi)Y,Y\rangle \geq \omega(u)|\xi|^2|Y|^2, \qquad \forall\, \xi \in \mathbb{R}^d,\; Y \in \mathbb{R}^{n-p},$$

where we define as usual

$$R(u;\xi) = \sum_{\alpha,\beta} \xi_\alpha \xi_\beta R^{\alpha\beta}(u).$$

4. *Strong dissipation is that R satisfies*

$$\langle R(u)F,F\rangle \geq \omega(u)|R(u)F|^2, \qquad \forall\, F \in \mathbf{M}_{(n-p)\times d}(\mathbb{R}).$$

Vanishing viscosity limit under weak dissipation: Suppose that the dissipation tensor has the form $\epsilon B(u)$, with $\epsilon > 0$ tending to zero. Suppose in addition that the corresponding solution u^ϵ of the Cauchy problem with initial data a converges boundedly almost everywhere to a limit $u(x,t)$. We know that u is a weak solution of the inviscid Cauchy problem for (23). If the system is strongly dissipative, we have seen that u satisfies the entropy inequality (34). Is it also true if the system is only weakly dissipative? To our knowledge, this is an open question.

Let us see where the difficulty lies. A straightforward calculation gives us the formula

$$\mathrm{D}^2\eta(B\nabla u,\nabla u) = \langle R\nabla z,\nabla z\rangle,$$

which yields the integral identity

$$\frac{d}{dt}\int_{\mathbb{R}^d} \eta(u)\,dx + \epsilon\int_{\mathbb{R}^d} \langle R(u)\nabla z,\nabla z\rangle\,dx = 0.$$

If the system is weakly dissipative, the last integral may not be positive. Of course, it must be greater than $\|\nabla z\|_{L^2}$ if R has constant coefficients. In the general case, we expect that it will be positive, up to a correction that can be controlled by the entropy itself. What we have in mind is a kind of Gårding inequality; a simple version would be

$$\int_{\mathbb{R}^d} \langle R(u)\nabla z,\nabla z\rangle\,dx \geq \omega\|\nabla z\|_{L^2} - \zeta\left(\int_{\mathbb{R}^d} \eta(u)\,dx\right)$$

for some increasing function ζ. If such an inequality holds true, then a Gronwall argument gives us a uniform estimate of

$$\int_{\mathbb{R}^d} \eta(u^\epsilon)\,dx, \qquad \epsilon\int_0^\infty\int_{\mathbb{R}^d} \langle R(u^\epsilon)\nabla z^\epsilon,\nabla z^\epsilon\rangle\,dx\,dt.$$

Because of the ellipticity of R, we have that $\epsilon^{1/2}\nabla z^\epsilon$ is bounded in $L^2_{x,t}$ and therefore converges weakly towards zero. Let us decompose

$$\epsilon\langle R(u^\epsilon)\nabla z^\epsilon, \nabla z^\epsilon\rangle = \epsilon\langle (R(u^\epsilon) - R(u))\nabla z^\epsilon, \nabla z^\epsilon\rangle + \epsilon\langle R(u)\nabla z^\epsilon, \nabla z^\epsilon\rangle.$$

Up to a subsequence, the last product has a non-negative limit in $\mathcal{D}'$ because of compensated compactness (see [26,37]). But even if u^ϵ converges boundedly almost everywhere to $u(x,t)$, we cannot infer that $\epsilon\langle (R(u^\epsilon) - R(u))\nabla z^\epsilon, \nabla z^\epsilon\rangle$ converges to zero; this was pointed out to me by L. Tartar. This convergence would hold true if $u^\epsilon \to u$ uniformly, but this is not an interesting case, since u would be continuous and therefore we should not need an admissibility condition.

5.2.1 Application to Gas Dynamics

Theorem 14 gives an answer to **Q.2**, which can be reformulated as follows.

QUESTION. Why is it the same quantity (the temperature) that occurs in Gibbs' relation

$$\theta \mathrm{d}s = \mathrm{d}e + p\mathrm{d}\frac{1}{\rho}$$

and in Fourier law

$$\text{heat flux} = -\kappa\nabla\theta \qquad ?$$

Answer. In full gas dynamics ($n = d + 2$), the conserved variables are

$$u_0 = \rho, \quad u_j = \rho \mathbf{v}_j, \quad u_{d+1} = \frac{1}{2}\rho|\mathbf{v}|^2 + \rho e.$$

The mathematical entropy is $\eta = -\rho s$ (s the physical entropy). It is an interesting exercise to check that, if θ is given by Gibb's relation, then

$$z_j = \frac{\partial \eta}{\partial u_j} = \frac{\mathbf{v}_j}{\theta} \quad (j = 1, \dots, d) \quad \text{and} \quad z_{d+1} = \frac{\partial \eta}{\partial u_{d+1}} = -\frac{1}{\theta}.$$

Therefore, Theorem 14 tells us that in NSF, the dissipation tensor $B(u)\nabla u$ must be a linear combination of the gradients of $\frac{\mathbf{v}}{\theta}$ and $\frac{1}{\theta}$, or equivalently of $\nabla\mathbf{v}$ and $\nabla\theta$. Newton's law of viscosity and Fourier's law of heat conduction agree with this conclusion.

For the Euler–Fourier system, the theorem implies instead an identity of the form

$$b(u)\nabla u = r(u)\nabla z_{d+1} = \kappa(u)\nabla\theta,$$

as predicted by the Fourier law.

We remark that the Euler system admits other entropies, of the form $\eta_h := -\rho h(s)$ where h is any numerical function. However, the 'temperature' associated with η_h is $\theta/h(s)$. Its gradient is not colinear to that of θ, unless[10] h is affine. Hence Theorem 14 implies that η_h is not dissipated. Only η can be dissipated.

Now that we know what the variables z are, we may calculate the tensor R. Tedious calculations yield

$$R^{\alpha\beta}_{\rm NSF}(u) = \theta^2 \begin{pmatrix} \frac{1}{\theta}\left[\mu(\delta_\alpha^\beta I_d + \mathbf{e}_\beta \mathbf{e}_\alpha^T) + \zeta \mathbf{e}_\alpha \mathbf{e}_\beta^T\right] & \mu(z_\alpha \mathbf{e}_\beta + \delta_\alpha^\beta z) + \zeta z_\beta \mathbf{e}_\alpha \\ \mu(z_\beta \mathbf{e}_\alpha^T + \delta_\alpha^\beta z^T) + \zeta z_\alpha \mathbf{e}_\beta^T & \kappa\delta_\alpha^\beta + \theta\left[\mu|z|^2\delta_\alpha^\beta + (\mu+\zeta)z_\alpha z_\beta\right] \end{pmatrix},$$

where the $\mathbf{e}_j$'s denote the vectors of the canonical basis of $\mathbb{R}^d$. We remark that these matrices satisfy

$$(R^{\alpha\beta})^T = R^{\beta\alpha}, \qquad \forall 1 \le \alpha, \beta \le d. \tag{62}$$

This amounts to saying that $R(u)$ is symmetric, as a linear operator over $\mathbf{M}_{(d+1)\times d}(\mathbb{R})$, endowed with the scalar product $\langle F, G\rangle := \mathrm{Tr}(F^T G)$; we have

$$\langle RF, G\rangle = \langle RG, F\rangle.$$

Decomposing the standard matrix $F \in \mathbf{M}_{(d+1)\times d}(\mathbb{R})$ into an upper block $M \in \mathbf{M}_d(\mathbb{R})$ and a lower row ℓ, and forming $H := M + \mathbf{v}\ell$, we have

$$\langle RF, F\rangle = \frac{\theta\mu}{2}\|H + H^T\|^2 + \theta\zeta(\mathrm{Tr}\, H)^2 + \theta^2\kappa\|\ell\|^2,$$

where the norms are Euclidean. The NSF system is weakly dissipative if and only if κ, μ and $2\mu + \zeta$ are non-negative. It is strongly dissipative if κ, μ and $2\mu + d\zeta$ are non-negative. If in addition $\kappa > 0$ and $2\mu + d\zeta > 0$, the kernel of R equals $\mathbf{Skew}_d \times \{0\}$; therefore R is never positive definite.

Isentropic Navier-Stokes. This is the system formed by (51, 52) where p is a function of the density only. There is no temperature. The dissipation tensor is just T. The "entropy" is played by the energy density

$$\eta := \frac{1}{2}\rho|\mathbf{v}|^2 + \rho e(\rho), \quad e(\rho) = \rho\int^\rho p'(s)\frac{ds}{s^2}\,.$$

The variable z is played by $\mathbf{v}$, thus R is given by T in the most direct way. Again R is symmetric and we have

[10] With the exception of the unphysical case $s = s(\theta)$, which means that $p = p(\rho)$ only.

$$\langle RF, F\rangle = \frac{\mu}{2}\|F + F^T\|^2 + \zeta(\operatorname{Tr} F^2).$$

Again, weak dissipation is that $\mu, 2\mu + \zeta \geq 0$, and strong dissipation occurs when $\mu \geq 0$ and $2\mu + d\zeta \geq 0$.

5.3 Onsager's Reciprocity Relations

Point 3 in Theorem 14 suggests that a symmetry property, which was irrelevant for B, could be relevant for the tensor R. This symmetry is the content of *Onsager's reciprocity relations*. At the origin, these relations concerned ordinary differential systems, such as those encountered in chemical kinetics, but they can be extended to every dissipative system endowed with a Lyapunov function. The calculations above give us a striking example, namely the Navier-Stokes–Fourier system.

When R is non-negative over $\mathbf{M}_{(n-p)\times d}(\mathbb{R})$, its symmetry implies the estimate

$$|R(u)F|^2 \leq \|R(u)\|\langle R(u)F, F\rangle, \qquad \forall F \in \mathbf{M}_{(n-p)\times d}(\mathbb{R}).$$

The non-negativity of R is thus related to the strong dissipation. However it does not imply that $R(u)$ is positive definite. For instance, we have seen that the kernel of $R_{\mathrm{NSF}}(u)$ is equal to $\mathbf{Skew}_d(\mathbb{R})\times\{0\}$. It will be important in the sequel that this kernel does not depend on $u \in \mathcal{U}$. We remark in passing that this forbids the (unphysical) case in NSF where $\kappa = 0$ but $2\mu + d\zeta > 0$ (Newtonian viscosity but no heat conduction), because then the kernel of R does depend upon the state. But it permits the Euler–Fourier case where $\mu = \zeta = 0$ and $\kappa > 0$ (heat conduction but no newtonian viscosity).

When the system is weakly dissipative only, R does not have a constant sign. We may only say that its isotropy cone intersects trivially the cone of rank-one matrices.

5.4 Normal Form and Local Well-Posedness

The normal form of (54) that is appropriate for the study of the Cauchy problem was identified by Kawashima in his PhD thesis [17], who derived it from the dissipativeness and block structure (H1) in [35].

Let us split our unkown u into two blocks v and w of respective sizes p and $n-p$. Theorem 14 tells us that the dissipative flux is linear in the gradient of $z := \mathrm{d}_w\eta$. This suggests to work with the new unknown

$$U := \begin{pmatrix} v \\ z \end{pmatrix}.$$

As pointed out in Theorem 14, the map $u \mapsto U$ is a change of variable. This follows from the strictly convexity[11] of η. We now introduce the symmetric positive definite[12] matrix

$$S_0(U) := \begin{pmatrix} D^2_{vv}\eta - D^2_{vw}\eta\,(D^2_{ww}\eta)^{-1}\,D^2_{wv}\eta & 0 \\ 0 & (D^2_{ww}\eta)^{-1} \end{pmatrix}.$$

The system (54) is equivalent to

$$S_0(U)\partial_t U + \sum_\alpha S_\alpha(U)\partial_\alpha U = \begin{pmatrix} 0 \\ \sum_{\alpha,\beta} \partial_\alpha(R^{\alpha\beta}\partial_\beta z) \end{pmatrix}, \tag{63}$$

where S_α is symmetric. This symmetrization is discussed in detail in [35]. It is at the core of the local well-posedness theory.

Local existence. The main result of [35] is

Theorem 15. *Consider a viscous system of conservation laws (54)*

$$\partial_t u + \sum_\alpha \partial_\alpha f^\alpha(u) = \sum_{\alpha,\beta} \partial_\alpha(B^{\alpha\beta}(u)\partial_\beta u).$$

Assume the following:

- *The maps $u \mapsto f^\alpha(u)$ and $u \mapsto B^{\alpha\beta}(u)$ are smooth over a convex open set $\mathcal{U}$ containing the origin.*
- *System (54) is strongly entropy-dissipative for some smooth strongly convex entropy η.*
- **(H1)** *the range of the symbol matrix $B(\xi;u)$ neither depends on $\xi \neq 0$ in $\mathbb{R}^d$, nor on the state u.*
- **(H2)** *the kernel of $R(u)$ is independent of u and $R(u)$ dominates its u-derivatives up to the order $[s]+1$.*

Then, given an initial data u_0 in $H^s(\mathbb{R}^d)$ with $s > 1 + d/2$, there exists a $T > 0$ and a unique solution in the class

$$u \in C(0,T;H^s), \quad \partial_t u \in L^2(0,T;H^{s-1}).$$

In addition, the component v belongs to $C^1(0,T;H^{s-1})$ and $R(u)\nabla z$ is in $L^2(0,T;H^s)$.

---

[11] Actually, only strict convexity with respect to w is needed here.

[12] The upper-left block is the Schur complement of $D^2_{ww}\eta$ in $D^2\eta$.

Comments:

- The local existence is actually proved for the more general class of systems of the form (63).
- In the latter context, this theorem is due to Kawashima [17], except for two aspects: we do not assume the symmetry of $R(u)$ and we are able to treat $H^{1+d/2+\epsilon}$-initial data, instead of $H^{2+d/2+\epsilon}$ in [17].
- The solution is not fully classical, because $\partial_t w$ might not be better than $L^2(0,T;H^{s-1})$; only $\partial_t v$ is $C(0,T;H^{s-1})$.

5.5 *Singular Limits*

Even if the symmetry of $R(u)$ was not necessary to establish the local well-posedness of the Cauchy problem, it is particularly useful for the study of singular limits. Such limits occur when the tensor B depends on one or several parameters, one of which being small (say ϵ), and when the rank of the limit

$$B_0 = \lim_{\epsilon \to 0} B_\epsilon$$

is strictly smaller than the rank of B_ϵ. The simplest example happens when $B_\epsilon = \epsilon B_1$ (then $B_0 \equiv 0$), which is the *vanishing viscosity limit*. Other examples happen in continuum mechanics when some dissipation processes are negligible and some others are not; the limit of NS–Fourier towards Euler–Fourier is such an example.

A fundamental question in the study of singular limits is whether the local existence of semi-classical solutions given in Theorem 15 is uniform with respect to ϵ. More precisely:

> QUESTION. For a given initial data, let T_ϵ be the existence time and u_ϵ the solution. Is T_ϵ bounded away from zero as $\epsilon \to 0$? Does u_ϵ admit a strong limit u, which will then be a solution of the limit Cauchy problem?

This question has a positive answer under natural assumptions, which are listed now.

Definition 6. Let us assume the symmetry (62) of $R_\epsilon(u)$, for every $\epsilon \in (0,1)$. We say that the family $(B_\epsilon)_{\epsilon\in[0,1]}$ is *stable* if

- The range of $B_\epsilon(\xi;u)$ does not depend upon $\epsilon > 0$ (however it may be, and in general is, different when $\epsilon = 0$),
- The kernel of $R_\epsilon(u)$ does not depend upon $\epsilon > 0$ either,
- The partial derivatives of R_ϵ are uniformly bounded in terms of R_ϵ itself: For every multi-index ℓ of length less than s (s the regularity considered in Theorem 15)

$$\|\partial_u^\ell R_\epsilon(u)F\| \le c_\ell(u)\|R_\epsilon(u)F\|, \tag{64}$$

with c_ℓ independent of ϵ and bounded over compact sets of $\mathcal{U}$.

Discussion

- In other words, we ask that the dissipativeness be satisfied uniformly in $\epsilon > 0$, as much as it could be.
- It should not be meaningful to compare the (fixed) kernel of $R_\epsilon(u)$ with that of $R_0(u)$. Because the range of $B_0(u)$ is strictly smaller than that of $B_\epsilon(u)$, $R_0(u)$ operates on a smaller matrix space than $R_\epsilon(u)$.
- Condition (64) amounts to saying that $\partial_u^\ell R_\epsilon(u) R_\epsilon(u)^\dagger$ remains bounded, where $R^\dagger$ denotes the Moore-Penrose inverse. Because $R(u)$ is symmetric and positive semi-definite, $R(u)^\dagger$ coincides with the usual inverse on the range of $R(u)$, and vanishes over $\ker R(u)$. Because of the symmetry of $R^\dagger$ and of $\partial_u^\ell R_\epsilon(u)$, and the fact that the norm of operators is unchanged under transposition, this is equivalent to saying that

$$\|R_\epsilon(u)^\dagger \partial_u^\ell R_\epsilon(u)\| \le c_\ell(u). \tag{65}$$

This notion yields the following stability result.

Theorem 16 ([36]). *Let us consider a system as in Theorem 15 with a viscous tensor $B = B_\epsilon(u)$ and a flux f independent of ϵ. We assume in addition that $R(u)$ is a symmetric tensor and that the family $(B_\epsilon)_{\epsilon\in[0,1]}$ is stable in the above mentioned sense.*

Let u_0 in $H^s(\mathbb{R}^d)$ with $s > 1 + d/2$ be a given initial data, independent of ϵ. Let u^ϵ denote the solution obtained in Theorem 15. Then there exists a $T > 0$ such that u^ϵ is defined over $(0, T)$ and the following sequences are bounded:

$$u^\epsilon \text{ in } C(0,T;H^s), \quad \partial_t u^\epsilon \text{ in } L^2(0,T;H^{s-1})$$

and

$$v^\epsilon \text{ in } C^1(0,T;H^{s-1}), \qquad R_\epsilon \nabla z^\epsilon \text{ in } L^2(0,T;H^s).$$

If in addition B_ϵ converges uniformly towards B as $\epsilon \to 0$, then u^ϵ converges towards the unique strong solution of the Cauchy problem associated to the viscous tensor B.

Comments.

- Because R_ϵ may behave badly as $\epsilon \to 0$, we do not expect the components z^ϵ to remain bounded in $L^2(0,T;H^{s+1})$.
- The time $T > 0$ mentionned in the theorem might be strictly smaller than $\liminf T_\epsilon$. For instance, if $B_\epsilon = \epsilon\Delta$, one often has $T_\epsilon = +\infty$ for every $\epsilon > 0$, but the classical solution of the inviscid Cauchy problem blows up in finite time.

5.6 Principal Sub-systems and Hyperbolic Modes

We consider here a different kind of singular limit, where

$$B_\epsilon(u) = \frac{1}{\epsilon} B(u), \qquad \epsilon \to 0^+.$$

This situation happens in the large-scale asymptotics. It is interesting in that it reveals that the full system (54) contains a smaller inviscid system of p equations in p unknowns (we recall that p is the dimension of $\ker B(\xi)$ when $\xi \neq 0$). In this way, one explains how the structure of the isothermal Euler system derives from the non-isothermal one.

The a priori estimate is now

$$\int_0^T \int_{\mathbb{R}^d} \omega(u^\epsilon)|B(u^\epsilon)\nabla u^\epsilon|^2 dx\, dt \leq \epsilon \int_{\mathbb{R}^d} \eta(a(x))dx.$$

Together with $B\nabla u = R\nabla z$ and the ellipticity of R, this suggests that the z-component tends to a constant $\bar{z}$ as $\epsilon \to 0$ ($\bar{z}$ is nothing but the value of z at infinity). If v^ϵ converges boundedly a.e., then passing to the limit in the non-viscous part of (54)

$$\partial_t v^\epsilon + \mathrm{Div}\, f_{(1,\dots,p)}(u^\epsilon) = 0$$

yields a *principal sub-system*

$$\partial_t v + \mathrm{Div} g(v) = 0, \tag{66}$$

where g is defined implicitly by

$$f_{(1,\dots,p)}(v,w) = g(v), \qquad \bar{z} = \mathrm{d}_w \eta(v,w).$$

In other words, g is obtained from $f_{(1,\dots,p)}$ by applying the inverse of the change of variable $u \mapsto U = (v,z)$ and then letting $z = \bar{z}$.

It is remarkable that the sub-system (66) is still symmetrizable hyperbolic:

Theorem 17 (Boillat–Ruggeri [5]). *The system (66) admits a convex entropy E, and is therefore symmetrizable hyperbolic.*

The Legendre transforms of E and η are related by

$$\mathrm{d}E^*(\mu) = \mathrm{d}_\mu \eta^*(\mu, \bar{z}).$$

In other words,

$$E(v) = \mu \cdot \mathrm{d}\eta^*(\mu,\bar{z}) - \eta^*(\mu,\bar{z}), \qquad \textit{where} \quad v := \mathrm{d}\eta^*(\mu,\bar{z}).$$

Equivalently, $E(v) = \eta(u) - \mathrm{d}_w\eta(u) \cdot w$*, where* $u = (v, w)$ *is determined by* $\mathrm{d}_w\eta(u) = \bar{z}$.

The Hessian of the new entropy E is given by the formula

$$\mathrm{D}^2 E = \mathrm{D}^2_{vv}\eta - \mathrm{D}^2_{vw}\eta(\mathrm{D}^2_{ww}\eta)^{-1}\mathrm{D}^2_{wv}\eta,$$

which is the *Schur complement* of the block $\mathrm{D}^2_{ww}\eta$ in $\mathrm{D}^2_{uu}\eta$.

Example. The Euler–Fourier system with thermal diffusivity $\frac{\kappa}{\epsilon}$ tends to the *isothermal* Euler system. Remember that $\eta = -\rho s$ and $z = \frac{1}{\theta}$. Then

$$E = \frac{1}{\bar{\theta}}\left(\frac{1}{2}\rho|\mathbf{v}|^2 + \rho(e - \theta s)\right).$$

The quantity $e_0 := e - \theta s$ is known as the Helmholtz *free energy.*

Interlacing property. Now that we have two inviscid systems (23) and (66), we therefore have two sets of wave velocities:

- Those of (23), denoted $\lambda_1(u;\xi) \le \cdots \le \lambda_n(u;\xi)$,
- Those of (66), denoted $a_1(v;\xi) \le \cdots \le a_p(v;\xi)$.

In a symmetrization $S_0\partial_t u + \sum_\alpha S^\alpha \partial_\alpha u = 0$ with $S_0 = \mathrm{D}^2\eta$, the λ_j's are the eigenvalues of the pair $(S_0(u), S(u;\xi))$:

$$\det(S(u;\xi) - \lambda_j(u;\xi)S_0(u)) = 0.$$

It turns out that the a_k's are the eigenvalues of the pair $(T_0(v), T(v;\xi))$, where $T_0(v)$ (respectively $T(v;\xi)$) is the upper-left $p \times p$ block of $S_0(u)$ (resp. of $S(u;\xi)$). Because of the characterization of eigenvalues as max min or min max, this yields the following fact.

Proposition 12. *For every* $j = 1, \dots, p$ *and whenever* $z(u) = \bar{z}$*, one has*

$$\lambda_j \le a_j \le \lambda_{j+n-p}. \tag{67}$$

The particular cases $\lambda \le a_1$ and $a_p \le \lambda_n$ are also known as the *subcharacteristic property*.

Example (continued). For full Euler ($n = d + 2$) and isothermal Euler ($p = d + 1$) systems, one has

$$\lambda_1 = \mathbf{v}\cdot\xi - c_{\mathrm{ful}}|\xi|, \quad \lambda_2 = \cdots = \lambda_{d+1} = \mathbf{v}\cdot\xi, \quad \lambda_{d+2} = \mathbf{v}\cdot\xi + c_{\mathrm{ful}}|\xi|$$

and

$$a_1 = \mathbf{v}\cdot\xi - c_{\mathrm{iso}}|\xi|, \quad a_2 = \cdots = a_d = \mathbf{v}\cdot\xi, \quad a_{d+1} = \mathbf{v}\cdot\xi + c_{\mathrm{iso}}|\xi|.$$

The interlacing property tells us that $c_{\mathrm{iso}} \le c_{\mathrm{ful}}$, and actually the inequality is strict: the sound velocity is smaller in the isothermal gas than in the non-isothermal one.

To understand this amazing statement, we point out that the a_j's are the velocities of the hyperbolic modes in the hyperbolic/parabolic system (54), whenever $z(u) = \bar{z}$. The a_k's are therefore velocities of hyperbolic waves (for instance, discontinuities in the gradient of u) that propagate in the viscous system, while the λ_j's are the velocities of so-called diffusion waves; the latter are smooth and are Gaussian, up to a twist due to the genuine non-linearity, their support spreads as $\sqrt{t}$ while their amplitude decays as $t^{-1/2}$. Finally, they carry a fixed total mass.

5.7 *Principal Sub-systems Without Entropy Structure: The Linear Case*

Even if the original system does not admit a convex entropy, something can be said about the relation between the well-posedness of the viscous and of the inviscid systems. The interpretation is the same as above. However, it is unclear whether we have an interlacing property such as (67).

Theorem 18 (Benzoni-Gavage and Serre [1]). *Assume that the forward Cauchy problem for the linear system with constant coefficients*

$$\partial_t v + \sum_\alpha A^\alpha \partial_\alpha v + \sum_\alpha C^\alpha \partial_\alpha w = 0,$$

$$\partial_t w + \sum_\alpha D^\alpha \partial_\alpha v + \sum_\alpha E^\alpha \partial_\alpha w = \sum_{\alpha,\beta} B^{\alpha\beta} \partial_\alpha \partial_\beta w,$$

is well-posed in $L^2(\mathbb{R}^d)^n$. Assume also that the symbol $B(\xi)$ is non-singular for $\xi \neq 0$.

Then the sub-system $\partial_t v + \sum_\alpha A^\alpha \partial_\alpha v = 0$ is hyperbolic.

Likewise, for *relaxation* models, the well-posedness of the non-damped part (the principal sub-system) is related to the asymptotic stability of the relaxed system.

Theorem 19 (Benzoni-Gavage and Serre [1]). *Let $B \in \mathbf{GL}_{n-p}(\mathbb{R})$ be given. Assume that the Cauchy problem for the linear system with constant coefficients*

$$\partial_t v + \sum_\alpha A^\alpha \partial_\alpha v + \sum_\alpha C^\alpha \partial_\alpha w = 0,$$

$$\partial_t w + \sum_\alpha D^\alpha \partial_\alpha v + \sum_\alpha E^\alpha \partial_\alpha w = Bw,$$

is well-posed in $L^2(\mathbb{R}^d)^n$, uniformly in forward time: the semi-group $(S_t)_{t>0}$ is uniformly bounded.

Then the sub-system $\partial_t v + \sum_\alpha A^\alpha \partial_\alpha v = 0$ is hyperbolic.

We remark that the well-posedness of the relaxed system is nothing but the hyperbolicity of its left-hand side. But the uniform boundedness for $t > 0$ depends on the choice of B.

References

1. S. Benzoni-Gavage, D. Serre, *Multi-dimensional Hyperbolic Partial Differential Equations. First Order Systems and Applications*. Oxford Mathematical Monographs (Oxford University Press, Oxford, 2007)
2. S. Benzoni-Gavage, D. Serre, K. Zumbrun, Alternate Evans functions and viscous shock waves. SIAM J. Math. Anal. **32**, 928–962 (2001)
3. S. Benzoni-Gavage, F. Rousset, D. Serre, K. Zumbrun, Generic types and transitions in hyperbolic initial-boundary value problems. Proc. R. Soc. Edinb. **132A**, 1073–1104 (2002)
4. G. Boillat, Sur l'existence et la recherche d'équations de conservation supplémentaires pour les systèmes hyperboliques. C. R. Acad. Sci. Paris Sér. A **278**, 909–912 (1974)
5. G. Boillat, T. Ruggeri, Hyperbolic principal subsystems: entropy convexity and subcharacteristic conditions. Arch. Ration. Mech. Anal. **137**, 305–320 (1997)
6. C. Dafermos, *Hyperbolic Conservation Laws in Continuum Physics*. Grundlehren der mathematischen Wissenschaften, vol. 325, 3rd edn. (Springer, Belin/Heidelberg, 2010)
7. J. Francheteau, G. Métivier, Existence de chocs faibles pour des systèmes quasi-linéaires hyperboliques multidimensionnels. Astérisque **268** (2000)
8. H. Freistühler, The persistence of ideal shock waves. Appl. Math. Lett. **7**, 1–5 (1994)
9. H. Freistühler, T.-P. Liu, Nonlinear stability of overcompressive shock waves in a rotationally invariant system of viscous conservation laws. Commun. Math. Phys. **153**, 147–158 (1993)
10. H. Freistühler, P. Szmolyan, Spectral stability of small shock waves. Arch. Ration. Mech. Anal. **164**, 287–309 (2002)
11. K.O. Friedrichs, P.D. Lax, Systems of conservation equations with a convex extension. Proc. Natl. Acad. Sci. U.S.A. **68**, 1686–1688 (1971)
12. R.A. Gardner, K. Zumbrun, The gap lemma and geometric criteria for instability of viscous shock profiles. Commun. Pure Appl. Math. **51**, 797–855 (1998)
13. S.P. Godunov, An interesting class of quasi-linear systems. Sov. Math. Doklady **2**, 947–949 (1961)
14. J.W. Helton, V. Vinnikov, Linear matrix inequality representation of sets. Commun. Pure Appl. Math. **60**, 654–674 (2007)
15. R. Hersh, Mixed problems in several variables. J. Math. Mech. **12**, 317–334 (1963)
16. K. Jenssen, G. Lyng, Evaluation of the Lopatinski condition for gas dynamics. Appendix to K. Zumbrun, Stability of large-amplitude shock waves of compressible Navier-Stokes equations, in *Handbook of Mathematical Fluid Dynamics*, ed. by S. Friedlander, D. Serre, vol. III (North-Holland, Amsterdam, 2004), pp. 507–524
17. S. Kawashima, Systems of a hyperbolic parabolic type with applications to the equations of magnetohydrodynamics. PhD thesis, Kyoto University, 1983
18. A. Knutson, T. Tao, The honeycomb model of $GL_n(\mathbb{C})$ tensor products. I. Proof of the saturation conjecture. J. Am. Math. Soc. **12**, 1055–1090 (1999)
19. H.-O. Kreiss, Initial boundary value problems for hyperbolic systems. Commun. Pure Appl. Math. **23**, 277–298 (1970)
20. P.D. Lax, Hyperbolic systems of conservation laws. II. Commun. Pure Appl. Math. **10**, 537–566 (1957)
21. P.D. Lax, Differential equations, difference equations and matrix theory. Commun. Pure Appl. Math. **11**, 175–194 (1958)

22. A. Majda, *The Stability of Multi-dimensional Shock Fronts*. Memoirs of the American Mathematical Society, vol. 41 (AMS, Providence, 1983), p. 275 and *The Existence of Multi-dimensional Shock Fronts*. Memoirs of the American Mathematical Society, vol. 41 (AMS, Providence, 1983), p. 281
23. A. Majda, R. Pego, Stable viscosity matrices for systems of conservation laws. J. Differ. Equ. **56**, 229–262 (1985)
24. G. Métivier, The block structure condition for symmetric hyperbolic systems. Bull. Lond. Math. Soc. **32**, 689–702 (2000)
25. G. Métivier, K. Zumbrun, Hyperbolic boundary value problems for symmetric systems with variable multiplicities. J. Differ. Equ. **211**(1), 61–134 (2005)
26. F. Murat, Compacité par compensation. Ann. Scuola Norm. Sup. Pisa Cl. Sci. **5**, 489–507 (1978)
27. R.L. Pego, Stable viscosities and shock profiles for systems of conservation laws. Trans. Am. Math. Soc. **282**, 749–763 (1984)
28. J. Rauch, Energy inequalities for hyperbolic initial boundary value problems. PhD thesis, New York University (1971); L^2 is a continuable initial condition for Kreiss' mixed problems. Commun. Pure Appl. Math. **25**, 265–285 (1972)
29. R. Sakamoto, *Hyperbolic Boundary Value Problems* (Cambridge University Press, Cambridge, 1982). Translated from the Japanese by Katsumi Miyahara
30. D.H. Sattinger, On the stability of waves of nonlinear parabolic systems. Adv. Math. **22**, 312–355 (1976)
31. D. Serre, La compacité par compensation pour les systèmes non linéaires de deux équations à une dimension d'espace. J. Math. Pures Appl. **65**, 423–468 (1987)
32. D. Serre, *Systems of Conservation Laws. Vol. I. Hyperbolicity, Entropies, Shock Waves; II. Geometric Structures, Oscillations, and Initial-Boundary Value Problems* (Cambridge University Press, Cambridge, 2000)
33. D. Serre, Weyl and Lidskiĭ inequalities for general hyperbolic polynomials. Chin. Ann. Math. Ser. B **30**, 785–802 (2009)
34. D. Serre, The structure of dissipative viscous system of conservation laws. Physica **D293**, 1381–1386 (2010)
35. D. Serre, Local existence for viscous system of conservation laws: H^s-data with $s > 1 + d/2$, in *Nonlinear PDEs and Hyperbolic Wave Phenomena*, ed. by H. Holden, K. Karlsen. Contemporary Mathematics, vol. 526 (AMS, Providence, 2010), pp. 339–358
36. D. Serre, Viscous system of conservation laws: Singular limits, in *IMA Volume Nonlinear Conservation Laws and Applications*, ed. by A. Bressan, G.-Q. Chen, M. Lewicka, D. Wang (Springer, New York, 2010)
37. L. Tartar, Compensated compactness and applications to partial differential equations, in *Nonlinear Analysis and Mechanics: Heriot-Watt Symposium*, vol. IV. Research Notes in Mathematics, vol. 39 (Pitman, London, 1979), pp. 136–212
38. B. Temple, Systems of conservation laws with invariant submanifolds. Trans. Am. Math. Soc. **280**, 781–795 (1983)
39. K. Zumbrun, Stability of large-amplitude shock waves of compressible Navier-Stokes equations, in *Handbook of Mathematical Fluid Dynamics*, ed. by S. Friedlander, D. Serre, vol. III (North-Holland, Amsterdam, 2004), pp. 311–533
40. K. Zumbrun, Instantaneous shock location and one-dimensional nonlinear stability of viscous shock waves. Q. Appl. Math. **69**, 177–202 (2011)
41. K. Zumbrun, D. Serre, Viscous and inviscid stability of multidimensional planar shock fronts. Indiana Univ. Math. J. **48**, 937–992 (1999)

The Nash-Moser Iteration Technique with Application to Characteristic Free-Boundary Problems

Ben Stevens

Abstract These notes are an overview of the Nash-Moser iteration technique for solving PDEs (or other non-linear problems) via linearisation, where the linearised equations admit estimates with a loss of regularity with respect to the source term, coefficients and/or boundary/initial data. We first introduce the abstract setting along with a version of the iteration scheme due to Hörmander (Arch Ration Mech Anal 62(1):1–52, 1976). We then introduce some modifications which allow the scheme to be applied to some characteristic free-boundary problems for hyperbolic conservation laws. We focus on the case of supersonic vortex sheets in 2D as considered by Coulombel and Secchi in Ann Sci Éc Norm Supér (4) 41(1):85–139, 2008.

2010 Mathematics Subject Classification 76N10 (35L65 35L67 35Q35 35R35)

1 Introduction

1.1 Summary

These notes are an overview of the Nash-Moser iteration technique for solving PDEs (or other non-linear problems) via linearisation, where the linearised equations admit estimates with a loss of regularity with respect to the source term, coefficients

B. Stevens (✉)
Mathematical Institute, University of Oxford, Oxford, OX1 3LB, UK
e-mail: ben.stevens@maths.ox.ac.uk

G.-Q.G. Chen et al. (eds.), *Hyperbolic Conservation Laws and Related Analysis with Applications*, Springer Proceedings in Mathematics & Statistics 49,
DOI 10.1007/978-3-642-39007-4_13, © Springer-Verlag Berlin Heidelberg 2014

and/or boundary/initial data. In these situations, Picard iteration (or the contraction mapping principle) fails, but a modified form of Newton-Raphson iteration, involving the application of smoothing operators to overcome the loss of regularity, may succeed in finding a solution for given data close to some special data for which a solution is known to exist. This technique is known as Nash-Moser iteration, or in some contexts as the Nash-Moser inverse function theorem. It was originally used by Nash in [21] for solving the isometric embedding problem. Moser in [20] and Schwartz in [23] simplified the method at the expense of a loss of regularity and showed how it could be applied in a more general setting. Hörmander, in his paper [15] on the boundary problems of physical geodesy, improved on Moser's scheme by reducing the loss of regularity, using a scheme more similar to Nash's original. More recently, Alinhac in [2] used a modified version of Hörmander's scheme to prove the short-time existence of rarefaction waves for a class of conservation laws and Coulombel and Secchi in [8] introduced an additional modification to prove the short-time existence of vortex sheets for the two dimensional isentropic Euler equations provided the Mach number is sufficiently large. A scheme similar to the one used by Coulombel and Secchi is also developed by Chen and Wang in [5] and [6] to prove the short-time existence of current-vortex sheets for three-dimensional MHD under certain stability assumptions.

We aim to provide an abstract setting for the technique, whilst keeping in mind that we want to apply it to PDE problems. Hopefully in an abstract setting it will be easier to see the key hypotheses needed on the equations to be solved than in specific situations, which may involve other complications. We first introduce the scheme used by Hörmander in [15], and detailed by Alinhac and Gérard in [3], which is closer than Moser's scheme to Nash's original technique except that Hörmander uses a discrete approximation scheme rather than one based on a continuous parameter t. Whilst Hörmander works in Hölder spaces, we work in more general Banach spaces, at the price of losing a small degree of regularity. We have in mind that the linearised equations are most likely to be estimated in Sobolev spaces (or weighted Sobolev spaces), probably with exponent two. This technique has the advantage over Moser's technique of obtaining a solution which is closer in regularity to the given data, but although Nash used his method to obtain optimal regularity, we are unlikely to obtain an optimal regularity result using this method in more complicated situations.

We then introduce a more complicated scheme which allows us to deal with difficulties in solving the linearised equations, inspired by the paper on 2D compressible vortex sheets by Coulombel and Secchi [8].

Following this, we give the construction of the smoothing operators used in Nash-Moser iteration on some Sobolev spaces which are used in practice, and some inequalities useful for obtaining the tame estimates used in the iteration scheme.

Finally, we show how the generalised scheme can be applied to the case considered by Coulombel and Secchi in [8], in a slightly simplified manner but at the expense of some loss of regularity.

1.2 Newton-Raphson Iteration, Picard Iteration, and Nash-Moser Iteration

Suppose we wish to solve the nonlinear equation $T(u) = f$ for the unknown $u \in X$, given $f \in Y$, where $T : X \to Y$. So as not to ask too much, let us look for a solution u close to u_0 of the equation $T(u) = T(u_0) + f$, where f is small. One of the most classical methods for solving such a nonlinear equation via linearisation is Newton-Raphson iteration. For $n \geq 1$, we set

$$u_{n+1} = u_n - L(u_n)(T(u_n) - T(u_0) - f)$$

where $L(u)$ is a right inverse of $DT(u)$. One can check by applying T to both sides and using Taylor's theorem that $T(u_{n+1}) = T(u_0) + f$ plus terms involving $u_{n+1} - u_n$ which one would hope to converge to zero. However, for this scheme to even make sense, we need an operator $L(u) : Y \to X$ which is a right inverse of $DT(u)$. The linearised equations $DT(u)v = g$ themselves may be difficult or impossible to solve for $v \in X$, hence we may not be able to find such an operator L.

As a possible remedy to this problem, we consider the contraction mapping theorem, or Picard iteration, which uses a slightly different kind of linearisation and may be able to solve equations where the operator L as above does not exist. For example, suppose we can write our equation in the form

$$S(u)u = 0$$

where, for fixed u, $S(u)$ is a linear operator. We seek the unknown $u \in X$, where X is a complete metric space, and we assume the initial/boundary conditions have been absorbed into the definition of X. We now define the map $F : X \to X$ by $F(u) = v$, where v is the solution to the linear equation

$$S(u)v = 0.$$

If we can prove that F is well-defined, and that F is a contraction, i.e. $d_X(F(u_1), F(u_2)) \leq \kappa d_X(u_1, u_2)$, where $\kappa < 1$, for all u_1, u_2 in X, then the contraction mapping theorem implies that F has a fixed point, w. By construction, w satisfies the original nonlinear equation we wished to solve.

Note that in order to apply this method, we require that the solution v of the linear equation be in the same space as u, on which v depends through the coefficients of the equation. This is a better situation than for Newton-Raphson iteration, which requires that the operator $L(u)$ regains the regularity lost by applying the operator T.

We can also write this method as an explicit iteration scheme (effectively re-proving the contraction mapping theorem). We pick $u_0 \in X$ and for $n \geq 0$ we define u_{n+1} as the solution of the linear equation

$$S(u_n)u_{n+1} = 0.$$

We then aim to show that, for $n \geq 1$, $d_X(u_{n+1}, u_n) \leq \kappa d_X(u_n, u_{n-1})$. This will ensure u_n is a Cauchy sequence which converges to a solution of the nonlinear equation. Using the explicit iteration scheme (known as Picard iteration) allows more scope for slight modification in specific cases. For example, Majda in [16], uses this iteration scheme, modified to include a smoothing of the initial data, to prove the short-time existence of classical solutions to multidimensional systems of conservation laws with a convex entropy.

However, it is possible that we cannot solve the linearised problem above for v in the same space as u, as required by Picard iteration. It may happen that we can solve the linear equation, but only for $v \in Z$, where $X \subset Z$. For example, perhaps, given $u \in C^k$, we can only prove that a solution v to the linearised equation exists in C^{k-1}. We refer to this as a loss of regularity in solving the linearised problem.

To overcome this, the key idea of Nash was to return to Newton-Raphson iteration, but to modify the scheme to include a smoothing operation at each step to compensate for the loss of regularity. Returning to the equation $T(u) = T(u_0) + f$, standard Newton-Raphson iteration may be written as follows. For $n \geq 0$, we set

$$u_{n+1} = u_n + \dot{u}_n.$$

The difference $\dot{u}_n$ is given by

$$\dot{u}_n = L(u_n)g_n$$

for

$$g_n = f + T(u_0) - T(u_n)$$

where $L(u)$ is a right-inverse of $DT(u)$.

Now let us suppose we have a family of smoothing operators S_n that regain the regularity lost by T and L, and such that $S_n \to \mathrm{id}$ as $n \to \infty$. Then there are two obvious ways we can modify the scheme.

The simplest is to set $u_{n+1} = u_n + S_n\dot{u}_n$, i.e. we smooth $\dot{u}_n$ after applying the operators T and L to u_n. Since $S_n \to \mathrm{id}$ as $n \to \infty$, this scheme looks like Newton-Raphson iteration for large n, so we might expect it to converge under certain conditions. This method is used by Moser in [20] and Schwartz in [23]. Whilst this is a very simple modification, it has the drawback that a solution u obtained by this method has a much lower degree of regularity than the given data f.

The other obvious modification is to smooth u_n before we apply the operators T and L. Thus we set

$$\dot{u}_n = L(S_n u_n)g_n.$$

We also adjust our choice of g_n (which should be smoothed) given this modification. This method is used by Hörmander in [15] and a continuous-parameter version was used by Nash in his original paper [21]. We motivate how to choose g_n in Sect. 3.1, which is based on the motivation given by Alinhac and Gérard in [3]. Again, the fact that $S_n \to \text{id}$ as $n \to \infty$ means the scheme looks like Newton-Raphson iteration for large n. The advantage of this method is that the solution u obtained can be quite close in regularity to the given data f, but generally the regularity obtained will not be optimal. In modifying Hörmander's method to deal with more general Banach spaces instead of just Hölder spaces, we lose an arbitrarily small degree of regularity if we can use fractional index spaces, or one degree of regularity if we are using integer index spaces. Other modifications to the scheme used in practice further reduce the degree of regularity obtained. Nevertheless, we may consider this an improvement over Moser's technique, which we can informally attribute to the fact that we have carefully constructed g_n to compensate for the introduction of the smoothing operators, whereas Moser's method involves no such modification.

1.3 Nash-Moser Iteration as an Inverse Function Theorem

It is instructive to consider a slightly different viewpoint, that is to consider Nash-Moser iteration as an inverse function theorem for a certain class of Fréchet spaces, which are a natural generalisation of Banach spaces.

Indeed, the standard version of the Inverse Function Theorem, which can be proved (under slightly stronger hypotheses than usual to make things simpler) by an application of the contraction mapping theorem with parameter, carries over analogously to an operator $T : X \to Y$ between Banach spaces. By this we mean that if the Fréchet derivative $DT(u)$ of T is invertible at a point $u \in X$, then T itself is invertible in a neighbourhood of u. Hence, if we wish to solve the equation $T(u) = T(u_0) + f$ for u near u_0, where f is small, we can simply apply the inverse function theorem.

However, it is possible in applications that we can only find an 'unbounded' inverse for $DT(u)$. For example, if we work with differential operators in the spaces C^k of k-times differentiable functions, then we might have $T : C^k \to C^{k-1}$, but we might only be able to find a right inverse $L(u)$ of $DT(u)$ on some subset of C^{k-1}, for example on C^k, so that $L(u) : C^k \to C^k$, or, even worse, on C^{k+1} so that $L(u) : C^{k+1} \to C^k$. This is solved if we work in the space $X = C^\infty$, since then $L(u)$ maps X to itself. However, this is no longer a Banach space, but a Fréchet space. Thus we are led to ask whether there is an inverse function theorem for Fréchet spaces. The answer is that if we assume the existence of a certain family of smoothing operators on our Fréchet space (which by no means exist in general, but do for most spaces of differentiable functions commonly used), then there is a sort of inverse function theorem. This requires that $DT(u)$ be invertible on a neighbourhood of u, not just at u itself.

This point of view is elegantly considered by Hamilton in [14], who refers to this special class of Fréchet spaces as 'tame' Fréchet spaces and the necessary estimates involved on the operator T as 'tame' estimates. The proof of this result uses Nash-Moser iteration, and Hamilton's proof in particular is quite close to

Nash's original method. The similarity with the usual inverse function theorem is why Nash-Moser iteration is sometimes referred to as the Nash-Moser inverse function theorem or the Nash-Moser implicit function theorem. See also the chapter 'Generalized Implicit Function Theorems' written by E. Zehnder in Nirenberg [22] for an introduction to Nash-Moser type theorems as generalisations of the standard inverse/implicit function theorem. Another implicit function theorem in the setting of Fréchet spaces is given by Ekeland in [10], whose approach does not rely on Newton-Raphson iteration but on Lebesgue's dominated convergence theorem and Ekeland's variational principle.

Whilst this viewpoint is conceptually simple, for actual applications to PDEs, working in Fréchet spaces is not necessary and complicates matters, and it is easier to consider a family of Banach spaces in which one has estimates for the linearised equations, for example $(C^k)_{k\in\mathbb{N}}$ or $(H^s)_{s\in\mathbb{R}_{\geq 0}}$.

1.4 Tame Estimates

The key estimates involved in Nash-Moser iteration are known as tame estimates. These are estimates of the following form. (Here we use the spaces C^k for definiteness.)

Let $T : C^\infty \to C^\infty$.

Then T satisfies a tame estimate if

$$||T(u)||_{C^k} \leq C_k(1 + ||u||_{C^{k+k_1}})$$

for some fixed integer k_1 and all u in some fixed bounded set $U \subset C^{k_0}$, for some k_0, where the constant $C_k > 0$ is independent of u.

The key point about this estimate is that it is affine in the norm of u on the right hand side with the variable index k.

Similarly, the second derivative of T, D^2T, is said to satisfy a tame estimate if

$$\begin{aligned}&\left|\left|D^2T(u)(v_1, v_2)\right|\right|_{C^k}\\ &\leq C_k(||v_1||_{C^{k_1+k}}\,||v_2||_{C^{k_2}} + ||v_1||_{C^{k_1}}\,||v_2||_{C^{k_2+k}} + ||v_1||_{C^{k_1}}\,||v_2||_{C^{k_2}}\,(1 + ||u||_{C^{k+k_3}}))\end{aligned}$$

for some fixed integers k_1, k_2, k_3 and all u in some fixed bounded set $U \subset C^{k_0}$, for some k_0, where the constant $C_k > 0$ is independent of u, v_1 and v_2.

Note that this estimate is also affine in the norms on the right hand side with the variable index k, and in addition it is quadratic (with no affine terms) in (v_1, v_2), which will be a key point in the iteration. The smoothing operators will control the large k norms in terms of lower ones at the price of poorer estimates and we require DT to be a good approximation for T to compensate.

Note that the framework of tame estimates fits differential operators well because of product estimates of the form

$$||fg||_{H^s} \leq C_s(||f||_{H^r}\,||g||_{H^s} + ||f||_{H^s}\,||g||_{H^r})$$

for $r > \frac{d}{2}$, where d is the dimension.

Similarly, we have estimates for compositions $G(x) = F(u(x))$ (sometimes called Moser-type inequalities) of the form

$$||\partial^\alpha G||_{L^2} \leq C_s\,||u||_{H^{|\alpha|}}$$

for u in an H^r-bounded set.

These estimates can be derived from the Sobolev embedding theorem for large index s, and details of these estimates for certain classes of Sobolev Spaces are given in Sect. 5.2.

2 The Abstract Setting

In order to describe Nash-Moser iteration in an abstract setting we will need to introduce some notation, as well as the idea of a derivative in this setting. We will simply use the notion of a directional derivative, since all we need is a linear approximation to an operator which satisfies Taylor's theorem.

2.1 Families of Banach Spaces and Differentiation

Definition 1. Let I be an interval in $\mathbb{R}$ or $\mathbb{Z}$ of the form $[0, a)$, $[0, a]$, or $[0, \infty)$, where $a > 0$.

We will say $\{X^s\}_{s \in I}$ is a *decreasing family of Banach spaces* if, for each $s \in I$, X^s is a Banach space with norm $||\cdot||_{X^s}$, and, for $s_1, s_2 \in I$ with $s_1 \leq s_2$, we have

$$X^{s_2} \subset X^{s_1} \text{ with } ||\cdot||_{X^{s_2}} \geq ||\cdot||_{X^{s_1}} \text{ on } X^{s_2}.$$

We will write

$$X^\infty = \cap_{s \in I} X^s$$

and

$$X^{\infty - m} = \cap_{s \in I, s \geq m} X^{s-m}$$

for $m \in I$.

Remark 1. Note that it is convenient to use the notation X^∞ for the intersection of all the Banach Spaces X^s with $s \in I$, even if I is a finite interval. In the case that $I = [0, \infty)$, $X^{\infty - m}$ as defined above is the same as X^∞, but if I is a finite interval then they are not the same.

Definition 2. Let $\{X^s\}_{s\in I}$ be a decreasing family of Banach spaces. Let $\alpha : U \to X^\infty$ where $U \subset \mathbb{R}$ is open, and let $t \in U$. We say α is *differentiable at* t if there exists a $w \in X^\infty$ such that

$$\left\| \frac{\alpha(t+h) - \alpha(t)}{h} - w \right\|_{X^s} \to 0 \text{ as } h \to 0\ (h \neq 0)$$

for all $s \in I$.

If such a w exists, we say w is the derivative of α at t, and write $\alpha'(t) = w$ or $\frac{\mathrm{d}\alpha}{\mathrm{d}t}(t) = w$.

We say α is *differentiable* if it is differentiable at t for all $t \in U$.

Definition 3. Let $\{X^s\}_{s\in I}$ and $\{Y^s\}_{s\in I}$ be two decreasing families of Banach spaces. Let $T : U \to Y^{\infty-m}$ for some $m \in I$, where $U \subset X^\infty$ is $||\cdot||_{X^r}$-open for some $r \in I$, and let $u \in U$. We say T is *differentiable at* u if, for each $v \in X^\infty$, the map $\alpha_v : (-\epsilon, \epsilon) \to Y^{\infty-m}$ defined on a small neighbourhood of 0 in $\mathbb{R}$ by

$$\alpha_v(t) = T(u + tv)$$

is differentiable at 0 in the sense of Definition 2, and

$$\alpha_v'(0) = DT(u)v$$

where $DT(u) : X^\infty \to Y^{\infty-m}$ is a linear map. We call $DT(u)$ the derivative of T at u.

We say T is *differentiable* if it is differentiable at u for all $u \in U$ and call DT the derivative of T.

For an integer $k \geq 2$, we say T is k*-times differentiable* with k-th derivative D^kT if the following inductive definition holds.

T is $k-1$ times differentiable with $(k-1)$-th derivative at u given by $D^{k-1}T(u) : (X^\infty)^{k-1} \to Y^{\infty-m}$ for each $u \in U$.

For each ordered set $(v_1, \dots, v_{k-1}) \in (X^\infty)^{k-1}$, the map $S : U \to Y^{\infty-m}$ defined by

$$S(u) = D^{k-1}(u)(v_1, \dots, v_{k-1})$$

is differentiable in the above sense.

Define the k-th derivative of T at $u \in U$ as $D^kT(u) : (X^\infty)^k \to Y^{\infty-m}$ where

$$D^kT(u)(v_1, \dots, v_k) = DS(u)v_k.$$

Remark 2. We will not need all the properties of standard derivatives. We merely require a linear approximation to within quadratic error of a nonlinear operator. Hence we give the above fairly weak definition of differentiability and don't worry about questions such as whether the partial derivatives commute.

Proposition 1. *Let* $\{X^s\}_{s\in I}$ *and* $\{Y^s\}_{s\in I}$ *be two decreasing families of Banach spaces. Let* $T : U \to Y^{\infty-m}$ *for some* $m \in I$*, where* $U \subset X^\infty$ *is* $||\cdot||_{X^r}$*-open* *for some $r \in I$. Then Taylor's theorem holds for T. More precisely, suppose T is k-times differentiable (in the sense of Definition 3) for some $k \geq 1$, let $u \in U$, $v \in X^\infty$, and suppose the line segment $[u, u+v]$ is contained in U. Then*

$$T(u+v) = T(u) + DT(u)v + \ldots + \frac{1}{(k-1)!} D^{k-1}T(u)(v, \ldots, v) + R_{k,u}(v)$$

where

$$||R_{k,u}(v)||_{Y^s} \leq \frac{1}{k!} \sup_{t \in [0,1]} \left|\left|D^k T(u+tv)(v, \ldots, v)\right|\right|_{Y^s}$$

for all $s \in I$ such that $s + m \in I$.

Proof. Fix $s \in I$ such that $s + m \in I$. Let $\phi \in (Y^s)^*$ be a continuous linear functional on Y^s.

Define $g : J \to \mathbb{R}$ by

$$g(t) = \phi \circ T(u + tv)$$

where J is an open interval in $\mathbb{R}$ containing $[0, 1]$.

Since ϕ is a continuous linear functional on Y^s, from the definition of differentiability we have that g is k-times differentiable with

$$g^{(k)}(t) = \phi \circ D^k T(u + tv)(v, \ldots, v).$$

Applying the one-dimensional Taylor's theorem to obtain an expansion for $g(1)$ about $g(0)$, we have

$$g(1) = g(0) + g'(0) + \ldots + \frac{1}{(k-1)!} g^{k-1}(0) + \frac{1}{k!} g^k(h) h^k$$

for some $h \in [0, 1]$ (which may depend on ϕ). Hence

$$\phi \circ T(u+v) =$$

$$\phi(T(u) + DT(u)v + \ldots + \frac{1}{(k-1)!} D^{k-1}T(u)(v, \ldots, v) + \frac{1}{k!} h^k D^k T(u + hv)(v, \ldots, v))$$

Rearranging, we have

$$\left|\phi(T(u+v) - (T(u) + DT(u)v + \ldots + \frac{1}{(k-1)!} D^{k-1}T(u)(v, \ldots, v)))\right|$$

$$\leq ||\phi||_{(Y^s)^*} \left|\left|\frac{1}{k!} h^k D^k T(u + hv)(v, \ldots, v)\right|\right|_{Y^s}$$

$$\leq ||\phi||_{(Y^s)^*} \frac{1}{k!} \sup_{t \in [0,1]} \left|\left|D^k T(u + tv)(v, \ldots, v)\right|\right|_{Y^s}.$$

Now use the Hahn-Banach theorem to pick $\phi \in (Y^s)^*$ with $||\phi||_{(Y^s)^*} = 1$ such that

$$\phi(T(u+v) - (T(u) + DT(u)v + \ldots + \frac{1}{(k-1)!}D^{k-1}T(u)(v,\ldots,v)))$$
$$= \left|\left|T(u+v) - (T(u) + DT(u)v + \ldots + \frac{1}{(k-1)!}D^{k-1}T(u)(v,\ldots,v))\right|\right|_{Y^s}.$$

We then obtain

$$\left|\left|T(u+v) - (T(u) + DT(u)v + \ldots + \frac{1}{(k-1)!}D^{k-1}T(u)(v,\ldots,v))\right|\right|_{Y^s}$$
$$\leq \frac{1}{k!}\sup_{t\in[0,1]}||D^kT(u+tv)(v,\ldots,v)||_{Y^s}.$$

This completes the proof.

Remark 3. Note that we can apply the above proposition when $\{X^s\}_{s\in I}$ is just $\{\mathbb{R}\}_{s\in I}$ to obtain Taylor's theorem for paths in Y^∞.

2.2 Definition of the Smoothing Operators

Definition 4. We will say a decreasing family of Banach spaces $\{X^s\}_{s\in I}$ *satisfies the smoothing hypothesis* if there exists a family of linear operators $\{S_\theta\}_{\theta\in\mathbb{R}_{\geq 1}}$ such that

$$S_\theta : X^0 \to X^\infty$$

and, for $u \in X^s$, we have

$$||S_\theta u||_{X^r} \leq C_{r,s}\theta^{(r-s)_+}\,||u||_{X^s} \text{ for all } r,s \in I \tag{1}$$
$$||u - S_\theta u||_{X^r} \leq C_{r,s}\theta^{-(s-r)}\,||u||_{X^s} \text{ for all } r,s \in I \text{ with } r \leq s \tag{2}$$
$$\left|\left|\frac{\mathrm{d}}{\mathrm{d}\theta}S_\theta u\right|\right|_{X^r} \leq C_{r,s}\theta^{r-s-1}\,||u||_{X^s} \text{ for all } r,s \in I \tag{3}$$

where the constant $C_{r,s} > 0$ remains bounded if r and s remain bounded.

Here $(a)_+$ denotes $\max\{a,0\}$ for $a \in \mathbb{R}$ or $a \in \mathbb{Z}$.

Note $\frac{\mathrm{d}}{\mathrm{d}\theta}S_\theta u$ is the derivative of the map $\theta \mapsto S_\theta u$ in the sense of Definition 2, which we require to exist for each $u \in X^0$.

3 Hörmander's Version of Nash-Moser Iteration

3.1 Motivation for the Iteration Scheme

Here we provide some motivation for the iteration scheme used by Hörmander in [15] by comparing it to Newton-Raphson iteration. This is unnecessary for the proof of the theorem, but the iteration scheme seems a little unmotivated without it. This motivation is partly based on the motivation given in Alinhac and Gérard [3].

3.1.1 Newton-Raphson Iteration

In order to solve the equation

$$T(u) = T(u_0) + f$$

the Newton-Raphson method uses the following iteration scheme.

$$u_{n+1} = u_n - L(u_n)(T(u_n) - (T(u_0) + f))$$

for L a right inverse of DT.

One way of justifying this is as follows.

We set

$$u_{n+1} = u_n + \dot{u}_n$$

where the increment $\dot{u}_n$ is to be determined. We then have

$$T(u_{n+1}) = T(u_n) + DT(u_n)\dot{u}_n + e_n$$

which defines the error e_n incurred by using the derivative of T to obtain a linear approximation to T. By Taylor's theorem, we expect this to be small when $\dot{u}_n$ is small.

Let us choose $\dot{u}_n$ such that

$$DT(u_n)\dot{u}_n = g_n$$

i.e.

$$\dot{u}_n = L(u_n)g_n$$

where g_n is to be determined so that u_n converges to a solution u of $T(u) = T(u_0) + f$.

From the equation

$$T(u_{n+1}) = T(u_n) + g_n + e_n$$

we obtain

$$\begin{aligned} T(u_{n+1}) &= T(u_0) + \sum_{m=0}^{n} g_m + \sum_{m=0}^{n} e_m \\ &= T(u_0) + \sum_{m=0}^{n} g_m + E_n + e_n \end{aligned}$$

where

$$E_n = \sum_{m=0}^{n-1} e_m.$$

Thus if we define g_n by

$$\sum_{m=0}^{n} g_m + E_n = f$$

we obtain

$$T(u_{n+1}) = T(u_0) + f + e_n$$

which we hope converges to $T(u_0) + f$ as $n \to \infty$ since $e_n \to 0$.

The formula for g_n implies $g_0 = f$ and

$$\begin{aligned} g_{n+1} &= -e_n \\ &= T(u_n) + g_n - T(u_{n+1}). \end{aligned}$$

Hence

$$g_{n+1} = T(u_0) + f - T(u_{n+1}).$$

Thus we obtain the iteration scheme

$$u_{n+1} = u_n - L(u_n)(T(u_n) - (T(u_0) + f))$$

3.1.2 Nash-Moser Iteration

We still wish to use an iteration scheme of the form

$$u_{n+1} = u_n + \dot{u}_n$$

but we are now concerned with the case when the application of the operator $L(u_n)$ to g_n causes a loss of regularity with respect to u_n and g_n. By this we mean that if u_n and g_n lie in X^s, then $L(u_n)g_n$ will lie in a larger space $X^{s'}$ for $s' < s$ so that for any fixed s the norm $||u_n||_{X^s}$ will blow up as $n \to \infty$. This loss of regularity is stated precisely in (5).

To overcome this, we apply smoothing operators S_n which allow us to control $||S_n u_n||_{X^s}$ for large s in terms of $||u_n||_{X^s}$ for small s. By choosing S_n to vary with n so that $S_n \to \mathrm{id}$ in some sense as $n \to \infty$, we hope to be able to overcome the error introduced by these smoothing operators. In this particular version of Nash-Moser iteration, we follow Hörmander in [15] and Alinhac and Gérard in [2] by choosing to apply smoothing operators before the application of the operator L. Hence we define

$$v_n = S_n u_n$$

and set

$$T(u_{n+1}) = T(u_n) + DT(v_n)\dot{u}_n + e_n$$

which defines the error e_n incurred by using the derivative of T, evaluated at v_n, to obtain a linear approximation to T. By Taylor's theorem, and the fact that $S_n \to \mathrm{id}$, we expect this to be small when $\dot{u}_n$ is small and n is large.

Following the same process as before, we define

$$\dot{u}_n = L(v_n)g_n$$

where g_n is to be determined so that u_n converges to a solution u of $T(u) = T(u_0) + f$, and g_n should be smoothed.

From the equation

$$T(u_{n+1}) = T(u_n) + g_n + e_n$$

we obtain

$$\begin{aligned} T(u_{n+1}) &= T(u_0) + \sum_{m=0}^{n} g_m + \sum_{m=0}^{n} e_m \\ &= T(u_0) + \sum_{m=0}^{n} g_m + E_n + e_n \end{aligned}$$

where

$$E_n = \sum_{m=0}^{n-1} e_m.$$

Before we defined g_n by

$$\sum_{m=0}^{n} g_m + E_n = f$$

but since we would like g_n to be smoothed, we define g_n by

$$\sum_{m=0}^{n} g_m = S_n(f - E_n).$$

From this, we obtain

$$T(u_{n+1}) = T(u_0) + S_n f + E_n - S_n E_n + e_n$$

which we hope converges to $T(u_0) + f$ as $n \to \infty$ since $e_n \to 0$ and $S_n \to \mathrm{id}$.

The formula for g_n implies $g_0 = S_0 f$ and

$$\begin{aligned} g_{n+1} &= S_{n+1}(f - E_{n+1}) - S_n(f - E_n) \\ &= (S_{n+1} - S_n)(f - E_n) - S_{n+1}e_n. \end{aligned}$$

Note that we may split the error e_n up into two parts,

$$e_n = e_n' + e_n''$$

where

$$e_n' = (DT(u_n) - DT(v_n))\dot{u}_n$$

is the error caused by replacing u_n by v_n and

$$e_n'' = T(u_{n+1}) - T(u_n) - DT(u_n)\dot{u}_n$$

is the standard quadratic error in the Newton-Raphson scheme.

3.2 Statement and Proof of the Theorem

Theorem 1. *Let $\{X^s\}_{s\in I}$ and $\{Y^s\}_{s\in I}$ be two decreasing families of Banach spaces, each satisfying the smoothing hypothesis. Let $u_0 \in X^\infty$ and let $T : U^{m_0} \to Y^0$ be continuous, where $U^{m_0} \subset X^{m_0}$ is a bounded open neighbourhood of u_0 in X^{m_0}, for some $m_0 \in I$. Suppose also $T : U \to Y^{\infty - m_1}$ for some fixed $m_1 \in I$, where $U := U^{m_0} \cap X^\infty$, and T satisfies the following conditions.*

1. *T is twice differentiable in the sense of Definition 3 and*

$$\begin{aligned} &||D^2T(u)(v_1, v_2)||_{Y^s} \\ &\le C_s^1(||v_1||_{X^{s+m_1}}\, ||v_2||_{X^{m_2}} + ||v_1||_{X^{m_2}}\, ||v_2||_{X^{s+m_1}} + ||v_1||_{X^{m_2}}\, ||v_2||_{X^{m_2}}\, (1 + ||u||_{X^{s+m_3}})) \end{aligned} \tag{4}$$

   for all $u \in U$, $v_1, v_2 \in X^\infty$ and $s \in I$ such that $s + m_1, s + m_3 \in I$, for some fixed numbers $m_1, m_2, m_3 \in I$, where the constant $C_s^1 > 0$ is bounded for s bounded.
2. *For each $u \in U$, there exists a linear map $L(u) : Y^\infty \to X^{\infty - \max\{l_1, m_4\}}$ such that*

$$DT(u)L(u) = \mathrm{id}$$

   and

$$||L(u)g||_{X^s} \le C_s^2(||g||_{Y^{s+l_1}} + ||g||_{Y^{l_1}}\, ||u||_{X^{s+m_4}}) \tag{5}$$

   for all $u \in U$, $g \in Y^\infty$ and $s \in I$ such that $s + l_1, s + m_4 \in I$, for some fixed numbers $l_1, m_4 \in I$, where the constant $C_s^2 > 0$ is bounded for s bounded.

Let $r_0 \in I$ with $r_0 > \max\{m_0, m_4, l_1 + m_1 + m_2, 2m_2, \frac{l_1+m_3}{2} + m_2\}$ and let $r_0 + 1 < s_1 \in I$ such that $s_1 + \max\{l_1, m_4\} \in I$ be sufficiently large depending on the constants m_i.

Then there exists a constant $0 < \epsilon \le 1$ such that if $f \in Y^{r_0+l_1}$ with

$$||f||_{Y^{r_0+l_1}} \le \epsilon$$

we can find $u \in U^{m_0}$ which solves the equation

$$T(u) = T(u_0) + f.$$

Moreover, let $J = \{r \in I : f \in Y^{r+l_1}, r \ge r_0\}$. Then for each $r \in J$ and $s \in I$ with $s < r$, assuming that $s_1 + r - r_0 + \max\{l_1, m_4\} \in I$, we have $u \in X^s$, and there exists a constant $K_{r,s}$ independent of f such that

$$||u - u_0||_{X^s} \le K_{r,s}\, ||f||_{Y^{r+l_1}}\, .$$

Proof.

Step 1 – Setup of the iteration scheme

Let $f \in Y^{r_0+l_1}$ be such that $||f||_{Y^{r_0+l_1}} \le \epsilon$, where $0 < \epsilon \le 1$ will be chosen later.

Denote the smoothing operators on $(X^s)_{s\in I}$ by $\{S_\theta^X\}_{\theta\ge1}$ and the smoothing operators on $(Y^s)_{s\in I}$ by $\{S_\theta^Y\}_{\theta\ge1}$.

We use an iteration scheme to construct a sequence $(u_n)_{n\ge0}$ in X^∞ which we aim to show converges to a solution $u \in U^{m_0}$ of $T(u) = T(u_0) + f$.

For $n \geq 0$, define

$$\theta_n = \theta_0 + n$$

where $\theta_0 > 1$ will be chosen later depending only on r_0, the constants m_i, l_1, and the constants in the smoothing hypothesis and in the inequalities satisfied by D^2T and L.

Note that

$$\theta_{n+1} \leq \theta_n + 1 \leq 2\theta_n.$$

We have dropped the parameter κ from the definition of θ_n in Hörmander's version since he introduced it to make e''_n as small as e'_n, but this will turn out to be automatically true under our hypotheses.

For $n \geq 0$, define

$$\begin{aligned} v_n &= S^X_{\theta_n} u_n \\ \dot{u}_n &= L(v_n) g_n \\ u_{n+1} &= u_n + \dot{u}_n \end{aligned}$$

where g_n is defined below.

Note that the overdot $\dot{}$ is simply notation indicating a sort of difference and does not denote differentiation.

For $n \geq 0$, define

$$\begin{aligned} g_0 &= S^Y_{\theta_0} f \\ g_{n+1} &= (S^Y_{\theta_{n+1}} - S^Y_{\theta_n})(f - E_n) - S^Y_{\theta_{n+1}} e_n \end{aligned}$$

where

$$E_n = \sum_{m=0}^{n-1} e_m$$

(so $E_0 = 0$), and the error e_n is defined below, for $n \geq 0$.

$$\begin{aligned} e'_n &= (DT(u_n) - DT(v_n))\dot{u}_n \\ e''_n &= T(u_n + \dot{u}_n) - T(u_n) - DT(u_n)\dot{u}_n \\ e_n &= e'_n + e''_n. \end{aligned}$$

Note that since g_0 is defined in terms of f only, and we are given u_0, from which v_0 is obtained immediately, the iteration scheme can be determined for $n \geq 0$ in the order $\dot{u}_n, u_{n+1}, v_{n+1}, e'_n, e''_n, e_n, E_n, g_{n+1}$.

Note that e_n is defined so that it measures how well $T(u_{n+1}) - T(u_n)$ is approximated by $DT(v_n)\dot{u}_n$, by which we mean

$$T(u_{n+1}) - T(u_n) = DT(v_n)\dot{u}_n + e_n$$
$$= g_n + e_n.$$

Also note that the formula for g_{n+1} can be rearranged to give

$$g_{n+1} = (S^Y_{\theta_{n+1}} f - S^Y_{\theta_n} f) - (S^Y_{\theta_{n+1}} E_{n+1} - S^Y_{\theta_n} E_n).$$

We thus obtain

$$\begin{aligned} T(u_{n+1}) - T(u_0) &= \sum_{m=0}^{n} (T(u_{m+1}) - T(u_m)) \\ &= \sum_{m=0}^{n} g_m + \sum_{m=0}^{n} e_m \\ &= S^Y_{\theta_n} f - S^Y_{\theta_n} E_n + E_{n+1} \\ &= S^Y_{\theta_n} f + (E_n - S^Y_{\theta_n} E_n) + e_n \end{aligned}$$

which we hope converges to f as $n \to \infty$, since, roughly speaking, $S^Y_{\theta_n} \to \text{id}$ and $e_n \to 0$.

Step 2 – Obtaining estimates for the iterates via induction

We will show the following inductive hypothesis holds.

$$||\dot{u}_n||_{X^s} \leq K\,||f||_{Y^{r_0+l_1}}\,\theta_n^{s-r_0-1} \qquad \text{for all } s \in [0, s_1] \qquad [H_n]$$

where the constant $K > 0$ will be chosen later, with K independent of n, f and ϵ, but depending on θ_0. We will choose ϵ sufficiently small such that $K\,||f||_{Y^{r_0+l_1}} \leq K\epsilon \leq 1$.

In what follows, $C_s > 0$ represents a constant, which is independent of n, f and ϵ, and is bounded for s bounded. It will also be independent of θ_0, which will allow us to choose θ_0 so that θ_n is large compared to C_s for s in a certain range. We will write $C > 0$ for a constant which is also independent of s.

Assume now that $[H_m]$ is true for all $0 \leq m \leq n$ and let us show that $[H_{n+1}]$ follows. (We will leave the proof of $[H_0]$ until later.)

Pick a real number $0 < \eta < 1$ such that $r_0 > \max\{m_0, m_4, l_1 + m_1 + m_2, 2m_2, \frac{l_1+m_3}{2} + m_2\} + 2\eta$.

For $s \in I$, define

$$P(s) = \begin{cases} (s - r_0)_+ & \text{for } |s - r_0| \geq \eta, \\ \eta & \text{for } |s - r_0| < \eta. \end{cases}$$

We claim that the following estimates for $0 \leq m \leq n+1$ follow directly from $[H_m]$ for $0 \leq m \leq n$.

$$||u_m - u_0||_{X^s} \leq C_s K \,||f||_{Y^{r_0+l_1}}\, \theta_m^{P(s)} \qquad \text{for } s \in [0, s_1], \tag{6}$$

$$\left|\left|S_{\theta_m}^X (u_m - u_0)\right|\right|_{X^s} \leq C_s K \,||f||_{Y^{r_0+l_1}}\, \theta_m^{P(s)} \qquad \text{for } s \in I, \tag{7}$$

$$\left|\left|(u_m - u_0) - S_{\theta_m}^X (u_m - u_0)\right|\right|_{X^s} \leq C_s K \,||f||_{Y^{r_0+l_1}}\, \theta_m^{(s-r_0)} \qquad \text{for } s \in [0, s_1], \tag{8}$$

$$||u_m - v_m||_{X^s} \leq C_s \theta_m^{s-r_0} \qquad \text{for } s \in [0, s_1], \tag{9}$$

$$||v_m||_{X^s} \leq C_s \theta_m^{P(s)} \qquad \text{for } s \in I, \tag{10}$$

$$||u_m||_{X^s} \leq C_s \theta_m^{P(s)} \qquad \text{for } s \in [0, s_1]. \tag{11}$$

Indeed, for $0 \leq m \leq n$, we have

$$\begin{aligned}
||u_{m+1} - u_0||_{X^s} &= \left|\left|\sum_{l=0}^{m} \dot{u}_l\right|\right|_{X^s} \\
&\leq \sum_{l=0}^{m} ||\dot{u}_l||_{X^s} \\
&\leq K \,||f||_{Y^{r_0+l_1}} \sum_{l=0}^{m} \theta_l^{s-r_0-1} \\
&= K \,||f||_{Y^{r_0+l_1}} \sum_{l=0}^{m} (\theta_0 + l)^{s-r_0-1} \\
&\leq K \,||f||_{Y^{r_0+l_1}} \sum_{l=0}^{m} (\theta_0 + l)^{Q(s)-1}
\end{aligned}$$

where

$$Q(s) = \begin{cases} s - r_0 & \text{for } |s - r_0| \geq \eta, \\ \eta & \text{for } |s - r_0| < \eta. \end{cases}$$

Set $h(x) = (\theta_0 + x)^{Q(s)-1}$ for $x \in [0, \infty)$. Then

$$\begin{aligned}
\sum_{l=0}^{m} (\theta_0 + l)^{Q(s)-1} &\leq \int_0^{m+1} h(x) \mathrm{d}x \\
&= \begin{cases} \frac{1}{s-r_0}((\theta_0 + m + 1)^{s-r_0} - \theta_0^{s-r_0}) & \text{for } |s - r_0| \geq \eta \\ \frac{1}{\eta}((\theta_0 + m + 1)^{\eta} - \theta_0^{\eta}) & \text{for } |s - r_0| < \eta \end{cases}
\end{aligned}$$

$$= \begin{cases} \frac{1}{s-r_0}(\theta_{m+1}^{s-r_0} - \theta_0^{s-r_0}) & \text{for } |s-r_0| \geq \eta \\ \frac{1}{\eta}(\theta_{m+1}^{\eta} - \theta_0^{\eta}) & \text{for } |s-r_0| < \eta \end{cases}$$

$$\leq \begin{cases} \frac{1}{s-r_0}\theta_{m+1}^{s-r_0} & \text{for } s-r_0 \geq \eta \\ \frac{1}{r_0-s}\theta_0^{-(r_0-s)} & \text{for } s-r_0 \leq -\eta \\ \frac{1}{\eta}\theta_{m+1}^{\eta} & \text{for } |s-r_0| < \eta \end{cases}$$

This implies (6), noting that the constant C_s remains bounded for s bounded. (We introduced η to avoid a constant involving $\frac{1}{s-r_0}$ which blows up as $s \to r_0$.)

For $s \geq r_0 + \eta$, use (1) from the smoothing hypothesis and (6) to obtain

$$\begin{aligned} \left|\left|S_{\theta_m}^X(u_m - u_0)\right|\right|_{X^s} &\leq C_s\theta_m^{s-r_0-\eta}\,||u_m - u_0||_{X^{r_0+\eta}} \\ &\leq C_s K\,||f||_{Y^{r_0+l_1}}\,\theta_m^{s-r_0-\eta}\theta_m^{\eta} \\ &\leq C_s K\,||f||_{Y^{r_0+l_1}}\,\theta_m^{s-r_0}. \end{aligned}$$

For $s < r_0 + \eta$, using (1) from the smoothing hypothesis and (6), we have

$$\left|\left|S_{\theta_m}^X(u_m - u_0)\right|\right|_{X^s} \leq C_s K\,||f||_{Y^{r_0+l_1}}\,\theta_m^{P(s)}.$$

This proves (7).

For $s \leq r_0 + \eta$, use (2) from the smoothing hypothesis and (6) to obtain

$$\begin{aligned} \left|\left|(u_m - u_0) - S_{\theta_m}^X(u_m - u_0)\right|\right|_{X^s} &\leq C_s\theta_m^{s-r_0-\eta}\,||u_m - u_0||_{X^{r_0+\eta}} \\ &\leq C_s K\,||f||_{Y^{r_0+l_1}}\,\theta_m^{s-r_0-\eta}\theta_m^{\eta} \\ &\leq C_s K\,||f||_{Y^{r_0+l_1}}\,\theta_m^{s-r_0}. \end{aligned}$$

For $r_0 + \eta < s \leq s_1$, using (6) and (7), we have

$$\left|\left|(u_m - u_0) - S_{\theta_m}^X(u_m - u_0)\right|\right|_{X^s} \leq C_s K\,||f||_{Y^{r_0+l_1}}\,\theta_m^{s-r_0}$$

as required. This proves (8).

Now

$$\begin{aligned} ||u_m - v_m||_{X^s} &= \left|\left|u_m - S_{\theta_m}^X u_m\right|\right|_{X^s} \\ &= \left|\left|(u_m - u_0) - S_{\theta_m}^X(u_m - u_0) + u_0 - S_{\theta_m}^X u_0\right|\right|_{X^s} \\ &\leq \left|\left|(u_m - u_0) - S_{\theta_m}^X(u_m - u_0)\right|\right|_{X^s} + \left|\left|u_0 - S_{\theta_m}^X u_0\right|\right|_{X^s} \\ &\leq C_s K\,||f||_{Y^{r_0+l_1}}\,\theta_m^{s-r_0} + C_s\theta_m^{s-r_0}\,||u_0||_{X^{\max\{r_0,s\}}} \end{aligned}$$

by applying (8) to the first term and (1) or (2) from the smoothing hypothesis to the second term. This proves (9). (Note $K\,||f||_{Y^{r_0+l_1}} \leq K\epsilon \leq 1$.)

Similarly,

$$\begin{aligned}||v_m||_{X^s} &= \left|\left|S^X_{\theta_m} u_m\right|\right|_{X^s} \\ &= \left|\left|S^X_{\theta_m}(u_m - u_0) + S^X_{\theta_m} u_0\right|\right|_{X^s} \\ &\leq \left|\left|S^X_{\theta_m}(u_m - u_0)\right|\right|_{X^s} + \left|\left|S^X_{\theta_m} u_0\right|\right|_{X^s} \\ &\leq \left|\left|S^X_{\theta_m}(u_m - u_0)\right|\right|_{X^s} + C_s \,||u_0||_{X^s}\end{aligned}$$

by (1) from the smoothing hypothesis. Now use (7) to obtain (10).

We have

$$||u_m||_{X^s} \leq ||u_m - u_0||_{X^s} + ||u_0||_{X^s} \,.$$

Now apply (6) to obtain (11).

This completes the proof of the claim.

Note that, using (6) and (9), we have

$$\begin{aligned}||v_m - u_0||_{X^{m_0}} &\leq ||v_m - u_m||_{X^{m_0}} + ||u_m - u_0||_{X^{m_0}} \\ &\leq C\theta_m^{m_0 - r_0} + CK\epsilon\theta_m^{P(m_0)} \\ &\leq C\theta_m^{m_0 - r_0} + CK\epsilon.\end{aligned}$$

Thus by taking ϵ sufficiently small depending on K and C, and θ_0 sufficiently large depending on C, we have $v_n,\ v_{n+1} \in U$. Also note that (6) in the case $s = m_0$ implies $u_n \in U$ for ϵ sufficiently small, and $[H_n]$ implies that $u_n + \dot{u}_n \in U$ for ϵ sufficiently small. This guarantees that e_n and $\dot{u}_{n+1}$ are well-defined. Note that the same argument also shows that the line segments $[u_n, u_n + \dot{u}_n]$ and $[u_n, v_n]$ are in U for ϵ sufficiently small.

Estimate of e'_n. We claim that for all $s \in [0, s_1 - \max\{m_1, m_3\}]$,

$$\left|\left|e'_n\right|\right|_{Y^s} \leq C_s K \,||f||_{Y^{r_0+l_1}}\, \theta_n^{M(s)-1+\eta}$$

where

$$M(s) = \max\{s + m_1 + m_2 - 2r_0, (s + m_3 - r_0)_+ + 2m_2 - 2r_0\}.$$

Indeed, we have

$$\begin{aligned}e'_n &= (DT(u_n) - DT(v_n))\dot{u}_n \\ &= (DT((u_n - v_n) + v_n) - DT(v_n))\dot{u}_n.\end{aligned}$$

Note that, since T is twice differentiable in the sense of Definition 3, the map

$$u \mapsto DT(u)\dot{u}_n$$

is differentiable in the sense of Definition 3 with derivative acting on v given by

$$D^2T(u)(\dot{u}_n, v).$$

Hence, applying Taylor's theorem, (4), $[H_n]$ and the estimates (9) and (10), we have, for $s \in [0, s_1 - \max\{m_1, m_3\}]$,

$$\begin{aligned}
||e'_n||_{Y^s} &= ||(DT((u_n - v_n) + v_n) - DT(v_n))\dot{u}_n||_{Y^s} \\
&\leq \sup_{t\in[0,1]} ||D^2T(t(u_n - v_n) + v_n)(\dot{u}_n, u_n - v_n)||_{Y^s} \\
&\leq C_s(||\dot{u}_n||_{X^{s+m_1}}\,||u_n - v_n||_{X^{m_2}} + ||\dot{u}_n||_{X^{m_2}}\,||u_n - v_n||_{X^{s+m_1}} \\
&+ ||\dot{u}_n||_{X^{m_2}}\,||u_n - v_n||_{X^{m_2}}\,(1 + \sup_{t\in[0,1]} ||v_n + t(u_n - v_n)||_{X^{s+m_3}})) \\
&\leq C_s(K\,||f||_{Y^{r_0+l_1}}\,\theta_n^{s+m_1-r_0-1}\theta_n^{m_2-r_0} + K\,||f||_{Y^{r_0+l_1}}\,\theta_n^{m_2-r_0-1}\theta_n^{s+m_1-r_0} \\
&+ K\,||f||_{Y^{r_0+l_1}}\,\theta_n^{m_2-r_0-1}\theta_n^{m_2-r_0}(1 + \theta_n^{P(s+m_3)} + \theta_n^{s+m_3-r_0})) \\
&\leq C_sK\,||f||_{Y^{r_0+l_1}}\,\theta_n^{M(s)-1+\eta}.
\end{aligned}$$

Estimate of e''_n. We claim that for all $s \in [0, s_1 - \max\{m_1, m_3\}]$,

$$||e''_n||_{Y^s} \leq C_sK\,||f||_{Y^{r_0+l_1}}\,\theta_n^{M(s)-1+\eta}.$$

Indeed, we have

$$e''_n = T(u_n + \dot{u}_n) - T(u_n) - DT(u_n)\dot{u}_n.$$

Hence, applying Taylor's theorem, (4), $[H_n]$ and the estimate (11), we have, for $s \in [0, s_1 - \max\{m_1, m_3\}]$,

$$\begin{aligned}
||e''_n||_{Y^s} &\leq \sup_{t\in[0,1]} ||D^2T(u_n + t\dot{u}_n)(\dot{u}_n, \dot{u}_n)||_{Y^s} \\
&\leq C_s(||\dot{u}_n||_{X^{s+m_1}}\,||\dot{u}_n||_{X^{m_2}} + ||\dot{u}_n||^2_{X^{m_2}}\,(1 + \sup_{t\in[0,1]} ||u_n + t\dot{u}_n||_{X^{s+m_3}})) \\
&\leq C_s(K\,||f||_{Y^{r_0+l_1}}\,\theta_n^{s+m_1-r_0-1}K\,||f||_{Y^{r_0+l_1}}\,\theta_n^{m_2-r_0-1} \\
&+ K^2\,||f||^2_{Y^{r_0+l_1}}\,\theta_n^{2m_2-2r_0-2}(1 + \theta_n^{P(s+m_3)} + K\,||f||_{Y^{r_0+l_1}}\,\theta_n^{s+m_3-r_0-1})) \\
&\leq \theta_n^{-1}C_sK\,||f||_{Y^{r_0+l_1}}\,\theta_n^{M(s)-1+\eta} \\
&\leq C_sK\,||f||_{Y^{r_0+l_1}}\,\theta_n^{M(s)-1+\eta}
\end{aligned}$$

where we have used $K\,||f||_{Y^{r_0+l_1}} \leq K\epsilon \leq 1$.

Estimate of e_n. Adding the estimates for e'_n and e''_n, we obtain

$$||e_n||_{Y^s} \le C_s K \, ||f||_{Y^{r_0+l_1}} \, \theta_n^{M(s)-1+\eta}$$

for all $s \in [0, s_1 - \max\{m_1, m_3\}]$.

Estimate of g_{n+1}. We claim that for all $s \in I$,

$$||g_{n+1}||_{Y^s} \le C_s(K \, ||f||_{Y^{r_0+l_1}} \, \theta_n^{M(s)-1+\eta} + ||f||_{Y^{r_0+l_1}} \, \theta_n^{s-r_0-l_1-1}).$$

Indeed, we have

$$g_{n+1} = (S^Y_{\theta_{n+1}} - S^Y_{\theta_n})(f - E_n) - S^Y_{\theta_{n+1}} e_n.$$

Note that for any $w \in Y^{s'}$,

$$\left|\left|(S^Y_{\theta_{n+1}} - S^Y_{\theta_n})w\right|\right|_{Y^s} \le C_{s',s} \theta_n^{s-s'-1} \, ||w||_{Y^{s'}}$$

by the smoothing hypothesis (3) and Taylor's theorem.

Setting $s' = r_0 + l_1$, we have

$$\left|\left|(S^Y_{\theta_{n+1}} - S^Y_{\theta_n})f\right|\right|_{Y^s} \le C_s \theta_n^{s-r_0-l_1-1} \, ||f||_{Y^{r_0+l_1}} \, .$$

We also have

$$\left|\left|(S^Y_{\theta_{n+1}} - S^Y_{\theta_n})E_n\right|\right|_{Y^s} \le C_{s',s} \theta_n^{s-s'-1} \, ||E_n||_{Y^{s'}} \, .$$

Now, for $s' \in [0, s_1 - \max\{m_1, m_3\}]$, we have, from the estimate for e_n,

$$\begin{aligned} ||E_n||_{Y^{s'}} &= \left|\left|\sum_{m=0}^{n-1} e_m\right|\right|_{Y^{s'}} \\ &\le C_{s'} K \, ||f||_{Y^{r_0+l_1}} \sum_{m=0}^{n-1} \theta_m^{M(s')-1+\eta} \\ &\le C_{s'} K \, ||f||_{Y^{r_0+l_1}} \, \theta_n^{M(s')+\eta} \end{aligned}$$

if $M(s') \ge 0$, by the integral comparison used before. Note that $M(s')$ has slope 1 for large enough s' depending on r_0 and the constants m_i, so to achieve $M(s') \ge 0$ it suffices to take s' large in relation to r_0 and the constants m_i. To do this we require s_1 sufficiently large in relation to r_0 and the constants m_i.

Hence

$$\left\|(S^Y_{\theta_{n+1}} - S^Y_{\theta_n})E_n\right\|_{Y^s} \leq C_{s'}C_{s',s}K\,||f||_{Y^{r_0+l_1}}\,\theta_n^{M(s')+s-s'-1+\eta}$$
$$\leq C_s K\,||f||_{Y^{r_0+l_1}}\,\theta_n^{M(s)-1+\eta}$$

by choosing s' sufficiently large compared to r_0 and the constants m_i so that $M(s)$ has slope 1 for $s \geq s'$. (Hence $M(s') - s' \leq M(s) - s$ for all s since $M(s) - s$ is decreasing for $s \leq s'$ and constant for $s \geq s'$.) Again, to do this we require s_1 sufficiently large in relation to r_0 and the constants m_i. This fixes s_1.

Similarly, for s' sufficiently large, we have

$$\left\|S^Y_{\theta_{n+1}} e_n\right\|_{Y^s} \leq C_{s',s}\theta_n^{s-s'}\,||e_n||_{Y^{s'}}$$
$$\leq C_{s',s}C_{s'}K\,||f||_{Y^{r_0+l_1}}\,\theta_n^{M(s')+s-s'-1+\eta}$$
$$\leq C_s K\,||f||_{Y^{r_0+l_1}}\,\theta_n^{M(s)-1+\eta}.$$

Hence the estimate for g_{n+1} holds.

Estimate of $\dot{u}_{n+1}$. We have

$$\dot{u}_{n+1} = L(v_{n+1})g_{n+1}.$$

Hence, for all $s \in I$ such that $s + l_1, s + m_4 \in I$, using (5), the estimate (10) and the estimate for g_{n+1}, we have

$$\begin{aligned}
||\dot{u}_{n+1}||_{X^s} &\leq C_s(||g_{n+1}||_{Y^{s+l_1}} + ||g_{n+1}||_{Y^{l_1}}\,(1 + ||v_{n+1}||_{X^{s+m_4}}))\\
&\leq C_s(K\,||f||_{Y^{r_0+l_1}}\,\theta_{n+1}^{M(s+l_1)-1+\eta} + ||f||_{Y^{r_0+l_1}}\,\theta_{n+1}^{s-r_0-1}\\
&\quad + (K\,||f||_{Y^{r_0+l_1}}\,\theta_{n+1}^{M(l_1)-1+\eta} + ||f||_{Y^{r_0+l_1}}\,\theta_{n+1}^{-r_0-1})(1 + \theta_{n+1}^{P(s+m_4)}))\\
&\leq C_s(K\,||f||_{Y^{r_0+l_1}}\,\theta_{n+1}^{M(l_1)+s-1+\eta} + ||f||_{Y^{r_0+l_1}}\,\theta_{n+1}^{s-r_0-1})
\end{aligned} \tag{12}$$

since $\theta_{n+1}^{P(s+m_4)} \leq \theta_{n+1}^s$ because $r_0 > m_4 + 2\eta$, and $M(l_1 + s) \leq M(l_1) + s$ because M has slope at most 1.

We want to obtain

$$||\dot{u}_{n+1}||_{X^s} \leq K\,||f||_{Y^{r_0+l_1}}\,\theta_{n+1}^{s-r_0-1}$$

for $s \in [0, s_1]$.

To make the first term sufficiently small, we require

$$-\gamma := M(l_1) + r_0 + \eta < 0.$$

Then we can choose θ_0 large enough so that

$$C_s \theta_{n+1}^{M(l_1)+s-1+\eta} = C_s \theta_{n+1}^{s-r_0-1} \theta_{n+1}^{-\gamma} \leq C_s \theta_{n+1}^{s-r_0-1} \theta_0^{-\gamma} \leq \frac{1}{2} \theta_{n+1}^{s-r_0-1}$$

for all $s \in [0, s_1]$.

We note that $M(l_1)+r_0+\eta < 0$ if and only if $r_0-\eta > l_1+m_1+m_2$, $r_0-\eta > 2m_2$ and $r_0 - \eta > m_2 + \frac{l_1+m_3}{2}$, which indeed hold by the choice of η.

To make the second term sufficiently small, we take $K \geq 2C_s$ for all $s \in [0, s_1]$.

This gives $[H_{n+1}]$.

Proof of $[H_0]$ We have

$$g_0 = S_{\theta_0}^Y f$$

and

$$v_0 = S_{\theta_0}^X u_0.$$

Hence

$$\begin{aligned}
||\dot{u}_0||_{X^s} &= \left|\left|L(S_{\theta_0}^X u_0) S_{\theta_0}^Y f\right|\right|_{X^s} \\
&\leq C_s(\left|\left|S_{\theta_0}^Y f\right|\right|_{Y^{s+l_1}} + \left|\left|S_{\theta_0}^Y f\right|\right|_{Y^{l_1}} (1 + \left|\left|S_{\theta_0}^X u_0\right|\right|_{X^{s+m_4}})) \\
&\leq C_s \left|\left|S_{\theta_0}^Y f\right|\right|_{Y^{s+l_1}} \\
&\leq C_s \,||f||_{Y^{r_0+l_1}}\, \theta_0^{(s-r_0)+} \qquad \text{by (1) and (2) from the smoothing hypothesis} \\
&\leq K \,||f||_{Y^{r_0+l_1}}\, \theta_0^{s-r_0-1}
\end{aligned}$$

for all $s \in [0, s_1]$, assuming that K is sufficiently large compared to θ_0 and C_s for $s \in [0, s_1]$.

This is $[H_0]$.

Step 3 – Better estimates if $f \in Y^{r+l_1}$ for $r > r_0$

Let $r \in J$, so that $f \in Y^{r+l_1}$, where $r \geq r_0$.

We will show that, for all $n \geq 0$ and for all $s \in I$ such that $s + \max\{m_1, m_3\} + \max\{l_1, m_4\} \in I$, we have

$$||\dot{u}_n||_{X^s} \leq C_{r,s} \,||f||_{Y^{r+l_1}}\, \theta_n^{s-r-1} \tag{13}$$

where the constant $C_{r,s} > 0$ is independent of n and f.

Firstly, note that we have proved $[H_n]$ for $n \geq 0$, and hence all the estimates from step 2 which were conditional on the inductive hypothesis are now valid, and we may use them as we wish.

We are going to prove the above statement by an induction argument, but not an induction on n. We are going to use the estimates from step 2 for each n separately to obtain the above inequality, and the constant will be independent of n because the constants from step 2 are independent of n.

We claim by induction on $k \geq 0$ that for all $s \in I$ such that $s + \max\{l_1, m_4\} \in I$, we have

$$||\dot{u}_n||_{X^s} \leq C_{k,r,s} \, ||f||_{Y^{r+l_1}} \, \theta_n^{s-r_0-\gamma_k-1} \qquad [G_k]$$

where the constant $C_{k,r,s} > 0$ is independent of n and f, and

$$\gamma_k = \min\{k\gamma, r - r_0\}.$$

Indeed, the estimate (12) for $\dot{u}_{n+1}$ in step 2 implies that

$$||\dot{u}_n||_{X^s} \leq C_s \, ||f||_{Y^{r_0+l_1}} \, \theta_n^{s-r_0-1} \tag{14}$$

for all $s \in I$ such that $s + \max\{l_1, m_4\} \in I$ (not just $s \in [0, s_1]$ which would follow directly from $[H_n]$).

Using this, we can obtain the following new versions of the estimates (9)–(11) for all $s \in I$ such that $s + \max\{l_1, m_4\} \in I$ (not just $s \in [0, s_1]$) via exactly the same calculations

$$||u_m - v_m||_{X^s} \leq C_s \theta_m^{s-r_0}, \tag{15}$$

$$||v_m||_{X^s} \leq C_s \theta_m^{P(s)}, \tag{16}$$

$$||u_m||_{X^s} \leq C_s \theta_m^{P(s)}. \tag{17}$$

Using the fact that $||f||_{Y^{r_0+l_1}} \leq ||f||_{Y^{r+l_1}}$, (14) immediately implies $[G_0]$.

Now we assume $[G_k]$ holds and aim to show $[G_{k+1}]$ holds.

Now we want to obtain new estimates for e'_n and e''_n. Note that in the estimates for both of these there was at least one factor involving $\dot{u}_n$ in each term. If we estimate this one factor using the new estimate given by $[G_k]$ and the other quantities using (14) and the slightly modified estimates (15)–(17), we obtain

$$||e_n||_{Y^s} \leq C_{k,r,s} \, ||f||_{Y^{r+l_1}} \, \theta_n^{M(s)-1+\eta-\gamma_k}$$

for all $s \in I$ such that $s + \max\{m_1, m_3\} + \max\{l_1, m_4\} \in I$. The constant $C_{k,r,s}$ is independent of f since we have only used the new estimate given by $[G_k]$ in one factor, and the other estimates we have used involve $||f||_{Y^{r_0+l_1}}$, which is bounded by $\epsilon \leq 1$.

This implies that for $s' \in I$ such that $s' + \max\{m_1, m_3\} + \max\{l_1, m_4\} \in I$, we have

$$\begin{aligned}
||E_n||_{Y^{s'}} &= \left|\left| \sum_{m=0}^{n-1} e_m \right|\right|_{Y^{s'}} \\
&\leq C_{k,r,s'} \, ||f||_{Y^{r+l_1}} \sum_{m=0}^{n-1} \theta_m^{M(s')-1+\eta-\gamma_k} \\
&\leq C_{k,r,s'} \, ||f||_{Y^{r+l_1}} \, \theta_n^{M(s')+\eta-\gamma_k}
\end{aligned} \tag{18}$$

as long as $M(s') \geq \gamma_k$. It is possible to pick such an s' if $s_1+r-r_0+\max\{l_1, m_4\} \in I$ given the fact that $M(s_1 - \max\{m_1, m_3\}) \geq 0$ and $M(s)$ has slope 1 for $s \geq s_1 - \max\{m_1, m_3\}$.

Hence

$$\left|\left|(S^Y_{\theta_{n+1}} - S^Y_{\theta_n})E_n\right|\right|_{Y^s} \leq C_{s',k} C_{k,r,s} \theta_n^{M(s')+s-s'-1+\eta-\gamma_k}$$
$$\leq C_{k,r,s} \,||f||_{Y^{r+l_1}}\, \theta_n^{M(s)-1+\eta-\gamma_k}$$

as long as $M(s') \geq \gamma_k$ and s' is sufficiently large compared to r_0 and the constants m_i so that $M(s)$ has slope 1 for $s \geq s'$.

We also have the estimate

$$\left|\left|S^Y_{\theta_{n+1}} e_n\right|\right|_{Y^s} \leq C_{k,r,s} \,||f||_{Y^{r+l_1}}\, \theta_n^{M(s)-1+\eta-\gamma_k}.$$

In addition we can use the new estimate

$$\left|\left|(S^Y_{\theta_{n+1}} - S^Y_{\theta_n})f\right|\right|_{Y^s} \leq C_{r,s} \theta_n^{s-r-l_1-1} \,||f||_{Y^{r+l_1}}\,.$$

This gives us the following new estimate for g_{n+1}, for all $s \in I$,

$$||g_{n+1}||_{Y^s} \leq C_{k,r,s} \,||f||_{Y^{r+l_1}}\, (\theta_n^{M(s)-1+\eta-\gamma_k} + \theta_n^{s-r-l_1-1}).$$

From this we obtain, for all $s \in I$ such that $s + \max\{l_1, m_4\} \in I$,

$$||\dot{u}_n||_{X^s} \leq C_{r,s} \,||f||_{Y^{r+l_1}}\, (\theta_n^{M(l_1)+s-1+\eta-\gamma_k} + \theta_n^{s-r-1})$$
$$\leq C_{r,s} \,||f||_{Y^{r+l_1}}\, (\theta_n^{s-r_0-1-\gamma_k-\gamma} + \theta_n^{s-r-1})$$
$$\leq C_{r,s} \,||f||_{Y^{r+l_1}}\, \theta_n^{s-r_0-\gamma_{k+1}-1}$$

where we have used the fact that $M(l_1) + r_0 + \eta = -\gamma$.

This is $[G_{k+1}]$.

For large enough k, we have $k\gamma \geq r - r_0$, so $\gamma_k = r - r_0$ and this gives (13).

Step 4 – Convergence to a solution

Let $r \in J$, so that $f \in Y^{r+l_1}$, where $r \geq r_0$.

Using (13), we have

$$\sum_{m=0}^{n} ||u_{m+1} - u_m||_{X^s} = \sum_{m=0}^{n} ||\dot{u}_m||_{X^s}$$
$$\leq C_{r,s} \,||f||_{Y^{r+l_1}}\, \theta_{n+1}^{(s-r)+}$$

for $r \neq s$.

Thus

$$\sum_{m=0}^{n} ||u_{m+1} - u_m||_{X^s}$$

converges as $n \to \infty$ for $s < r$. Hence, by completeness, $u_n \to u$ in X^s as $n \to \infty$, for all $s < r$, for some $u \in \bigcap_{0 \le s < r} X^s$.

Note the above calculation also implies that

$$||u_n - u_0||_{X^s} \le C_{r,s} \, ||f||_{Y^{r+l_1}}$$

for $s < r$, so we have

$$||u - u_0||_{X^s} \le C_{r,s} \, ||f||_{Y^{r+l_1}} \, .$$

Next we claim that

$$T(u_{n+1}) - T(u_0) \to f$$

in X^s as $n \to \infty$, for all $s < r$.

Indeed,

$$T(u_{n+1}) - T(u_0) = S^Y_{\theta_n} f + (E_n - S^Y_{\theta_n} E_n) + e_n$$

so

$$T(u_{n+1}) - T(u_0) - f = (S^Y_{\theta_n} f - f) + (E_n - S^Y_{\theta_n} E_n) + e_n.$$

By (2) from the smoothing hypothesis, we have

$$\left|\left|S^Y_{\theta_n} f - f\right|\right|_{Y^{s+l_1}} \le C_{r,s} \theta_n^{s-r} \, ||f||_{Y^{r+l_1}} \to 0 \text{ as } n \to \infty.$$

Also,

$$\begin{aligned}
\left|\left|E_n - S^Y_{\theta_n} E_n\right|\right|_{Y^{s+l_1}} &\le C_{s,s'} \theta_n^{s-s'} \, ||E_n||_{Y^{s'+l_1}} \qquad \text{for } s' \ge s \\
&\le C_{s,s'} \theta_n^{s-s'} C_{r,s} \theta_n^{M(s'+l_1)+\eta-(r-r_0)} \, ||f||_{Y^{r+l_1}} \\
&\quad \text{using (18), for } s' \text{ large enough such that } M(s'+l_1) \ge r - r_0 \\
&\le C_{r,s} \theta_n^{M(s'+l_1)+s-s'+\eta-(r-r_0)} \, ||f||_{Y^{r+l_1}} \\
&\le C_{r,s} \theta_n^{s-r} \, ||f||_{Y^{r+l_1}} \to 0 \text{ as } n \to \infty
\end{aligned}$$

since $M(s' + l_1) + \eta + r_0 \le M(l_1) + \eta + r_0 + s' < s'$.

Finally,

$$||e_n||_{Y^{s+l_1}} \leq C_{r,s}\theta_n^{M(s+l_1)+\eta-(r-r_0)-1}||f||_{Y^{r+l_1}} \to 0 \text{ as } n \to \infty$$

since $M(s+l_1)+\eta+r_0 \leq M(l_1)+\eta+r_0+s < s$.

This proves the claim.

Now since $T : U \to Y^0$ is continuous as a map from X^{m_0} to Y^0, and $r_0 > m_0$, so $u_n \to u$ in X^{m_0}, we have that $T(u_n) \to T(u)$ in Y^0, hence $T(u) = T(u_0) + f$ as required.

This completes the proof.

Remark 4. We make a remark here on the rate of convergence of the above scheme as compared to the Newton-Raphson scheme. Since we have in mind applying the result in existence proofs in PDE problems, we have made no effort to optimise the rate of convergence in the above scheme in any way. One of the key features of the Newton-Raphson scheme is that the rate of convergence is quadratic, i.e. the error at step $n+1$ is proportional to the square of the error at step n. However, we can see in the above scheme that the error $||T(u_{n+1}) - T(u_0) - f||_{X^s}$ is proportional to θ_n^{s-r} where $r > s$ is such that $f \in Y^{r+l_1}$, and θ_n increases like n. Thus according to the crude bounds we have in the above proof, the ratio of the errors at steps n and $n+1$ may tend to 1 as $n \to \infty$, although it may be possible to better by being more careful.

4 Modified Version of Nash-Moser Iteration

4.1 Changes from Hörmander's Iteration Scheme

Here, we introduce two modifications to Hörmander's scheme which will allow it to be applied as in Coulombel and Secchi [8]. The basic principle is that the error $T(u_n)-T(u_0)-f$ in the above scheme tends to zero, so we may introduce additional approximations into the scheme that can be controlled in terms of this error. One disadvantage is that we lose regularity with respect to f since we need this error to be controlled to high order.

Firstly, we note that it may be inconvenient to solve the linearised system

$$DT(u)v = g.$$

It may in fact be more convenient to solve the system

$$A(u)v = g$$

where the operator $A(u)$ is approximately equal to $DT(u)$, such that $A(u) - DT(u)$ can be controlled in terms of the error $T(u) - T(u_0) - f$. This modification was made by Alinhac in [2] when he introduced the 'good unknown'.

Secondly, it may only be possible to solve the linearised system

$$A(u)v = g$$

under certain constraints on u which are not preserved by the iteration scheme, which was the problem encountered by Coulombel and Secchi in [8]. Abstractly, we suppose that the linear system can only be solved for $u \in V$, whereas the iteration scheme only preserves $u \in U$. In fact under the iteration scheme we are trying to solve the problem

$$A(v_n)\dot{u}_n = g_n$$

where

$$v_n = S_n u_n.$$

Therefore we denote by R an operator that maps U to V and set $w_n = R(v_n)$ and solve the system

$$A(w_n)\dot{u}_n = g_n.$$

This will require that $R(u) - u$ is controlled in terms of the error $T(u) - T(u_0) - f$ and also that R and the smoothing operators satisfy some commutation estimates.

4.2 Statement and Proof of the Theorem

Theorem 2. *Let $\{X^s\}_{s\in I}$ and $\{Y^s\}_{s\in I}$ be two decreasing families of Banach spaces, each satisfying the smoothing hypothesis. Let $u_0 \in X^\infty$ and let $T : U^{m_0} \to Y^0$ be continuous, where $U^{m_0} \subset X^{m_0}$ is a bounded open neighbourhood of u_0 in X^{m_0}, for some $m_0 \in I$. Suppose also $T : U \to Y^{\infty - m_1}$ for some fixed $m_1 \in I$, where $U := U^{m_0} \cap X^\infty$. Let $f \in Y^{s_1 - \max\{m_1, m_3\}}$ with $||f||_{Y^{s_1-\max\{m_1,m_3\}}} \leq C^0$, where $s_1, m_3 \in I$ are defined below and C^0 is a constant. Assume the following conditions are satisfied, where the constants are independent of f (at least for $||f||_{Y^{s_1-\max\{m_1,m_3\}}} \leq C^0$).*

1. *T is twice differentiable in the sense of Definition 3 and*

$$\begin{aligned} &||D^2T(u)(v_1, v_2)||_{Y^s} \\ &\leq C_s^1(||v_1||_{X^{s+m_1}}\,||v_2||_{X^{m_2}} + ||v_1||_{X^{m_2}}\,||v_2||_{X^{s+m_1}} + ||v_1||_{X^{m_2}}\,||v_2||_{X^{m_2}}\,(1 + ||u||_{X^{s+m_3}})) \end{aligned} \tag{19}$$

for all $u \in U$, $v_1, v_2 \in X^\infty$ and $s \in I$ such that $s + m_1, s + m_3 \in I$, for some fixed numbers $m_1, m_2, m_3 \in I$, where we assume $\max\{m_1, m_3\} > 0$, and the constant $C_s^1 > 0$ is bounded for s bounded. Also,

$$||DT(u)v||_{Y^s} \leq C_s^2(||v||_{X^{s+m_1}} + ||v||_{X^{m_2}}\,(1 + ||u||_{X^{s+m_3}})) \tag{20}$$

for all $u \in U$, $v \in X^\infty$ *and* $s \in I$ *such that* $s + m_1, s + m_3 \in I$, *where the constant* $C_s^2 > 0$ *is bounded for s bounded.*

2. *For each* $u \in U$, *there exists an operator* $A(u) : X^\infty \to Y^{\infty - m_1}$ *such that*

$$\begin{aligned} &||(A(u) - DT(u))v||_{Y^s} \\ &\leq C_s^3(||v||_{X^{s+m_5}} ||T(u) - T(u_0) - f||_{Y^{l_3}} + ||v||_{X^{m_6}} ||T(u) - T(u_0) - f||_{Y^{s+l_4}} \\ &+ ||v||_{X^{m_6}} ||T(u) - T(u_0) - f||_{Y^{l_3}} (1 + ||u||_{X^{s+m_9}}) \end{aligned} \tag{21}$$

for all $v \in X^\infty$ *and* $s \in I$ *such that* $s + m_5, s + m_9 \in I$, $s + l_4 + \max\{m_1, m_3\} \leq s_1$, *for some fixed numbers* $m_5, m_6, m_9, l_3, l_4 \in I$, *where the constant* $C_s^3 > 0$ *is bounded for s bounded.*

Also, for each $v \in X^\infty$ *that map defined on* U *by* $A_v : u \mapsto A(u)v$ *is differentiable with*

$$\begin{aligned} &||DA_v(u)h||_{Y^s} \\ &\leq C_s^4(||h||_{X^{s+m_1}} ||v||_{X^{m_2}} + ||h||_{X^{m_2}} ||v||_{X^{s+m_1}} + ||h||_{X^{m_2}} ||v||_{X^{m_2}} (1 + ||u||_{X^{s+m_3}})) \end{aligned} \tag{22}$$

for all $h \in X^\infty$ *and* $s \in I$ *such that* $s + m_1, s + m_3 \in I$, *where the constant* $C_s^4 > 0$ *is bounded for s bounded.*

3. *For each* $u \in V$, *where* $u_0 \in V \subset X^{\infty - m_7}$, *there exists a linear map* $B(u) : Y^\infty \to X^{\infty - \max\{l_1, m_4 + m_7\}}$ *such that*

$$A(u)B(u) = \text{id}$$

and

$$||B(u)g||_{X^s} \leq C_s^5(||g||_{Y^{s+l_1}} + ||g||_{Y^{l_1}} ||u||_{X^{s+m_4}}) \tag{23}$$

for all $u \in V$, $g \in Y^\infty$ *and* $s \in I$ *such that* $s + l_1, s + m_4 + m_7 \in I$, *for some fixed numbers* $l_1, m_4, m_7 \in I$, *where the constant* $C_s^5 > 0$ *is bounded for s bounded.*

4. *There exists an operator* $R : U \to V$ *such that*

$$||R(u) - u||_{X^0} \leq C \, ||T(u) - T(u_0) - f||_{Y^{l_2}} \tag{24}$$

for some fixed number $l_2 \in I$, *where we assume* $l_2 \leq l_1$ *(else increase* l_1*), and some constant* $C > 0$. *In addition*

$$||R(u)||_{X^s} \leq C_s^6(1 + ||u||_{X^{m_8}})(1 + ||u||_{X^{s+m_7}}) \tag{25}$$

for all $u \in U$ *and* $s \in I$ *such that* $s + m_7 \in I$, *for some fixed number* $m_8 \in I$, *where* $\{S_\theta^X\}_{\theta \geq 1}$ *are the smoothing operators on* $(X^s)_{s \in I}$, *and the constant* $C_s^6 > 0$ *is bounded for s bounded.*

We also assume the commutator estimate

$$\begin{aligned}&||R(S_\theta^X u) - S_\theta^X R(u)||_{X^s}\\&\le C_{r',r,s}(\theta^{s-r}(1+||u||_{X^{m_8}})(1+||u||_{X^{r+m_7}}) + \theta^{-r'}(1+||u||_{X^{s+m_7}})(1+||u||_{X^{r'+m_8}}))\end{aligned} \tag{26}$$

for all $u \in U$ and $r', r, s \in I$ such that $r + m_7, s + m_7, r' + m_8 \in I$, where $\{S_\theta^X\}_{\theta \ge 1}$ are the smoothing operators on $(X^s)_{s\in I}$, and the constant $C_{r',r,s} > 0$ is bounded for r', r, s bounded.

Let $r_0 \in I$ with $r_0 > \max\{m_0+\max\{m_7, m_8\}, m_4, m_9, l_1+m_1+m_2+\max\{m_7, m_8\}, 2m_2+2\max\{m_7, m_8\}, \frac{l_1+m_3}{2}+m_2+\max\{m_7, m_8\}, l_1+\max\{m_5, m_6\}+(l_3-l_1)_+, l_1+m_6+\max\{m_1, m_3\}+l_4\}$ and let $s_1 \in I$ with $r_0+1 < s_1$, $r_0+\max\{m_1, m_3\}+l_1 \le s_1$ and $s_1 + \max\{l_1, m_4+m_7\} \in I$ be sufficiently large depending on the constants m_i, l_i.

Then there exists a constant $0 < \epsilon \le 1$ such that if

$$||f||_{Y^{r_0+l_1}} \le \epsilon$$

we can find $u \in U^{m_0}$ which solves the equation

$$T(u) = T(u_0) + f.$$

Moreover, suppose that $f \in Y^{s_2-\max\{m_1,m_3\}}$ where $s_2 \in I$ with $s_2 \ge s_1$ and $s_2 + \max\{l_1, m_4+m_7\} \in I$, and suppose $||f||_{Y^{s_2-\max\{m_1,m_3\}}} \le C_{s_2}$. Assume also that the estimate (21) holds for all $s \in [0, s_2 - l_4 - \max\{m_1, m_3\}]$. Then for each $r \in [r_0, s_2 - \max\{m_1, m_3\} - l_1]$ and $s \in I$ with $s < r$, assuming that $s_1 + r - r_0 + \max\{l_1, m_4+m_7\} \in I$, we have $u \in X^s$, and there exists a constant $K_{r,s}$, possibly increasing with C_{s_2}, but otherwise independent of f, such that

$$||u - u_0||_{X^s} \le K_{r,s}||f||_{Y^{r+l_1}}$$

Proof.

Step 1 – Setup of the iteration scheme

Assume that $||f||_{Y^{r_0+l_1}} \le \epsilon$, where $0 < \epsilon \le 1$ will be chosen later.

Denote the smoothing operators on $(X^s)_{s\in I}$ by $\{S_\theta^X\}_{\theta\ge 1}$ and the smoothing operators on $(Y^s)_{s\in I}$ by $\{S_\theta^Y\}_{\theta\ge 1}$.

We use an iteration scheme to construct a sequence $(u_n)_{n\ge 0}$ in X^∞ which we aim to show converges to a solution $u \in U^{m_0}$ of $T(u) = T(u_0) + f$.

For $n \ge 0$, define

$$\theta_n = \theta_0 + n$$

where $\theta_0 > 1$ will be chosen later depending only on r_0, the constants m_i,l_i and the constants in the smoothing hypothesis and in the inequalities satisfied by DT, D^2T, A, B and R.

Note that

$$\theta_{n+1} \le \theta_n + 1 \le 2\theta_n.$$

For $n \ge 0$, define

$$\begin{aligned} v_n &= S^X_{\theta_n} u_n \\ w_n &= R(v_n) \\ \dot{u}_n &= B(w_n) g_n \\ u_{n+1} &= u_n + \dot{u}_n \end{aligned}$$

where g_n is defined below.

Note that the overdot $\dot{}$ is simply notation indicating a sort of difference and does not denote differentiation.

For $n \ge 0$, define

$$\begin{aligned} g_0 &= S^Y_{\theta_0} f \\ g_{n+1} &= (S^Y_{\theta_{n+1}} - S^Y_{\theta_n})(f - E_n) - S^Y_{\theta_{n+1}} e_n \end{aligned}$$

where

$$E_n = \sum_{m=0}^{n-1} e_m$$

(so $E_0 = 0$), and the error e_n is defined below, for $n \ge 0$,

$$\begin{aligned} e'_n &= (A(u_n) - A(w_n))\dot{u}_n, \\ e''_n &= T(u_n + \dot{u}_n) - T(u_n) - A(u_n)\dot{u}_n, \\ e_n &= e'_n + e''_n. \end{aligned}$$

Note that since g_0 is defined in terms of f only, and we are given u_0, from which v_0 is obtained immediately, the iteration scheme can be determined for $n \ge 0$ in the order $\dot{u}_n$, u_{n+1}, v_{n+1}, w_{n+1}, e'_n, e''_n, e_n, E_n, g_{n+1}.

Note that e_n is defined so that it measures how well $T(u_{n+1}) - T(u_n)$ is approximated by $A(w_n)\dot{u}_n$, by which we mean

$$\begin{aligned} T(u_{n+1}) - T(u_n) &= A(w_n)\dot{u}_n + e_n \\ &= g_n + e_n. \end{aligned}$$

Also note that the formula for g_{n+1} can be rearranged to give

$$g_{n+1} = (S^Y_{\theta_{n+1}} f - S^Y_{\theta_n} f) - (S^Y_{\theta_{n+1}} E_{n+1} - S^Y_{\theta_n} E_n).$$

We thus obtain

$$\begin{aligned}
T(u_{n+1}) - T(u_0) &= \sum_{m=0}^{n} (T(u_{m+1}) - T(u_m)) \\
&= \sum_{m=0}^{n} g_m + \sum_{m=0}^{n} e_m \\
&= S^Y_{\theta_n} f - S^Y_{\theta_n} E_n + E_{n+1} \\
&= S^Y_{\theta_n} f + (E_n - S^Y_{\theta_n} E_n) + e_n
\end{aligned}$$

which we hope converges to f as $n \to \infty$, since, roughly speaking, $S^Y_{\theta_n} \to \text{id}$ and $e_n \to 0$.

Step 2 – Obtaining estimates for the iterates via induction

We will show the following inductive hypothesis, $[H_n]$, holds.

$$||\dot{u}_n||_{X^s} \leq K_1 \, ||f||_{Y^{r_0+l_1}} \, \theta_n^{s-r_0-1} \qquad \text{for all } s \in [0, s_1]$$

$$||T(u_n) - T(u_0) - f||_{Y^{s+l_1}} \leq K_2 \, ||f||_{Y^{r_0+l_1}} \, \theta_n^{s-r_0} \qquad \text{for } s \in [0, r_0]$$

where the constants $K_1, K_2 > 0$ will be chosen later, with K_1, K_2 independent of n, f and ϵ, but depending on θ_0, and with K_2 depending on K_1. We will choose ϵ sufficiently small such that $K_1 \, ||f||_{Y^{r_0+l_1}} \leq K_1 \epsilon \leq 1$ and $K_2 \, ||f||_{Y^{r_0+l_1}} \leq K_2 \epsilon \leq 1$.

In what follows, $C_s > 0$ represents a constant, which is independent of n, f and ϵ, and is bounded for s bounded. It will also be independent of θ_0, which will allow us to choose θ_0 so that θ_n is large compared to C_s for s in a certain range. We will write $C > 0$ for a constant which is also independent of s.

Assume now that $[H_m]$ is true for all $0 \leq m \leq n$ and let us show that $[H_{n+1}]$ follows. (We will leave the proof of $[H_0]$ until later.)

Pick a real number $0 < \eta < 1$ such that $r_0 > \max\{m_0 + \max\{m_7, m_8\}, m_4, m_9, l_1 + m_1 + m_2 + \max\{m_7, m_8\}, 2m_2 + 2\max\{m_7, m_8\}, \frac{l_1+m_3}{2} + m_2 + \max\{m_7, m_8\}, l_1 + \max\{m_5, m_6\} + (l_3 - l_1)_+, l_1 + m_6 + \max\{m_1, m_3\} + l_4\} + 2\eta$ and $\eta < \max\{m_1, m_3\}$.

For $s \in I$, define

$$P(s) = \begin{cases} (s - r_0)_+ & \text{for } |s - r_0| \geq \eta, \\ \eta & \text{for } |s - r_0| < \eta. \end{cases}$$

We claim that the following estimates for $0 \leq m \leq n+1$ follow directly from $[H_m]$ for $0 \leq m \leq n$.

$$||u_m - u_0||_{X^s} \le C_s K_1 \,||f||_{Y^{r_0+l_1}}\, \theta_m^{P(s)} \qquad \text{for } s \in [0, s_1], \tag{27}$$

$$\left|\left|S^X_{\theta_m}(u_m - u_0)\right|\right|_{X^s} \le C_s K_1 \,||f||_{Y^{r_0+l_1}}\, \theta_m^{P(s)} \qquad \text{for } s \in I, \tag{28}$$

$$\left|\left|(u_m - u_0) - S^X_{\theta_m}(u_m - u_0)\right|\right|_{X^s} \le C_s K_1 \,||f||_{Y^{r_0+l_1}}\, \theta_m^{(s-r_0)} \qquad \text{for } s \in [0, s_1], \tag{29}$$

$$||u_m - v_m||_{X^s} \le C_s \theta_m^{s-r_0} \qquad \text{for } s \in [0, s_1], \tag{30}$$

$$||v_m||_{X^s} \le C_s \theta_m^{P(s)} \qquad \text{for } s \in I, \tag{31}$$

$$||u_m||_{X^s} \le C_s \theta_m^{P(s)} \qquad \text{for } s \in [0, s_1]. \tag{32}$$

Indeed, the proofs of (27)–(32) are exactly the same as the proofs of (6)–(11).

We also claim that, for $0 \le m \le n$, we have

$$||v_m - w_m||_{X^s} \le C_s \theta_m^{s+\max\{m_7,m_8\}-r_0} \qquad \text{for } s \in I \text{ such that } s + m_7 \in I, \tag{33}$$

$$||u_m - w_m||_{X^s} \le C_s \theta_m^{s+\max\{m_7,m_8\}-r_0} \qquad \text{for } s \in [0, s_1], \tag{34}$$

$$||w_m||_{X^s} \le C_s \theta_m^{\max\{P(s), s+\max\{m_7,m_8\}-r_0\}} \qquad \text{for } s \in I \text{ such that } s + m_7 \in I. \tag{35}$$

Indeed, first we assume $s \le r_0 + \eta$. We have

$$\begin{aligned} &||v_m - w_m||_{X^s} \\ &= \left|\left|S^X_{\theta_m} u_m - R(S^X_{\theta_m} u_m)\right|\right|_{X^s} \\ &\le \left|\left|S^X_{\theta_m}(u_m - R(u_m))\right|\right|_{X^s} + \left|\left|S^X_{\theta_m} R(u_m) - R(S^X_{\theta_m} u_m)\right|\right|_{X^s}. \end{aligned}$$

Now, using the smoothing hypothesis, estimate (24) and $[H_n]$, we obtain

$$\begin{aligned} \left|\left|S^X_{\theta_m}(u_m - R(u_m))\right|\right|_{X^s} &\le C_s \theta_m^s \,||u_m - R(u_m)||_{X^0} \\ &\le C_s \theta_m^s \,||T(u_m) - T(u_0) - f||_{Y^{l_2}} \\ &\le C_s \theta_m^{s-r_0} K_2 \,||f||_{Y^{r_0+l_1}} \\ &\le C_s \theta_m^{s-r_0} \end{aligned}$$

(where we have used $K_2 \,||f||_{Y^{r_0+l_1}} \le K_2 \epsilon \le 1$).

For the second term, using the estimates (26) and (32), we obtain, choosing $r, r' \geq r_0 + \eta$,

$$\begin{aligned}
&\left|\left|S^X_{\theta_m} R(u_m) - R(S^X_{\theta_m} u_m)\right|\right|_{X^s} \\
&\leq C_s(\theta_m^{s-r}(1 + ||u_m||_{X^{m_8}})(1 + ||u_m||_{X^{r+m_7}}) \\
&\qquad + \theta_m^{-r'}(1 + ||u_m||_{X^{s+m_7}})(1 + ||u_m||_{X^{r'+m_8}})) \\
&\leq C_s(\theta_m^{s-r}(1 + \theta_m^{r+m_7-r_0}) + \theta_m^{-r'}\theta_m^{P(s+m_7)}(1 + \theta_m^{r'+m_8-r_0})) \\
&\leq C_s \theta_m^{s+\max\{m_7,m_8\}-r_0}.
\end{aligned}$$

If $s \geq r_0 + \eta$, then we can directly estimate, using (32) and (25),

$$\begin{aligned}
&||v_m - w_m||_{X^s} \\
&= \left|\left|S^X_{\theta_m} u_m - R(S^X_{\theta_m} u_m)\right|\right|_{X^s} \\
&\leq \left|\left|S^X_{\theta_m} u_m\right|\right|_{X^s} + \left|\left|R(S^X_{\theta_m} u_m)\right|\right|_{X^s} \\
&\leq C_s\, ||u_m||_{X^s} + C_s(1 + ||u_m||_{X^{m_8}})(1 + ||u_m||_{X^{s+m_7}}) \\
&\leq C_s \theta_m^{s-r_0} + C_s \theta_m^{s+m_7-r_0} \\
&\leq C_s \theta_m^{s+m_7-r_0}.
\end{aligned}$$

This proves (33). Now, using (33) and (30), for $s \in [0, s_1]$, we have

$$\begin{aligned}
||u_m - w_m||_{X^s} &\leq ||w_m - v_m||_{X^s} + ||v_m - u_m||_{X^s} \\
&\leq C_s \theta_m^{s+\max\{m_7,m_8\}-r_0}.
\end{aligned}$$

This proves (34).

Using (33) and (31), for $s \in I$, we have

$$\begin{aligned}
||w_m||_{X^s} &\leq ||w_m - v_m||_{X^s} + ||v_m||_{X^s} \\
&\leq C_s \theta_m^{\max\{P(s), s+\max\{m_7,m_8\}-r_0\}}.
\end{aligned}$$

This proves (35).

This completes the proof of the claim.

Note that, using (27) and (30), we have

$$\begin{aligned}
||v_m - u_0||_{X^{m_0}} &\leq ||v_m - u_m||_{X^{m_0}} + ||u_m - u_0||_{X^{m_0}} \\
&\leq C\theta_m^{m_0-r_0} + CK_1\epsilon\theta_m^{P(m_0)} \\
&\leq C\theta_m^{m_0-r_0} + CK_1\epsilon.
\end{aligned}$$

Thus by taking ϵ sufficiently small depending on K_1 and C, and θ_0 sufficiently large depending on C, we have $v_n, v_{n+1} \in U$. Similarly we can ensure $w_n \in U$ using (34). Also note that (6) in the case $s = m_0$ implies $u_n \in U$ for ϵ sufficiently small, and $[H_n]$ implies that $u_n + \dot{u}_n \in U$ for ϵ sufficiently small. Note that the same argument also shows that the line segments $[u_n, u_n + \dot{u}_n]$ and $[u_n, w_n]$ are in U for ϵ sufficiently small.

We claim that the following estimate holds.

$$||T(u_n) - T(u_0) - f||_{Y^s} \leq C_s \theta_n^{s+\max\{m_1,m_3\}-r_0} \quad \text{for } s \in [0, s_1 - \max\{m_1, m_3\}]. \tag{36}$$

Indeed, for $s \in [r_0, s_1 - \max\{m_1, m_3\}]$, using Taylor's theorem, (20) and (27), we have

$$\begin{aligned}
&||T(u_n) - T(u_0) - f||_{Y^s} \\
&\leq ||T(u_n) - T(u_0)||_{Y^s} + ||f||_{Y^s} \\
&\leq \left|\left| \sup_{t\in[0,1]} DT(u_0 + t(u_n - u_0))(u_n - u_0) \right|\right|_{Y^s} + C^0 \\
&\leq C_s(||u_n - u_0||_{X^{s+m_1}} + ||u_n - u_0||_{X^{m_2}} (1 + ||u_n - u_0||_{X^{s+m_3}})) + C^0 \\
&\leq C_s \theta_n^{s+\max\{m_1,m_3\}-r_0}
\end{aligned}$$

(assuming that $\max\{m_1, m_3\} \geq \eta$). We combine this with $[H_n]$ for $s \in [0, r_0]$ to get

$$||T(u_n) - T(u_0) - f||_{Y^s} \leq C_s \theta_n^{s+\max\{m_1,m_3\}-r_0}$$

for all $s \in [0, s_1 - \max\{m_1, m_3\}]$.

Estimate of e_n'. We claim that for all $s \in [0, s_1 - \max\{m_1, m_3\}]$,

$$||e_n'||_{Y^s} \leq C_s K_1 \, ||f||_{Y^{r_0+l_1}} \, \theta_n^{M'(s)-1+\eta}$$

where

$$\begin{aligned}
M'(s) = \max\{&s + m_1 + m_2 + \max\{m_7, m_8\} - 2r_0, \\
&(s + m_3 - r_0)_+ + 2\max\{m_7, m_8\} + 2m_2 - 2r_0\}.
\end{aligned}$$

Applying the estimate (22) together with Taylor's theorem, $[H_n]$ and the estimates (34) and (35), we have, for $s \in [0, s_1 - \max\{m_1, m_3\}]$,

$$\begin{aligned}
&||e_n'||_{Y^s} \\
&= ||(A((u_n - w_n) + w_n) - A(w_n))\dot{u}_n||_{Y^s} \\
&\leq C_s(||\dot{u}_n||_{X^{s+m_1}} \, ||u_n - w_n||_{X^{m_2}} + ||\dot{u}_n||_{X^{m_2}} \, ||u_n - w_n||_{X^{s+m_1}}
\end{aligned}$$

$$+ ||\dot{u}_n||_{X^{m_2}} ||u_n - w_n||_{X^{m_2}} (1 + ||w_n||_{X^{s+m_3}} + ||u_n - w_n||_{X^{s+m_3}}))$$
$$\leq C_s K_1 ||f||_{Y^{r_0+l_1}} (\theta_n^{s+m_1-r_0-1} \theta_n^{m_2-r_0+\max\{m_7,m_8\}} + \theta_n^{m_2-r_0-1} \theta_n^{s+\max\{m_7,m_8\}+m_1-r_0}$$
$$+ \theta_n^{m_2-r_0-1} \theta_n^{m_2+\max\{m_7,m_8\}-r_0} (1 + \theta_n^{P(s+m_3)+\max\{m_7,m_8\}} + \theta_n^{s+\max\{m_7,m_8\}+m_3-r_0}))$$
$$\leq C_s K_1 ||f||_{Y^{r_0+l_1}} \theta_n^{M'(s)-1+\eta}.$$

Estimate of e_n''. We claim that for all $s \in [0, s_1 - \max\{m_1 + l_4, m_3 + l_4, m_5, m_9\}]$,

$$||e_n''||_{Y^s} \leq C_s K_1 ||f||_{Y^{r_0+l_1}} \theta_n^{M(s)-1+\eta}$$

where

$$M(s) = \max\{s + m_1 + m_2 + \max\{m_7, m_8\} - 2r_0,$$
$$(s + m_3 - r_0)_+ + 2\max\{m_7, m_8\} + 2m_2 - 2r_0,$$
$$s + \max\{m_5, m_6\} + (l_3 - l_1)_+ - 2r_0, s + m_6 + \max\{m_1, m_3\} + l_4 - 2r_0\}.$$

Indeed, we have

$$e_n'' = T(u_n + \dot{u}_n) - T(u_n) - A(u_n)\dot{u}_n$$
$$= T(u_n + \dot{u}_n) - T(u_n) - DT(u_n)\dot{u}_n + (A(u_n) - DT(u_n))\dot{u}_n.$$

Applying Taylor's theorem, (19), $[H_n]$ and the estimate (32), we have, for $s \in [0, s_1 - \max\{m_1, m_3\}]$,

$$||T(u_n + \dot{u}_n) - T(u_n) - DT(u_n)\dot{u}_n||_{Y^s}$$
$$\leq \sup_{t\in[0,1]} ||D^2T(u_n + t\dot{u}_n)(\dot{u}_n, \dot{u}_n)||_{Y^s}$$
$$\leq C_s(||\dot{u}_n||_{X^{s+m_1}} ||\dot{u}_n||_{X^{m_2}} + ||\dot{u}_n||^2_{X^{m_2}} (1 + \sup_{t\in[0,1]} ||u_n + t\dot{u}_n||_{X^{s+m_3}}))$$
$$\leq C_s(K_1 ||f||_{Y^{r_0+l_1}} \theta_n^{s+m_1-r_0-1} K_1 ||f||_{Y^{r_0+l_1}} \theta_n^{m_2-r_0-1}$$
$$+ K_1^2 ||f||^2_{Y^{r_0+l_1}} \theta_n^{2m_2-2r_0-2}(1 + \theta_n^{P(s+m_3)} + K_1 ||f||_{Y^{r_0+l_1}} \theta_n^{s+m_3-r_0-1}))$$
$$\leq \theta_n^{-1} C_s K_1 ||f||_{Y^{r_0+l_1}} \theta_n^{M'(s)-1+\eta}$$
$$\leq C_s K_1 ||f||_{Y^{r_0+l_1}} \theta_n^{M'(s)-1+\eta}$$

where we have used $K_1 ||f||_{Y^{r_0+l_1}} \leq K_1\epsilon \leq 1$.

For $s \in [0, s_1 - \max\{m_5, m_9, m_1 + l_4, m_3 + l_4\}]$, we have, using (21), $[H_n]$, and (36),

$$\begin{aligned}
&||(A(u_n) - DT(u_n))\dot{u}_n||_{Y^s} \\
&\leq C_s(||\dot{u}_n||_{X^{s+m_5}} ||T(u_n) - T(u_0) - f||_{Y^{l_3}} + ||\dot{u}_n||_{X^{m_6}} ||T(u_n) - T(u_0) - f||_{Y^{s+l_4}} \\
&+ ||\dot{u}_n||_{X^{m_6}} ||T(u_n) - T(u_0) - f||_{Y^{l_3}} (1 + ||u||_{X^{s+m_9}})) \\
&\leq C_s K_1 ||f||_{Y^{r_0+l_1}} (\theta_n^{s+m_5-r_0-1} \theta_n^{(l_3-l_1)_+-r_0} + \theta_n^{m_6-r_0-1} \theta_n^{s+\max\{m_1,m_3\}+l_4-r_0} \\
&+ \theta_n^{m_6-r_0-1} \theta_n^{(l_3-l_1)_+-r_0} \theta_n^{(s+m_9-r_0)_++\eta}) \\
&\leq C_s K_1 ||f||_{Y^{r_0+l_1}} \theta_n^{M''(s)-1+\eta}
\end{aligned}$$

where

$$\begin{aligned}
M''(s) = \max\{&s + m_5 + (l_3 - l_1)_+ - 2r_0, s + m_6 + \max\{m_1, m_3\} + l_4 - 2r_0, \\
&s + m_6 + (l_3 - l_1)_+ - 2r_0\}
\end{aligned}$$

where we have used $r_0 \geq m_9$.

Adding the two above estimates yields the estimate for e''_n.

Estimate of e_n. Adding the estimates for e'_n and e''_n, we obtain

$$||e_n||_{Y^s} \leq C_s K_1 ||f||_{Y^{r_0+l_1}} \theta_n^{M(s)-1+\eta}$$

for all $s \in [0, s_1 - \max\{m_1 + l_4, m_3 + l_4, m_5, m_9\}]$.

Estimate of g_{n+1}. We claim that for all $s \in I$,

$$||g_{n+1}||_{Y^s} \leq C_s(K_1 ||f||_{Y^{r_0+l_1}} \theta_n^{M(s)-1+\eta} + ||f||_{Y^{r_0+l_1}} \theta_n^{s-r_0-l_1-1}).$$

Indeed, we have

$$g_{n+1} = (S^Y_{\theta_{n+1}} - S^Y_{\theta_n})(f - E_n) - S^Y_{\theta_{n+1}} e_n.$$

Note that for any $z \in Y^{s'}$,

$$\left|\left|(S^Y_{\theta_{n+1}} - S^Y_{\theta_n})z\right|\right|_{Y^s} \leq C_{s',s} \theta_n^{s-s'-1} ||z||_{Y^{s'}}$$

by the smoothing hypothesis (3) and Taylor's theorem.

Setting $s' = r_0 + l_1$, we have

$$\left|\left|(S^Y_{\theta_{n+1}} - S^Y_{\theta_n})f\right|\right|_{Y^s} \leq C_s \theta_n^{s-r_0-l_1-1} ||f||_{Y^{r_0+l_1}} .$$

We also have

$$\left\|(S^Y_{\theta_{n+1}} - S^Y_{\theta_n})E_n\right\|_{Y^s} \le C_{s',s}\theta_n^{s-s'-1}\,||E_n||_{Y^{s'}}\,.$$

Now, for $s' \in [0, s_1 - \max\{m_1 + l_4, m_3 + l_4, m_5, m_9\}]$, we have, from the estimate for e_n,

$$\begin{aligned}||E_n||_{Y^{s'}} &= \left\|\sum_{m=0}^{n-1} e_m\right\|_{Y^{s'}} \\ &\le C_{s'}K_1\,||f||_{Y^{r_0+l_1}}\sum_{m=0}^{n-1}\theta_m^{M(s')-1+\eta} \\ &\le C_{s'}K_1\,||f||_{Y^{r_0+l_1}}\,\theta_n^{M(s')+\eta} \end{aligned} \tag{37}$$

if $M(s') \ge 0$, by the integral comparison used before. Note that $M(s')$ has slope 1 for large enough s' depending on r_0 and the constants m_i, l_i, so to achieve $M(s') \ge 0$ it suffices to take s' large in relation to r_0 and the constants m_i, l_i. To do this we require s_1 sufficiently large in relation to r_0 and the constants m_i, l_i.

Hence

$$\begin{aligned}\left\|(S^Y_{\theta_{n+1}} - S^Y_{\theta_n})E_n\right\|_{Y^s} &\le C_{s'}C_{s',s}K_1\,||f||_{Y^{r_0+l_1}}\,\theta_n^{M(s')+s-s'-1+\eta} \\ &\le C_sK_1\,||f||_{Y^{r_0+l_1}}\,\theta_n^{M(s)-1+\eta}\end{aligned}$$

by choosing s' sufficiently large compared to r_0 and the constants m_i so that $M(s)$ has slope 1 for $s \ge s'$. (Hence $M(s') - s' \le M(s) - s$ for all s since $M(s) - s$ is decreasing for $s \le s'$ and constant for $s \ge s'$.) Again, to do this we require s_1 sufficiently large in relation to r_0 and the constants m_i, l_i. This fixes s_1.

Similarly, for s' sufficiently large, we have

$$\begin{aligned}\left\|S^Y_{\theta_{n+1}}e_n\right\|_{Y^s} &\le C_{s',s}\theta_n^{s-s'}\,||e_n||_{Y^{s'}} \\ &\le C_{s',s}C_{s'}K_1\,||f||_{Y^{r_0+l_1}}\,\theta_n^{M(s')+s-s'-1+\eta} \\ &\le C_sK_1\,||f||_{Y^{r_0+l_1}}\,\theta_n^{M(s)-1+\eta}.\end{aligned}$$

Hence the estimate for g_{n+1} holds.

Estimate of $T(u_{n+1}) - T(u_0) - f$ We have

$$T(u_{n+1}) - T(u_0) - f = (S^Y_{\theta_n}f - f) + (E_n - S^Y_{\theta_n}E_n) + e_n.$$

Let $s \in [0, r_0]$.

By (2) from the smoothing hypothesis, we have

$$\left|\left|S^{Y}_{\theta_n} f - f\right|\right|_{Y^{s+l_1}} \leq C_s \theta_n^{s-r_0} \,||f||_{Y^{r_0+l_1}} \,.$$

Also,

$$\begin{aligned}\left|\left|E_n - S^{Y}_{\theta_n} E_n\right|\right|_{Y^{s+l_1}} &\leq C_{s,s'} \theta_n^{s-s'} \,||E_n||_{Y^{s'+l_1}} \qquad \text{for } s' \geq s\\ &\leq C_{s,s'} \theta_n^{s-s'} C_s \theta_n^{M(s'+l_1)+\eta} K_1 \,||f||_{Y^{r_0+l_1}}\\ &\quad \text{using (37), for } s' \text{ large enough such that } M(s'+l_1) \geq 0\\ &\leq C_s \theta_n^{M(s'+l_1)+s-s'+\eta} K_1 \,||f||_{Y^{r_0+l_1}}\\ &\leq C_s \theta_n^{s-r_0} K_1 \,||f||_{Y^{r_0+l_1}}\end{aligned}$$

since $M(s'+l_1)+\eta \leq M(l_1)+\eta+s' < s'-r_0$.

Finally,

$$\begin{aligned}||e_n||_{Y^{s+l_1}} &\leq C_s \theta_n^{M(s+l_1)+\eta-1} K_1 \,||f||_{Y^{r_0+l_1}}\\ &\leq C_s \theta_n^{s-r_0-1} K_1 \,||f||_{Y^{r_0+l_1}}\end{aligned}$$

since $M(s+l_1)+\eta \leq M(l_1)+\eta+s < s-r_0$.

Hence we have

$$||T(u_{n+1}) - T(u_0) - f||_{Y^{s+l_1}} \leq C_s \theta_n^{s-r_0} K_1 \,||f||_{Y^{r_0+l_1}}$$

for $s \in [0, r_0]$. Thus, by choosing K_2 sufficiently large depending on K_1 and C_s for $s \in [0, r_0]$, we have

$$||T(u_{n+1}) - T(u_0) - f||_{Y^{s+l_1}} \leq K_2 \,||f||_{Y^{r_0+l_1}} \,\theta_{n+1}^{s-r_0} \tag{38}$$

for $s \in [0, r_0]$.

The estimates

$$||v_{n+1} - w_{n+1}||_{X^s} \leq C_s \theta_{n+1}^{s-r_0+\max\{m_7,m_8\}} \qquad \text{for } s \in I \text{ such that } s+m_7 \in I \tag{39}$$

$$||u_{n+1} - w_{n+1}||_{X^s} \leq C_s \theta_{n+1}^{s-r_0+\max\{m_7,m_8\}} \qquad \text{for } s \in [0, s_1] \tag{40}$$

$$||w_{n+1}||_{X^s} \leq C_s \theta_{n+1}^{\max\{P(s),s+\max\{m_7,m_8\}-r_0\}} \qquad \text{for } s \in I \text{ such that } s+m_7 \in I \tag{41}$$

now hold, and are proved exactly as for the estimates (33)–(35) using the estimate (38) to go from n to $n+1$.

Estimate of $\dot{u}_{n+1}$. We have

$$\dot{u}_{n+1} = B(w_{n+1})g_{n+1}.$$

Hence, for all $s \in I$ such that $s + l_1, s + m_4 + m_7 \in I$, using (23), the estimate (35) and the estimate for g_{n+1}, we have

$$\begin{aligned}
&||\dot{u}_{n+1}||_{X^s} \\
&\leq C_s(||g_{n+1}||_{Y^{s+l_1}} + ||g_{n+1}||_{Y^{l_1}} (1 + ||w_{n+1}||_{X^{s+m_4}})) \\
&\leq C_s(K_1 \,||f||_{Y^{r_0+l_1}} \,\theta_{n+1}^{M(s+l_1)-1+\eta} + ||f||_{Y^{r_0+l_1}} \,\theta_{n+1}^{s-r_0-1} + \\
&(K_1 \,||f||_{Y^{r_0+l_1}} \,\theta_{n+1}^{M(l_1)-1+\eta} + ||f||_{Y^{r_0+l_1}} \,\theta_{n+1}^{-r_0-1})(1 + \theta_{n+1}^{\max\{P(s+m_4), s+\max\{m_7,m_8\}-r_0\}})) \\
&\leq C_s(K_1 \,||f||_{Y^{r_0+l_1}} \,\theta_{n+1}^{M(l_1)+s-1+\eta} + ||f||_{Y^{r_0+l_1}} \,\theta_{n+1}^{s-r_0-1}) \qquad (42)
\end{aligned}$$

since $P(s + m_4) \leq s$ because $r_0 > m_4 + 2\eta$ and $s + \max\{m_7, m_8\} - r_0 \leq s$ because $r_0 > \max\{m_7, m_8\}$, and $M(l_1 + s) \leq M(l_1) + s$ because M has slope at most 1.

We want to obtain

$$||\dot{u}_{n+1}||_{X^s} \leq K_1 \,||f||_{Y^{r_0+l_1}} \,\theta_{n+1}^{s-r_0-1}$$

for $s \in [0, s_1]$.

To make the first term sufficiently small, we require

$$-\gamma := M(l_1) + r_0 + \eta < 0.$$

Then we can choose θ_0 large enough so that

$$C_s\theta_{n+1}^{M(l_1)+s-1+\eta} = C_s\theta_{n+1}^{s-r_0-1}\theta_{n+1}^{-\gamma} \leq C_s\theta_{n+1}^{s-r_0-1}\,\theta_0^{-\gamma} \leq \frac{1}{2}\theta_{n+1}^{s-r_0-1}$$

for all $s \in [0, s_1]$.

We note that $M(l_1) + r_0 + \eta < 0$ if and only if $r_0 - \eta > l_1 + m_1 + m_2 + \max\{m_7, m_8\}$, $r_0 - \eta > 2m_2 + 2\max\{m_7, m_8\}$, $r_0 - \eta > m_2 + \max\{m_7, m_8\} + \frac{l_1+m_3}{2}$, $r_0 - \eta > l_1 + \max\{m_5, m_6\} + (l_3 - l_1)_+$ and $r_0 - \eta > l_1 + m_6 + \max\{m_1, m_3\} + l_4$, which indeed hold by the choice of r_0 and η.

To make the second term sufficiently small, we take $K_1 \geq 2C_s$ for all $s \in [0, s_1]$. This gives $[H_{n+1}]$.

Proof of $[H_0]$ We have

$$g_0 = S_{\theta_0}^Y f$$

and

$$v_0 = S^X_{\theta_0} u_0$$

and

$$w_0 = R(v_0).$$

Hence

$$\begin{aligned}
||\dot{u}_0||_{X^s} &= ||B(R(S^X_{\theta_0} u_0)) S^Y_{\theta_0} f||_{X^s} \\
&\leq C_s(||S^Y_{\theta_0} f||_{Y^{s+l_1}} + ||S^Y_{\theta_0} f||_{Y^{l_1}} (1 + ||R(S^X_{\theta_0} u_0)||_{X^{s+m_4}})) \\
&\leq C_s ||S^Y_{\theta_0} f||_{Y^{s+l_1}} \\
&\leq C_s ||f||_{Y^{r_0+l_1}} \theta_0^{(s-r_0)+} \qquad \text{by (1) and (2) from the smoothing hypothesis} \\
&\leq K_1 ||f||_{Y^{r_0+l_1}} \theta_0^{s-r_0-1}
\end{aligned}$$

for all $s \in [0, s_1]$, assuming that K_1 is sufficiently large compared to θ_0 and C_s for $s \in [0, s_1]$.

Now for $s \in [0, r_0]$,

$$\begin{aligned}
||T(u_0) - T(u_0) - f||_{Y^{s+l_1}} &= ||f||_{Y^{s+l_1}} \\
&\leq K_2 ||f||_{Y^{r_0+l_1}} \theta_0^{s-r_0}
\end{aligned}$$

for all $s \in [0, r_0]$, assuming that K_2 is sufficiently large compared to θ_0.

This proves $[H_0]$.

Step 3 – Better estimates if $f \in Y^{s_2-\max\{m_1,m_3\}}$ for $s_2 \geq s_1$

Assume $f \in Y^{s_2-\max\{m_1,m_3\}}$ where $s_2 \in I$ with $s_2 \geq s_1$ and $s_2 + \max\{l_1, m_4 + m_7\} \in I$, and suppose $||f||_{Y^{s_2-\max\{m_1,m_3\}}} \leq C_{s_2}$. Let $r \in I$ with $r \geq r_0$ be such that $s_1 + r - r_0 + \max\{l_1, m_4 + m_7\} \in I$. We will show that, for all $n \geq 0$ and for all $s \in [0, s_2]$, we have

$$||\dot{u}_n||_{X^s} \leq C_{r,s} ||f||_{Y^{r+l_1}} \theta_n^{s-r-1} \tag{43}$$

where the constant $C_{r,s} > 0$ is independent of n and f, except that it may increase with $||f||_{Y^{s_2-\max\{m_1,m_3\}}}$.

Firstly, note that we have proved $[H_n]$ for $n \geq 0$, and hence all the estimates from step 2 which were conditional on the inductive hypothesis are now valid, and we may use them as we wish.

We are going to prove the above statement by an induction argument, but not an induction on n. We are going to use the estimates from step 2 for each n separately to obtain the above inequality, and the constant will be independent of n because the constants from step 2 are independent of n.

We claim by induction on $k \geq 0$ that for all $s \in [0, s_2]$, we have

$$||\dot{u}_n||_{X^s} \leq C_{k,r,s}\, ||f||_{Y^{r+l_1}}\, \theta_n^{s-r_0-\gamma_k-1} \qquad [G_k]$$

where the constant $C_{k,r,s} > 0$ is independent of n and f, and

$$\gamma_k = \min\{k\gamma, r - r_0\}.$$

Indeed, the estimate (42) for $\dot{u}_{n+1}$ in step 2 implies that

$$||\dot{u}_n||_{X^s} \leq C_s\, ||f||_{Y^{r_0+l_1}}\, \theta_n^{s-r_0-1} \tag{44}$$

for all $s \in I$ such that $s + \max\{l_1, m_4 + m_7\} \in I$ (not just $s \in [0, s_1]$ which would follow directly from $[H_n]$).

Using this, we can obtain the following new versions of the estimates (27), (30)–(32) for all $s \in [0, s_2]$ (not just $s \in [0, s_1]$) via exactly the same calculations

$$||u_m - u_0||_{X^s} \leq C_s \theta_m^{s-r_0}, \tag{45}$$

$$||u_m - v_m||_{X^s} \leq C_s \theta_m^{s-r_0}, \tag{46}$$

$$||v_m||_{X^s} \leq C_s \theta_m^{P(s)}, \tag{47}$$

$$||u_m||_{X^s} \leq C_s \theta_m^{P(s)}. \tag{48}$$

We then obtain, for all $s \in [0, s_2]$, the estimates

$$||w_m - v_m||_{X^s} \leq C_s \theta_m^{s+\max\{m_7,m_8\}-r_0}, \tag{49}$$

$$||w_m - u_m||_{X^s} \leq C_s \theta_m^{s+\max\{m_7,m_8\}-r_0}, \tag{50}$$

$$||w_m||_{X^s} \leq C_s \theta_m^{\max\{P(s), s+\max\{m_7,m_8\}-r_0\}}. \tag{51}$$

Using the fact that $||f||_{Y^{r_0+l_1}} \leq ||f||_{Y^{r+l_1}}$, (44) immediately implies $[G_0]$.

Now we assume $[G_k]$ holds and aim to show $[G_{k+1}]$ holds.

Now we want to obtain new estimates for e'_n and e''_n.

First we estimate e'_n. Note that in the estimate for e'_n there was at least one factor involving $\dot{u}_n$ in each term. If we estimate this one factor using the new estimate given by $[G_k]$ and the other quantities using (44) and the slightly modified estimates (45)–(48), we obtain

$$\left|\left|e'_n\right|\right|_{Y^s} \leq C_{k,r,s}\, ||f||_{Y^{r+l_1}}\, \theta_n^{M'(s)-1+\eta-\gamma_k}$$

for all $s \in [0, s_2 - \max\{m_1, m_3\}]$. The constant $C_{k,r,s}$ is independent of f since we have only used the new estimate given by $[G_k]$ in one factor, and the other estimates we have used involve $||f||_{Y^{r_0+l_1}}$, which is bounded by $\epsilon \leq 1$.

Now we estimate e_n''. The first part of the estimate can be modified in exactly the same way as above, to obtain

$$||T(u_n + \dot{u}_n) - T(u_n) - DT(u_n)\dot{u}_n||_{Y^s} \leq C_{k,r,s}\, ||f||_{Y^{r+l_1}}\, \theta_n^{M'(s)-1+\eta-\gamma_k}$$

for all $s \in [0, s_2 - \max\{m_1, m_3\}]$.

We proceed similarly for the second part of the estimate of e_n'' to obtain

$$||(A(u_n) - DT(u_n))\dot{u}_n||_{Y^s} \leq C_{k,r,s}\, ||f||_{Y^{r+l_1}}\, \theta_n^{M''(s)-1+\eta-\gamma_k}$$

for all $s \in [0, s_2 - \max\{m_5, m_9, m_1 + l_4, m_3 + l_4\}]$.

Here, the constant depends on $||f||_{Y^{s_2-\max\{m_1,m_3\}}}$, and we need to assume that the estimate (21) holds for all $s \in [0, s_2 - l_4 - \max\{m_1, m_3\}]$.

Thus we obtain the estimate

$$||e_n||_{Y^s} \leq C_{k,r,s}\, ||f||_{Y^{r+l_1}}\, \theta_n^{M(s)-1+\eta-\gamma_k}$$

for all $s \in [0, s_2 - \max\{m_5, m_9, m_1 + l_4, m_3 + l_4\}]$.

This implies that for $s' \in [0, s_2 - \max\{m_5, m_1 + l_4, m_3 + l_4\}]$, we have

$$\begin{aligned} ||E_n||_{Y^{s'}} &= \left|\left|\sum_{m=0}^{n-1} e_m\right|\right|_{Y^{s'}} \\ &\leq C_{k,r,s'}\, ||f||_{Y^{r+l_1}} \sum_{m=0}^{n-1} \theta_m^{M(s')-1+\eta-\gamma_k} \\ &\leq C_{k,r,s'}\, ||f||_{Y^{r+l_1}}\, \theta_n^{M(s')+\eta-\gamma_k} \end{aligned} \tag{52}$$

as long as $M(s') \geq \gamma_k$. It is possible to pick such an s' if $s_1 + r - r_0 + \max\{l_1, m_4\} \in I$ given the fact that $M(s_1 - \max\{m_5, m_9, m_1 + l_4, m_3 + l_4\}) \geq 0$ and $M(s)$ has slope 1 for $s \geq s_1 - \max\{m_5, m_9, m_1 + l_4, m_3 + l_4\}$.

Hence

$$\begin{aligned} \left|\left|(S^Y_{\theta_{n+1}} - S^Y_{\theta_n})E_n\right|\right|_{Y^s} &\leq C_{s',k} C_{k,r,s} \theta_n^{M(s')+s-s'-1+\eta-\gamma_k} \\ &\leq C_{k,r,s}\, ||f||_{Y^{r+l_1}}\, \theta_n^{M(s)-1+\eta-\gamma_k} \end{aligned}$$

as long as $M(s') \geq \gamma_k$ and s' is sufficiently large compared to r_0 and the constants m_i, l_i so that $M(s)$ has slope 1 for $s \geq s'$.

We also have the estimate

$$\left|\left|S^Y_{\theta_{n+1}} e_n\right|\right|_{Y^s} \leq C_{k,r,s}\, ||f||_{Y^{r+l_1}}\, \theta_n^{M(s)-1+\eta-\gamma_k}.$$

In addition we can use the new estimate

$$\left|\left|(S^Y_{\theta_{n+1}} - S^Y_{\theta_n})f\right|\right|_{Y^s} \leq C_{r,s}\theta_n^{s-r-l_1-1} \, ||f||_{Y^{r+l_1}} \, .$$

This gives us the following new estimate for g_{n+1}, for all $s \in I$,

$$||g_{n+1}||_{Y^s} \leq C_{k,r,s} \, ||f||_{Y^{r+l_1}} \, (\theta_n^{M(s)-1+\eta-\gamma_k} + \theta_n^{s-r-l_1-1}).$$

From this we obtain,for all $s \in [0, s_2]$,

$$\begin{aligned} ||\dot{u}_n||_{X^s} &\leq C_{r,s} \, ||f||_{Y^{r+l_1}} \, (\theta_n^{M(l_1)+s-1+\eta-\gamma_k} + \theta_n^{s-r-1}) \\ &\leq C_{r,s} \, ||f||_{Y^{r+l_1}} \, (\theta_n^{s-r_0-1-\gamma_k-\gamma} + \theta_n^{s-r-1}) \\ &\leq C_{r,s} \, ||f||_{Y^{r+l_1}} \, \theta_n^{s-r_0-\gamma_{k+1}-1} \end{aligned}$$

where we have used the fact that $M(l_1) + r_0 + \eta = -\gamma$.

This is $[G_{k+1}]$.

For large enough k, we have $k\gamma \geq r - r_0$, so $\gamma_k = r - r_0$ and this gives (43).

Step 4 – Convergence to a solution

Assume as above that $f \in Y^{s_2-\max\{m_1,m_3\}}$ where $s_2 \in I$ with $s_2 \geq s_1$ and $s_2 + \max\{l_1, m_4, m_7\} \in I$, and suppose $||f||_{Y^{s_2-\max\{m_1,m_3\}}} \leq C_{s_2}$. Let $r \geq r_0$.

Using (13), we have

$$\begin{aligned} \sum_{m=0}^{n} ||u_{m+1} - u_m||_{X^s} &= \sum_{m=0}^{n} ||\dot{u}_m||_{X^s} \\ &\leq C_{r,s} \, ||f||_{Y^{r+l_1}} \, \theta_{n+1}^{(s-r)+} \end{aligned}$$

for $r \neq s$, with $r, s \in [0, s_2]$.

Thus

$$\sum_{m=0}^{n} ||u_{m+1} - u_m||_{X^s}$$

converges as $n \to \infty$ for $s < r$. Hence, by completeness, $u_n \to u$ in X^s as $n \to \infty$, for all $s < r$, for some $u \in \bigcap_{0 \leq s < r} X^s$.

Note the above calculation also implies that

$$||u_n - u_0||_{X^s} \leq C_{r,s} \, ||f||_{Y^{r+l_1}}$$

for $s < r$, so we have

$$||u - u_0||_{X^s} \leq C_{r,s} \, ||f||_{Y^{r+l_1}} \, .$$

Next we claim that

$$T(u_{n+1}) - T(u_0) \to f$$

in X^s as $n \to \infty$, for all $s < r$.

Indeed,

$$T(u_{n+1}) - T(u_0) = S^Y_{\theta_n} f + (E_n - S^Y_{\theta_n} E_n) + e_n$$

so

$$T(u_{n+1}) - T(u_0) - f = (S^Y_{\theta_n} f - f) + (E_n - S^Y_{\theta_n} E_n) + e_n.$$

By (2) from the smoothing hypothesis, we have

$$\left|\left|S^Y_{\theta_n} f - f\right|\right|_{Y^{s+l_1}} \le C_{r,s}\theta_n^{s-r}\,||f||_{Y^{r+l_1}} \to 0 \text{ as } n \to \infty.$$

Also,

$$\begin{aligned}
\left|\left|E_n - S^Y_{\theta_n} E_n\right|\right|_{Y^{s+l_1}} &\le C_{s,s'}\theta_n^{s-s'}\,||E_n||_{Y^{s'+l_1}} \qquad \text{for } s' \ge s \\
&\le C_{s,s'}\theta_n^{s-s'} C_{r,s}\theta_n^{M(s'+l_1)+\eta-(r-r_0)}\,||f||_{Y^{r+l_1}} \\
&\quad \text{using (18), for } s' \text{ large enough such that } M(s'+l_1) \ge r - r_0 \\
&\le C_{r,s}\theta_n^{M(s'+l_1)+s-s'+\eta-(r-r_0)}\,||f||_{Y^{r+l_1}} \\
&\le C_{r,s}\theta_n^{s-r}\,||f||_{Y^{r+l_1}} \to 0 \text{ as } n \to \infty
\end{aligned}$$

since $M(s'+l_1) + \eta + r_0 \le M(l_1) + \eta + r_0 + s' < s'$.

Finally,

$$||e_n||_{Y^{s+l_1}} \le C_{r,s}\theta_n^{M(s+l_1)+\eta-(r-r_0)-1}\,||f||_{Y^{r+l_1}} \to 0 \text{ as } n \to \infty$$

since $M(s + l_1) + \eta + r_0 \le M(l_1) + \eta + r_0 + s < s$.

This proves the claim.

Now since $T : U \to Y^0$ is continuous as a map from X^{m_0} to Y^0, and $r_0 > m_0$, so $u_n \to u$ in X^{m_0}, we have that $T(u_n) \to T(u)$ in Y^0, hence $T(u) = T(u_0) + f$ as required.

This completes the proof.

5 Applying the Theorem in Sobolev Spaces

This section assumes familiarity with the standard Sobolev spaces $W^{k,p}(\Omega)$ of functions on the domain Ω with weak derivatives up to order k in $L^p(\Omega)$, and Sobolev embedding theorems – see for example the chapter of Evans [11] entitled

'Sobolev Spaces', or see Adams and Fournier [1] for a more complete reference. We do however give the definition of fractional Sobolev spaces below, since these are slightly less standard. See, for example, Adams and Fournier [1] for much more detail.

5.1 The Smoothing Operators in H^s

Definition 5. For $d \in \mathbb{N}$ and $0 \le s \in \mathbb{R}$ we define the *Sobolev space of order* s, $H^s(\mathbb{R}^d)$, by

$$H^s(\mathbb{R}^d) = \{u \in L^2(\mathbb{R}^d) : (1 + |\xi|^2)^{\frac{s}{2}} \hat{u}(\xi) \in L^2(\mathbb{R}^d)\}$$

where $\hat{u}$ denotes the Fourier transform of u, which we also denote by $\mathscr{F}[u]$. We endow H^s with norm $||\cdot||_{H^s}$ given by

$$||u||_{H^s} = \left|\left|(1 + |\xi|^2)^{\frac{s}{2}} \hat{u}(\xi)\right|\right|_{L^2}.$$

Then $H^s(\mathbb{R}^d)$ is a Banach space for each s and $(H^s(\mathbb{R}^d), ||\cdot||_{H^s})_{s \ge 0}$ is a decreasing family of Banach spaces, in the sense of Definition 1.

Notation. For $\phi \in C^\infty(\mathbb{R}^d)$ (with values in $\mathbb{R}$), write $\phi_\epsilon = \epsilon^{-d} \phi(\frac{x}{\epsilon})$.

Notation. We write $\mathscr{S}(\mathbb{R}^d)$ for the Schwartz space of smooth functions which decay faster than the reciprocal of any polynomial, and use the well-known fact that the Fourier transform is an automorphism of $\mathscr{S}(\mathbb{R}^d)$.

Proposition 2. *The decreasing family of Banach spaces* $(H^s(\mathbb{R}^d), ||\cdot||_{H^s})_{s \ge 0}$ *satisfies the smoothing hypothesis 4. Moreover, the smoothing operators can be taken as* $S_\theta u = \rho_{\frac{1}{\theta}} * u$ *for* $\theta \ge 1$, *where* $\rho \in \mathscr{S}(\mathbb{R}^d)$ *is a specially constructed mollifier.*

Proof. Let $\hat{\rho} \in C_c^\infty(\mathbb{R}^d)$ with $0 \le \hat{\rho} \le 1$ be an even function such that $\hat{\rho} = 1$ on $B_{\frac{1}{2}}(0)$ and $\hat{\rho} = 0$ outside $B_1(0)$, where $B_r(x)$ denotes the closed ball of radius r about x.

Define ρ to be the inverse Fourier transform of $\hat{\rho}$, which is real since $\hat{\rho}$ is even, and $\rho \in \mathscr{S}(\mathbb{R}^d)$, since $\hat{\rho} \in \mathscr{S}(\mathbb{R}^d)$.

For $u \in H^0(\mathbb{R}^d) = L^2(\mathbb{R}^d)$, we define

$$S_\theta u = \rho_{\frac{1}{\theta}} * u.$$

Let $0 \le r, s \in \mathbb{R}$ and $u \in H^s(\mathbb{R}^d)$.

Note that, by properties of the Fourier transform,

$$\begin{aligned}\widehat{S_\theta u}(\xi) &= \widehat{\rho_{\frac{1}{\theta}}}(\xi)\hat{u}(\xi)\\ &= \hat{\rho}(\frac{\xi}{\theta})\hat{u}(\xi).\end{aligned}$$

Hence

$$\begin{aligned}\int_{\mathbb{R}^d}(1+|\xi|^2)^r\left|\widehat{S_\theta u}(\xi)\right|^2 \mathrm{d}\xi &= \int_{\mathbb{R}^d}(1+|\xi|^2)^r\hat{\rho}(\frac{\xi}{\theta})^2\hat{u}(\xi)^2\mathrm{d}\xi\\ &= \int_{\mathbb{R}^d}(1+|\xi|^2)^{r-s}\hat{\rho}(\frac{\xi}{\theta})^2(1+|\xi|^2)^s\hat{u}(\xi)^2\mathrm{d}\xi\\ &\leq ||u||^2_{H^s}\sup_{\xi\in\mathbb{R}^d}(1+|\xi|^2)^{r-s}\hat{\rho}(\frac{\xi}{\theta})^2\\ &\leq ||u||^2_{H^s}(1+\theta^2)^{(r-s)_+}\\ &\leq C_{r,s}\,||u||^2_{H^s}\,\theta^{2(r-s)_+}\end{aligned}$$

since $0 \leq \hat{\rho} \leq 1$ and $\hat{\rho}(\frac{\xi}{\theta}) = 0$ for $\xi \geq \theta$.

This proves (1), and also that $S_\theta : H^0(\mathbb{R}^d) \to \cap_{s\geq 0} H^s(\mathbb{R}^d)$.

Now

$$\begin{aligned}\int_{\mathbb{R}^d}(1+|\xi|^2)^r\left|\widehat{u-S_\theta u}(\xi)\right|^2 \mathrm{d}\xi &= \int_{\mathbb{R}^d}(1+|\xi|^2)^r(1-\hat{\rho}(\frac{\xi}{\theta}))^2\hat{u}(\xi)^2\mathrm{d}\xi\\ &= \int_{\mathbb{R}^d}(1+|\xi|^2)^{r-s}(1-\hat{\rho}(\frac{\xi}{\theta}))^2(1+|\xi|^2)^s\hat{u}(\xi)^2\mathrm{d}\xi\\ &\leq ||u||^2_{H^s}\sup_{\xi\in\mathbb{R}^d}(1+|\xi|^2)^{r-s}(1-\hat{\rho}(\frac{\xi}{\theta}))^2\\ &\leq ||u||^2_{H^s}(1+(\frac{\theta}{2})^2)^{(r-s)}\\ &\leq C_{r,s}\,||u||^2_{H^s}\,\theta^{2(r-s)}\end{aligned}$$

assuming $r \leq s$, since $0 \leq \hat{\rho} \leq 1$ and $1-\hat{\rho}(\frac{\xi}{\theta}) = 0$ for $\xi \leq \frac{\theta}{2}$.

This proves (2).

Finally, for small $h \in \mathbb{R}$, we have

$$\begin{aligned}\mathscr{F}\left[\frac{S_{\theta+h}u - S_\theta u}{h}\right](\xi) &= \frac{\hat{\rho}(\frac{\xi}{\theta+h}) - \hat{\rho}(\frac{\xi}{\theta})}{h}\hat{u}(\xi)\\ &= \left(-\frac{1}{\theta^2}\sum_{i=1}^{d}\xi_i\partial_i\hat{\rho}(\frac{\xi}{\theta}) + R(h,\theta,\xi)\right)\hat{u}(\xi)\end{aligned}$$

by Taylor's theorem, where

$$|R(h,\theta,\xi)| \leq h \sup_{\theta \leq \phi \leq \theta+h} \frac{\mathrm{d}^2}{\mathrm{d}\phi^2}\hat{\rho}(\frac{\xi}{\phi}).$$

This implies

$$\int_{\mathbb{R}^d} (1+|\xi|^2)^r \, |R(h,\theta,\xi)|^2 \, |\hat{u}(\xi)|^2 \to 0 \text{ as } h \to 0$$

so that $S_\theta u$ is differentiable with respect to θ with derivative the inverse Fourier transform of

$$-\frac{1}{\theta^2} \sum_{i=1}^{d} \xi_i \, \partial_i \hat{\rho}(\frac{\xi}{\theta}) \hat{u}(\xi).$$

We also see that

$$\begin{aligned}
\left|\left| \frac{\mathrm{d}}{\mathrm{d}\theta} S_\theta u \right|\right|_{H^r}^2 &= \int_{\mathbb{R}^d} (1+|\xi|^2)^r (\frac{1}{\theta^2} \sum_{i=1}^{d} \xi_i \, \partial_i \hat{\rho}(\frac{\xi}{\theta}))^2 \, |\hat{u}(\xi)|^2 \\
&\leq ||u||_{H^s}^2 \sup_{\xi \in \mathbb{R}^d} (1+|\xi|^2)^{r-s} (\frac{1}{\theta^2} \sum_{i=1}^{d} \xi_i \, \partial_i \hat{\rho}(\frac{\xi}{\theta}))^2 \\
&\leq C_{r,s} \, ||u||_{H^s}^2 \, \theta^{2(r-s-1)}
\end{aligned}$$

since $\partial_i \hat{\rho}(\frac{\xi}{\theta})$ is zero for $\xi \leq \frac{\theta}{2}$ and $\xi \geq \theta$.

This proves (3).

5.2 *Tame Estimates in Sobolev Spaces*

The results in this section are fairly standard, and are based on standard Sobolev embeddings. Results of this type can be found in classical references on Sobolev spaces, for example Adams and Fournier [1]. However, we try and formulate them in a form which is most useful for obtaining tame estimates in the applications we have in mind.

The following lemma is very useful for proving chain and product rules in Sobolev spaces.

Lemma 1. *Let $p \in [1,\infty]$, $\Omega \subset \mathbb{R}^d$, for $d \geq 1$, be a domain where the standard Sobolev embedding holds and let $m > \frac{d}{p}$ be an integer. Let $0 \leq m_i \leq m$ be integers for $1 \leq i \leq n$ with $\sum_{i=1}^n m_i \geq (n-1)m$ and let $u_i \in W^{m_i,p}(\Omega)$. Then $\prod_{i=1}^n u_i \in L^p(\Omega)$ and*

$$\left\|\prod_{i=1}^{n} u_i\right\|_{L^p(\Omega)} \leq C \prod_{i=1}^{n} ||u_i||_{W^{m_i,p}(\Omega)} \, .$$

Proof. For $p = \infty$ the result is obvious and in fact only requires $m \geq 0$, so we will assume $p < \infty$.

We will use the following Sobolev embeddings. Let $k \geq 1$ be an integer and $u \in W^{k,p}(\Omega)$. Then for $q \geq p$,

$$||u||_{L^q(\Omega)} \leq C \, ||u||_{W^{k,p}(\Omega)}$$

provided

$$\frac{1}{q} > \frac{1}{p} - \frac{k}{d}$$

and $kp \leq d$. (Note it is the case $kp = d$ that requires the inequality to be strict.) If $kp > d$ then

$$||u||_{L^\infty(\Omega)} \leq C \, ||u||_{W^{k,p}(\Omega)} \, .$$

Suppose $m_i p > d$ for some i. By renumbering if necessary, we may assume $m_n p > d$. Then

$$\begin{aligned}\left\|\prod_{i=1}^{n} u_i\right\|_{L^p(\Omega)} &\leq \left\|\prod_{i=1}^{n-1} u_i\right\|_{L^p(\Omega)} ||u_n||_{L^\infty(\Omega)} \\ &\leq C \left\|\prod_{i=1}^{n-1} u_i\right\|_{L^p(\Omega)} ||u_n||_{W^{m_n,p}(\Omega)} \, .\end{aligned}$$

Also note that since $m_n \leq m$, we have $\sum_{i=1}^{n-1} m_i \geq (n-2)m$. Hence we are reduced to proving the result with n replaced by $n-1$. Thus we may assume $m_i p \leq d$ for all i.

Suppose $m_i = 0$ for some i. By renumbering if necessary, we may assume $m_n = 0$. Then $\sum_{i=1}^{n-1} m_i \geq (n-1)m$ and $0 \leq m_i \leq m$ implies $m_i = m > \frac{d}{p}$ for all $i < n$, hence

$$\begin{aligned}\left\|\prod_{i=1}^{n} u_i\right\|_{L^p(\Omega)} &\leq \prod_{i=1}^{n-1} ||u_i||_{L^\infty(\Omega)} \, ||u_n||_{L^p(\Omega)} \\ &\leq C \prod_{i=1}^{n-1} ||u_i||_{W^{m_i,p}(\Omega)} \, ||u_n||_{W^{m_n,p}(\Omega)} \, .\end{aligned}$$

Thus we may assume $m_i > 0$ for all i.

Now, using Hölder's inequality,

$$\left\| \prod_{i=1}^{n} u_i \right\|_{L^p(\Omega)} \leq \prod_{i=1}^{n} ||u_i||_{L^{\frac{p}{\lambda_i}}(\Omega)}$$

where $\sum_{i=1}^{n} \lambda_i = 1$ and $0 \leq \lambda_i \leq 1$ for all i. Hence, using Sobolev embedding, we have

$$\left\| \prod_{i=1}^{n} u_i \right\|_{L^p(\Omega)} \leq C \prod_{i=1}^{n} ||u_i||_{W^{m_i,p}(\Omega)}$$

provided

$$\frac{\lambda_i}{p} > \frac{1}{p} - \frac{m_i}{d}$$

for all i. But, summing the above inequalities, it is possible, assuming $0 < m_i \leq \frac{d}{p}$, to choose such $0 \leq \lambda_i \leq 1$ with $\sum_{i=1}^{n} \lambda_i = 1$ if and only if

$$\frac{n}{p} - \frac{\sum_{i=1}^{n} m_i}{d} < \frac{1}{p} \iff \sum_{i=1}^{n} m_i > (n-1)\frac{d}{p}.$$

But this does indeed hold since $\sum_{i=1}^{n} m_i \geq (n-1)m$ and $m > \frac{d}{p}$.

Corollary 1 (Leibniz's Rule or The Product Rule). *Let $p \in [1, \infty]$, $\Omega \subset \mathbb{R}^d$, for $d \geq 1$, be a domain where the standard Sobolev embedding holds and let $m > \frac{d}{p}$ be an integer. Let $0 \leq m_i \leq m$ be integers for $1 \leq i \leq n$ and $0 \leq k \leq m$ be an integer, with $\sum_{i=1}^{n} m_i \geq (n-1)m + k$. Let $u_i \in W^{m_i,p}(\Omega)$. Then $\prod_{i=1}^{n} u_i \in W^{k,p}(\Omega)$ with weak derivatives given by the classical Leibniz rule and*

$$\left\| \prod_{i=1}^{n} u_i \right\|_{W^{k,p}(\Omega)} \leq C \prod_{i=1}^{n} ||u_i||_{W^{m_i,p}(\Omega)} .$$

Proof. Let γ^i be multi-indices with $\sum_{i=1}^{n} \gamma_i = \alpha$, where $|\alpha| \leq k$. Note that $\sum_{i=1}^{n}(m_i - |\gamma_i|) \geq (n-1)m$, hence we may apply the above result to obtain

$$\left\| \prod_{i=1}^{n} \partial^{\gamma^i} u_i \right\|_{L^p(\Omega)} \leq C \prod_{i=1}^{n} ||u_i||_{W^{m_i,p}(\Omega)} .$$

Assuming u_i are smooth, we immediately obtain the result, since $\partial^\alpha \prod_{i=1}^n u_i$ is a sum of terms of the form $\prod_{i=1}^n \partial^{\gamma^i} u_i$ by the classical chain rule. For non-smooth u_i we use approximation by smooth functions together with this inequality.

Corollary 2 (The Chain Rule). *Let $p \in [1, \infty]$, $\Omega \subset \mathbb{R}^d$, for $d \geq 1$, be a domain where the standard Sobolev embedding holds and let $m > \frac{d}{p}$ be an integer. Let $F \in C_b^m(\mathbb{R}^{d'})$ and $u : \Omega \to \mathbb{R}^{d'}$ with $u \in W^{m,p}(\Omega)$. Let α be a multi-index with $1 \leq |\alpha| \leq m$. Let $0 < \beta \leq \alpha, 0 < \gamma^j \leq \alpha$ $(1 \leq j \leq |\beta|)$, be multi-indices with $\sum_{j=1}^{|\beta|} |\gamma^j| = |\alpha|$. Then*

$$\left|\left|(\partial^\beta F)(u) \prod_{j=1}^{|\beta|} \partial^{\gamma^j} u_{i_j}\right|\right|_{L^p(\Omega)} \leq C \, ||u||_{W^{m,p}(\Omega)}^{|\beta|}$$

where u_{i_j} denotes a component of u depending on j. Moreover, the function $F(u) \in L^\infty(\Omega)$ has a weak α-derivative in $L^p(\Omega)$ given as in the classical chain rule by sums of terms of the above form which satisfies the inequality

$$||\partial^\alpha (F(u))||_{L^p(\Omega)} \leq C \, ||u||_{W^{m,p}(\Omega)} \left(1 + ||u||_{W^{m,p}(\Omega)}\right)^{m-1}.$$

In addition, if $F(0) = 0$, then $F(u) \in W^{m,p}(\Omega)$ with

$$||F(u)||_{W^{m,p}(\Omega)} \leq C \, ||u||_{W^{m,p}(\Omega)} \left(1 + ||u||_{W^{m,p}(\Omega)}\right)^{m-1}.$$

Proof. Note that $\sum_{j=1}^{\beta} (m - |\gamma^j|) = |\beta| \, m - |\alpha| \geq (|\beta| - 1)m$, hence we may use the above result and the fact that $\partial^\beta F$ is bounded to obtain

$$\begin{aligned}\left|\left|(\partial^\beta F)(u) \prod_{j=1}^{|\beta|} \partial^{\gamma^j} u_{i_j}\right|\right|_{L^p(\Omega)} &\leq C \prod_{j=1}^{|\beta|} \left|\left|\partial^{\gamma^j} u_{i_j}\right|\right|_{W^{m-|\gamma^j|,p}(\Omega)} \\ &\leq C \, ||u||_{W^{m,p}(\Omega)}^{|\beta|} .\end{aligned}$$

Assuming u_i are smooth, we immediately obtain the required inequalities, since $\partial^\alpha(F(u))$ is a sum of terms of the form $(\partial^\beta F)(u) \prod_{j=1}^{|\beta|} \partial^{\gamma^j} u_{i_j}$ by the classical product and chain rules. For non-smooth u_i we use approximation by smooth functions together with this inequality.

Finally, if $F(0) = 0$, then we have

$$\begin{aligned}|F(u)| &= \left|\int_0^1 DF(tu)u \, \mathrm{d}t\right| \\ &\leq C \, |u|\end{aligned}$$

since DF is bounded. Thus $F(u) \in L^p(\Omega)$ with $||F(u)||_{L^p(\Omega)} \leq C\,||u||_{L^p(\Omega)}$. Together with the previous part, this implies the final statement of the result.

Corollary 3. *Let $D, D' \subset \mathbb{R}^{d'}$ be open with $D' \subset\subset D$. Let $F \in C^m(D)$ and $u : \Omega \to D'$ with $u \in W^{m,p}(\Omega)$. Then the above chain rule holds with these new F and u.*

Proof. Since u takes values in D', we may modify F outside D' by multiplying by a smooth cut-off function which is identically 1 on D' and 0 outside D'' for some $D'' \subset\subset D$, so we may assume $F \in C_b^m(\mathbb{R}^{d'})$, and then we can apply the above result.

Proposition 3 (The Derivative of a Differential Operator on Sobolev Spaces). *Let $p \in [1,\infty]$, $\Omega \subset \mathbb{R}^d$, for $d \geq 1$, be a domain where the standard Sobolev embedding holds and let $m \geq 0$ be an integer. Let I be a subinterval of $\mathbb{N}_0$ containing 0 and $l+m$, where $l > \frac{d}{p}$ is an integer, and set $X^s = W^{s,p}(\Omega,\mathbb{R}^{d'})$ for $s \in I$ and $Y^s = W^{s,p}(\Omega,\mathbb{R}^{d''})$. Let $U \subset X^\infty$ be $||\cdot||_{X^r}$-open for some $r \in I$ with $r \geq l+m$ and assume $0 \in U$. Define*

$$T : U \to Y^{\infty - m}$$

by

$$T(u)(x) = F(\{\partial^\alpha u(x) : 0 \leq |\alpha| \leq m\})$$

where $F : \mathbb{R}^{d'} \times \cdots \times \mathbb{R}^{d'} \to \mathbb{R}^{d''}$ is smooth and bounded with bounded derivatives on the range of $\{\partial^\alpha u : 0 \leq |\alpha| \leq m\}$ for $u \in U$ (so we may assume F is smooth and bounded with bounded derivatives), and $F(0) = 0$. The above rather complicated notation is merely a convenient way of expressing that $F(\{\partial^\alpha u : 0 \leq |\alpha| \leq m\})$ is a smooth function of u and its partial derivatives up to order m, which can be evaluated at x to give a function of x.

Write v^i_α for the argument of F which is evaluated at $\partial^\alpha u^i(x)$ in the above formula.

Then T is twice differentiable with derivatives given by

$$(DT(u)h)(x) = \sum_{0 \leq i \leq d'} \sum_{0 \leq \beta \leq m} \partial^\beta h^i(x) \frac{\partial F}{\partial v^i_\beta}(\{\partial^\alpha u(x)\})$$

$$D^2T(u)(h,h')(x) = \sum_{0 \leq i,j \leq d'} \sum_{0 \leq \beta,\gamma \leq m} \partial^\beta h^i(x) \partial^\gamma h'^j(x) \frac{\partial^2 F}{\partial v^j_\gamma \partial v^i_\beta}(\{\partial^\alpha u(x)\})$$

and the following inequalities hold.

$$||DT(u)h||_{Y^s} \leq C_s(||h||_{X^{s+m}} + ||h||_{X^l}(1 + ||u||_{X^{s+m}}))$$

$$\left|\left|D^2T(u)(h,h')\right|\right|_{Y^s}$$

$$\leq C_s(||h||_{X^{s+m}}\left|\left|h'\right|\right|_{X^l} + ||h||_{X^l}\left|\left|h'\right|\right|_{X^{s+m}} + ||h||_{X^l}\left|\left|h'\right|\right|_{X^l}(1 + ||u||_{X^{s+m}}))$$

for all $u \in U$, $h, h' \in X^\infty$ and $s \in I$ such that $s+m \in I$, where the constant $C_s > 0$ is bounded for s bounded.

Proof. First we assume all functions are smooth, or else we can use approximation by smooth functions. Note that by the chain rule, $||F(\{\partial^\alpha u(x) : 0 \le |\alpha| \le m\})||_{Y^s} \le C_s\, ||u||_{X^{s+m}}$, since $r > \frac{d}{p}$ (and the constant depends on U), hence T is well-defined. Using Taylor's Theorem, for $u \in U$, $t \in (-1,1)$ and $||h||_{X^r}$ small enough such that the line segment $[u-h, u+h]$ lies in U, we have

$$\begin{aligned}
&\frac{1}{t}(T(u+th) - T(u))(x)\\
&= \frac{1}{t}(F(\{\partial^\alpha u(x) + \partial^\alpha h(x)\}) - F(\{\partial^\alpha u(x)\}))\\
&= \sum_{0\le i\le d'}\sum_{0\le\beta\le m} \frac{\partial F}{\partial v^i_\beta}(\{\partial^\alpha u(x)\})\partial^\beta h^i(x)\\
&\quad + t\sum_{0\le i,j\le d'}\sum_{0\le\beta,\gamma\le m} \partial^\beta h^i(x)\partial^\gamma h^j(x)\int_0^1 (1-\tau)\frac{\partial^2 F}{\partial v^j_\gamma \partial v^i_\beta}(\{\partial^\alpha u(x) + \tau\partial^\alpha h(x)\})\mathrm{d}\tau.
\end{aligned}$$

Applying the chain rule to $\frac{1}{t}$ times the last term, which may be thought of as a function of (u,h), we see that $\frac{1}{t}$ times the last term is in Y^s for $s \in I$ such that $s+m \in I$ hence the last term converges to zero in Y^s as $t \to 0$. Similarly

$$\begin{aligned}
&\frac{1}{t}(DT(u+th')h - DT(u)h)(x)\\
&= \frac{1}{t}\Big(\sum_{0\le i\le d'}\sum_{0\le\beta\le m} \partial^\beta h^i(x)\frac{\partial F}{\partial v^i_\beta}(\{\partial^\alpha u(x) + t\partial^\alpha h'(x)\})\\
&\qquad - \sum_{0\le i\le d'}\sum_{0\le\beta\le m} \partial^\beta h^i(x)\frac{\partial F}{\partial v^i_\beta}(\{\partial^\alpha u(x)\})\Big)\\
&= \sum_{0\le i,j\le d'}\sum_{0\le\beta,\gamma\le m} \partial^\beta h^i(x)\partial^\gamma h'^j(x)\frac{\partial^2 F}{\partial v^j_\gamma \partial v^i_\beta}(\{\partial^\alpha u(x)\})\\
&\quad + t\sum_{0\le i,j,k\le d'}\sum_{0\le\beta,\gamma,\delta\le m} \partial^\beta h^i(x)\partial^\gamma h'^j(x)\partial^\delta h'^k(x)\\
&\qquad \times \int_0^1 (1-\tau)\frac{\partial^3 F}{\partial v^k_\delta \partial v^j_\gamma \partial v^i_\beta}(\{\partial^\alpha u(x) + \tau\partial^\alpha h'(x)\})\mathrm{d}\tau.
\end{aligned}$$

Applying the same argument we see the last term converges to zero as $t \to 0$.

Now, using the chain rule we have

$$\left|\frac{\partial F}{\partial v^i_\beta}(\{\partial^\alpha u(x)\})\right|_{W^{s,p}(\Omega)} \leq C_s \,||u||_{W^{s+m,p}(\Omega)}$$

for integer $s \geq 1$ and $u \in U$, where $|\cdot|_{W^{s,p}(\Omega)}$ denotes the Sobolev semi-norm of order s (the sum of the L^p norms of the weak derivatives of order s).

Define

$$H(x) = \frac{\partial F}{\partial v^i_\beta}(\{\partial^\alpha u(x)\}).$$

For integer $s \geq 0$, using the product rule and the above, we have

$$\begin{aligned}
&\left|\frac{\partial F}{\partial v^i_\beta}(\{\partial^\alpha u(x)\})\partial^\beta h^i(x)\right|_{W^{s,p}(\Omega)} \\
&\leq C_s\Big(\sum_{1\leq\delta\leq s} ||\partial^\delta H \partial^{s-\delta}\partial^\beta h^i(x)||_{L^p(\Omega)} + ||H\partial^s\partial^\beta h^i(x)||_{L^p(\Omega)}\Big) \\
&\leq C_s(||DH||_{W^{l-1,p}(\Omega)}\,||h||_{W^{s+m,p}(\Omega)} + ||DH||_{W^{s-1,p}(\Omega)}\,||h||_{W^{l,p}(\Omega)} \\
&\qquad + ||H||_{L^\infty(\Omega)}\,||h||_{W^{s+m,p}(\Omega)}) \\
&\leq C_s(||h||_{W^{s+m,p}(\Omega)} + ||h||_{W^{l,p}(\Omega)}\,(1 + ||u||_{W^{s+m,p}(\Omega)}))
\end{aligned}$$

for any $h \in X^\infty$ and $u \in U$, where we have used $r \geq l + m$.

In a similar manner, we obtain the inequality for the second derivative of T.

6 Application to Compressible Vortex Sheets in 2D

Here we show how the paper [8] of Coulombel and Secchi fits into the above framework. In fact the above framework is specifically devised to fit this case and the original ideas are contained in the paper by Coulombel and Secchi and earlier papers. For the sake of brevity, to follow this section it is necessary to refer to their paper. Note though that a significant portion of the work of the full result of Coulombel and Secchi is in solving the linearised equations with an appropriate energy estimate, which can be found in [7]. We believe that the abstract framework below should also fit the scheme used by Trakhinin in [25], since his scheme is very similar to the one used by Coulombel and Secchi.

We make some simplifications to the scheme of Coulombel and Secchi – firstly we take the boundary condition for the continuity of density (which is a linear condition) as part of the definition of the function spaces. Secondly, we treat the

Eikonal equations in a slightly simpler way which is less optimal with respect to regularity. It appears that although we need more regularity on the approximate solution, we only require it to be small in a lower-order Sobolev space.

The aim of their paper is to show short-time structural stability of plane vortex sheets for the 2D isentropic Euler equations of gas dynamics. This means the following. We start with two constant states $\bar{U}^+ = (\bar{\rho}, \bar{v}^+, 0)$, $\bar{U}^- = (\bar{\rho}, \bar{v}^-, 0)$ with pressure given by $\bar{p} = p(\bar{\rho})$ and sound speed given by $\bar{c} = \sqrt{p'(\bar{\rho})}$. When patched together either side of $\{x_2 = 0\}$ these form a weak solution of the 2D isentropic Euler equations equal to $\bar{U}^+$ in $\{x_2 > 0\}$ and equal to $\bar{U}^-$ in $\{x_2 < 0\}$, since the Rankine-Hugoniot jump conditions are satisfied across $\{x_2 = 0\}$. Since the normal velocity is continuous whereas the tangential velocity jumps this is called a vortex-sheet solution and it is characteristic in the sense that the boundary matrix for the system evaluated at this state is singular. We then impose smooth initial data close to this state (satisfying the Rankine-Hugoniot conditions with continuous normal velocity) which includes perturbing the discontinuity slightly so it is the graph of a function. The aim is to show the short-time existence of a solution with the same structure – that is, smooth either side of a surface of discontinuity across which the Rankine-Hugoniot conditions are satisfied with continuous normal velocity. This requires a stability assumption on the background state, $\left|\bar{u}^+ - \bar{u}^-\right| > 2\sqrt{2}\bar{c}$, and also a smallness assumption on the initial data. After some reductions the problem is reduced to finding a local inverse of a nonlinear operator, so that Nash-Moser iteration may be applied. The preliminary work includes changing coordinates to fix the free surface, which involves adding the Eikonal equations to the system to be solved, and introducing an approximate solution so that the initial data can be taken as zero. The main work is then to obtain a tame estimate for the linearised equations, after which a modified version of Nash-Moser iteration as above can be applied.

Notation. We will use the notation of [8], and to avoid conflict of notation with the above we will write u, v, w, f, g used in the above in bold face as $\mathbf{u}, \mathbf{v}, \mathbf{w}, \mathbf{f}, \mathbf{g}$. We will also write $\mathscr{U}$ and $\mathscr{V}$ instead of U and V used above.

6.1 The Function Spaces

For $T > 0$, define

$$\Omega_T = \{(t, x_1, x_2) \in \mathbb{R}^3 : t < T, x_2 > 0\}$$
$$\omega_T = \{(t, x_1) \in \mathbb{R}^2 : t < T\}.$$

For integer $s \geq 0$ and real $\gamma \geq 1$ define the weighted Sobolev space

$$H^s_\gamma(\Omega_T) = \{\exp(\gamma t)v : v \in H^m(\Omega_T)\}$$

where $H^s(\Omega_T)$ is the usual Sobolev space of order s. We define $H^s_\gamma(\omega_T)$ similarly. The norm on $u \in H^s_\gamma(\Omega_T)$ is given by

$$||u||_{H^s_\gamma(\Omega_T)} = ||\exp(-\gamma t)u||_{H^s(\Omega_T)}\,.$$

Next, we define, for integer $s \geq 0$,

$$\mathscr{F}^s_\gamma(\Omega_T) = \{u \in H^s_\gamma(\Omega_T) : u = 0 \text{ for } t < 0\}$$

and we define $\mathscr{F}^s_\gamma(\omega_T)$ similarly. Now, adapting the notation of [8] to our framework, we define

$$\begin{aligned} X^s = \{\mathbf{u} \in (\mathscr{F}^{s+3}_\gamma(\Omega_T))^3 \times (\mathscr{F}^{s+3}_\gamma(\Omega_T))^3 \times \mathscr{F}^{s+3}_\gamma(\Omega_T) \times \mathscr{F}^{s+3}_\gamma(\Omega_T) \\ : \Psi^+|_{x_2=0} = \Psi^-|_{x_2=0}, \rho^+|_{x_2=0} = \rho^-|_{x_2=0}\} \end{aligned}$$

where we write

$$\mathbf{u} = (V^+, V^-, \Psi^+, \Psi^-)$$

and

$$V = (\rho, v, u)$$

and define

$$\psi := \Psi^+|_{x_2=0} = \Psi^-|_{x_2=0}.$$

Note that we omit the superscripts $^+$ and $^-$ in formulae which apply to both. We have chosen X^0 to consist of products of Sobolev spaces of order 3 because of the embedding $H^s(\mathbb{R}^d) \subset W^{1,\infty}(\mathbb{R}^d)$ for $s > \frac{d}{2} + 1$, and in this case the dimension d is 3 (two space and one time).

We define the norm $||\cdot||_{X^s}$ on X^s as the usual product norm (the sum of the norms of the components). Then $\{X^s\}_{s\in I}$ is a decreasing family of Banach spaces, where $I = [0, s_3]$ is an interval in $\mathbb{N}_0$, for integer $s_3 > 0$ which we will fix later sufficiently large.

Similarly, we define

$$Y^s = \{\mathbf{g} \in (\mathscr{F}^{s+3}_\gamma(\Omega_T))^3 \times (\mathscr{F}^{s+3}_\gamma(\Omega_T))^3 \times \mathscr{F}^{s+3}_\gamma(\Omega_T) \times \mathscr{F}^{s+3}_\gamma(\Omega_T)\}$$

where we write

$$\mathbf{g} = (f^+, f^-, h^+, h^-)$$

and

$$f = (f_1, f_2, f_3).$$

6.2 The Smoothing Operators

Note that in order to define the smoothing operators on $\{X^s\}_{s\in I}$ (which can then be used on $\{Y^s\}_{s\in I}$ as well), we must make some modifications from those on $H^s(\mathbb{R}^d)$. Firstly, we must replace $\mathbb{R}^d$ by a domain with a Lipschitz boundary with finite covering, which is easily done via an extension operator. Next, we must ensure that the property $\mathbf{u} = 0$ for $t < 0$ is preserved under the action of the smoothing operators, which was done by Alinhac in [2], and finally we must ensure that the two properties $\Psi^+|_{x_2=0} = \Psi^-|_{x_2=0} = \psi$ and $\rho^+|_{x_2=0} = \rho^-|_{x_2=0}$ are preserved. See [8] for the details of this construction using a lifting operator.

6.3 The Background Solution and the Approximate Solution

Although we will not introduce the original problem considered in [8] (since we wish to show the use of Nash-Moser iteration only), we need to introduce the background or stationary solution and approximate solution for reference.

The background solution is given in the form

$$(\overline{U}^{\pm} = (\overline{\rho}^{\pm} = \overline{\rho}, \pm\overline{v}, \overline{u}^{\pm} = 0), \overline{\Phi}^{\pm} = \pm x_2)$$

where $\overline{\rho}, \overline{v}$ are constants with $\overline{\rho} > 0$.

We assume the existence of an 'approximate solution' $(U^{a+}, U^{a-}, \Phi^{a+}, \Phi^{a-})$ with $U^a - \overline{U}, \Phi^a - \overline{\Phi} \in H^{s_4+3}(\Omega_T)$ having compact support, which has the following properties. Here, s_4 is a sufficiently large integer with $s_4 \geq s_3 + 2$. In fact $s_4 = s_3 + 2$ will do.

$$\partial_t^j \mathbb{L}(U^a, \Phi^a)|_{t=0} = 0 \text{ for } 0 \leq j \leq s_3 + 3$$

$$\partial_t \Phi^a + v^a \partial_{x_1} \Phi^a - u^a = 0$$

$$\Phi^{a+}|_{x_2=0} = \Phi^{a-}|_{x_2=0} =: \phi^a$$

$$\rho^{a+} - \rho^{a-} = 0$$

$$\partial_{x_2} \Phi^{a+} \geq \frac{3}{4}$$

$$\partial_{x_2} \Phi^{a-} \leq -\frac{3}{4}$$

$$\rho^{a\pm} \geq \delta_0$$

$$||U^a - \overline{U}||_{H^7(\Omega_T)} + ||\Phi^a - \overline{\Phi}||_{H^7(\Omega_T)} \leq \delta_1$$

for some $\delta_0 > 0$, where we are allowed to choose constant $\delta_1 > 0$ as small as we like (which restricts the size of the initial data in the original problem). The first order differential operator $\mathbb{L}$ is defined in the next section.

6.4 *The Nonlinear Operator and Equations*

6.4.1 The Operator T and the Set $\mathscr{U}$

We set $m_0 = 4$ and define

$$\mathscr{U}^4 = \{\mathbf{u} \in X^4 : ||\mathbf{u}||_{X^4} \leq \delta_2\}$$

where $\delta_2 > 0$ is chosen sufficiently small. In particular, we need

$$||\Psi^{\pm}||_{W^{1,\infty}(\Omega_T)} \leq \frac{1}{2}$$
$$||\rho^{\pm}||_{L^{\infty}(\Omega_T)} \leq \frac{\delta_0}{2}$$

which is possible via Sobolev embedding. This ensures that $\partial_{x_2}(\Phi^{a\pm} + \Psi^{\pm})$ and $\rho^{a\pm} + \rho^{\pm}$ are bounded away from zero.

We define the operator $T : \mathscr{U}^4 \to Y^0$ by

$$T(\mathbf{u}) = \begin{pmatrix} \mathscr{L}(V^+, \Psi^+) \\ \mathscr{L}(V^-, \Psi^-) \\ \mathscr{E}(V^+, \Psi^+) \\ \mathscr{E}(V^-, \Psi^-) \end{pmatrix}.$$

Here,

$$\mathscr{L}(V, \Psi) = \mathbb{L}(U^a + V, \Phi^a + \Psi) - \mathbb{L}(U^a, \Phi^a)$$

and

$$\mathbb{L}(U, \Phi) = \partial_t U + A_1(U)\partial_{x_1} U + \frac{1}{\partial_{x_2}\Phi}(A_2(U) - \partial_t \Phi - \partial_{x_1}\Phi A_1(U))\partial_{x_2} U.$$

The matrices $A_1(U)$ and $A_2(U)$ are smooth functions of U for $U_1 > 0$, where U_1 is the first component of U (the 'ρ' component). See [8] for the exact expressions of these matrices. Also,

$$\mathscr{E}(V, \Psi) = \partial_t \Psi + (v^a + v)\partial_{x_1}\Psi - u + v\partial_{x_1}\Phi^a.$$

We note also that $T : \mathscr{U} \to Y^{\infty - 1}$, where $\mathscr{U} = \mathscr{U}^4 \cap X^{\infty}$.

6.4.2 The Equations

Define

$$f^a = \begin{cases} -\mathbb{L}(U^a, \Phi^a) \text{ for } t \geq 0, \\ 0 \text{ for } t < 0. \end{cases}$$

Then by the properties of the approximate solution, we have $f^a \in \mathscr{F}_\gamma^{s_3+3}(\Omega_T)$ and together with the definition of $\mathbb{L}$ we obtain

$$||f^a||_{Y^{s_3}} \leq C\delta_1 =: \epsilon.$$

Set

$$\mathbf{f} = \begin{pmatrix} f^{a+} \\ f^{a-} \\ 0 \\ 0 \end{pmatrix}.$$

For ϵ sufficiently small, we wish to solve the equation

$$T(\mathbf{u}) = \mathbf{f}$$

which is equivalent to

$$T(\mathbf{u}) = T(\mathbf{u_0}) + \mathbf{f}$$

if we set

$$\mathbf{u_0} = 0$$

since $T(0) = 0$.

6.5 *The Linearised Operator, Modified Linearised Operator, Modified State and Linearised Equations*

6.5.1 The Operator DT

Notation. To make the notation easier, let us use $\tilde{\mathbf{u}}$ instead of $\mathbf{v}$ to represent a vector to which we apply $DT(\mathbf{u})$, with the obvious notation

$$\tilde{\mathbf{u}} = (\tilde{V}^+, \tilde{V}^-, \tilde{\Psi}^+, \tilde{\Psi}^-)$$

and $\tilde{\Psi}^\pm|_{x_2=0} = \tilde{\psi}$.

Then we have

$$DT(\mathbf{u})\tilde{\mathbf{u}} = \begin{pmatrix} \mathcal{L}'(V^+,\Psi^+)(\tilde{V}^+,\tilde{\Psi}^+) \\ \mathcal{L}'(V^-,\Psi^-)(\tilde{V}^-,\tilde{\Psi}^-) \\ \mathcal{E}'(V^+,\Psi^+)(\tilde{V}^+,\tilde{\Psi}^+) \\ \mathcal{E}'(V^-,\Psi^-)(\tilde{V}^-,\tilde{\Psi}^-) \end{pmatrix}$$

where $\mathcal{L}'$ is the derivative of $\mathcal{L}$ and $\mathcal{E}'$ is the derivative of $\mathcal{E}$. Calculating these, we obtain

$$\mathcal{L}'(V,\Psi)(\tilde{V},\tilde{\Psi}) = \mathbb{L}'(U^a + V, \Phi^a + \Psi)(\tilde{V},\tilde{\Psi})$$

where $\mathbb{L}'$ is the derivative of $\mathbb{L}$ and is given by

$$\begin{aligned}\mathbb{L}'(U,\Phi)(\tilde{V},\tilde{\Psi}) = {}& \partial_t \tilde{V} + A_1(U)\partial_{x_1}\tilde{V} + \frac{1}{\partial_{x_2}\Phi}(A_2(U) - \partial_t\Phi - \partial_{x_1}\Phi A_1(U))\partial_{x_2}\tilde{V} \\ & +(DA_1(U)\tilde{V})\partial_{x_1}U - \frac{\partial_{x_2}\tilde{\Psi}}{(\partial_{x_2}\Phi)^2}(A_2(U) - \partial_t\Phi - \partial_{x_1}\Phi A_1(U))\partial_{x_2}U \\ & +\frac{1}{\partial_{x_2}\Phi}(DA_2(U)\tilde{V} - \partial_t\tilde{\Psi} - \partial_{x_1}\tilde{\Psi}A_1(U) - \partial_{x_1}\Phi DA_1(U)\tilde{V})\partial_{x_2}U.\end{aligned}$$

Also,

$$\mathcal{E}'(V,\Psi)(\tilde{V},\tilde{\Psi}) = \partial_t\tilde{\Psi} + (v^a + v)\partial_{x_1}\tilde{\Psi} - \tilde{u} + \tilde{v}\partial_{x_1}\Phi^a + \tilde{v}\partial_{x_1}\Psi.$$

6.5.2 The Operator A

We define

$$A(\mathbf{u})\tilde{\mathbf{u}} = \begin{pmatrix} \mathbb{L}'_e(U^{a+} + V^+, \Phi^{a+} + \Psi^+)\check{V}^+ \\ \mathbb{L}'_e(U^{a-} + V^-, \Phi^{a-} + \Psi^-)\check{V}^- \\ \mathcal{E}'(V^+,\Psi^+)(\tilde{V}^+,\tilde{\Psi}^+) \\ \mathcal{E}'(V^-,\Psi^-)(\tilde{V}^-,\tilde{\Psi}^-) \end{pmatrix}$$

where, as in [8], we have introduced the ‘good unknown’, which we denote by $\check{V}$ instead of $\dot{V}$ to avoid conflict of notation, as

$$\check{V} = \tilde{V} - \frac{\tilde{\Psi}}{\partial_{x_2}(\Phi^a + \Psi)}\partial_{x_2}(U^a + V).$$

The operator $\mathbb{L}'_e$ is defined as

$$\mathbb{L}'_e(U,\Phi)\check{V} = \partial_t \check{V} + A_1(U)\partial_{x_1}\check{V} + \frac{1}{\partial_{x_2}\Phi}(A_2(U) - \partial_t\Phi - \partial_{x_1}\Phi A_1(U))\partial_{x_2}\check{V}$$
$$+(DA_1(U)\check{V})\partial_{x_1}U + \frac{1}{\partial_{x_2}\Phi}(DA_2(U)\check{V} - \partial_{x_1}\Phi DA_1(U)\check{V})\partial_{x_2}U.$$

Note that, with $(U,\Phi) = (U^a + V, \Phi^a + \Psi)$, we have

$$\mathbb{L}'(U,\Phi)(\tilde{V},\tilde{\Psi}) - \mathbb{L}'_e(U,\Phi)\check{V} = \frac{\tilde{\Psi}}{\partial_{x_2}\Phi}\partial_{x_2}(\mathbb{L}(U,\Phi))$$
$$= \frac{\tilde{\Psi}}{\partial_{x_2}\Phi}\partial_{x_2}(\mathscr{L}(V,\Psi) - f^a).$$

6.5.3 The Set $\mathscr{V}$ and the Operator R

We set $m_7 = 1$ and define

$$\mathscr{V} = \{\mathbf{u} \in X^{\infty-1} : \mathscr{E}(V^+,\Psi^+) = 0, \mathscr{E}(V^-,\Psi^-) = 0, ||\mathbf{u}||_{X^3} \leq \delta_5\}$$

where $0 < \delta_5$ is to be chosen sufficiently small.

We define the operator $R : \mathscr{U} \to \mathscr{V}$ by

$$R(\mathbf{u}) = \begin{pmatrix} \rho^+ \\ v^+ \\ \partial_t\Psi^+ + (v^{a+} + v^+)\partial_{x_1}\Psi^+ + v^+\partial_{x_1}\Phi^{a+} \\ \rho^- \\ v^- \\ \partial_t\Psi^- + (v^{a-} + v^-)\partial_{x_1}\Psi^- + v^-\partial_{x_1}\Phi^{a-} \\ \psi^+ \\ \psi^- \\ \psi \end{pmatrix}.$$

One can check that indeed $R(\mathbf{u}) \in \mathscr{V}$. In particular, one can see that $||R(\mathbf{u})||_{X^3}$ can be controlled in terms of $||\mathbf{u}||_{X^4} \leq \delta_2$ for $\mathbf{u} \in \mathscr{U}$.

We then calculate

$$R(\mathbf{u}) - \mathbf{u} = \begin{pmatrix} 0 \\ 0 \\ \mathscr{E}(V^+, \Psi^+) \\ 0 \\ 0 \\ \mathscr{E}(V^-, \Psi^-) \\ 0 \\ 0 \\ 0 \end{pmatrix}.$$

6.5.4 The Linearised Equations

Given $\mathbf{u} \in \mathscr{V}$ and $\mathbf{g} \in Y^\infty$, we wish to solve the equation

$$A(\mathbf{u})\tilde{\mathbf{u}} = \mathbf{g}$$

for $\tilde{\mathbf{u}} \in X^{\infty - \max\{l_1, m_4 + m_7\}}$. Let us write

$$\mathbf{g} = \begin{pmatrix} f^+ \\ f^- \\ h^+ \\ h^- \end{pmatrix}$$

where $h^+|_{x_2=0} = h^-_{x_2=0} = g$. Then we want to solve the system

$$\begin{pmatrix} \mathbb{L}'_e(U^{a+} + V^+, \Phi^{a+} + \Psi^+)\check{V}^+ \\ \mathbb{L}'_e(U^{a-} + V^-, \Phi^{a-} + \Psi^-)\check{V}^- \\ \mathscr{E}'(V^+, \Psi^+)(\check{V}^+ + \frac{\tilde{\Psi}^+}{\partial_{x_2}(\Phi^{a+} + \Psi^+)}\partial_{x_2}(U^{a+} + V^+), \tilde{\Psi}^+) \\ \mathscr{E}'(V^-, \Psi^-)(\check{V}^- + \frac{\tilde{\Psi}^-}{\partial_{x_2}(\Phi^{a-} + \Psi^-)}\partial_{x_2}(U^{a-} + V^-), \tilde{\Psi}^-) \end{pmatrix} = \begin{pmatrix} f^+ \\ f^- \\ h^+ \\ h^- \end{pmatrix}$$

where in the last two equations we have written $\tilde{V}^+$ in terms of the 'good unknown' $\check{V}$ and $\tilde{\Psi}$. The introduction of the 'good unknown' allows us to split the solution of this system into two steps. First we solve the system

$$\mathbb{L}'_e(U^{a\pm} + V^\pm, \Phi^{a\pm} + \Psi^\pm)\check{V}^\pm = f^\pm \tag{53}$$

with boundary conditions

$$\begin{aligned} &\check{\rho}^+|_{x_2=0} + \frac{\tilde{\Psi}}{\partial_{x_2}(\Phi^{a+} + \Psi^+)|_{x_2=0}}\partial_{x_2}(\rho^{a+} + \rho^+)|_{x_2=0} \\ -&\check{\rho}^-|_{x_2=0} - \frac{\tilde{\Psi}}{\partial_{x_2}(\Phi^{a-} + \Psi^-)|_{x_2=0}}\partial_{x_2}(\rho^{a-} + \rho^-)|_{x_2=0} = 0 \end{aligned} \tag{54}$$

$$\partial_t \tilde{\psi} + (v^{a\pm} + v^{\pm})|_{x_2=0} \partial_{x_1} \tilde{\psi}$$
$$-(\check{u}^{\pm}|_{x_2=0} + \frac{\tilde{\psi}}{\partial_{x_2}(\Phi^{a\pm} + \Psi^{\pm})|_{x_2=0}} \partial_{x_2}(u^{a\pm} + u^{\pm})|_{x_2=0})$$
$$+(\check{v}^{\pm}|_{x_2=0} + \frac{\tilde{\psi}}{\partial_{x_2}(\Phi^{a\pm} + \Psi^{\pm})|_{x_2=0}} \partial_{x_2}(v^{a\pm} + v^{\pm})|_{x_2=0}) \partial_{x_1}(\psi^a + \psi)$$
$$= h^{\pm}|_{x_2=0} \qquad (55)$$

for the unknowns $(\check{V}^{\pm}, \tilde{\psi})$. Note the first boundary condition is $\tilde{\rho}^{+}|_{x_2=0} - \tilde{\rho}^{-}|_{x_2=0} = 0$ written in terms of the 'good unknown', and the second boundary condition is a rewriting of

$\mathscr{E}'(V^{\pm}, \Psi^{\pm})(\tilde{V}^{\pm}, \tilde{\Psi}^{\pm})\big|_{x_2=0} = h^{\pm}|_{x_2=0}$ in terms of the 'good unknown', where we replace $\tilde{\Psi}^{\pm}|_{x_2=0}$ with $\tilde{\psi}$.

Secondly, having solved the above system for $(\check{V}^{\pm}, \tilde{\psi})$, we solve the two separate equations

$$\mathscr{E}'(V^{\pm}, \Psi^{\pm})(\check{V}^{\pm} + \frac{\tilde{\Psi}^{\pm}}{\partial_{x_2}(\Phi^{a\pm} + \Psi^{\pm})} \partial_{x_2}(U^{a\pm} + V^{\pm}), \tilde{\Psi}^{\pm}) = h^{\pm} \qquad (56)$$

for $\tilde{\Psi}^{\pm}$. By restricting to $\{x_2 = 0\}$, we see that $\tilde{\Psi}^{\pm}|_{x_2=0}$ satisfy the same equations as $\tilde{\psi}$ given in the boundary conditions above, hence by uniqueness of solutions we have $\tilde{\Psi}^{\pm}|_{x_2=0} = \tilde{\psi}$.

Finally, we can rearrange to obtain $\tilde{V}$ from $\check{V}$ and $\tilde{\Psi}$.

6.6 Solution of the Linearised Equations

Assume $\mathbf{u} \in \mathscr{V}$ and $\mathbf{g} \in Y^{\infty}$. We wish to solve the equation

$$A(\mathbf{u})\tilde{\mathbf{u}} = \mathbf{g}$$

for $\tilde{\mathbf{u}}$, using the steps described above.

The key to the whole iteration scheme is the solution of the linearised problem (53)–(55).

We have the following result, stated in [8]. Assume that the stationary solution satisfies the supersonic condition

$$\overline{v} > \sqrt{2}c(\overline{\rho}).$$

Assume that U, Φ are such that $U - \overline{U}, \Phi - \overline{\Phi} \in H^{s+3}_{\gamma}(\Omega_T)$ for integer $s \in [3, s_3]$ with

$$||(U - \overline{U}, \nabla(\Phi - \overline{\Phi}))||_{H^5_\gamma(\Omega_T)} + ||(U - \overline{U}, \partial_{x_2} U, \nabla(\Phi - \overline{\Phi}))|_{x_2=0}||_{H^4_\gamma(\omega_T)} \leq \delta_4 \tag{57}$$

for some $\delta_4 > 0$, where $\Phi^+|_{x_2=0} = \Phi_-|_{x_2=0} = \phi$.

Assume also that (U, Φ) satisfy the eikonal equation

$$\partial_t \Phi + v\partial_{x_1}\Phi - u = 0.$$

Assume in addition that the coefficients $(U - \overline{U}, \Phi - \overline{\Phi})$ have fixed compact support – a technical condition which can be achieved by truncating the coefficients without affecting the solution due to the finite speed of propagation of the Euler equations.

Then if δ_4 is sufficiently small, given

$$(f^{\pm}, g^{\pm}) \in \mathscr{F}^{s+1}_{\gamma}(\Omega_T) \times \mathscr{F}^{s+1}_{\gamma}(\omega_T)$$

we have a unique solution

$$(\check{V}^{\pm}, \tilde{\psi}) \in \mathscr{F}^{s}_{\gamma}(\Omega_T) \times \mathscr{F}^{s+1}_{\gamma}(\omega_T)$$

of (53)–(55), replacing $h^{\pm}|_{x_2=0}$ with $g^{\pm}$, provided $\gamma \geq 1$ is sufficiently large depending on s_3. Moreover, the following estimate holds, for some constant $C_s > 0$,

$$\begin{aligned}
&\left|\left|\check{V}\right|\right|_{H^s_\gamma(\Omega_T)} + ||\tilde{\psi}||_{H^{s+1}_\gamma(\omega_T)} \\
&\leq C_s(||f||_{H^{s+1}_\gamma(\Omega_T)} + ||g||_{H^{s+1}_\gamma(\omega_T)} \\
&\qquad + (||f||_{H^4_\gamma(\Omega_T)} + ||g||_{H^4_\gamma(\omega_T)})\,||(U - \overline{U}, \Phi - \overline{\Phi})||_{H^{s+3}_\gamma(\Omega_T)}).
\end{aligned}$$

Here, we set $U = U^a + V$, $\Phi = \Phi^a + \Psi$, where $(U, \Psi) \in \mathscr{V}$. Note that the smallness condition (57) holds provided δ_5 and δ_1 are sufficiently small. Also note that the Eikonal equation holds since the approximate solution satisfies the Eikonal equation and by the definition of $\mathscr{V}$. We are given f and h and set $g^{\pm} = h^{\pm}|_{x_2=0}$. Unfortunately this method, which is slightly simpler than the one described in [8], results in a further loss of regularity due to taking the trace of h. So in fact given

$$(f^{\pm}, h^{\pm}) \in \mathscr{F}^{s+1}_{\gamma}(\Omega_T) \times \mathscr{F}^{s+2}_{\gamma}(\omega_T)$$

we have a unique solution

$$(\check{V}^{\pm}, \tilde{\psi}) \in \mathscr{F}^s_\gamma(\Omega_T) \times \mathscr{F}^{s+1}_\gamma(\omega_T)$$

satisfying the estimate

$$\left|\left|\check{V}\right|\right|_{H^s_\gamma(\Omega_T)} + ||\tilde{\psi}||_{H^{s+1}_\gamma(\Omega_T)} \leq C_s(||f||_{H^{s+1}_\gamma(\Omega_T)} + ||h||_{H^{s+2}_\gamma(\omega_T)}$$

$$+(||f||_{H^4_\gamma(\Omega_T)} + ||h||_{H^5_\gamma(\Omega_T)})(||(U^a - \overline{U}, \Phi^a - \overline{\Phi})||_{H^{s+3}_\gamma(\Omega_T)} + ||(V, \Psi)||_{H^{s+3}_\gamma(\Omega_T)})).$$

Having solved this system, it remains to solve the Eqs. (56) for $\tilde{\Psi}^{\pm}$. Each of these equations is a first order scalar linear equation, so has a unique solution (for smooth enough coefficients and source term). More precisely, assuming that

$$||(U^a + V, \Phi^a + \Psi)||_{H^3_\gamma(\Omega_T)}$$

is small enough (which is guaranteed by taking δ_4 small enough), we have a unique solution

$$\tilde{\Psi} \in \mathscr{F}^s_\gamma(\Omega_T)$$

of (56). Moreover, the following estimate holds, for some constant $C_s > 0$ (which may depend on the bound on $||(U^a + V, \Phi^a + \Psi)||_{H^3_\gamma(\Omega_T)}$),

$$||\tilde{\Psi}||_{H^s_\gamma(\Omega_T)} \leq$$

$$C_s(||h||_{H^s_\gamma(\Omega_T)} + \left|\left|\check{V}\right|\right|_{H^s_\gamma(\Omega_T)} + \left|\left|\check{V}\right|\right|_{H^3_\gamma(\Omega_T)}(||\Phi^a - \overline{\Phi}||_{H^{s+1}_\gamma(\Omega_T)} + ||\Psi||_{H^{s+1}_\gamma(\Omega_T)})$$

$$+ ||\tilde{\Psi}||_{H^3_\gamma(\Omega_T)}(||(U^a - \overline{U}, \Phi^a - \overline{\Phi})||_{H^{s+1}_\gamma(\Omega_T)} + ||(V, \Psi)||_{H^{s+1}_\gamma(\Omega_T)})).$$

Taking $s = 3$ and assuming δ_4 is sufficiently small, we obtain

$$||\tilde{\Psi}||_{H^3_\gamma(\Omega_T)} \leq C(||h||_{H^3_\gamma(\Omega_T)} + \left|\left|\check{V}\right|\right|_{H^3_\gamma(\Omega_T)}).$$

Thus

$$||\tilde{\Psi}||_{H^s_\gamma(\Omega_T)} \leq C_s(||h||_{H^s_\gamma(\Omega_T)} + \left|\left|\check{V}\right|\right|_{H^s_\gamma(\Omega_T)}$$

$$+(||h||_{H^3_\gamma(\Omega_T)} + \left|\left|\check{V}\right|\right|_{H^3_\gamma(\Omega_T)})(||(U^a - \overline{U}, \Phi^a - \overline{\Phi})||_{H^{s+1}_\gamma(\Omega_T)} + ||(V, \Psi)||_{H^{s+1}_\gamma(\Omega_T)})).$$

From the previous estimate for $\check{V}$ with $s = 3$, we obtain

$$\left|\left|\check{V}\right|\right|_{H^3_\gamma(\Omega_T)} \leq C_s(||f||_{H^4_\gamma(\Omega_T)} + ||h||_{H^5_\gamma(\omega_T)})$$

(where the constant will depend on $\left|\left|(U^a - \overline{U}, \Phi^a - \overline{\Phi})\right|\right|_{H^6_\gamma(\Omega_T)} + ||(V, \Psi)||_{H^6_\gamma(\Omega_T)} \leq \delta_1$). Thus we obtain

$$||\tilde{\Psi}||_{H^s_\gamma(\Omega_T)} \leq$$
$$C_s(||f||_{H^{s+1}_\gamma(\Omega_T)} + ||h||_{H^{s+2}_\gamma(\Omega_T)} + (||f||_{H^4_\gamma(\Omega_T)} + ||h||_{H^5_\gamma(\Omega_T)})(1 + ||(V, \Psi)||_{H^{s+3}_\gamma(\Omega_T)}))$$

(where the constant will depend on $\left|\left|(U^a - \overline{U}, \Phi^a - \overline{\Phi})\right|\right|_{H^{s+3}_\gamma(\Omega_T)}$). Combining and writing $\tilde{V}$ in terms of $\check{V}$ and $\tilde{\Psi}$, we obtain

$$||\tilde{V}||_{H^s_\gamma(\Omega_T)} + ||\tilde{\Psi}||_{H^s_\gamma(\Omega_T)} + ||\tilde{\psi}||_{H^{s+1}_\gamma(\omega_T)} \leq$$
$$C_s(||f||_{H^{s+1}_\gamma(\Omega_T)} + ||h||_{H^{s+2}_\gamma(\Omega_T)} + (||f||_{H^4_\gamma(\Omega_T)} + ||h||_{H^5_\gamma(\Omega_T)})(1 + ||(V, \Psi)||_{H^{s+3}_\gamma(\Omega_T)})).$$

Hence, for $\mathbf{u} \in \mathcal{V}$, we have $B(\mathbf{u}) : Y^\infty \to X^{\infty-4}$ and

$$||B(\mathbf{u})\mathbf{g}||_{X^s} \leq C_s(||\mathbf{g}||_{Y^{s+2}} + ||\mathbf{g}||_{Y^2}\,(1 + ||\mathbf{u}||_{X^{s+3}}))$$

for all s such that $s + 4 \in I$. Thus we have $l_1 = 2$ and $m_4 = 3$.

6.7 *Estimates of the Operators*

6.7.1 Estimate of R

Clearly from the definition of R and T, we have

$$||R(\mathbf{u}) - \mathbf{u}||_{X^0} \leq ||T(\mathbf{u}) - T(\mathbf{u_0}) - \mathbf{f}||_{Y^0}\,.$$

Thus $l_2 = 0$.

Also, using Sobolev embedding and that R is a first order differential operator, we have the tame estimate

$$||R(\mathbf{u})||_{X^s} \leq C_s(1 + ||u||_{X^0})(1 + ||u||_{X^{s+1}})$$

for $s \in [0, s_3 - 1]$. Thus $m_8 = 0$, and as we have already stated, $m_7 = 1$.

Now we estimate the commutator

$$||R(S_\theta^X \mathbf{u}) - S_\theta^X R(\mathbf{u})||_{X^s}$$

for $\mathbf{u} \in \mathscr{U}$.

We have

$$\begin{aligned}
&\mathscr{E}(S_\theta V, S_\theta \Psi) - S_\theta \mathscr{E}(V, \Psi)\\
&= \partial_t (S_\theta \Psi) + (v^a + S_\theta v)\partial_{x_1}(S_\theta \Psi) + (S_\theta v)\partial_{x_1}\Phi^a\\
&- S_\theta \partial_t \Psi - S_\theta(v^a \partial_{x_1}\Psi) - S_\theta(v\partial_{x_1}\Psi) - S_\theta(v\partial_{x_1}\Phi)\\
&= \partial_t(S_\theta \Psi - \Psi) + (\partial_t \Psi - S_\theta \partial_t \Psi)\\
&+ (v^a \partial_{x_1}\Psi - S_\theta(v^a \partial_{x_1}\Psi)) + v^a \partial_{x_1}(S_\theta \Psi - \Psi)\\
&+ (v\partial_{x_1}\Psi - S_\theta(v\partial_{x_1}\Psi)) + v\partial_{x_1}(S_\theta \Psi - \Psi) + (S_\theta v - v)\partial_{x_1} S_\theta \Psi\\
&+ (v\partial_{x_1}\Phi^a - S_\theta(v\partial_{x_1}\Phi^a)) + (S_\theta v - v)\partial_{x_1}\Phi^a.
\end{aligned}$$

Hence, using the property (2) of the smoothing operators and product estimates for Sobolev norms, we have, for $r-3, s-3 \in [0, s_3 - 1]$ with $r \geq s$ and $r' \in [3, s_3]$,

$$\begin{aligned}
&||\mathscr{E}(S_\theta V, S_\theta \Psi) - S_\theta \mathscr{E}(V, \Psi)||_{H_\gamma^s(\Omega_T)}\\
&\leq C_{r,s}(\theta^{s-r}\,||\Psi||_{H_\gamma^{r+1}(\Omega_T)}\\
&+ \theta^{s-r}((||v^a - \overline{v}||_{H_\gamma^2(\Omega_T)} + 1)\,||\Psi||_{H_\gamma^{r+1}(\Omega_T)} + ||v^a - \overline{v}||_{H_\gamma^r(\Omega_T)}\,||\Psi||_{H_\gamma^3(\Omega_T)})\\
&+ (||v^a - \overline{v}||_{H_\gamma^s(\Omega_T)} + 1)\theta^{3-r'}\,||\Psi||_{H_\gamma^{r'}(\Omega_T)}\\
&+ \theta^{s-r}(||v||_{H_\gamma^2(\Omega_T)}\,||\Psi||_{H_\gamma^{r+1}(\Omega_T)} + ||v||_{H_\gamma^r(\Omega_T)}\,||\Psi||_{H_\gamma^3(\Omega_T)})\\
&+ ||v||_{H_\gamma^s(\Omega_T)}\,\theta^{3-r'}\,||\Psi||_{H_\gamma^{r'}(\Omega_T)}\\
&+ \theta^{s-r}\,||v||_{H_\gamma^r(\Omega_T)}\,||\Psi||_{H_\gamma^3(\Omega_T)} + \theta^{2-r'}\,||v||_{H_\gamma^{r'}(\Omega_T)}\,||\Psi||_{H_\gamma^{s+1}(\Omega_T)}\\
&+ \theta^{s-r}(||v||_{H_\gamma^2(\Omega_T)}\,||\Phi^a - \overline{\Phi}||_{H_\gamma^{r+1}(\Omega_T)} + ||v||_{H_\gamma^r(\Omega_T)}\,||\Phi^a - \overline{\Phi}||_{H_\gamma^3(\Omega_T)})\\
&+ \theta^{s-r}\,||v||_{H_\gamma^r(\Omega_T)}\,||\Phi^a - \overline{\Phi}||_{H_\gamma^3(\Omega_T)} + \theta^{2-r'}\,||v||_{H_\gamma^{r'}(\Omega_T)}\,||\Phi^a - \overline{\Phi}||_{H_\gamma^{s+1}(\Omega_T)})\\
&\leq C_{r,s}(\theta^{s-r}(1 + ||v||_{H_\gamma^2(\Omega_T)} + ||\Psi||_{H_\gamma^3(\Omega_T)})(1 + ||v||_{H_\gamma^r(\Omega_T)} + ||\Psi||_{H_\gamma^{r+1}(\Omega_T)})\\
&+ \theta^{3-r'}(1 + ||v||_{H_\gamma^s(\Omega_T)} + ||\Psi||_{H_\gamma^{s+1}(\Omega_T)})(||v||_{H_\gamma^{r'}(\Omega_T)} + ||\Psi||_{H_\gamma^{r'}(\Omega_T)})).
\end{aligned}$$

Hence, for $r', r, s \in I$ with $r \geq s$,

$$\|R(S_\theta^X \mathbf{u}) - S_\theta^X R(\mathbf{u})\|_{X^s}$$
$$\leq C_{r,s}(\theta^{s-r}(1+\|\mathbf{u}\|_{X^0})(1+\|\mathbf{u}\|_{X^{r+1}}) + \theta^{-r'}(1+\|\mathbf{u}\|_{X^{s+1}})(1+\|\mathbf{u}\|_{X^{r'}})).$$

6.7.2 Estimate of the Derivatives of T

Since T is a first order differential operator, that is, $T(\mathbf{u})$ can be written as a smooth bounded function of $\mathbf{u}$ and its first order derivatives for $\mathbf{u} \in \mathscr{U}^1$, we immediately see that $T : \mathscr{U}^1 \to Y^0$ is continuous and it satisfies (20) and (19) with $m_1 \geq 1, m_2 \geq 0, m_3 \geq 1$. Note that we have used the Sobolev embedding $H^3_\gamma(\Omega_T) \subset W^{1,\infty}(\Omega_T)$. We will in fact need to estimate the derivative of A before we fix m_1, m_2, m_3.

6.7.3 Estimate of $A - DT$

We estimate

$$\left\|\frac{\tilde{\Psi}}{\partial_{x_2}(\Phi^a + \Psi)}\partial_{x_2}(\mathscr{L}(V,\Psi) - f^a)\right\|_{H^s_\gamma(\Omega_T)} \leq$$
$$C_s \|\tilde{\Psi}\|_{H^s_\gamma(\Omega_T)} \|\mathscr{L}(V,\Psi) - f^a\|_{H^3_\gamma(\Omega_T)} + C_s \|\tilde{\Psi}\|_{H^2_\gamma(\Omega_T)} \|\mathscr{L}(V,\Psi) - f^a\|_{H^{s+1}_\gamma(\Omega_T)}$$
$$+ C_s \|\tilde{\Psi}\|_{H^2_\gamma(\Omega_T)} \|\mathscr{L}(V,\Psi) - f^a\|_{H^3_\gamma(\Omega_T)} (1 + \|\Phi^a + \Psi - \overline{\Phi}\|_{H^{s+1}_\gamma(\Omega_T)}).$$

Hence

$$\|(A(\mathbf{u}) - DT(\mathbf{u}))\tilde{\mathbf{u}}\|_{Y^s}$$
$$\leq C_s(\|\tilde{\mathbf{u}}\|_{X^s} \|T(\mathbf{u}) - T(\mathbf{u_0}) - \mathbf{f}\|_{Y^0} + \|\tilde{\mathbf{u}}\|_{X^0} \|T(\mathbf{u}) - T(\mathbf{u_0}) - \mathbf{f}\|_{Y^{s+1}}$$
$$+ \|\tilde{\mathbf{u}}\|_{X^0} \|T(\mathbf{u}) - T(\mathbf{u_0}) - \mathbf{f}\|_{Y^0} (1 + \|\mathbf{u}\|_{X^{s+1}}))$$

where the constant C_s depends on $\|\Phi^a - \overline{\Phi}\|_{H^{s+1}_\gamma(\Omega_T)}$. Thus $m_5 = 0$, $m_6 = 0$, $m_9 = 1, l_3 = 0, l_4 = 1$.

6.7.4 Estimate of the Derivative of A

Note that

$$\mathbb{L}'_e(U,\Phi)\check{V} = \mathbb{L}'(U,\Phi)(\tilde{V},\tilde{\Psi}) - \frac{\tilde{\Psi}}{\partial_{x_2}\Phi}\partial_{x_2}(\mathbb{L}(U,\Phi))$$

where $(U, \Phi) = (U^a + V, \Phi^a + \Psi)$. The first term is a component of DT. For fixed $(\tilde{U}, \tilde{\Psi})$, the second term is $\tilde{\Psi}$ multiplied by a differential operator of order 2. Hence

$$\begin{aligned}
&||DA(\mathbf{u})_{\tilde{\mathbf{u}}}\mathbf{h}||_{Y^s} \\
&\leq C_s(||\mathbf{h}||_{X^{s+1}}\,||\tilde{\mathbf{u}}||_{X^0} + ||\mathbf{h}||_{X^0}\,||\tilde{\mathbf{u}}||_{X^{s+1}} + ||\mathbf{h}||_{X^0}\,||\tilde{\mathbf{u}}||_{X^0}\,(1 + ||\mathbf{u}||_{X^{s+1}}) \\
&+ ||\tilde{\mathbf{u}}||_{X^0}\,(||\mathbf{h}||_{X^{s+2}} + ||\mathbf{h}||_{X^1}\,(1 + ||\mathbf{u}||_{X^{s+2}})) + ||\tilde{\mathbf{u}}||_{X^s}\,||\mathbf{h}||_{X^1}) \\
&\leq C_s(||\mathbf{h}||_{X^{s+2}}\,||\tilde{\mathbf{u}}||_{X^1} + ||\mathbf{h}||_{X^1}\,||\tilde{\mathbf{u}}||_{X^{s+2}} + ||\mathbf{h}||_{X^1}\,||\tilde{\mathbf{u}}||_{X^1}\,(1 + ||\mathbf{u}||_{X^{s+2}})).
\end{aligned}$$

Thus we fix $m_1 = 2, m_2 = 1, m_3 = 2$.

6.8 Conclusion

We have seen that the hypotheses of the theorem are satisfied with $m_0 = 4, m_1 = 2, m_2 = 1, m_3 = 2, m_4 = 3, m_5 = 0, m_6 = 0, m_7 = 1, m_8 = 0, m_9 = 1, l_1 = 2, l_2 = 0, l_3 = 0, l_4 = 1$. Hence we may take $r_0 = 6$. Note that in the proof we required $s_1 > r_0 + 1$, $s_1 \geq r_0 + \max\{m_1, m_3\} + l_1$ and $M(s_1 - \max\{m_1 + l_4, m_3 + l_4, m_5, m_9\}) \geq 0$ (with slope 1 which is satisfied for $s_1 > r_0 + 1$ automatically). One can check that $M(s) = s - 8$ hence we require $s_1 - 3 \geq 8$, so $s_1 \geq 11$. Now we require $s_1 + \max\{l_1, m_4 + m_7\} \in I$, hence $s_3 \geq 11 + 4 = 15$, and thus $s_4 \geq 17$ will do.

Thus we conclude that if we are given the approximate solution $(U^{a+}, U^{a-}, \Phi^{a+}, \Phi^{a-})$ with $U^a - \overline{U}, \Phi^a - \overline{\Phi} \in H^{20}(\Omega_T)$ which satisfies the conditions described above, with

$$||U^a - \overline{U}||_{H^7(\Omega_T)} + ||\Phi^a - \overline{\Phi}||_{H^7(\Omega_T)}$$

sufficiently small, then we have a unique solution $(V^+, V^-, \Psi^+, \Psi^-) \in \mathscr{F}^7_\gamma(\Omega_T)$ to the following equations (for both $^+$ and $^-$ components),

$$\mathbb{L}(U^a + V, \Phi^a + \Psi) = 0$$
$$\partial_t(\Phi^a + \Psi) + (v^a + v)\partial_{x_1}(\Phi^a + \Psi) - (u^a + u) = 0.$$

In fact, since $f^a \in Y^{s_2 - 2}$, where $s_2 = 12 \leq s_3 - 3$, we may use the last part of the theorem to conclude that $(V^+, V^-, \Psi^+, \Psi^-) \in \mathscr{F}^{11}_\gamma(\Omega_T)$.

7 Further Applications and Open Problems

There are several other situations involving characteristic discontinuities for the Euler equations or the equations of ideal magnetohydrodynamics where it may be possible to obtain a tame estimate for the linearised equations, and thus apply the above Nash-Moser iteration scheme. In these contexts a characteristic discontinuity is a surface of discontinuity in the fluid across which the Rankine-Hugoniot jump conditions are satisfied with zero mass transfer. The first step is usually to perform a normal modes analysis by linearising about a background state (constant either side of a plane across which the Rankine-Hugoniot jump conditions are satisfied) and to determine criteria which rule out exponentially growing solutions. The aim is then to show short-time existence of solutions with the same structure as the background state (that is, smooth either side of a surface of discontinuity across which the Rankine-Hugoniot jump conditions are satisfied) where the initial data is a small perturbation of the background state, under the assumption that the background state satisfies the stability criteria. We call this structural stability.

One obvious open problem is to extend the above result by Coulombel and Secchi in [8] on the 2D isentropic Euler equations to the 2D full Euler equations. Miles showed in [17] that the stability criterion on the background solution $\bar{U}^{\pm}$ (using notation as above) in this case is

$$|[\bar{u}]| > ((\bar{c}^{+})^{\frac{2}{3}} + (\bar{c}^{-})^{\frac{2}{3}})^{\frac{3}{2}}$$

(where $[u] = u^{+} - u^{-}$) under the simplifying assumption

$$\bar{\rho}^{+}(\bar{c}^{+})^{2} = \bar{\rho}^{-}(\bar{c}^{-})^{2}.$$

The main difficulty is to solve, and to deduce a tame estimate for, the linearised equations, assuming this stability criterion, after which we would expect the application of Nash-Moser iteration to be similar. In fact Morando and Trebeschi have obtained an L^2 estimate with derivative loss for the linearised equations under this stability criterion – see [18]. We note that vortex sheets in 3D Euler are always unstable according to normal modes analysis – see Miles and Fejer [13].

A modification of the Nash-Moser scheme similar to the one above has been used successfully by Chen and Wang in [5] and [6] for current-vortex sheets in ideal compressible magnetohydrodynamics under the assumption that the jump in the non-parallel component of the magnetic field dominates the jump in tangential velocity. This stability criterion was first found by Trakhinin by forming a new symmetric form of the equations – see [24] – although it is almost certainly stricter than necessary. One of the key observations made by Chen and Wang is that, using this new symmetric form of the equations, the linearised problem for current-vortex sheets is endowed with a well-structured decoupled formulation into a standard initial-boundary value problem for a symmetric hyperbolic system and a separate scalar PDE for the front. Chen and Wang then modify the iteration scheme to reconstruct the extensions of the front, $\Psi^{\pm}$, with $\Psi^{+} = \Psi^{-}$ on the boundary, which is why their scheme does not exactly fit into the above framework, but would require a small modification. In fact Trakhinin in [25] obtained the same result on current-vortex sheets, but instead of modifying significantly the iteration scheme of

Coulombel and Secchi, he solved the original linearised equations having used his new symmetric form only to help with the treatment of the linearised equations, which results in his approach being longer, although it should fit into the above framework. The normal modes analysis to determine the expected weakest possible stability criteria for current-vortex sheets in compressible magnetohydrodynamics leads to high order algebraic equations which seem impossible to solve analytically, and is detailed by Fejer in [12], where some special cases are considered.

The stability criterion for current-vortex sheets in incompressible magnetohydrodynamics is easier to determine – see e.g. Axford [4]. In 2D, the condition is

$$2(|\bar{H}_+|^2 + |\bar{H}_-|^2) > |[\bar{u}]|^2 .$$

In 3D, there are two conditions

$$\begin{aligned} 2(|\bar{H}_+|^2 + |\bar{H}_-|^2) &> |[\bar{u}]|^2 \\ 2\,|\bar{H}_+ \times \bar{H}_-|^2 &> |\bar{H}_+ \times [\bar{u}]|^2 + |\bar{H}_- \times [\bar{u}]|^2 \end{aligned}$$

although in fact the first follows from the second under the additional assumption $\bar{H}_+ \times \bar{H}_- \neq 0$.

Given these stability criteria, one would hope to be able to obtain a tame estimate for the linearised equations and then use Nash-Moser iteration as above to prove nonlinear structural stability of incompressible current-vortex sheets. In [19], Morando, Trakhinin and Trebeschi obtain an energy estimate for the linearised 3D equations under the above stability criteria. Also, using a different approach, Coulombel et al. [9] have derived a priori high order energy estimates directly for the nonlinear equations in 3D, using the incompressible version of Trakhinin's stability criterion – see Coulombel et al. [9]. However, the full problem of nonlinear structural stability of incompressible current-vortex sheets is still open.

The case of current-vortex sheets in 2D isentropic magnetohydrodynamics, where the magnetic fields are parallel on either side of the discontinuity, has been considered by Wang and Yu in [26]. They obtain a low order energy estimate for the linearised equations with loss of derivatives, under some quite restrictive assumptions to simplify the algebra and make the treatment similar to that of 2D isentropic Euler.

Another open problem is the case of current-entropy waves for the full magnetohydrodynamics equations, where the normal component of the magnetic field is no longer zero on the surface of discontinuity, but the velocity and magnetic field are continuous, with only the pressure, entropy and density experiencing a jump. There are strong indications that such waves ought to be stable under certain conditions, but the normal modes analysis again results in high-order algebraic equations which are difficult to study analytically.

Acknowledgements My research is supported by a UK EPSRC grant to the Department of Mathematics at Oxford University. I would like to thank my supervisor, Gui-Qiang G. Chen, for helpful discussions on this problem.

References

1. R.A. Adams, J.J.F. Fournier, *Sobolev Spaces*. Volume 140 of Pure and Applied Mathematics (Amsterdam), 2nd edn. (Elsevier/Academic, Amsterdam, 2003)
2. S. Alinhac, Existence d'ondes de raréfaction pour des systèmes quasi-linéaires hyperboliques multidimensionnels. Commun. Partial Differ. Equ. **14**(2), 173–230 (1989)
3. S. Alinhac, P. Gérard, *Pseudo-differential Operators and the Nash-Moser Theorem* (American Mathematical Society, Providence, 2007)
4. W.I. Axford, The stability of plane current-vortex sheets. Q. J. Mech. Appl. Math. **13**(3), 314–324 (1960)
5. G.-Q. Chen, Y.-G. Wang, Existence and stability of compressible current-vortex sheets in three-dimensional magnetohydrodynamics. Arch. Ration. Mech. Anal. **187**(3), 369–408 (2008)
6. G.-Q. Chen, Y.-G. Wang, Characteristic discontinuities and free boundary problems for hyperbolic conservation laws, in *Nonlinear Partial Differential Equations – The Abel Symposium 2010*, Oslo, ed. by H. Holden, K. Karlsen. Volume 7 of Abel Symposia (Springer, 2012)
7. J.-F. Coulombel, P. Secchi, The stability of compressible vortex sheets in two space dimensions. Indiana Univ. Math. J. **53**(4), 941–1012 (2004)
8. J.-F. Coulombel, P. Secchi, Nonlinear compressible vortex sheets in two space dimensions. Ann. Sci. Éc. Norm. Supér. (4) **41**(1), 85–139 (2008)
9. J.-F. Coulombel, A. Morando, P. Secchi, P. Trebeschi, A priori estimates for 3D incompressible current-vortex sheets. Commun. Math. Phys. **311**(1), 247–275 (2012)
10. I. Ekeland, An inverse function theorem in Fréchet spaces. Ann. Inst. H. Poincaré Anal. Non Linéaire **28**(1), 91–105 (2011)
11. L.C. Evans, *Partial Differential Equations*. Volume 19 of Graduate Studies in Mathematics, 2nd edn. (American Mathematical Society, Providence, 2010)
12. J.A. Fejer, Hydromagnetic stability at a fluid velocity discontinuity between compressible fluids. Phys. Fluids **7**, 499–503 (1964)
13. J.A. Fejer, J.W. Miles, On the stability of a plane vortex sheet with respect to three-dimensional disturbances. J. Fluid Mech. **15**, 335–336 (1963)
14. R.S. Hamilton, The inverse function theorem of Nash and Moser. Bull. Am. Math. Soc. (N.S.) **7**(1), 65–222 (1982)
15. L. Hörmander, The boundary problems of physical geodesy. Arch. Ration. Mech. Anal. **62**(1), 1–52 (1976)
16. A. Majda, *Compressible Fluid Flow and Systems of Conservation Laws in Several Space Variables* (Springer, New York, 1984)
17. J.W. Miles, On the disturbed motion of a plane vortex sheet. J. Fluid Mech. **4**, 538–552 (1958)
18. A. Morando, P. Trebeschi, Two-dimensional vortex sheets for the nonisentropic Euler equations: linear stability. J. Hyperbolic Differ. Equ. **5**(3), 487–518 (2008)
19. A. Morando, Y. Trakhinin, P. Trebeschi, Stability of incompressible current-vortex sheets. J. Math. Anal. Appl. **347**(2), 502–520 (2008)
20. J. Moser, A new technique for the construction of solutions of nonlinear differential equations. Proc. Natl. Acad. Sci. USA **47**, 1824–1831 (1961)
21. J. Nash, The imbedding problem for Riemannian manifolds. Ann. Math. (2) **63**, 20–63 (1956)
22. L. Nirenberg, *Topics in Nonlinear Functional Analysis*. Volume 6 of Courant Lecture Notes in Mathematics (New York University Courant Institute of Mathematical Sciences, New York, 2001). Chapter 6 by E. Zehnder, Notes by R. A. Artino, Revised reprint of the 1974 original.

23. J. Schwartz, *Nonlinear Functional Analysis* (Gordon and Breach, New York, 1969)
24. Y. Trakhinin, Existence of compressible current-vortex sheets: variable coefficients linear analysis. Arch. Ration. Mech. Anal. **177**(3), 331–366 (2005)
25. Y. Trakhinin, The existence of current-vortex sheets in ideal compressible magnetohydrodynamics. Arch. Ration. Mech. Anal. **191**(2), 245–310 (2009)
26. Y.-G. Wang, F. Yu, Stabilization effect of magnetic fields on two-dimensional compressible current-vortex sheets. Arch. Ration. Mech. Anal. (2013). doi:10.1007/s00205-012-0601-9

MIX
Papier aus verantwortungsvollen Quellen
Paper from responsible sources
FSC® C105338

If you have any concerns about our products,
you can contact us on
ProductSafety@springernature.com

In case Publisher is established outside the EU,
the EU authorized representative is:
Springer Nature Customer Service Center GmbH
Europaplatz 3, 69115 Heidelberg, Germany

Printed by Libri Plureos GmbH
in Hamburg, Germany

BIN TRAVERLER FORM

Cut By: #30 **Qty** 24 **Date** 10/16/20

Scanned By: ____________ **Qty** ______ **Date** ________

Scanned Batch ID's

______________ ______________ ______________

Notes / Exceptions

____